Physiological Perspectives on Food Safety: Exploring the Intersection of Health and Nutrition

Tanmay Sarkar • Ahmed Hamad
Ayan Chatterjee
Editors

Physiological Perspectives on Food Safety: Exploring the Intersection of Health and Nutrition

Springer

Editors
Tanmay Sarkar
Malda Polytechnic
Department of Food Processing Technology
Malda, West Bengal, India

Ahmed Hamad
Department of Food Hygiene and Control
Benha University
Toukh city, Egypt

Ayan Chatterjee
Medhavi Skills University
School of Health Science and Technology
East Sikkim, Sikkim, India

ISBN 978-3-031-94581-6 ISBN 978-3-031-94582-3 (eBook)
https://doi.org/10.1007/978-3-031-94582-3

This Springer imprint is published by the registered company Springer Nature Switzerland AG
The registered company address is: Gewerbestrasse 11, 6330 Cham, Switzerland

To my beloved mother, Arati Sarkar, and father, Bholanath Sarkar,and my sister, Tanusree Sarkar,whose unwavering love, support, and encouragement have shaped my journey.

—Dr. Tanmay Sarkar

To my beloved mother, Baisakhi Chatterjee, and my father, Ashis Chatterjee, and my wife, Parna Chatterjee.

—Dr. Ayan Chatterjee

This book is dedicated to my beloved father and mother.

Your unwavering love, sacrifice, and guidance that have shaped me into who I am today.
Your endless encouragement and belief in my dreams have been my greatest inspiration.
This book is a testament to the values you instilled in me and the unwavering support you've given me every step of the way.
With all my love and gratitude,

Dr. Ahmed Hamad

And to God, for the strength and guidance throughout every step of the way.

Preface

In the nexus of nutrition and health, food safety is a crucial element in global health and wellness policies. The publication *Physiological Perspectives on Food Safety: Exploring the Intersection of Health and Nutrition* offers a comprehensive analysis of the intricate physiological mechanisms underlying food safety, spanning from gut health to hormone regulation. By synthesizing global insights, this study presents a multidisciplinary view of food safety issues through various physiological perspectives. This book aims at an interdisciplinary analysis related to food safety problems through distinct physiological lenses with the integration of expertise from around the world. There are various subjects in different chapters of the book. This book begins with a basic understanding of the physiological components of food safety, researching how the immune and digestive systems interact with foodborne pathogens to affect human health. Later chapters explore metabolic, hormonal, and neurophysiological responses to contaminants in food and their roles in immune responses, hunger regulation, and nutrient absorption. It also discusses how the physiology of aging and maternal-fetus interactions influences susceptibility to foodborne diseases and newer areas of concern, such as the impact of toxins in food on the lungs, kidneys, and heart. Discussions of chronobiology, sensory perception, and developmental physiology bring readers state-of-the-art thinking about how individuals respond to issues of food safety and quality. To address such complex issues, the book also discusses chapters on food safety management and its integrative physiology. This book is intended for professionals, scholars, and students in public health, food science, and related disciplines. It also serves as a call for action with the hope of stimulating new physiology-based methods of performing food safety and as an instructional aid. In the hope that this book may stimulate valuable discussions and progress in food safety research and practice, the editors would like to extend their sincere thanks to the contributors for their dedication and expertise.

Malda, West Bengal, India — Tanmay Sarkar
Toukh City, Egypt — Ahmed Hamad
East Sikkim, Sikkim, India — Ayan Chatterjee

Contents

About the Editors

Tanmay Sarkar is currently working as a Lecturer in the Government of West Bengal. He has received his Bachelor of Technology, Master of Technology, and PhD in Engineering from Jadavpur University, Kolkata, India. With 9 years of experience in both teaching and research, his areas of interest include bioactive components from natural sources, ethnic foods, mathematical modeling, and phytochemicals. He is an editor for several SCI, SCIE, and Scopus-indexed journals and has edited 12 books with Elsevier, Springer, and Wiley. He has been classified among the top 2% of global highly cited researchers published by Stanford University, USA

Ahmed Hamad is a highly accomplished food scientist, microbiologist, and academician from Egypt, specializing in food safety, quality control, and molecular food hygiene. With a PhD in Food Hygiene and Control from Benha University, he has established himself as a leader in research and education, focusing on the molecular characterization of pathogens in food systems and innovative strategies for foodborne pathogen control. He currently serves as a Lecturer at the Faculty of Veterinary Medicine, Benha University, where he leads courses and research initiatives in food safety and quality. His expertise extends to consulting for international brands and institutions on food safety management systems, Hazard Analysis and Critical Control Point (HACCP) implementation, and advanced sanitation practices. Over the years, he has conducted more than 2500 food safety audits and spearheaded numerous consulting projects in the hospitality and food manufacturing industries. An author and editor of several influential books, including works on 3D printing in food manufacturing and cellular agriculture, he is a passionate advocate for innovation in sustainable food systems. His research has been widely published in peer-reviewed journals, addressing critical topics such as food microbiology, food science, one health, biofilm control, food safety in processing plants, and the implications of climate change on foodborne pathogens. He is a member of prestigious organizations like the Institute of Food Science and Technology (UK), the International Association for Food Protection (USA), and the American Meat Science Association. His dedication to advancing food science has earned him multiple awards, including the Scientific Excellence Award from Benha University and recognition as an outstanding peer reviewer by leading scientific publishers. Beyond academia, he is deeply committed to bridging the gap between science and industry through consulting, training, and mentorship, empowering professionals to implement cutting-edge practices in food safety and sustainability.

Ayan Chatterjee is currently working as Associate Professor and Associate Dean of School of Health Science and Technology, Medhavi Skills University, Sikkim, India. He also has teaching experience in Human Physiology, Biochemistry, and allied subjects at undergraduate, postgraduate, and diploma levels for 12 years. He also served as Member of Editorial Board of three UGC CARE listed journals, Reviewer of three Scopus indexed, and three UGC CARE listed journals. He also served as editor of five books of different publication houses (Elsevier, Springer Nature, IGI Global USA).

Chapter 1
Introduction to Physiological Aspects of Food Safety

W. Vinu, Dilshith A. Kabeer, Rameer Thaslim Khan, N. K. Ashif Ali, P. M. Ajmal, V. C. Fuad Sharaf, and Borem Siri Reddy

1.1 Introduction

Food safety is a global multifaceted concern and is defined as a scientific discipline ensuring that the food will not cause any adverse health effects when prepared and consumed according to its intended use. Food safety involves the handling, preparation, storage and distribution of food to prevent foodborne illnesses (Motarjemi and Smithers 2024; Witczak and Pokorska-Niewiada 2023). Producers, processors, distributors and retailers possess important roles in ensuring the safe consumption of food with sufficient support from government, consumers, and other parties such as academia and non-governmental bodies (Pray and Yaktine 2009). The concept of safe food is always a divisive issue when it comes to consumers, who are the last link to food supply chain services, from production to distribution. Some of the key concepts about safe food are when food is prepared on sanitized surfaces and served on sanitized plates and utensils, it is safe to eat. Some people would rather consume safe food that is pesticide-free but must contain sufficient nutrients such as vitamins and minerals, food that has not gone bad and has been distributed and kept in a

W. Vinu (✉) · D. A. Kabeer (✉) · N. K. Ashif Ali · P. M. Ajmal · V. C. Fuad Sharaf
Department of Physical Education & Sports, Pondicherry University,
Kalapet, Puducherry, India
e-mail: vinu@pondiuni.ac.in; dilshith.dpe@pondiuni.ac.in; ashifalink4@gmail.com; ajmalmusthafa89@gmail.com; vcfuad@gmail.com

R. T. Khan
Department of Food Sciences & Technology, Pondicherry University,
Kalapet, Puducherry, India
e-mail: rameercht@gmail.com

B. S. Reddy
Conestoga College, Kitchener, ON, Canada
e-mail: sborem9797@conestogac.on.ca

T. Sarkar et al. (eds.), *Physiological Perspectives on Food Safety: Exploring the Intersection of Health and Nutrition*, https://doi.org/10.1007/978-3-031-94582-3_1

temperature-controlled environment is considered safe, some people are aware of the concept of contamination and think that safe food will not have any contaminants. Some others will refer to healthy food in more pragmatic terms, like food that does not cause illness, while some others think that anything that smells or looks bad should not be eaten because your senses will warn you that it is dangerous (Schmidt and Rodrick 2003). The mentioned concepts indicate that, in practical terms, food safety is far more complex and encompasses the means and methods used throughout the food chain to ensure that the food consumed is safe. Food safety measures typically consist of ensuring better quality in food production and distribution, proper food handling and storage techniques, monitoring for environmental, physical, chemical hazards and biological contamination, adhering to government food safety compliances, and educating the public on foodborne illness and prevention.

The increasing complexity of food safety chains, environmental pollution and urbanization, contribute to the presence of harmful chemical compounds and microbiological hazards in animal and plant-based foods, posing a major challenge in ensuring safe consumption of food and highlights the significance of preventive actions and risk analysis in ensuring food safety (Bertolatti and Theobald 2011; Webb 2015). Thus, all the food service providers, including those engaged in pre-preparation, processing/preparation, packing and serving, must adhere to good manufacturing principles in order to guarantee food safety.

1.1.1 Physiological Relevance of Food Safety

The human body relies on a sufficient amount of safe and nutritious food for better living and promoting health. Food contaminants such as viruses, bacteria, parasites, chemicals and other toxic agents are the reason for more than 200 diseases found in human race, ranging from diarrhoea to cancer. Contaminated food usually contains detrimental microorganisms such as bacteria, parasites and viruses, some common pathogens among them such as Salmonella, E. coli, Giardia, Aspergillus and Listera can even cause a range of ailments from delicate gastrointestinal discomfort to severe life-threatening conditions (Mohammed and Elseory 2024; Mountzouris 2007). Pesticides, heavy metals, and mycotoxins, which are listed under chemical contaminants, can lead to poisoning and long-term health issues. Some common examples of such chemicals are aflatoxin and deoxynivalenol (DON) which causes gastrointestinal diseases, immune suppression and liver cancer (Gonya et al. 2024). A sedentary lifestyle, along with unhealthy food practices which involve harmful additives, high levels of unhealthy fats and sugars in processed food, can contribute to widespread chronic health challenges such as obesity, diabetes and cardiovascular diseases (Kinsey 2005). Nutritional deficiency, another important concern, arises from unsafe food which can result in the malabsorption of essential nutrients that affect overall health, especially in children. Chronic exposure to some mycotoxins can impair nutrient absorption and may lead to growth failure in children (Gonya et al. 2024). Consumption of trace elements which are essential for biological functions in an excessive manner, can be toxic and will cause severe consequences in human body (Miano et al. 2014). Genotoxic and cytotoxic compounds which are formed by the lipid oxidation in stored processed foods, can damage cells and may lead to cancer. In conclusion food safety has profound and diverse effects on human physiology ranging from mild ailments to chronic health conditions. Ensuring food safety is very crucial for better living with optimum physiological conditions and reducing foodborne disease-associated economic burden. Physiological concerns of food safety involve understanding how various factors, including microbial, chemical and physical hazards, impact human health and how these risks can be governed throughout the food chain.

1.1.2 The Importance of Food Safety in Public Health

Food safety is a very crucial concern of daily life that usually goes unnoticed until a problem occurs. Foodborne diseases affect over 600 million people globally and causing approximately 420,000 deaths annually, which denotes that, every year 10% of global population became ill through foodborne diseases. Each year 12,500 children under the age of five die from foodborne diseases and make about 40% of the burden of the foodborne diseases (Ghazanfar et al. 2024; Song et al. 2020). The most common disease caused by unsafe food are diarrheal infections, which impacts over 550 million people every year and results in 230,000 deaths. In particular unsafe food affects infants, young children, the elderly and the sick resulting in a vicious cycle of illness and malnutrition. Common symptoms of foodborne illness include vomiting, diarrhea (with or without blood), fever, abdominal cramping, headache, dehydration, myalgia and arthralgias (Switaj et al. 2015). A safe food supply promotes sustainable development by fostering national economies, trade and tourism in addition to food and nutrition security. The globalisation of food trade, a growing overall population, climate change and constantly evolving food systems have an impact on the safety of food (WHO 2025).

Food safety regulations and standards are very crucial for ensuring the safety and quality of food consumed globally. These are developed for the sake of public health, to prevent food born illness and to facilitate international food trades. Developed countries always possess high end standards for food safety while developing countries usually struggle with foodborne risks even for basic issues like access to safe water (Singh and Puniya 2024). Intake of unsafe food will result in far reaching health consequences including acute and chronic diseases, economic losses and increased health care costs. Countries must assure quality of imported food entering in the international market as well as implementing effective surveillance system and enforcing food safety regulations in a strict manner along with raising awareness and about the importance of food safety (Singh and Puniya 2024), is very crucial for preventing foodborne diseases and protect public health. Health concerns are becoming more widespread as food supply chains transcend multiple national boundaries (Gizaw 2019).

1.1.3 Historical Evolution of Food Safety Practices

Food safety has always been a top priority for human race, hence the knowledge of it and the techniques for keeping food safe have evolved significantly over time. Food safety regulations were rudimentary in ancient societies, including Egypt, Greece and Rome. In order to prolong the shelf life of perishable products, people would examine food for obvious signs of rotting and employ preservation methods including drying, salting and fermenting. The earliest framed food regulation dates back to Babylon, where provisions pertaining to food safety and quality were incorporated in a code of laws known as the Hammurabi (Bonny et al. 2023). During the middle-ages the outbreak of plague raised awareness about the importance of food safety as people realised the connection between the outbreak of challenging diseases and unsanitary food practices. During nineteenth century. The noticeable changes brought by industrial revolution in the production and distribution of food followed by the invention of role of microorganisms in food spoilage and illness laid the foundation for advanced modern food microbiology and safety measures (Fung et al. 2018). The emergence of regulatory authorities like WHO and U.S. Food and Drug Administration (FDA) in the twentieth century brought up food safety regulations and standards on a global scale. Codex alimentarius commission is one of the prominent international frameworks founded by FAO and WHO that devolopes harmonized international food standards, guidelines and codes of practice to protect consumer health and ensure fair trade practices (Boutrif 2014). Food Safety and Standards Authority of India (FSSAI) controls food safety regulation, integrating multiple existing laws under the food safety and standards act (Vasanthi and Bhat 2018) in India since 2006. Food safety measures require continuous renovation and adaptation to emerging challenges and technologies to protect public health effectively.

1.2 Physiological Systems and Food Safety

Food safety is a critical aspect of public health closely connected with physiological systems and overall human health. Contaminated food can affect multiple physiological systems, leading to broad range of health consequences. The body's defence mechanism's ranging from sensory systems that detect spoilage of food to advanced immune responses and detoxifying processes in the liver and gastrointestinal tract plays really crucial role in minimizing the hazards associated with contaminated food. Food contaminants, including food pathogens, microplastics, heavy metals, endocrine disruptors and other toxic substances, possess significant impact over most of the physiological systems in human body especially, gastrointestinal system, immune system, nervous system, endocrine system and metabolic health. Understanding the physiological impacts of food safety is very crucial for grasping how human body responds to various food contaminants aids in developing effective food safety measures and reducing health risks.

1.2.1 Digestive System

Eating is a complex physiological process that maintains energy balance by combining homeostatic and hedonic stimuli. It might be thought of as a cycle of hunger, fullness and contentment (Cifuentes and Acosta 2021). Intake of safe and nutritious food is a crucial concern for human society as it promotes optimum health and better living. The food contains ingredients that can have either positive or negative effects on human health. The phases and structure of consumed foods can vary according to their production process, formulas and places of origin. The digestive system comprises a highly complex union of adaptable and multiple-scale physicochemical methods. The digestive tract extending from mouth to anus will break down the food in to its basic components which then absorbed in to the blood stream, transported to different parts of the body and cleansing the waste, which altogether results in optimum physiological functioning of human body (Sensoy 2021).

Intake of contaminated food will definitely affect the human digestive system adversely which will lead to several health consequences. Contaminated food can carry several bacteria (Salmonella, E. coli), Viruses (Nora virus, Rotavirus), Fungi (Aspergillus, Penicillium) and Parasites (Giardia lamblia, Toxoplasma gondii) which may drive to food borne illness. These agents can cause health consequences through the production of toxins either produced in the food or composed after ingestion (Cangemi 2011), which ranging from delicate gastrointestinal discomfort, vomiting and diarrhoea to chronic health consequences such as gastroenteritis and inflammatory bowel disorder (IBDs) which involves the inflammation of stomach and intestine (Dimitrakopoulou et al. 2024; Mohammed and Elseory 2024). The food ingredients, including its fat and protein content, can influence in the survival of foodborne pathogens during digestion. As an example, high protein content can inactivate some pathogens during the gastric phase of digestion while fat can secure the pathogens along with helping them to survive through the gastrointestinal tract (Akritidou et al. 2022a, b). As we mentioned above, contaminated food can drive us to a range of gastrointestinal issues ranging from acute symptoms to chronic conditions. The survival and effect of pathogens and contaminants are influenced by several factors which involves food composition, digestive parameter and individual susceptibility. Understanding and taking adequate food safety measures will help to enhance the overall performance of digestive system.

1.2.2 Immune System

The immune system is a sophisticated web of cells, tissues, and organs that collaborate to protect the body from harmful pathogens and other foreign substances (Choudhury et al. 2007). Contaminated food refers to food that contains harmful substances such as pathogens like bacteria, viruses, fungi, parasites, chemicals such as pesticides, heavy metals, and other toxic substances (Singh et al. 2019). This is where the importance of the immune system in our body comes forward as this identifies and distinguishes these harmful pathogens from the body's own cells (Cota and Midwinter 2015). When pathogens are present in contaminated food, the immune system is triggered to activate diverse cells and mechanisms to neutralize and destroy the threat. There are various components of immune system include innate immune system, adaptive immune system. Both systems will work together which result in effective protection and proper homeostasis (Medzhitov 2007; Mukherjee et al. 2023; Stambas et al. 2020). The innate immune system is the body's initial response to pathogens, acting within minutes to hours after exposure. It recognizes broad patterns associated with pathogens through pattern recognition receptors. Upon detection, these receptors trigger signalling pathways that lead to the production of inflammatory molecules, which help in recruiting immune cells to the site of infection. The adaptive immune response is another critical component of the body's defence against foodborne pathogen. It is also known as acquired immunity, is a sophisticated and highly specific defence mechanism that develops over an organism's lifetime in response to exposure to pathogens. It involves the activation of lymphocytes, specifically T and B cells, which recognise and response to specific antigens. This system is characterized by its ability to remember past encounters with pathogens, allowing for a more rapid and efficient response upon subsequent exposures.

Contaminated food will undoubtfully affect the human immune system and can have negative health consequences in a variety of ways. Especially food borne diseases can cause severe infections and higher mortality rates in Individuals with compromised immune systems, like infants, elderly, women and those who are suffering from different kind of diseases (Obadina 2023; Rocourt 2014). Some of the foodborne pathogens can regulate the immune response which may drive to chronic inflammatory conditions. For example, yeast antigens can perpetuate inflammatory process in patients with Crohn's disease (Seibold 2005). Contaminated food can disrupt the immune homeostasis by altering the composition of gut microbiota which plays an important role in its immune homeostasis. Ruptures in the gut microbiota can lead to immune dysfunctions and increased exposure to infections (Hachimura 2022; Lamas et al. 2020). In fact, the effect of contaminated food on the immune system is well-rounded with different contaminants and food substances potentially drive to immune reactivity, disruptions in immune function and other long term health effects.

1.2.3 Endocrine System

The endocrine system is a complex network of hormones and related glands that regulate necessary bodily functions. It works in close association with nervous and immune systems to maintain homeostasis and effective response to changes. Understanding the basic components, functioning and controlling mechanisms is very important for diagnosing and treating related disorders (Akritidou et al. 2022a; Crespi et al. 2022; Ehlers and Malone 2024; Gorman and Chiasera 2013). Endocrine system possesses a significant role in growth, development, metabolism and reproductive process (Fink 2007; Yuan et al. 2021). Some of the key glands are the hypothalamus, pituitary gland, thyroid gland, adrenal gland, pancreas and gonads. The gland will secrete the hormones into the bloodstream and it will then travel to target organs or tissues where the bind to specific receptors and trigger some sections (Ehlers and Malone 2024; Galligan et al. 2023; Zia and Mukhopadhyay 2016).

The connection between endocrine system and food safety is primarily focused around the presence and effect of endocrine disrupting chemicals (EDC's) in food. They are the substances which can interfere with the normal functioning of the endocrine system, leading to diverse health issues. EDCs can enter the food chain through various ways, including the environment (e.g. fertilizer and feeds), direct use in food manufacture (e.g. pesticides) and discharge from food contact materials like bisphenol A (BPA) and phthalates (Mantovani 2016; Wang et al. 2024b). The impact of EDCs on human health is depends up on the duration of exposure and concentration of the substance as well as the life stage of the individual. For instance, prenatal and postnatal exposure can cause significant long-term adversities such as reproductive, neurological, immunological, metabolic disorders and even some types of cancer (Dhiman and Dureja 2020; Hamlin 2013; Mantovani 2016). The modulation of EDCs is very challenging as they having diverse nature and very complex interactions with the endocrine system. Regulatory boards should address the identification, testing and risk assessment of these chemicals (Dhiman and Dureja 2020; Manibusan and Touart 2018). Ensuring food safety in relation to the endocrine system needs inclusive risk assessment and governing measures to manage the presence and effect of EDCs in the food chain.

1.2.4 Nervous System

The nervous system is a complex network which transmits signals and harmonizing bodily functions. It includes two main components named central nervous system (CNS) and peripheral nervous system (PNS). The CNS includes the brain and spinal cord while PNS comprise cranial and peripheral nerves along with their associated structures (Banerjee et al. 2023; Wu et al. 2021). Despite the topographical classification, the CNS and PNS are closely connected and function in an integrated and harmonized manner. Neurons are the structural and functional unit of nervous system which is responsible for transmitting signals. The PNS connects the CNS to other body organs and tissues which results in functioning as a signal pathway. It is very crucial for human intelligence, behaviour, movement and sensation (Bazira 2021, 2024; Zochodne 2020).

It's very important to keep stringent food safety regulations to prevent adverse effects of food contamination on nervous system as intake of unsafe food will results in numerous health consequences. Heavy metals, pesticides, mycotoxins, BPA, acrylamide and micro plastic are all implicated in causing detrimental effects on nervous system. Exposure to common toxic metals like lead, cadmium, mercury and arsenic which are prominent in the food chain through industrial activities, can cause serious neuro developmental toxicities. Mercury, in particular is famous for its detrimental effects on the nervous system (Muñoz and Arango-Ruiz 2020; Witczak and Pokorska-Niewiada 2023). Organophosphates along with organochlorine compounds like DDT can cause neuro developmental issues in children. These chemicals can cause adverse effects on developing brain and prominently listed by the regulatory authorities due to its widespread use in agriculture (Borchers et al. 2010). Emerging research in the area suggests that microplastics, found in various foods can affect the neurological system. The complex nature of micro plastics makes them to require intensive studies to understand their complete health implications (de Carvalho et al. 2024).

1.2.5 Metabolism

Efficient digestion, absorption and metabolism of nutrients are essential for maintaining overall health. Factors such as food structure, nutrient interaction, intestinal health and genetics of the individual have very important roles in these processes (Cahill and El-Sohemy 2012; Zietek et al. 2020). Metabolism refers to the complete set of chemical reactions that take place inside a living organism to maintain life. These reactions are divided into two main categories named as catabolism and anabolism. Catabolism includes the break-down of molecules to produce energy while Anabolism involves the synthesis of all compounds which is essential for a living cell (Baart and Martens 2012; Evans and Heather 2016; Liao et al. 2021; Xu et al. 2016). Metabolic pathways are designed in the manner that they can produce energy and biomass while maintaining an equilibrium. Governance of metabolic pathways are very crucial for maintaining homeostasis and disruptions can cause various health problems including diabetes, obesity and cancer (De and Tomar 2012; Evans and Heather 2016, 2022; Xu et al. 2016) enzymes possess a very important role in metabolism by catalysing chemical reactions. Such enzymes are regulated by various factors which include substrate availability and feedback mechanisms (Di Filippo et al. 2022; Xu et al. 2016).

To understand how contaminated food affects the process of metabolism, it is essential to understand the various ways contaminants interact with nutrients and the gastrointestinal system (GI) system. Contaminants in the consumed food can alter the nutritional content and metabolism by interacting with nutrients at various stages, including digestion, absorption and excretion. For example, anti-nutritional factors (ANFs) present in contaminated food can inhibit the absorption of nutrients and digestion, which may lead to reduced bioavailability of nutrients (Kaur et al. 2021; Xu 2022). Heavy metals (e.g. cadmium and nickel), microplastics and Pathogens like E. coli can cause gastro intestinal illness that disrupts the guts ability to absorb nutrients effectively which may result in malnutrition and other health issues (Griffin et al. 2008; Khan et al. 2024; Shahid et al. 2017). Exposure to a mixture of food contaminants like pesticides, antibiotics, heavy metals and persistent organic pollutants can cause damage to gut homeostasis along with altering glucose tolerance, insulin sensitivity and adipose tissue inflammation, especially in the presence of high-fat, high sucrose diet, which may result in serious health adversities such as diabetes and obesity (Elmassry et al. 2022; Li et al. 2022; Naville et al. 2015). In fact, contaminated food can have severe and significant implications for metabolic health primarily through gut microbiota disruption, endocrine interference and may have direct impacts on lipid and energy metabolism. Addressing these issues needs comprehensive strategies to minimize exposure and mitigate the detrimental effects of these food contaminants.

In fact, food safety is not merely an entity of external control but also a biological concern deeply rooted in human physiology. A thorough understanding of these physiological aspects strengthens our ability to design safer food systems, establish targeted interventions and educate people about the significance of both personal and systemic food safety measures. This integrated approach is very crucial for protecting public health and reducing the burden of food borne disease globally.

1.3 Foodborne Pathogens and Their Effects

1.3.1 Introduction to Foodborne Pathogens

Microorganisms that cause disease through the consumption of contaminated water or food known as foodborne pathogens. These include bacteria, parasites, viruses and fungi each contributing different types of diseases also known as food poisoning, the symptoms can be range from mild discomfort to severe health complications and sometimes leads to death in vulnerable populations such as children, immunocompromised and old aged people *Hepatitis A virus* and *Norwalk-like caliciviruses* are the most common type foodborne viruses and *Escherichia coli, Campylobacter jejuni, Clostridium perfringens, Salmonella spp*. are the common bacteria in foodborne illnesses, while parasites such as *Cryptosporidium parvum, Giardia duodenails* and *Toxoplasma gondii* are also can cause foodborne diseases (Boughattas and Salehi 2014).

1.3.2 Classification of Foodborne Pathogens

Based on the nature and the mode of transmission foodborne pathogens are classified such as bacteria, viruses, parasites and fungi. Bacterial pathogens, such as *Escherichia coli*, *Salmonella* and *Listeria* are responsible for major foodborne diseases and transmits through undercooked meats, diary and other contaminated produces. *Norovirus* and *Hepatitis A* are the highly contagious viral pathogens which can be transmitted through shellfish, contaminated water or improper handling of food. Parasitic pathogens, such as *Giardia lamblia* and *Toxoplasma gondii* are not that common but still they cause serious health issues, especially transmits through contaminated water and undercooked meats.

1.3.2.1 Bacterial Foodborne Pathogens

Bacterial foodborne pathogens are causing numerous diseases and deaths each year and it's a significant public health concern globally. Bacterial pathogens are typically transmitted through the contaminated water and food, consumption of such food items leads to foodborne infections (Reshma et al. 2023). *Salmonella spp., Escherichia coli, Listeria monocytogens, Campylobacter jejuni, Staphylococcus aureus, Clostridium botulinum, Shigella spp., Vibrio spp.* and *Bacillus cereus* are the common bacterial foodborne pathogens (Bintsis 2017; Hu et al. 2016).

1.3.2.1.1 *Salmonella spp.*

Salmonella is a bacteria genus which is ranks among the common foodborne pathogens globally. They are scientifically classified under the *Enterobacteriaceae* family and the genus consist of two species which are *Salmonella bongori* and *Salmonella enterica* also the *Salmonella enterica* further subdivides into six subspecies. The ability of *S. enterica* to infect warm-blooded mammals, including humans makes them most clinically significant pathogen (Cohn et al. 2021). There are 1500 serovars within this subspecies and they vary in their capacity to cause disease (Cohn et al. 2021). The ability to contaminate diverse food sources makes *Salmonella spp.* is responsible for approximately 53.4% of global foodborne outbreaks (Liu et al. 2018). *Salmonella* is transmitted primarily through intake of contaminated water or food. Contaminated animal products such as undercooked or raw meat, milk and eggs are major source of transmission. Fresh produce can also be contaminated by irrigation water which contains *Salmonella* which may originated from the surface water such as rivers and ponds which are exposed to animal feces (Liu et al. 2018). Unwashed hands, utensils and cutting boards can leads to cross contamination during food preparation also the direct contact with infected animals can transmit pathogens (Silva et al. 2014).

Physiological impact on the body

Gastrointestinal tract is the primary infection target of *salmonella*, where it induces physiological disruptions by damaging the intestinal epithelium. They can survive the acidic environment of the stomach and colonize in the small intestine. The pathogen once entered inside epithelial cells they resides in a specialized compartment called the *Salmonella*-containing vacuole (SCV), where the pathogens replicates and evades host immune responses (Sun et al. 2020). This results in the increased intestinal permeability, fluid secretion, and neutrophil recruitment, which collectively shows as diarrhoea and abdominal cramps (Patel and McCormick 2014). In severe *Salmonella* infection cases, the pathogen breaches the intestinal barrier and spreads through the bloodstream with the help of the bacterial effectors such as SpvB which can disrupt epithelial integrity (Sun et al. 2020). The pathogens migrate to the organs such as liver and spleen by using phagocytic immune cells, and this systemic spread can lead to the condition like bacteraemia, sepsis or other localized infections in immunocompromised individuals (Dai et al. 2024; Worley 2023). *Salmonella* can cause log-term damage to the intestinal lining which can lead to the post-infectious irritable bowel syndrome (IBS) or worst condition like inflammatory bowel disease (IBD) (Griffin and McSorley 2011). Additionally systemic *Salmonells* infection can trigger reactive arthritis in the individuals who are genetically predisposed, this is an autoimmune condition which develops weeks after the infection and it characterized by joint pain, eye inflammation, also the chronic carriers can asymptomatically shed bacteria in their feces for months or years (Li 2022).

1.3.2.1.2 *Escherichia coli* (*E. coli*)

Escherichia coli is a rod shaped, gram-negative bacterium that plays a significant role in digestion of warm-blooded animals and resides in their gastrointestinal tract (Martinson and Walk 2020). Initially the name was *Bacillus coli* but later changed to *Escherichia coli* to honor its discoverer Theodor Escherich who discovered it in 1885 (Foster-Nyarko and Pallen 2022). Mostly *E. coli* strains are harmless but certain pathogenic variants can cause gastrointedstinal and systemic diseases due to their ability to produce toxins, such variant like Shiga toxin-producing *E. coli* (STEC) (Ruiz and Silhavy 2022). Also, *E. coli* used as a model organism for microbiological and molecular biology research due to its genetic diversity and adaptability.

Physiological impact on the body

Shiga toxin-producing *E. coli* infection leads to severe gastrointestinal damages because of Shiga toxin bind to globotriaosylceramide (Gb3) receptors on intestinal epithelial cells and disrupting protein synthesis which leads to bloody diarrhoea (Havira et al. 2020). After entering to the systemic circulation, the toxins released from bacteria targets glomerular endothelial cells and podocytes in the kidneys, where the Gb3 receptors are abundant, leading to thrombotic microangiopathy- a key characteristic of haemolytic uremic syndrome (HUS), also it leads to acute tubular necrosis and intestinal inflammation due to the renal damage by toxin-mediated inhibition of ribosomal function (Trachtman et al. 2012). This leads to 40% of the STEC-HUS patients develop acute kidney injury and due to the endothelial injury and thrombotic occlusion of glomerular capillaries which leads to renal replacement therapy (Trachtman et al. 2012). Shiga toxins can amplify endothelial dysfunction and delaying tissue repair by upregulate pro-inflammatory cytokines like tumour necrosis factor-alpha (TNF-α) (Havira et al. 2020). The toxin's effect on vascular integrity resulting in affecting multiple organs, mostly kidney (Trachtman et al. 2012).

1.3.2.1.3 *Listeria monocytogenes*

Listeria monocytogenes is a facultative intracellular Gram-positive pathogen responsible for listeriosis, which is a serious foodborne illness with high mortality rate of 20–30% mortality rate globally (Quereda et al. 2021). They show high mortality rate, particularly among vulnerable populations such as pregnant women, newborns, old age people and immunocompromised individuals, also they can adapt extreme conditions such as refrigeration temperature and high salt which helps them to survive in food production environments and contaminating the ready-to-eat food items (Osek et al. 2022). Transmission of this pathogen occurs through the consumption of contaminated food products including deli meats, unpasteurized dairy and ready-to-eat meals.

Physiological impact on the body

The physiological impact depends on the severity of *Listeria monocytogenes* infection and the host's immune response. *Listeria* can survive and replicate within host cells and leading to systemic infections, mostly in vulnerable populations such as immunocompromised and pregnant women. In pregnant women they can cross the placental barrier and make fetal infection and causing damage to fetal tissues which can lead to miscarriage or still birth (Pizarro-Cerdá and Cossart 2018). *Listeria* causes meningitis or encephalitis once they reached central nerve system, severe headaches, stiff neck seizures and confusion are the symptoms (Sibanda and Buys 2022).

1.3.2.1.4 *Campylobacter jejuni*

Campylobacter jejuni is a spiral shaped gram-negative microaerophilic bacterium and plays major role in bacterial gastroenteritis globally. It infects and colonizes poultry intestines and through the contaminated food it transmits to humans and causing campylobacterosis, symptoms are inflammatory diarrhoea, abdominal cramps and fever (Dasti et al. 2010; Jorgensen et al. 2022; Sibanda and Buys 2022).

Physiological impact on the body

Campylobacter jejuni invades intestinal epithelial cells and with help of protease like HtrA it disrupts cell junction leading to bloody diarrhoea and mucosal inflammation (Dasti et al. 2010). The cytolethal distending toxin blocking the cell cycle and induces host cell apoptosis and exacerbating tissue damage (Jorgensen et al. 2022). Molecular mimicry between bacterial lipooligosaccharides and host nerve gangliosides triggers systemic complications include Gullain-Barre syndrome (GBS) resulting in autoimmune neuropathy (Heimesaat et al. 2021). In immunocompromised individuals septicemia and bacterecemia can occur with high mortality risks (Jorgensen et al. 2022).

1.3.2.1.5 *Clostridium perfringens*

Clostridium perfringens is a spore forming, anaerobic Gram-positive bacterium found in soil mostly, decaying vegetation and the gastrointestinal tract of humans and animals (De la Mora et al. 2020). It is one of the major cause of foodborne disease and gas gangrene with its five toxinotypes ranging from A to E linked to diverse pathologies (De la Mora et al. 2020; Uzal et al. 2014). Its toxin production and rapid sporulation shows its role in necrotic enteritis and traumatic infections (Uzal et al. 2014).

Physiological impact on the body
Clostridium disrupts host cell membranes using the α-toxin such as phospholipase C and inducing tissue necrosis, haemolysis and vascular damage in gas gangrene, while its enterotoxin forms pores in intestinal cells, causing diarrhoea and cramping (Ranjan Sahoo and Deb 2022). Bacteraemia, septicaemia and multiorgan failure causes by systemic spread in severe cases (Ranjan Sahoo and Deb 2022).

1.3.2.1.6 *Clostridium botulinum*

Clostridium botulinum is a rod shaped Gram-positive anaerobic, spore forming bacterium which producing botulinum toxin, one of the most potent toxins known. It is responsible for botulism, a severe neuroparalytic disease characterized by descending muscle paralysis. Botulism can occur through wound infections, infant exposure to spores or foodborne transmission. The spores can survive in harsh environments making them significant health concern in improperly preserved foods (Smith et al. 2014).

Physiological impact on the body
Clostridium botulinum has ability to cleave the SNARE proteins in presynaptic neurons, preventing neurotransmitter release and causing paralysis, its early symptoms include blurred vision, ptosis, and difficulty in swallowing, also in severe cases respiratory muscle paralysis (Bradshaw et al. 2009). The systemic effects of toxin can also disrupt autonomic functions, causing dry mouth and constipation, the recovery requires nerve regeneration and it can take weeks to months (Caya et al. 2004).

1.3.2.1.7 *Staphylococcus aureus*

Staphylococcus aureus is a Gram-positive, facultative anaerobic bacterium that mostly found on human skin and mucosal surfaces, 30% of the population carries it asymptomatically (Cheung et al. 2021). *S. aureus* is a versatile pathogen responsible for a wide range of infections such as skin and soft tissue infections, Bactericemia, pneumonia, and endocarditis (Thomer et al. 2016). Its pathogenicity contains numerous virulence factors such as coagulases, toxins and immune evasion strategies makes them a major cause of both health-care and community associated infections (Tong et al. 2015). The antibiotic resistant strains, mainly methicillin-resistant *S. aureus* (MRSA) further complicates treatment and becomes a threat for public health (Balasubramanian et al. 2017).

Physiological impact on the body
The impact of *S. aureus* infections depends on the site and severity of the infection, such as in blood stream infection the pathogen secretes coagulases that activate prothrombin to form fibrin clots, leading to abscess formation (Thomer et al. 2016). *S. aureus* produces toxins such as α-toxin and Panton-Valentine leucocidin (PVL), which lyse host cells and contribute to tissue damage (Cheung et al. 2021). *S. aureus* can cause systemic inflammation, septic shock or multiorgan failure in severe cases (Tong et al. 2015). Biofilm formation on medical devices can lead to chronic infections and leading to persistent inflammation and treatment resistance (Tong et al. 2015).

1.3.2.1.8 *Shigella spp.*

Shigella spp. facultative anaerobic Gram-negative bacteria responsible for shigellosis, which is a highly contagious diarrhoeal disease causing mortality worldwide (Sahl et al. 2015). The genus includes species such as *S. dysteriae, S. flexneri, Sboyndii and S. sonnei,* among these *S. flexneri* is most prevalent in developing countries and in industrialized regions *S. sonnei* is dominating (Anderson et al. 2016). Transmission of *Shigella* pathogen occurs via the fecal-oral route, also through the contaminated food and water, 10–100 organisms are required to cause the infection (Pakbin et al. 2023).

Physiological impact on the body
Shigella can invade intestinal epithelial cells using type III secretion system (T3SS), which injects effector proteins into host cells and promoting bacterial uptake and immune evasion (Schnupf and Sansonetti 2019). Once inside the host cell *Shigella* multiplies and spreads between cells via actin polymerization which leading to epithelial damage and inflammation (Mattock and Blocker 2017). Also, *S. dysenteries* produces Shiga toxin which inhibits protein synthesis in host cells and it causes severe complications such as haemolytic uremic syndrome (HUS) with the symptoms of bloody diarrhoea, fever and abdominal pain due to inflammation and mucosal ulceration (Halimeh et al. 2021; Pakbin et al. 2023).

1.3.2.1.9 *Vibrio spp.*

Vibrio spp. are rod-shaped Gram-negative bacteria commonly found in marine and estuarine environments, which contains 70 species identified, there are several human pathogenic species such as *Vibrio cholerae, Vibrio parahaemolyticus, and Vibrio vulnificu,* and they causes disease ranging from mild gastroenteritis to life threatening septicemia (Marques et al. 2022). These are facultative anaerobic bacteria and motile sue to a single polar flagellum, which jhelps them to thrive in aquatic ecosystems and adapt to various hosts including marine organisms and humans (Onohuean et al. 2022; Xiao et al. 2022).

Physiological impact on the body
The physiological impact of *Vibrio spp.* varies by the species, like *V. cholerae* produces cholera toxin (CTX), which disrupts ion transport in intestinal cells which leading to severe watery diarrhoea and dehydration (Marques et al. 2022; Thompson et al. 2004). *V. parahaemolyticus* infection results in gastroenteritis with diarrhoea, abdominal cramps and fever due to thermostable direct hemolysin (TDH) and TDH-related hemolysin (TRH) (Sultana et al. 2024; Xiao et al. 2022). *Vibrio vulnificus* severe infection can lead to production of metalloproteases and leads to septicemia or necrotizing fasciitis (Onohuean et al. 2022; Sultana et al. 2024).

1.3.2.1.10 *Bacillus cereus*

Bacillus cereus is a rod shaped, Gam-positive facultative anaerobic bacterium mostly found in environment, particularly in soil and vegetation. It produces protective endospores which allows them to survive in adverse environmental conditions, it is well known for foodborne illness with two distinct syndromes which are emetic and diarrhoeal, in this the emetic syndrome is caused by cereulidetoxin which is produced in contaminated food whereas the enterotoxin produced in the intestine intestine after ingestion causes diarrhoeal syndrome (Ehling-Schulz et al. 2019).

Physiological impact on the body
The cereulidetoxin disrupts mitochondrial fumnction, leading to nausea and vomiting within 1–6 hours after ingestion (Ehling-Schulz et al. 2019). The diarrhoeal syndrome caused by enterotoxins like hemolysin and nonhemolytic enterotoxin, these toxins can cause intestinal epithelial cell damage leads to watery diarrhoea and abdominal cramps after 8–16 hours of consumption (Ehling-Schulz et al. 2019).

1.3.2.2 Viral Foodborne Pathogens

Viral foodborne pathogens are causing several foodborne illnesses worldwide such as gastroenteritis and all, viral pathogens primarily transmitted through contaminated food, water or close contact with infected individuals (Petrović and D'Agostino 2016). These viruses will be inactive in food without multiplying but gets activated once entered to the host, the transmission often through poor hygene, inadequate sanitation or unhygienic food handling practice (Seymour and Appleton 2008). Norovirus, Hepatitis A Virus, Rotavirus are the common foodborne viruses which are responsible for major outbreaks globally (Bintsis 2017).

1.3.2.2.1 Norovirus

Norovirus is an RNA virus which is highly contagious belonging under the family *Caliciviridae.* Norovirus is one of the leading causes of gastroenteritis globally which is responsible for significant morbidity (Ogunsakin et al. 2022). This virus was first identified in 1968 in an outbreak in Norwalk, and that's why it got the name Norwalk virus (Lucero et al. 2021). It requires a very small infectious dose and transmitted through contaminated food, water or person-to-person contact (Robilotti et al. 2015). It causes symptoms such as diarrhoea, vomiting, nausea and abdominal cramps (Winder et al. 2022).

Physiological impact on the body
Norovirus disrupting the epithelial cells after infecting in small intestine, which leading to acute gastroenteritis. The symptoms starts within 12–48 hours of exposure and it last of 1–3 days (Robilotti et al. 2015). It can trigger immune responses but doesn't confer long-term immunity. Dehydration can happen in severe cases, particularly in vulnerable populations such as children and the elderly (Winder et al. 2022).

1.3.2.2.2 Hepatitis A Virus (HAV)

Hepatitis A virus is a non-enveloped single stranded RNA virus belongs to thee *Picornaviridae* family, the pathogen is primarily transmitted via the fecal-oral route through contaminated water, food, or contact with infected individuals (Kunanitthaworn et al. 2023; Nagashima et al. 2021). This virus is highly stable in the environment, enabling its survival on surfaces and non-porous environments for extended periods (Nagashima et al. 2021). Asymptomatic infection in children under five while older children and adults experiences the symptoms such as jaundice, fever, nausea and abdominal cramps (Kunanitthaworn et al. 2023). HAV leads to acute liver inflammation but does not cause chronic infection.

Physiological impact on the body
Hepatitis A virus infection impacts on host body by triggers hepatocyte damage through direct cytopathic effects and immune-mediated responses and which is resulting in elevated serum aminotransferases and bilirubin (Kunanitthaworn et al. 2023). The virus shed in feces by excreted into bile via transcytosis, facilitated by hepatocyte-speific transport mechanisms (Counihan and Anderson 2016). In severe cases, prolonged cholestasis exacerbating liver dysfunction and immune responses including IgA-mediated neutralization plays a dual role such as limiting viremia while adding viral ecretion (Counihan and Anderson 2016).

1.3.2.2.3 Rotavirus

Rotavirus is a highly contagious and couse of gastroenteritis, especially in young children and infants (Rotavirus (for Parents) | Nemours KidsHealth n.d.). The pathogen porimarily spread through fecal-oral route often due to poor hygiene (Rotavirus (for Parents) | Nemours KidsHealth n.d.). The infection begins within 2 days of exposure to the virus, symptoms shown such as fever, vomiting, watery diarrhoea leads to dehydration.

Physiological impact on the body
Rotavirus infection leads to acute gastroenteritis with symptoms such as watery diarrhoea, vomiting, and fever. Enterocyte damage and fluid loss leads to severe dehydration and leads to malabsorption (Parashar et al. 2013). Some cases will show neurological features such as seizures (Parashar et al. 2013). Rotavirus can increase the pro-inflammatory cytokines in the body, contributing to systemic inflammation and complications (Dian et al. 2021).

1.3.2.3 Parasite Foodborne Pathogens

Foodborne parasite pathogens are capable of causing severe illness through contaminated water and food. They are ranging from microscopic protozoa to larger helminths and often have complex life cycle involving multiple hosts. Among the parasites the most important and common foodborne parasite pathogens are *Toxoplasma gondii, Cryptosporidium spp. and Giardia lamblia.* They show symptoms like diarrhoea and gastrointestinal distress.

1.3.2.3.1 *Toxoplasma gondii*

Toxoplasma gondii is a protozoan parasite responsible for toxoplasmosis, it's a zoonotic disease transmitted through foodborne routes. Consumption of undercooked meat containing tissue cysts or ingesting water or other foods contaminated with oocytes leads to toxoplasmosis (Almeria and Dubey 2021). Mostly the infections are asymptomatic but still they can couse severe neurological, ocular and systemic complications in immunocompromised patients and infants (Almeria and Dubey 2021; Shapiro et al. 2019).

Physiological impact on the body
There are two type infections happens with *Toxoplasma gondii* acute infection and chronic infection, in acute infection flu-like symptoms, transient inflammation in the lungs, heart muscle or retina and immune-driven tissue damage particularly in the central nervous system (CNS) (Almeria anttd Dubey 2021). In chronic infection it correlates with neurotransmitter dysregulation, structural brain alterations, cognitive and psychiatric effects like impaired memory executive function decline, schizophrenia like symptoms are shown (Almeria and Dubey 2021; Shapiro et al. 2019).

1.3.2.3.2 *Cryptosporidium spp.*

Cryptosporidium spp. are the parasites responsible for cryptosporidiosis which is a diarrhoeal disease transmitted through contaminated food, water or direct contact with infected hosts. These pathogens infect intestinal epithelial cells and it causing self-limiting gastroenteritis in immunocompetent individuals and chronic illness in immunocompromised individuals (Hassan et al. 2021).

Physiological impact on the body
After the ingestion *Cryptosporidium* oocysts excyst in the intestine and releasing sporozoites that invade the epithelial cells of the gastrointestinal tract. The parasite can do both sexual and asexual reproduction inside the cells, it leads to production of new oocysts that are shed in feces (Leitch and He 2011). Acute gastroenteritis with self-limiting and severe outcomes happens with the immunocompromised individuals.

1.3.2.3.3 *Giardia lamblia*

Giardia lamblia is a flagellated protozoan parasite that primarily inhabits the small intestine of humans and other mammals. It causes giardiasis, a diarrhoeal disease characterized by gastrointestinal distress. The pathogen exists in in two forms, the motile trophozite which can adher the epithilim and the other onr dormant cyst which can survive outside the host (Adam 2001).

Physiological impact on the body
Symptoms such as watery diarrhoea, abdominal cramps, boating, nausea and fatigue are shown in acute infection for 1–3 weeks post-infection and in chronic infection symptoms like chronic diarrhoea and malabsorption. It results in significant weight loss and nutritional deficiencies (Adam 2001).

1.3.2.4 Fungal Foodborne Pathogens

Foodborne fungal pathogens are threat in food safety because of their ability to produce harmful toxins by contaminating the food. These fungi pathogens can infect food items during production, storage, or processing due to the conditions such as high humidity. Among the most notable foodborne fungal pathogens aspergillus, fusarium and penicillium species are most common, they can produce toxins such as aflatoxin, fumnisins, ochratoxins and ergot alkaloids. These toxins can lead to several health effects such as neurotoxicity, carcinogenicity and immunosuppression.

1.3.2.4.1 *Aspergillus*

Aspergillus is a filamentous fungi genus that includes several hundred species, they play a crucial role in the decomposition of organic matter and nutrient cycling but also, they have pathogenic potential too. Among the most significant species *Aspergillus fumigates, A. flavus,* and *A. Niger* are having pathogenic potential and they can particularly infect immunocompromised individuals. While many of them are harmless or beneficial, certain strains can cause serious diseases. Including respiratory infections and systemic diseases such as aspergillosis (Latgé and Chamilos 2020).

Physiological impact on the body
Aspergillus primarily effect the respiratory system but depending on the immune status of the individual it can widespread. Its spores, conidia can cause allergic reactions such as allergic bronchopulmonary aspergillosis (ABPA), resulting to inflammation wheezing and asthma-like symptoms. In case of immunocompromised individuals, the spores can germinate into hyphae, on lung tissue and causing invasive aspergillosis, it's a life-threatening condition which can spread to other organs like brain, kidney or skin. It can result into symptoms like fever, chest pain, coughing blood and systemic organ damage (Latgé and Chamilos 2020; Segal and Walsh 2012).

1.3.2.4.2 *Fusarium*

Fusarium species are filamentous fungi which can produce harmful mycotoxins that contaminate various food products particularly grains. These fungi primarily infect crops such as maize, wheat and barley, leading to diseases like ear rot and head blight. Deoxynivalenol (DON), fumonisin and trichothecenes are the toxins produced by fusarium (Hof and Schrecker 2024).

Physiological impact on the body
Fusarium mycotoxins can lead to a range of adverse health effects after consumption of contaminated food. Gastrointestinal disturbance including vomiting and nausea. Chronic exposure can result in organ damage and immunosuppression. In severe cases scute mycotoxicosis can occur leading to symptoms such as fever, abdominal pain and sometimes instant death (Hafez et al. 2022; Qiu et al. 2020).

1.3.2.4.3 *Penicillium*

Penicillium is a fungi known for food spoilage and they can produce mycotoxin. Also, they used in biotechnology for the production of antibiotic like penicillin. It commonly found in soil, decaying organic matter, and some are found in spoilt fruits producing mycotoxin in it (Wang et al. 2017).

Physiological impact on the body
After contaminating the food, they can produce toxin inside the food that pose significant health risks. *Penicillium expansum* produces patulin, which is associated with gastrointestinal disturbances and has potential carcinogenic effects (Houbraken et al. 2012; Wang et al. 2017). Symptoms such as nausea, vomiting, and abdominal cramps, some species can cause allergic reactions in sensitive individuals, leading to respiratory issues or skin irritations (Houbraken et al. 2012).

1.4 Long-Term Health Consequences of Unsafe Food

Food is the most basic and indispensable need for life. Hence, Food safety is a primary right of human beings. Both physiological demand and caloric requirements have been the primary drivers of food intake historically (Fung et al. 2018). Beyond basic needs and only a source of nourishment, it contributes necessary nutrients for physical health, cognitive development and disease prevention. Still, with a global rise in food malpractices, public perception has elevated the concern about food safety. Food that has been infected by dangerous bacteria, viruses, chemicals, or poisons from sources such as incorrect storage, contaminated water, or pesticides remains is referred to as unsafe food (Singh and Puniya 2024). Unsafe food creates a threat to billions of people worldwide.

Consuming unsafe food can bring about severe outcomes, stretching from mild sickness to serious health hazards. Every year, millions of people fall ill, and hundreds of thousands of people lose their lives because of this unsafe food consumption. Contaminated food affects millions of people all globally, resulting in high mortality rates and chronic sickness, accentuating the importance of stringent food safety laws and hygienic quality, according to WHO (Rodríguez-Melcón et al. 2024). Therefore, the nutritious and healthy consumption of food leads to improved individual and societal health, and regional prosperity is revamped when food safety is encouraged and executed. As long as the quality of life is refined to a certain point, there will be a concern over food safety, and potential toxic substances will persist as an important health issue. Indeed, every customer looks forward to safe and quality foodstuffs since they are the basic energy source and nutrients. Personal, chemical, biological and environmental events are among the challenges and setbacks in food safety (Fung et al. 2018).

1.4.1 Chronic Diseases from Unsafe Food

1.4.1.1 Kidney and Liver Damage

Chronic kidney disease (CKD) can be caused by consuming unsafe food infected with heavy metals, including Lead, Arsenic and Cadmium, which accumulate in the body over time and damage kidney function. A major threat to liver cancer is keeping long exposure to aflatoxins, which are harmful toxicants made by moulds in poorly stored grains and nuts. Hepatocellular carcinoma, a severe type of liver cancer, and cirrhosis are affected by these poisons that interfere with liver function (Jalili et al. 2021).

1.4.1.2 Neurological Disorders

Extended nerve damage and cognitive disability can result from foodborne diseases caused by germs such as Clostridium botulinum and Listeria monocytogenes. These contaminations affect the nervous system and lead to different problems like paralysis and meningitis (Adhikari and Kharel 2018; Batz et al. 2013; Kanaan and Tarek 2020).

Furthermore, memory loss, childhood developmental abnormalities, and degenerative neurological diseases like Alzheimer's and Parkinson's have been connected to pesticide remnants in fruits and vegetables and mercury pollution in seafood (Joyce et al. 2022; Llop et al. 2013; Suresh et al. 2023).

1.4.1.3 Cancer Risks

Certain substances contained in unsafe food might cause cancer. Processed meat contains N-nitroso compounds, which dramatically raise colon cancer (Cross and Sinha 2016; Crowe et al. 2019). Extended exposure to pesticides through tainted produce is connected to pancreatic and breast cancer due to hormonal imbalance and genetic deformities (Tayour et al. 2019).

1.4.2 Metabolic and Endocrine Disorders

1.4.2.1 Endocrine Disrupters

The endocrine system gets disturbed by numerous chemicals found in food, including pesticide residue, phthalates and bisphenol A (BPA), which upset the balance of hormones. These drugs cause different issues like early puberty, metabolic anomalies and problems with reproduction by emulating or blocking natural hormones. Additionally, prepackaged food chemicals and preservatives can change metabolism and take up the risk of obesity and diabetes (Tayour et al. 2019).

1.4.2.2 Obesity and Diabetes

The metabolic syndrome, which is related to obesity, insulin resistance and Type 2 diabetes, is aggravated by unhealthy food ingredients such as trans fats, artificial sweeteners, and high fructose corn syrup (Mukhopadhyay et al. 2020; Sheptulina et al. 2023). Consuming these nutrients over time affects lipid metabolism and digestion (Moszak et al. 2020). Furthermore, eating contaminated food changes the gut microbiota, which is necessary for metabolism and digestion. Insulin resistance can result from changes by formulating gut flora, increasing the risk of diabetes (Haneishi et al. 2024).

1.4.2.3 Thyroid Dysfunction

Thyroid hormones get affected by unsafe or contaminated food that is tainted with fluoride and heavy metals, which cause Hypothyroidism and other metabolic diseases (Hall et al. 2023). Thyroid function gets further disturbed by goitrogenic chemicals included in several meals, such as cruciferous vegetables and soy, which may intervene with iodine intake (Muzzaffar et al. 2022). These disturbances may result in weight change, exhaustion, and developmental issues (Hall et al. 2023).

1.4.3 Neurodegenerative Diseases and Food Safety

1.4.3.1 Alzheimer's and Parkinson's Disease

According to scientific studies, degenerative neurological diseases and food toxins appear to be closely related because aluminium can develop in the brain and aid in the making of plaque, which has been connected to Alzheimer's disease. Aluminium is often present in processed foods, packaging and cookware. (Nie 2023). Furthermore, it has been evident that pesticides and herbicides, including paraquat, raise oxidative stress and brain damage, which are considered threat factors for Parkinson's disease (See et al. 2024).

1.4.3.2 Heavy Metal Toxicity and Brain Health

Due to redox imbalance and neurotoxicity, heavy metals such as lead, arsenic, and mercury are exposed through tainted food and water can affect cognitive abilities. Loss of memory, cognitive decline, and developmental disability can result from the long-term creation of these metals. Furthermore, bacterial toxins formed in food, which are produced by clostridium botulinum, can aggravate brain inflammation (Althobaiti 2025; Wang et al. 2024a).

1.5 Chronic Diseases Linked to Foodborne Pathogens

Microorganisms like viruses, bacteria, parasites, and fungi that pollute food and cause both acute and chronic diseases are known as foodborne pathogens (Batz et al. 2013). Some microorganisms from food lead to chronic infection, while most foodborne infections only result in transient gastrointestinal symptoms. This long-term exposure to chemicals and immune response causes diseases that include cancer, neurological illness, liver damage, and kidney damage (Reshma et al. 2023).

Contaminated food can bring out dangerous bacteria like Salmonella, Listeria, and E. coli, which remain inside the body, triggering long-term inflammation and autoimmune disease (Dimitrakopoulou et al. 2024; Munekata et al. 2020). Over a period of time, various toxins can infect organs, including aflatoxins from mouldy cereals and heavy metals found in seafood (Reshma et al. 2023). The World Health Organization (WHO) estimates that foodborne diseases kill around 4 lakh people globally each year and impact 600 million of the population (Pires and Devleesschauwer 2021; Yin et al. 2017). The most weekend vulnerable communities include children, the elderly and those with compromised immune systems.

1.5.1 Overview of Chronic Conditions Like Kidney Disease, Liver Damage, and Neurological Issues

1.5.1.1 Chronic Kidney Disease and Liver Damage

Contaminated food and water can instigate toxicants, heavy metals, and infectious agents that lead to sustained organ damage (Khalaf et al. 2024). Chronic exposure to these toxic substances can cause progressive kidney damage and extreme liver disorders, including hepatitis and cancer (Perera et al. 2024). Therefore, kidney and liver health are closely associated with food safety (Bokhtiar et al. 2023).

The main function of the kidney is to filter harmful poisons from the bloodstream, but continuous exposure to heavy metals like cadmium and lead, which are most frequently found in polluted food and water, can generate nephrotoxicity. Ultimately, these toxic substances or elements limit the function by accumulating inside the kidney (Lin et al. 2023). Furthermore, foodborne E. coli infections, especially from spoiled meat and dairy, can lead to hemolytic uremic syndrome, a serious disease that ultimately results in kidney failure over an extended time period (Lavrek et al. 2018; Pomajzl et al. 2009).

The primary function of the liver is to absorb poisons and toxic components. However, prolonged exposure to dangerous foodborne substances can cause liver dysfunction. Aflatoxin, which is a strong carcinogen that leads to liver cancer, has been created by fungi in decayed grains and nuts (Groopman et al. 2008). Additionally, chronic liver inflammation can be caused by foodborne hepatitis A and E viruses through polluted food and water, which in turn poses a threat to liver cirrhosis and chronic damage. Making certain strong food safety standards, limiting the tainting of heavy metals, and promoting hygienic food handling are all essential factors in preventing kidney and liver diseases caused by contaminated food (Deon et al. 2014; Roya and Akram 2016).

1.5.1.2 Neurological Disorders from Foodborne Pathogens

Foodborne agents and toxicants can severely impact or harm brain health, resulting in neuroinflammation, peripheral nerve damage, and cognitive disablement (Ciobanu et al. 2012). Long-term neurological disorders are caused by bacteria, toxins, and heavy metals in infected food, which makes food safety a critical public health concern. A foodborne contaminant like Listeria monocytogenes is mostly found in unpasteurised dairy, raw vegetables, and processed meat. It has the unique ability to cross the blood-brain barrier (it protects the brain from harmful substances), unlike other bacteria, which results in Listeriosis (Rossi et al. 2008). This infection causes meningitis (infection on the protective membrane around the brain and spinal cord) and encephalitis (inflammation of cerebral tissue), leading to brain inflammation and adverse neurological impacts like seizures, cognitive loss and also a deficiency in movements. Patients with immunodeficiency, pregnant women, and elderly people are more prone to experience these kinds of dangerous results (Senay et al. 2020). Clostridium botulinum, which is found inside an inadequate tinned or conserved food.

Produce one of the most effective neurotoxins, Botulinum toxin. This highly potent toxin inhibits or blocks neurotransmitter production in the neuromuscular junction, which leads to paralysis and persistent nerve damage (Liu et al. 2023), which completely depends on the gravity of the exposure. Botulism is one of the severe conditions caused by toxins in the nervous system that can induce respiratory failure, which leads to difficulty breathing because the respiratory muscles get paralysed. It demands prolonged medication. Recovery might be slow, even after the treatment, since the nerve regeneration is gradual and impacted people have constant and permanent muscle debility and cognitive disability (Habib 2014).

Prolonged exposure to mercury and arsenic through contaminated seafood and groundwater results in neurotoxicity, which impacts brain function. Mercury affects dopaminergic neurons, raising the risk of neurodegenerative disorders like Parkinson's disease. Meanwhile, arsenic hinders memory, cognition and learning ability (Pyatha et al. 2022). Chronic exposure has been related to neuroinflammation, developmental problems and mood changes, especially in children and adults. A neurotransmitter balance is disrupted by these heavy metals, leading to long-term neurological effects (Tyler and Allan 2014).

1.5.1.3 Cancer Risks Associated with Foodborne Toxins

Food contaminants and cancer risks
Food pollutants have always been related to various types of cancers, with some compounds playing a role in carcinogenesis. The prevalence of aflatoxins, N-nitro compounds, and pesticide residues in food poses a significant public health concern.

Aflatoxin and Liver cancer
Aspergillus fungi produce aflatoxins, a toxic component that pollutes grains, nuts and legumes when stored in a muggy environment. Chronic exposure via food

consumption leads to DNA mutation, liver infection and, eventually, hepatocellular cancer (Marchetti et al. 2013). Aflatoxin B1 is a harmful substance and a toxin created by certain fungi that grow on food when stored improperly and interact with DNA to produce adducts that stimulate mutation in important tumour suppressor genes like p53, which normally help prevent cancer. Still, here, it increases liver cancer, particularly in places like Asia and Africa, where food preservation practices are poor. Extended exposure to aflatoxins through contaminated food is a major health concern in this region (Ferreira et al. 2019).

N-Nitroso compounds and colorectal cancer
N-nitroso compounds are chemicals present in processed meat like bacon, sausages, and smoked salmon. When these substances are exposed to high heat or an acidic stomach environment, they produce harmful substances called carcinogenic nitrosamines. Nitrosamines are known to cause cancer, especially in the digestive system. A high intake of NOC has been linked to an increased risk of colorectal cancer since these compounds can disrupt DNA through a process called alkylation, which causes genetic abnormalities and abnormal cell growth. WHO has classified them as GROUP 1 carcinogens because they are proven to cause colorectal cancer (Turesky 2018).

Pesticide Residues and Cancer
Prolonged exposure to pesticide remains in food such as fruits, vegetables, and cereals has been connected to cancers like pancreatic, breast and prostate cancer. Some pesticides include endocrine-disrupting chemicals (EDCs), which interfere with hormone balance in the body and gradually foster unrestrained cell growth, eventually increasing the risk of cancer over time (Kumar et al. 2023). Organochlorine pesticides like DDT have been associated with breast cancer, which can disintegrate hormonal balance and support tumour growth due to their estrogen effect. Furthermore, studies suggest that there is a relation between pesticide exposure and pancreatic cancer is likely because of chronic inflammation and oxidative stress., which result in DNA damage and cancer growth (Miao et al. 2021; Valerón et al. 2009) (Table 1.1).

1.6 Metabolic and Endocrine Disorders

Endocrine-disrupting chemicals (EDCs) are molecules that interfere with the body's hormonal balance. Hormones regulate specific functions like metabolism, growth, reproduction, and immune reactions (Encarnação et al. 2019). These compounds (EDCs) can imitate or block natural hormones, causing the body to respond inappropriately and interrupt normal physiological functions (Stroustrup and Swan 2024). These chemicals can induce metabolic disorders like obesity, diabetes and thyroid dysfunction, as well as reproductive and developmental difficulties when interfering with hormonal balance (Ganakumar et al. 2024). EDCs are commonly detected in several food sources, leading to contamination from industrial processes, covering materials, and farming methods (Vandenberg 2014).

Table 1.1 Causes, examples of long-term health consequences of unsafe food

Health consequences	Causes	Examples	References
Chronic kidney disease	Heavy metal contamination (lead, cadmium)	Contaminated water, polluted food	Khalaf et al. (2024)
Liver cancer	Aflatoxin exposure from moldy grain and nuts	Poor food storage, contaminated crops	Perera et al. (2024)
Neurological disorder	Listeria, Clostridium infection, mercury exposure	Contaminated seafood, dairy products	Pyatha et al. (2022), Rossi et al. (2008), Senay et al. (2020)
Colorectal cancer	N-nitroso compounds in processed meats	Excessive consumption of processed food	Perera et al. (2024)
Metabolic syndrome	Artificial sweeteners, trans fats, BPA exposure	Packaged and processed foods	Perera et al. (2024)
Thyroid dysfunction	Fluoride contamination, goitrogen	Polluted water, cruciferous vegetables	Muzzaffar et al. (2022)

Key sources include:

1. *Pesticides:* Many fertilisers, including organophosphate insecticides, function as endocrine disrupters by interfering with the hormonal system. These pollutants commonly remain in farming but can persist as residue on fruits, vegetables, and grains, causing a threat to consumer's health.
2. *Plastic contaminants:* Certain chemicals contained in plastic covering, such as Bisphenol A (BPA) and Phthalates, seep into food and beverages, especially when heated or extended storage (Janda et al. 2021). BPA mimics estrogen, a key hormone that potentially leads to hormonal imbalances, reproductive issues, and metabolic disorders (Martínez-Ibarra et al. 2021).
3. *Hormone Residues in Meat:* Livestock are frequently provided with additional synthetic hormones such as estrogen, progesterone, and growth stimulants to enhance growth and milk production (Poladia et al. 2007). This residue in meat and all dairy products can alter the endocrine system in humans, eventually increasing the risk of metabolic diseases (Pleadin and Samardžija 2019).

1.6.1 Role in Metabolic Disorders

EDCs are associated with metabolic disorders such as obesity, diabetes, and metabolic syndrome. They improvise fat storage, damage insulin function, and support chronic inflammation. Eventually, everything impacts metabolism (Bokobza et al. 2021; Migliarini et al. 2011). Prolonged exposure to EDCs results in hormonal imbalance that lasts for generations, which points to the need for proper and stricter control and secure dietary practices (Ahn and Jeung 2023).

1.6.2 Impact on Obesity and Metabolism

The relationship between food safety and disorders like obesity, diabetes, and thyroid dysfunction and certain dietary additives like Monosodium glutamate (MSG) and artificial sweeteners and additives like aspartame and saccharin may also contribute to disorders like obesity (Udom et al. 2025). Monosodium glutamate (MSG) is a flavour enhancer often used in processed and packed foods, which causes excessive drive and desire by influencing the hypothalamus, the part of the brain responsible for regulating and managing hunger (Fernandez-Tresguerres Hernández 2005). By having interference with the signals can lead to an excessive and increasing desire for food and increased calorie intake, further contributing to obesity. Meanwhile, artificial sweeteners may interrupt gut microbiota, changing metabolic activity and the increased fat accumulation (Pearlman et al. 2017). Despite being considered a low-calorie alternative to sugar. However, these additives result in severe and unexpected health issues. Additionally, it can trigger resistance and elevate your cravings for sweet foods, resulting in weight gain (Yunker et al. 2020). These hazards can be decreased by reducing your intake of processed foods and choosing complete, nutrient-content foods (Nettleton et al. 2016).

1.6.3 Diabetes and Unsafe Food

Arsenic is a naturally occurring toxicant that is produced or occurs in the environment. It is commonly found in groundwater and can accumulate in foods like rice, seafood, and leafy greens (Saxena and Kumar 2020). Long-term exposure to this toxic element arsenic, especially through drinking water or food, can cause serious health issues, including an increased chance of developing conditions like Type 2 diabetes and cancer (Díaz-Villaseñor et al. 2007). Arsenic inhibits insulin signalling by disrupting pancreatic beta cells, which are essential for producing insulin. Furthermore, it leads to oxidant stress and inflammation, contributing to insulin resistance over time, these effects hinder the body's ability to modulate glucose levels, eventually increasing the risk of developing Type 2 diabetes (Lu et al. 2011). Hence, prolonged arsenic exposure can significantly affect metabolic health by damaging glucose metabolism, causing high blood sugar levels (Giri et al. 2018).

Trans fats are the type of artificial fats mainly found in fried or processed foods, baked goods, and processed snacks. They are produced by the process that converts liquid oil into solid fats, which is known as hydrogenation. Extending the product lifespan (Shoaib et al. 2024). However, these trans fats are more dangerous to metabolic health because they cause inflammation and damage cell membrane function, making cells less responsive to insulin, and this contributes to insulin resistance and turndown glucose absorption, which are the basic characteristics of Type 2 diabetes (Odegaard and Pereira 2006; Zhao et al. 2016). Furthermore, trans fats elevate LDL (bad) cholesterol while decreasing HDL (good) cholesterol, increasing the risk of circulatory disease. Avoiding packaged or processed food and choosing healthy fats like those found in seeds, nuts, and olive oil can help alleviate these risks (Shoaib et al. 2024).

The gut microbiome contains billions of microorganisms that are pivotal in controlling metabolism and overall health. A well-balanced microbiome maintains proper glucose metabolism by producing short-chain fatty acids (SCFAs) that improve insulin sensitivity (Caesar 2024). However, foods that are processed and contain sweets and unhealthy or trans fats can definitely change gut flora composition, which eventually leads to dysbiosis (Miclotte and Van de Wiele 2022), an imbalance that creates inflammation and insulin resistance (Ispas et al. 2023). Furthermore, a disordered microbiome can hinder the generation of critical metabolites that control blood sugar levels (Abuqwider et al. 2023). A food that is high in fibre, fermented and probiotics. These can help to regulate glucose and lower the chance of developing Type 2 diabetes (Howard et al. 2022).

1.6.4 Thyroid Dysfunction and Food Safety

The thyroid gland manages and controls metabolism, growth, and energy generation by secreting hormones. However, some of the environmental and dietary variables might delay its function, leading to thyroid-related illness (Dev et al. 2016). Goitrogens are naturally found chemicals in foods like soy, broccoli, cabbage, and other cruciferous vegetables (Muzzaffar et al. 2022). This chemical interferes with the consumption of thyroid gland iodine intake, which is required for thyroid hormone synthesis. In iodine-deficient populations, excessive consumption of goitrogenic foods may cause goitre (Thyroid enlargement) (Chandra 2010). However, reducing the goitrogenic effects in these foods by cooking these veggies, and making a moderate intake is generally safe (Chandra et al. 2005).

Heavy metals like lead and cadmium are environmental toxicants that can build up in the body and affect and damage the thyroid function (Kim et al. 2022). These metals interfere with the endocrine system and damage thyroid cells by disrupting thyroid hormone production (Brix et al. 2023). Chronic exposure leads to redox imbalance and inflammation, which turn down the thyroid gland to secrete hormones (Bushra et al. 2024). Tainted food, cigarette smoking, and industrial pollution are the notable sources of heavy metal exposure, which increase the threat of hypothyroidism and other thyroid issues (Sonia and Seth 2019).

Fluoride is a naturally occurring mineral found mostly in soil, water and certain foods, often added to drinkable water for oral hygiene which can eventually affect the thyroid function (Mobeen and Kumar 2017). Fluoride ions may hinder iodine absorption in areas with high fluoride levels, causing decreased production of thyroid hormone production (Malin et al. 2018). If the iodine consumption is less, then this disruption can cause hypothyroidism (O'Neill and Quinn 2013). Monitoring fluoride content in drinking water and ensuring iodine adequate iodine presence can help reduce these risks (Neeti and Singh 2023). Adopting preventive measures like including a balanced diet and regulating environmental exposure can help safeguard thyroid function and enhance overall well-being (Brent 2010).

1.7 Neurodegenerative Diseases and Food Safety

1.7.1 Neurotoxins in Food

Heavy metals like lead, mercury and arsenic settle in the body over a longer period of time, which gradually results in oxidative stress and inflammation, which are connected to neurodegenerative illnesses like Alzheimer's and Parkinson's diseases. Pesticide remains found in vegetables, fruits, and cereals can pass the blood-brain barrier, leading to neuronal injury and cognitive damage (Althobaiti 2025). Furthermore, bacterial toxicants, such as botulinum and mycotoxins present in tainted food, can damage neurotransmission, resulting in neurological symptoms (Gatto et al. 2021).

1.7.2 Alzheimer's Diseases and Unsafe Food

Aluminum, mercury, artificial food colours and food preservatives are common pollutants in packed or processed foods that can ultimately harm brain health (Althobaiti 2025; Brent 2010; Gatto et al. 2021). Aluminum, which is prevalent in processed foods, baking powders, and packaging materials, has been related to an elevated risk of Alzheimer's disease (Nie 2023). Mercury, which is commonly found in seafood such as tuna, swordfish, and mackerel, is a highly toxicant. Long-term exposure can damage cognitive functioning, reduce memory capacity, and eventually lead to developmental problems in children (Choi et al. 2014). Mercury's tendency to pass the blood-brain barrier makes it especially toxic and even dangerous, contributing to neurological illness and cognitive damage (Takahashi and Shimohata 2019). Artificial food dyes and preservatives are always used to improve colour, flavour and durability and can potentially damage brain function. Certain dyes, such as Red 40 and Yellow 5, are the reasons for disrupting neurotransmitter action, which leads to memory loss and brain dysfunction (Damotharan et al. 2024). Furthermore, preservatives such as sodium benzoate and nitrates may obstruct brain signalling and increase oxidative stress (Khoshnoud et al. 2018).

1.7.3 Parkinson's Disease and Food Contaminants

Parkinson's disease (PD) is a progressive neurological condition that is characterised by motor disease, tremors, rigidity, and poor stability (Yaliman and Şen 2011). Emerging scientific studies suggest that chronic exposure to particular dietary toxins may raise the chance of developing Parkinson's disease (PD) (Ghosh et al. 2024). Pesticide remains are most widely found in fruits, vegetables, and cereals (Hadian et al. 2019) and contain neurotoxic chemicals that can impede dopamine-secreting neurons in the brain (Selvasubramaniam et al. 2008). There is an increased risk of Parkinson's disease due to extended exposure to organophosphate and paraquat.

Heavy metals, namely Lead, Cadmium and Mercury, mostly present in polluted water, seafood and processed foods, can pass the blood-brain barrier, contributing to oxidative stress and neurological infection (Rehman et al. 2018). These metals advance neuronal degeneration by disrupting mitochondrial function, leading to dopamine depletion, which is a hallmark of Parkinson's disease (Tian et al. 2021). Furthermore, foodborne toxins like bacterial endotoxins and fungal mycotoxins can stimulate systemic inflammation, aggravating neurodegeneration. Chronic exposure to these toxins may hasten the development of Parkinson's symptoms (Brown 2019).

1.7.4 Impact of Metals and Foodborne Toxins on the Nervous System

Lead exposure from food packaging can considerably damage children's brain development (Canfield and Jusko 2008). Lead can flow directly into food through contaminated packaging, mainly through cans or containers with Lead-based solder (Diviš et al. 2017). Lead accumulates in the body, pierces the blood-brain barrier and disturbs cognitive function (Santa Maria et al. 2019). This metal gets exposed among children, and due to its neurotoxic effect, it might result in lowered IQ, learning disability and behavioural problems (Gatzke-Kopp et al. 2024).

Another important concern is arsenic contamination in specific areas where groundwater and rice are an important source of food (Tiwari 2019). Rice is a major source of arsenic exposure and is absorbed more quickly from soil and water (Bhardwaj et al. 2019). Chronic exposure can cause neurological damage, which gradually leads to cognitive impairment, memory loss, and an increased risk of neurological sickness (Zarazúa et al. 2013). In addition, foodborne bacteria present in contaminated fruit, dairy, or undercooked meat led to chronic inflammation. Some bacteria, like Listeria monocytogenes and Escherichia coli, secrete endotoxins that operate the immunological response (Pierre 2014). While the body's defence mechanism fights infection, long-term inflammation may result in neurological illness, eventually increasing the chance of neurological disruptions, including Alzheimer's and Parkinson's (Skaper et al. 2018).

To label this distress, proper and strict food safety standards, reduce environmental pollution, and promote public awareness must be executed (Gałkowska et al. 2023). Furthermore, taking up safer packaging alternatives and preserving a clean water supply are very important in reducing exposure and protecting mental well-being (Schwendimann et al. 2021). Unsafe food has serious and extended health consequences, involving chronic disease, Neurological deformities and metabolic conditions. When the food is contaminated due to hazardous substances like heavy metals, pesticides, and foodborne viruses (Amouei et al. 2024; Bintsis 2017; Feng et al. 2019). These toxic substances can accumulate in the body, which eventually results in oxidative stress, inflammation, and cellular damage (Mesa-Garcia et al. 2018). To reduce these health hazards, strong food hygiene laws are essential (Crawford 2014). Governing bodies or regulatory agencies must implement strict food manufacturing, storage and (Nollet and Ali 2024) criteria (Nollet and Ali 2024). Additionally, consumer awareness is a crucial factor in preventing foodborne infections (Lin 2014a). Educating the general public on food hygiene, proper food handling techniques and recognising pollution symptoms can seriously reduce health issues (Fanaselle et al. 2022).

Therefore, global cooperation in food safety monitoring, policy change and transparent supply chain management is critical. By intensifying the food safety framework and promoting public education, the government and other important authorities can ensure safer, healthier and hygienic food for everybody (Kumar et al. 2021).

1.8 Foodborne Disease: Control and Prevention Methods

Foodborne diseases, or food poisoning, are any sickness caused by eating contaminated food or drinking polluted water. Contaminants can include pathogenic bacteria, viruses, parasites, or toxin substances (Adley and Ryan 2016; Kureljušić et al. 2021). Foodborne illness is a major worldwide health issue. According to the World Health Organization (WHO), nearly one in ten people fall ill each year due to foodborne pathogens, with 33 million healthy life years lost. There are about 6 million cases of foodborne illness each year, leading to 420,000 deaths. Particularly children under the age of five, who account for 30% of all foodborne diseases-related deaths. These troubling figures highlighting the need for better food safety practices around the world (Muthusamy et al. 2024).

1.8.1 Global Governance for Food Safety

Food safety regulations are essential worldwide to ensure that food is safe for consumption and to prevent contamination and health risks. Different countries and international organizations create these specific guidelines to regulate food production, processing, and distribution. Global food safety regulations created by international organizations such as the World Health Organization (WHO) and the Codex Alimentarius Commission are vital for making sure food is safe, high-quality, and trustworthy across global supply chains (Hill 2014; Jackson and Jansen 2010). Therefore, The Codex Alimentarius, developed together by WHO and the Food and Agriculture Organization (FAO), offers a set of internationally recognized standards and guidelines designed to reduce foodborne risks and promote fair trade practice (Boutrif 2014).

1.8.1.1 How Important Are Food Safety Standards for Public Well-Being?

Adhering to food safety standards is crucial for safeguarding public health, building consumer trust, and ensuring safe international trade. These standards are key to preventing foodborne diseases, particularly in vulnerable populations such as children and the elderly (Madilo et al. 2024). Various regulations exist to manage food safety, requiring cooperation among food manufacturers, regulatory agencies, and consumers. For example, the European Union aligns its regulations with international guidelines from the Codex Alimentarius (Lindner 2008). The COVID-19 pandemic has emphasized the need for stringent safety measures within the food industry, prompting enhancements like real-time monitoring systems to minimize human error (Pepito et al. 2023). Regular inspections play an essential role in addressing compliance issues and preventing outbreaks of foodborne diseases (Chatzimpyrou et al. 2025).

1.8.1.2 Challenges in Maintaining Food Safety Implementation

1.8.1.2.1 Differences in Food Safety Laws Across Nations

Food safety regulations differ greatly from country to country because of various factors such as legal systems, risk assessment approaches, and how well these laws are enforced. Although international organizations like the World Health Organization (WHO) and Codex Alimentarius provide general guidelines for food safety, each country decides how to adopt and use these standards based on its own policies. This fluctuation eventually led to a challenge to maintain consistent food safety practices beyond the global distribution.

- Developed nations such as the USA (Humphrey 2012), EU (Rusconi 2016) and Japan (Kirchsteiger-Meier and Baumgartner 2014) implement strict and stringent food safety regulations such as HACCP and EFSA standards (EU), with strict penalties for non-compliance (Kirchsteiger-Meier and Baumgartner 2014; May 2010; Rusconi 2016).
- Developing countries often have weakened enforcement due to limited resources, inadequate testing infrastructure and corruption. Therefore, frequently lack of necessary financial and qualified human resources to support regulatory authorities efficiently (Arblaster and Hooper 2015).
- Harmonization problems arise in worldwide trade because food exports must follow different rules and regulations in each market. This can reason trade clashes and make economic barriers (Bittisnich 2024).

1.8.1.2.2 Barriers to Regulations Enforcement in Developing Nations

In many developing nations, institutions that enforce laws often make it hard for them to function efficiently. Political instability along with bureaucratic issues further weakens these governance structures (Mussagulova 2021). Also, high levels of corruption within government and law or regulation enforcement erode public trust and hinder the enforcement of laws (Otairu et al. 2013). Low public awareness and cultural norms contribute to non-compliance, especially in non-western countries where people often reject imposed rules (Araujo and de Souza 2010; Buehn and Schneider 2012). Strict rules and regulations may push businesses into the informal economy to prevent burden (Araujo and de Souza 2010; Buehn and Schneider 2012). Extended case proceedings and ineffective agency associations lead to lengthy case resolutions and undermine trust in the judicial system (Contini 2023; Jiang 2006; Kamil Lakho 2021). Corporate lobbying and international pressure also affect enforcement priorities. To improve governance, its essential to strengthen institutions and simplify regulations, boosting economic growth and reducing the informal economy (Anaya-Muñoz 2024; Choi et al. 2015; Sadrieh and Annavarjula 2005). Effective governance must tackle cultural, economic, and institutional factors to enhance compliance.

1.8.1.2.3 Globalization and Rising Supply Chain Complexity

Globalization has added complexity to supply chains through factors like varied customer needs, a wider range of products, and uncertainty about suppliers. While this complexity often negatively impacts operational performance, it can enhance innovation and financial results. Increased connection among supply chain parts means that problems at any point can lead to risks throughout the entire system, companies are integrating their supply chains to better manage this complexity, through benefits can vary based on each country's logistics capacities. Additionally, globalization introduces new risks, making strong risk management practices crucial for ensuring a smooth flow of goods during disruptions. Companies must concentrate on key factors such as supplier reliability, accurate forecasting, and effective information sharing.

1.8.1.3 The Future Trends in Food Safety Regulations: Innovations and Shifts

Food safety rules are emerging because consumers are more aware, new technologies are developing, and global health concerns are rising, controlling agencies are making stricter rules and regulations to ensure food safety, transparency, and quality while also encouraging innovation in the industry, Issues like climate change and disruptions in supply chains are leading to the need for more adaptable rules.

Key Emerging Standards in Food Safety Regulations:

- *Stricter Ingredient Regulations:* This means protecting consumers from harmful food ingredients and contaminants. They help ensure food is safe, reducing the risk of disease and health issues (Aruoma 2006, 2014; Barney et al. 2023).
- *Global Harmonization of Standards:* Organizations like Iso and Code Alimentarius are working to unify global food safety rules and regulations (Heilandt et al. 2014; Moy and Godefroy 2024; Stankovic 2016).
- *Technological Integration:* New advancements in AI, blockchain, and IoT are improving food traceability and detecting contamination (Liu et al. 2025; Pal and Kant 2019; Sreevathsan et al. 2023).
- *Alternative Food Regulations:* New safety assessments are required for lab-grown meat and plant-based proteins due to potential risks and the need for comprehensive regulatory frameworks (Bennie et al. 2025; Bogueva and McClements 2023; Hadi and Brightwell 2021).

1.8.2 Role of Food Handling in Contamination Prevention

Maintaining proper hygiene in food preparation and storage is crucial in foodborne diseases and ensuring food quality. Effective hygiene practices include personal hygiene safe handling of food, and correct storage techniques to reduce contamination risks.

1.8.2.1 The Importance of Personal Hygiene in Preventing Contamination

Hand Hygiene: Regular hand cleaning with soap helps prevent microbial transfer and contamination.

Protective Clothing: Wearing clean clothes, gloves, and hairnets lowers the risk of contamination.

Good Manufacturing Practices: Ensures hygienic workstations and equipment are sanitized and staff are trained properly.

1.8.2.2 Safe Handling Practices for Raw and Cooked Food

Separation of Raw and Cooked Food: Keep raw and cooked food apart using different utensils and storage to avoid cross-contamination.

Temperature control: Cooking at a safe temperature kills pathogens while storing them properly to prevent them from being in the "danger zone" (5–60 °C).

HACCP Implementation: Identifies and Controls hazards at critical control points to manage risks during food processing.

1.8.2.3 Best Strategies for Food Safety and Quality

Temperature & Humidity Control: Keep refrigeration below 5 °C and freezing at −18 °C to slow bacterial growth.

Advanced packing: Use Vacuum sealing and modified atmosphere packing (MAP) to extend shelf life by reducing oxygen exposure.

FIFO & Expiration Management: Use older stocks first to prevent waste and ensure food safety.

1.8.2.3.1 Ensuring Food Safety: Critical Control Points in the Food Supply Chain

Food safety is essential for preventing health hazards related to contaminated foods or products, requiring more vigilance from production to consumption. The Hazard Analysis and Critical Control Point (HACCP) system is a crucial role in identifying and managing hazards at critical control points (CCPs) where risks can be effectively reduced. Complementing HACCP, Good Manufacturing Practices (GMP) provides a framework of fundamental principles that make sure consistent quality throughout food production, including maintaining sanitation and proper hygiene. Together, these systems foster a proactive approach to food safety, emphasizing the need for regular training and monitoring among all personnel involved in the food supply chain. By effectively implementing HACCP and GMP, food businesses not only protect public health but also support their reputations and instil greater consumer confidence in the safety of their products. This comprehensive focus on safety and quality ultimately contributes to a healthier food environment to everyone.

1.8.2.3.2 Identifying Key Contamination Risks from Farm-to-Table Journey

Food contamination can happen at different stages from farming to eating. On the farm, risks come from polluted soil, contaminated water, and pesticide residues. During processing, lack of sanitation and cross-contamination can lead to bacterial growth (Radford 2019; Ummulkhair Salamah et al. 2024). Supply issue problems, such as temperature abuse and poor packaging, further threaten safety. At home, improper storage, handling, and preparation increase the risk of contamination. To check food safety, following strict protocols, complying with laws and regulations, and raising consumer awareness is essential (Ahmad et al. 2019; Ummulkhair Salamah et al. 2024).

1.8.2.3.3 Hazard Analysis and Critical Control Points (HACCP) Systematic Approach

The Hazard Analysis and Critical Control Points (HACCP) system is a science-based approach to food safety that identifies and controls potential hazards in the food production process. This proactive method focuses on preventing biological, chemical, and physical hazards rather than relying on end-product testing.

Key Principles of HACCP

- *Analysis of Hazards:* Analysis of probable hazards at the stages of food production.
- *Identification of Critical Control Points (CCPs):* Determining stages where risks can be successfully stopped.
- *Monitoring of The Procedures:* Realizing measuring to track CCP compliance.
- *Takes Corrective Actions:* Identifying deviations to prevent unsafe food.
- *Analysis or Verifications of Procedure:* Ensuring the system functions efficiently through audits and reviews.
- *Record-keeping and Documentation:* Maintaining records to track regulations.

1.8.2.3.4 Roles of Good Manufacturing Practices in Food Processing

Good Manufacturing Practices (GMP) ensure food safety by maintaining hygiene while enforcing sanitation protocols and standardizing processing, packing, and storage (Turatti 2007; Walia et al. 2023). Employee training further minimizes contamination hazards. Integrating HACCP and GMP strengthens hazard control while ensuring regulatory compliance and consumer protection (de Oliveira et al. 2016; Varzakas 2016).

1.8.2.4 Foodborne Pathogens and Modes of Transmission

Foodborne pathogens, including bacteria, viruses and parasites, pose serious hazards to public health conditions. Common pathogens like Salmonella, Escherichia coli and Listeria monocytogenes can cause severe infections. Viral pathogens, including Norovirus and Hepatitis A, spread quickly through contaminated water and foods (Liang et al. 2022; Pakdel et al. 2023). Parasitic infections from Toxoplasma gondii and Giardia also contribute to foodborne diseases. Food contamination can occur through various pathways, such as direct contact with contaminated surfaces, poor hygiene practices, cross-contamination during food processing, and improper cooking or storage. The primary sources of these pathogens like water, soil, and raw food products (Kwon et al. 2025; Liang et al. 2022; Nerín et al. 2016; Pakdel et al. 2023).

1.8.2.4.1 Emerging Foodborne Pathogens: Risks and Mitigation Strategies

Foodborne infections are changing due to globalization, climate change, and intensive food production, they are becoming more harmful and adaptable. This rise of zoonotic diseases that transfer from animals to humans, novel bacterial strains, and rapid viral changes make the complexity of prevention and control harder. Antimicrobial resistance (AMR) is a major growing issue caused by overusing antibiotics in farming and medicine. Resistant strains of E. coli, Salmonella, and Campylobacter lead to more severe infections and larger outbreaks. Combating AMR, we require better surveillance, responsible antibiotic use, and enhanced food safety practices to measure to protect public health.

1.8.2.5 The Key Roles of Food Packaging in Preventing Contamination

Food packing is essential in the food industry. Improvements in packing technologies, such as vacuum packing and eco-friendly materials, these enhance functions (Silvestre and Cimmino 2013). Smart packing provides information about food freshness, also increasing safety and consumer trust. Overall, advancement in food packing balance safety quality, and environmental responsibility, meeting the needs both consumers and producers. It protects and preserves food, keeping fresh and safe from spoilage and contamination.

1.8.2.5.1 Importance of Food Grade Packing Materials

Food-grade materials are designed to be safe for direct contact with food, preventing harmful chemicals from migration and contamination. They must meet regulatory standards to ensure consumer safety and maintain product quality.

Non-toxic and chemically stable: Prevents harm and unwanted substances from entering food.

Moisture and gas barrier: keep moisture and lower spoilage by controlling exposure to oxygen and humidity.

Temperature-resistant: Maintains food stable in various storage conditions.

Durability and hygiene: Prevent against physical damage and microbial contamination.

1.8.2.5.2 Role of Intelligent and Active Packaging in Food Safety

The role of intelligent and active packaging in food safety is becoming increasingly significant. Intelligent packaging is equipped with sensors that monitor the freshness of food and can detect spoilage, providing real-time information about the product's quality. On the other hand, active packaging works by releasing antimicrobial agents and controlling moisture levels to extend the shelf life of food products. Together, these technologies not only enhance food safety but also help in reducing waste and improving overall quality control in the food supply chain.

1.8.2.5.3 Sustainable Packaging Solutions and Challenges of Plastic Waste

The excessive reliance on plastic in food packaging creates serious environmental challenges, highlighting the need for sustainable options. Biodegradable materials decompose naturally, which helps cut down pollution. In addition, recyclable packaging promotes reuse and decreases waste. Edible coatings and films offer a protective layer for food without contributing to plastic waste (Table 1.2).

Table 1.2 Common foodborne pathogen and their preventive measures

Pathogen	Common food sources	Symptoms	Preventive measures	Reference
Salmonella spp.	Dairy products, eggs, poultry, meat	Abdominal cramps, diarrhea, vomiting, fever	Good hygiene practices, proper food cooking, avoid cross-contamination	Reshma et al. (2023), Gourama (2020), Rai and Bai (2014)
Escherichia coli	Contaminated water, raw milk, undercooked beef	Abdominal pain, vomiting, fever, severe diarrhea	Wash hands properly, through cooking of meat, pasteurization	Gourama (2020), Rai and Bai (2014), Reshma et al. (2023)
Bacillus cereus	Pasta, rice, starchy foods	Vomiting, nausea, diarrhea	Proper storge of cooked foods, reheating throughly	Gourama (2020), Reshma et al. (2023)
Norovirus	Shellfish, contaminated water	Stomach pain, vomiting, diarrhea, nausea	Safe water sources, good hygiene practices, proper cooking of shellfish	Gourama (2020), Rai and Bai (2014), Reshma et al. (2023)
Vibrio spp.	Raw or undercooked seafood	Chills, abdominal cramping, vomiting, fever, diarrhea	Proper cooking of meat, good hygiene practices, avoiding contaminated water	Gourama (2020), Rai and Bai (2014), Reshma et al. (2023)
Campylobacter jejuni	Unpasteurized milk, poultry, water	Bloody diarrhea, fever, abdominal cramping, vomiting	Avoiding raw milk, safe water sources, proper cooking of poultry	Gourama (2020), Rai and Bai (2014), Reshma et al. (2023)
Listeria monocytogenes	Unpasteurized dairy, ready-to-eat meats	Muscle aches, severe cases can lead to meningitis	Cooking ready-to-eat meats, proper refrigeration, unpasteurized dairy	Gourama (2020), Reitano et al. (2024), Reshma et al. (2023)

1.8.3 Innovative Technologies in Food Safety

1.8.3.1 Evolving Technologies in Food Safety

Evolving technologies in food safety are fundamentally transforming our approach to ensuring the quality and safety of food products. Advancements in detection, monitoring, and traceability are paving the way for innovations like nanotechnology, blockchain, and artificial intelligence (AI). These technologies significantly enhance food quality control by reducing the risks of contamination and bolstering consumer confidence in the safety of the foods they consume. As these innovations continue to develop, they promise to create a more secure and trustworthy food supply chain.

1.8.3.2 Nanotechnology for Pathogen Detection

Nano-sensors have unique physical and chemical properties at the nanoscale, enhancing their sensitivity and specificity in detecting microbial contamination (Awlqadr et al. 2025; Bülbül et al. 2015; Singh et al. 2024; Stephen Inbaraj and Chen 2016). They provide quick results, which is essential for preventing foodborne diseases. By integrating with bio sensing techniques, nano-sensors enable real-time monitoring of pathogens such as Escherichia coli and salmonella, as well as toxins and other contaminants, including toxins and pesticides (Kuswandi et al. 2017; Lin 2014b). Additionally, Nano-sensors have important applications in food safety, helping assess freshness, shelf life, and authenticity, and are integrated into smart packaging for ongoing monitoring during storage and transport (Awlqadr et al. 2025; Ghosh et al. 2022; Nath et al. 2023; Tiwari et al. 2023). They are portable and easy to use, allowing for on-site testing without needing complex lab equipment. These sensors also support the simultaneous detection of multiple contaminants, making them more efficient than traditional methods (Awlqadr et al. 2025; Kuswandi et al. 2017). However, challenges like commercialization and regulatory approval still exist, requiring further research to improve their performance and safety. Future innovations are expected to come from new nanomaterials and the integration of technologies like artificial intelligence (AI), they are boosting the role of nano-sensors in food safety management.

1.8.3.3 Antimicrobial Applications

Nanoparticles, such as silver (AgNPs), Zin oxide (ZnONps), and Chitosan (ChNPs), exhibit significant antimicrobial properties, making them highly effective in reducing the risk of foodborne diseases by inhibiting bacterial growth on food surfaces and packing. Nanotechnology's integration into food safety enhances contamination control, reduces spoilage, and minimizes chemical preservative use, ensuring safer and higher-quality food products (Darshan et al. 2024; Ghosh et al. 2022; Koilparambil et al. 2020; Lin 2014b; Nakazato et al. 2017; Steffi et al. 2024; Yoo et al. 2021).

1.8.3.4 Blockchain for Supply Chain Integrity and Fraud Detection: Revolutionizing Food Safety

Blockchain technology is transforming food safety by enhancing transparency and traceability in the supply chain. This minimizes risks of contamination and fraud, enabling the identification of fake or misbranded products, such as those with false expiration dates. Consumers gain access to detailed product information, fostering trust in food brands. It provides a secure digital record of each transaction, allowing real-time tracking of food from origin to distribution.

1.8.3.5 Machine and AI Learning in Food Safety Monitoring

Machine learning (ML) and artificial intelligence (AI) are transforming real-time food safety monitoring. AI rapidly identifies spoilage and harmful microbes, enabling faster response. Therefore, these technologies analyze data from sensors and supply chains to detect contamination and ensure compliance. Automation in food processing reduces human errors and maintains hygiene. Machine learning improves detection by recognizing patterns and predicting risks. AI enhances food traceability and transparency when used with blockchain and the Internet of Things (IoT), allowing quick responses to contamination. As AI continues to develop, it will further improve food quality, reduce foodborne diseases and public health problems.

1.8.3.6 Rapid Detection and Biosensors Techniques

Biosensors are innovative devices that use biological components and physical sensors to quickly identify various substances, including pathogens, contaminations, and toxins, in real-time. They have more benefits compared to older detection techniques, such as quicker results, greater sensitivity and capability to analyze samples on-site without needing well trained staff or personnel (Rah 2000; Rasooly and Herold 2006; Sharma and Mutharasan 2013). Real-time monitoring allows for quick action to reduce contamination risks. Portable testing kits with biosensors help business and consumers check food safety on-site, without lab equipment. This simplifies monitoring for restaurants, manufacturers, and regulators. Integrating with AI and Internet of Things (IoT) improves tracking. Biosensors are an efficient and cost-effective solution for food safety and quality control.

1.8.3.7 Automation and Robotics in Food Processing

AI quality control checks for defects and monitors ingredient quality to meet safety standards. Machine vision and deep learning enable real-time contamination detection, making food processing safer and more efficient. Advanced methods such as UV light and ozone sanitize equipment and food, killing germs and cutting chemical use (Rozno 1959; Singh et al. 2021). Automation and robotics enhance food safety by reducing human contact, lowering contamination issues, and increase efficacy. Automated systems accurately sort, cut and package food, maintaining cleanliness and consistency.

1.8.3.8 Emerging Trends in Food Safety Pathogen Detection and Monitoring

Smart food labels provide real-time information about food quality. These labels change color or use sensors to indicate spoilage and expiration, helping reduce waste and ensure safety. The Internet of Things (IoT) enhances food safety monitoring with connected sensors and automatic alerts. IoT devices track storage conditions and identify contaminations, increasing transparency and reducing risks in the supply chain. Predictive analytics improves food safety by providing outbreaks using data from supply chains, weather, and consumer reports. AI Models identify risks early, allowing for preventive actions.

1.8.4 Consumer Responsibility in Food Safety

1.8.4.1 Safe Food Practices at Home

To keep food safe at home, clean hands, utensils, and surfaces before and after food preparation. Cook food at the right temperature to kill harmful bacteria and refrigerate it to slow the growth of germs. Handle raw meat, dairy, and fresh produce carefully by keeping raw meat separate from other foods, storing dairy at the correct temperature, and washing produce thoroughly. Moreover, watch for signs of spoilage, such as changes in texture, color, or smell, and always check expiration dates. By following these steps, you can decrease health hazards and make sure food safety at home.

1.8.4.2 Role of Unknown to Known in Food Safety

Food safety education is important to reducing and preventing foodborne illnesses (Murray et al. 2017). Effective programs can provide practice and knowledge among target groups, such as low-income families (Murray et al. 2017). Measuring success through surveys before and after the programs supports tracking changes in behavior (Milton and Mullan 2010). Most digital tools have to provide access to food safety information very easily at any time. Apps such as MyWarung© help users learn better practices to avoid food poisoning. Success depends on good data addressing and dressing privacy concerns. Government agencies and NGOs raise awareness through programs like fight BAC and partnerships with organizations like the WHO and FAO. Private and Public partnerships also help set and enforce food safety standards, while campaigns educated consumer about safe food handling practices (Fraser and Miller 2014; Gerardi 2023; Nichols and McConnell 2012; Qian et al. 2023; Redmond and Griffith 2006; Zaujan et al. 2021).

1.8.4.3 Costumer Awareness and Role of Food Labelling

Food labelling is essential for safety and awareness to public. Labels provide critical information on expiry dates; those ingredients and they often miss nutritional information. Term like "organic" and "non-organic" can confuse consumers. An effective label depends on shopper understanding, and many are willing to pay more for eco-labelled products due to food safety concerns. In Europe, strict rules and regulations require clear allergen and nutrition information. Public awareness campaigns are essential to help consumers interpret labels, ultimately fostering trust in food safety.

1.8.4.4 Ensuring Food Safety Through Sustainable Consumer Practices

Reducing food waste is essential for sustainability and food security. We can do this through meal planning minimizing leftovers, and freezing food promptly. While some eco-friendly actions, such as eating expired food or using unwashed reusable bags, may pose food safety risks, public agencies should combine sustainability and food safety guidance to address these issues (Massari et al. 2022). Consumers increasingly prefer sustainable and ethically produced options, like organic, local, and plant-based foods, which lower environmental impact and improve health (O'Neill et al. 2023). The eco-friendly coating can help keep fruits and vegetables fresh for long, which reduces waste. It's also important to practice good home sterilization by cleaning reusable bags and containers. This prevents cross-contamination and makes sure food is safe to eat (Ammar et al. 2025).

1.8.5 The Social and Economic Impact of Foodborne Diseases

1.8.5.1 Economic Impacts of Foodborne Illnesses

Foodborne diseases create a major economic burden on businesses and the healthcare system, especially in developing countries. For example, the 2008 listeriosis outbreak in Canada cost about $242 million CAD (Sanitation and HACCP 2013;

Thomas et al. 2015). In the U.S., these illnesses cause around 48 million cases, 128,000 hospitalizations, and 3000 deaths annually, putting pressure on healthcare and leading to lost workdays (Sanitation and HACCP 2013). Developing countries pose even greater challenges due to weaker safety regulations and higher disease rates; these often lack the resources to manage outbreaks, which impacts public-income countries, the economic burden can exceed $100 million a year, affecting the most vulnerable (Grace 2023). Overall, foodborne illnesses burden healthcare, productivity and economic stability, especially in developing countries.

1.8.5.2 Psychological and Social Burden of Foodborne Illnesses

Foodborne diseases can seriously impair social trust and consumer confidence in the food industry, as seen in the BSE outbreak in Europe, which created extensive fear about food safety (Lemyre et al. 2009). Ongoing safety issues and food fraud worsen this mistrust, leading to more regulations (Kendall et al. 2019). These outbreaks can cause anxiety and emotional distress, similar to the increased fear experienced during COVID-19 pandemic (Lu and Lin 2021). Long-term health effects, like cancer and heart disease from harmful substances such as arsenic, add to the problem (Oberoi et al. 2019). Additionally, conditions like irritable bowel syndrome (IBS) can develop after childhood gastroenteritis (Cremon et al. 2014). These issues underscore the need for strong food safety measures and clear communication to restore public trust and alleviate anxiety (Reitano et al. 2024).

1.8.5.3 Government and Industry Strategies for Reducing Foodborne Diseases

Government and industry strategies for reducing foodborne illnesses encompass several key areas, including strengthening surveillance and reporting systems to quickly detect and respond to outbreaks, which has been supported by combining food safety programs and creating a clear grading system that improves public health. Public-private partnerships enhance food safety by sharing resources and expertise, with programs like Hazard Analysis Critical Control Points (HACCP) helping to reduce food contamination. Additionally, global crisis management and quick response strategies are vital for handling cross-border foodborne diseases through the establishment of international standards and risk-based methods that improve food safety; together, these efforts ensure food system can effectively prevent and respond to outbreaks.

1.9 Conclusion and Future Directions

Food safety is inevitable for public health as foodborne contaminants, toxic components and inappropriate handling can adversely affect physiological functions, especially those with digestive, immunological and endocrine systems. These vulnerabilities result in a series of health issues ranging from acute gastrointestinal discomfort to chronic diseases such as metabolic syndrome, neurodegenerative conditions and cancer. Food safety is inevitable for public health as foodborne contaminants, toxic components and inappropriate handling can adversely affect physiological functions, especially those with digestive, immunological and endocrine systems. These vulnerabilities result in a series of health issues ranging from acute gastrointestinal discomfort to chronic diseases such as metabolic syndrome, neurodegenerative conditions and cancer. This change in gut microbiota caused by toxicants represents the complex relationship between food safety and general health. Prolonged exposure to dietary contaminants such as pesticides, heavy metals, and industrial chemicals highlights the essential need for improved public knowledge and regulation of this food safety among public via policy-making, which eventually mitigate the health hazards and protect health outcomes. Negligence of food safety violations can cause immediate health concerns which lead to chronic health problems. This interconnection emphasises the requirement for a collaborative approach among all representatives and authorities in food production and consumption, which makes food safety development a collective responsibility that deeply impacts societal health and demands urgent action to safeguard public well-being.

Food safety is gradually challenged by significant issues like climate change, globalisation of food supply chains, and the alarming rise of antimicrobial resistance among pathogens that commonly cause foodborne diseases. Each of these interconnected issues has the potential to compromise public. In response to these obstacles, innovative technologies make an appearance as a promising solutions for labelling these problems. Examples like nanotechnology help to improve preservation and food safety, meanwhile blockchain technology enhance traceability and transparency in the food supply chain. Furthermore, rapid pathogen detection systems permit for faster identification of dangerous bacteria and viruses in our food, which leads to faster and more effective breakout responses and ultimately protects consumers.

Ongoing research plays a critical role in understanding the long-term health effects of foodborne contaminants and in developing effective interventions designed to prevent foodborne illnesses. As scientific knowledge continues to grow and evolve, collaboration among public health authorities, food industries, and consumers becomes absolutely essential for success. By fostering greater vigilance, promoting education, and utilizing technological advancements, we can work together towards a safer and healthier future for global food safety. This collective effort is vital for protecting public health and ensuring that everyone has reliable access to safe food, thereby contributing to the well-being of communities as a whole. Ongoing research plays a critical role in understanding the long-term health effects of foodborne contaminants and in developing effective interventions designed to prevent foodborne illnesses. As scientific knowledge continues to grow and evolve, collaboration among public health authorities, food industries, and consumers becomes absolutely essential for success. By fostering greater vigilance, promoting education, and utilizing technological advancements, we can work together towards a safer and healthier future for global food safety. This collective effort is vital for protecting public health and ensuring that everyone has reliable access to safe food, thereby contributing to the well-being of communities as a whole.

References

Abuqwider J, Corrado A, Scidà G, Lupoli R, Costabile G, Mauriello G, Bozzetto L (2023) Gut microbiome and blood glucose control in type 1 diabetes: a systematic review. Front Endocrinol 14. https://doi.org/10.3389/fendo.2023.1265696

Adam RD (2001) Biology of Giardia lamblia. Clin Microbiol Rev 14(3):447. https://doi.org/10.1128/CMR.14.3.447-475.2001

Adhikari A, Kharel K (2018) Safe food handling for successful pregnancy outcomes. In: Nutrition and health (United Kingdom), vol Part F3934. https://doi.org/10.1007/978-3-319-90988-2_7

Adley CC, Ryan MP (2016) The nature and extent of foodborne disease. In: Antimicrobial food packaging. Elsevier, pp 1–10. https://doi.org/10.1016/B978-0-12-800723-5.00001-2

Ahmad S, Masood F, Khatoon K, Malik A (2019) Risk management of chemical hazards arising during food manufacturing. In: Health and safety aspects of food processing technologies. Springer International Publishing, pp 403–418. https://doi.org/10.1007/978-3-030-24903-8_13

Ahn C, Jeung E-B (2023) Endocrine-disrupting chemicals and disease endpoints. Int J Mol Sci 24(6). https://doi.org/10.3390/ijms24065342

Akritidou T, Akkermans S, Gaspari S, Azraini ND, Smet C, Van de Wiele T, Van Impe JFM (2022a) Effect of gastric pH and bile acids on the survival of Listeria monocytogenes and Salmonella Typhimurium during simulated gastrointestinal digestion. Innov Food Sci Emerg Technol 82:103161. https://doi.org/10.1016/j.ifset.2022.103161

Akritidou T, Akkermans S, Smet C, Delens V, Van Impe JFM (2022b) Influence of food composition on the survival of invading pathogens during simulated digestion. In: 12th International conference on simulation and modelling in the food and bio-industry, FOODSIM 2022, pp 173–177

Almeria S, Dubey JP (2021) Foodborne transmission of Toxoplasma gondii infection in the last decade. An overview. Res Vet Sci 135:371–385. https://doi.org/10.1016/J.RVSC.2020.10.019

Althobaiti NA (2025) Heavy metals exposure and Alzheimer's disease: underlying mechanisms and advancing therapeutic approaches. Behav Brain Res 476. https://doi.org/10.1016/j.bbr.2024.115212

Ammar EE, Zou X, Ghosh S, Onyeaka H, Elmasry SA, Alkeay AM, AL-Farga A, Rady HA, EL-Shershaby NA, Sallam AS (2025) Fresh futures: cutting-edge eco-friendly coating techniques for fruits. J Food Process Preserv 2025(1). https://doi.org/10.1155/jfpp/5201632

Amouei A, Naghibi EF, Fallah H, Asgharnia H, Khalilpour A, Tabarinia H (2024) Health risk assessment of heavy metals in vegetables consumed in the central part of Mazandaran province. Environ Health Eng Manag 11(3):349–359. https://doi.org/10.34172/EHEM.2024.34

Anaya-Muñoz A (2024) Willingness, capacities and (non)compliance with human rights norms. Int J Hum Rights:1–20. https://doi.org/10.1080/13642987.2024.2432960

Anderson M, Sansonetti PJ, Marteyn BS (2016) Shigella diversity and changing landscape: insights for the twenty-first century. Front Cell Infect Microbiol 6(APR):184919. https://doi.org/10.3389/FCIMB.2016.00045/BIBTEX

Araujo RA, de Souza NA (2010) An evolutionary game theory approach to the dynamics of the labour market: a formal and informal perspective. Struct Chang Econ Dyn 21(2):101–110. https://doi.org/10.1016/j.strueco.2009.11.002

Arblaster M, Hooper P (2015) Light handed regulation–can it play a role in the developing world? Transp Policy 43:32–41. https://doi.org/10.1016/j.tranpol.2015.05.014

Aruoma OI (2006) The impact of food regulation on the food supply chain. Toxicology 221(1):119–127. https://doi.org/10.1016/j.tox.2005.12.024
Aruoma OI (2014) World Trade Organization and food regulation. In: Nutraceutical and functional food regulations in the United States and around the world. Elsevier, pp 479–494. https://doi.org/10.1016/B978-0-12-405870-5.00026-8
Awlqadr FH, Altemimi AB, Qadir SA, Hama Salih TA, Alkanan ZT, AlKaisy QH, Mohammed OA, Hesarinejad MA (2025) Emerging trends in nano-sensors: a new frontier in food safety and quality assurance. Heliyon 11(1):e41181. https://doi.org/10.1016/j.heliyon.2024.e41181
Baart GJE, Martens DE (2012) Genome-scale metabolic models: reconstruction and analysis. Methods Mol Biol 799:107–126. https://doi.org/10.1007/978-1-61779-346-2_7
Balasubramanian D, Harper L, Shopsin B, Torres VJ (2017) Staphylococcus aureus pathogenesis in diverse host environments. Pathog Dis 75(1):ftx005. https://doi.org/10.1093/FEMSPD/FTX005
Banerjee D, Das PK, Mukherjee J (2023) Nervous system. In: Textbook of veterinary physiology. Springer Nature, pp 265–293. https://doi.org/10.1007/978-981-19-9410-4_11
Barney J, Campbell M, Schwartz C, Osband AJ (2023) Foods. In: Translational surgery. Elsevier, pp 547–549. https://doi.org/10.1016/B978-0-323-90300-4.00008-2
Batz MB, Henke E, Kowalcyk B (2013) Long-term consequences of foodborne infections. Infect Dis Clin N Am 27(3):599–616. https://doi.org/10.1016/j.idc.2013.05.003
Bazira PJ (2021) An overview of the nervous system. Surgery (United Kingdom) 39(8):451–462. https://doi.org/10.1016/j.mpsur.2021.06.012
Bazira PJ (2024) An overview of the nervous system. Surgery (United Kingdom) 42(8):525–535. https://doi.org/10.1016/j.mpsur.2024.05.013
Bennie RZ, Ogilvie OJ, Loo LSW, Zhou H, Ng SK, Jin A, Trlin HJF, Wan A, Yu H, Domigan LJ, Dobson RCJ (2025) A risk-based approach can guide safe cell line development and cell banking for scaled-up cultivated meat production. Nat Food 6(1):25–30. https://doi.org/10.1038/s43016-024-01085-9
Bertolatti D, Theobald C (2011) Food safety and risk analysis. In: Encyclopedia of environmental health, vol 2. Elsevier, pp 792–802. https://doi.org/10.1016/B978-0-444-52272-6.00620-6
Bhardwaj A, Misra K, Rajput R (2019) Status of arsenic remediation in India. In: Advances in water purification techniques: meeting the needs of developed and developing countries. https://doi.org/10.1016/B978-0-12-814790-0.00009-0
Bintsis T (2017) Foodborne pathogens. AIMS Microbiol 3(3):529–563. https://doi.org/10.3934/microbiol.2017.3.529
Bittisnich D (2024) Food safety in global trade: opportunities and challenges to trade harmonization. In: Encyclopedia of food safety. Elsevier, pp 474–481. https://doi.org/10.1016/B978-0-12-822521-9.00009-5
Bogueva D, McClements DJ (2023) Safety and nutritional risks associated with plant-based meat alternatives. Sustainability 15(19):14336. https://doi.org/10.3390/su151914336
Bokhtiar SM, Islam MR, Ahmed MJ, Rahman A, Rafiq K (2023) Assessment of heavy metals contamination and antimicrobial drugs residue in broiler edible tissues in Bangladesh. Antibiotics 12(4). https://doi.org/10.3390/antibiotics12040662
Bokobza E, Hinault C, Tiroille V, Clavel S, Bost F, Chevalier N (2021) The adipose tissue at the crosstalk between EDCs and cancer development. Front Endocrinol 12. https://doi.org/10.3389/fendo.2021.691658
Bonny P, Joshua OO, Nishimwe K, Marole TA, Omari R, Essia Ngang JJ (2023) Food safety management in developing countries. In: Encyclopedia of food safety, vol 1–4, 2nd edn, pp V4-170–V4-177. https://doi.org/10.1016/B978-0-12-822521-9.00205-7
Borchers A, Teuber SS, Keen CL, Gershwin ME (2010) Food safety. Clin Rev Allergy Immunol 39(2):95–141. https://doi.org/10.1007/s12016-009-8176-4
Boughattas S, Salehi R (2014) Molecular approaches for detection and identification of foodborne pathogens. J Food Qual Haz Control 1(1):1–6

Boutrif E (2014) Institutions involved in food safety: Food and Agriculture Organization of the United Nations (FAO). In: Encyclopedia of food safety, vol 4. Elsevier, pp 354–358. https://doi.org/10.1016/B978-0-12-378612-8.00007-X
Bradshaw M, Marshall KM, Heap JT, Tepp WH, Minton NP, Johnson EA (2009) Construction of a nontoxigenic Clostridium botulinum strain for food challenge studies. Appl Environ Microbiol 76(2):387. https://doi.org/10.1128/AEM.02005-09
Brent GA (2010) Environmental exposures and autoimmune thyroid disease. Thyroid 20(7):755–761. https://doi.org/10.1089/thy.2010.1636
Brix KV, Baken S, Poland CA, Blust R, Pope LJ, Tyler CR (2023) Challenges and recommendations in assessing potential endocrine-disrupting properties of metals in aquatic organisms. Environ Toxicol Chem 42(12):2564–2579. https://doi.org/10.1002/etc.5741
Brown GC (2019) The endotoxin hypothesis of neurodegeneration. J Neuroinflammation 16(1):180. https://doi.org/10.1186/s12974-019-1564-7
Buehn A, Schneider F (2012) Shadow economies around the world: novel insights, accepted knowledge, and new estimates. Int Tax Public Financ 19(1):139–171. https://doi.org/10.1007/s10797-011-9187-7
Bülbül G, Hayat A, Andreescu S (2015) Portable nanoparticle-based sensors for food safety assessment. Sensors 15(12):30736–30758. https://doi.org/10.3390/s151229826
Bushra, Kapoor S, Mahmood T (2024) Study of oxidative stress and inflammatory status in thyroid dysfunction: a systematic review. Int J Drug Deliv Technol 14(1):515–522. https://doi.org/10.25258/ijddt.14.1.71
Caesar R (2024) The impact of novel probiotics isolated from the human gut on the gut microbiota and health. Diabetes Obes Metab. https://doi.org/10.1111/dom.16129
Cahill LE, El-Sohemy A (2012) Genetic variation and nutrient metabolism. In: Present knowledge in nutrition, 10th edn, Wiley-Blackwell, pp 27–37. https://doi.org/10.1002/9781119946045.ch3
Canfield RL, Jusko TA (2008) Lead poisoning. In: Encyclopedia of infant and early childhood development, vol 1–3. https://doi.org/10.1016/B978-012370877-9.00091-8
Cangemi JR (2011) Food poisoning and diarrhea: small intestine effects. Curr Gastroenterol Rep 13(5):442–448. https://doi.org/10.1007/s11894-011-0209-5
Caya JG, Agni R, Miller JE (2004) Clostridium botulinum and the clinical laboratorian: a detailed review of botulism, including biological warfare ramifications of botulinum toxin. Arch Pathol Lab Med 128(6):653–662. https://doi.org/10.5858/2004-128-653-CBATCL
Chandra AK (2010) Goitrogen in food: cyanogenic and flavonoids containing plant foods in the development of goiter. In: Bioactive foods in promoting health. https://doi.org/10.1016/B978-0-12-374628-3.00042-6
Chandra AK, Lahari D, Mukhopadhyay S (2005) Goitrogen content of cyanogenic plant foods of Indian origin. J Food Sci Technol 42(1):95–98
Chatzimpyrou O, Chaidoutis E, Keramydas D, Papalexis P, Thomaidis NS, Pitiriga VC, Langi P, Koutsiari F, Drikos I, Giannari M, Chelidonis G, Lazaris AC, Kavantzas N (2025) Health inspections of restaurant establishments in the Attica region, Greece. Non-compliance data within the food hygiene sector. J Food Prot 88(2):100452. https://doi.org/10.1016/j.jfp.2025.100452
Cheung GYC, Bae JS, Otto M (2021) Pathogenicity and virulence of Staphylococcus aureus. Virulence 12(1):547. https://doi.org/10.1080/21505594.2021.1878688
Choi AL, Mogensen UB, Bjerve KS, Debes F, Weihe P, Grandjean P, Budtz-Jørgensen E (2014) Negative confounding by essential fatty acids in methylmercury neurotoxicity associations. Neurotoxicol Teratol 42:85–92. https://doi.org/10.1016/j.ntt.2014.02.003
Choi S-J, Jia N, Lu J (2015) The structure of political institutions and effectiveness of corporate political lobbying. Organ Sci 26(1):158–179. https://doi.org/10.1287/orsc.2014.0936
Choudhury SA, Asefa B, Webb A, Ramotar D, Chow TYK (2007) Functional and genetic analysis of the Saccharomyces cerevisiae RNC1/TRM2: evidences for its involvement in DNA double-strand break repair. Mol Cell Biochem 300(1–2):215–226. https://doi.org/10.1007/s11010-006-9386-1

Cifuentes L, Acosta A (2021) Homeostatic regulation of food intake. Clin Res Hepatol Gastroenterol 46(2):101794. https://doi.org/10.1016/J.CLINRE.2021.101794

Ciobanu C, Slencu BG, Cuciureanu R (2012) Estimation of dietary intake of cadmium and lead through food consumption. Revista Medico-Chirurgicală a Societăţii de Medici Şṃi Naturalişṃti Din Iaşṃi 116(2):617–623

Cohn AR, Cheng RA, Orsi RH, Wiedmann M (2021) Moving past species classifications for risk-based approaches to food safety: Salmonella as a case study. Front Sustain Food Syst 5. https://doi.org/10.3389/fsufs.2021.652132

Contini F (2023) The never-ending crisis of Italian justice: role and responsibility of its governance system. Law Technol Hum 5(1):153–177. https://doi.org/10.5204/lthj.2695

Cota AM, Midwinter MJ (2015) The immune system. Anaesth Intensive Care Med 16(7):353–355. https://doi.org/10.1016/J.MPAIC.2015.04.006

Counihan NA, Anderson DA (2016) Specific IgA enhances the transcytosis and excretion of hepatitis A virus. Sci Rep 6(1):1–9. https://doi.org/10.1038/srep21855

Crawford LM (2014) Public health measures: challenges of industrialized countries in food safety management. In: Encyclopedia of food safety, vol 4. https://doi.org/10.1016/B978-0-12-378612-8.00311-5

Cremon C, Stanghellini V, Pallotti F, Fogacci E, Bellacosa L, Morselli-Labate AM, Paccapelo A, Di Nardo G, Cogliandro RF, De Giorgio R, Corinaldesi R, Barbara G (2014) Salmonella gastroenteritis during childhood is a risk factor for irritable bowel syndrome in adulthood. Gastroenterology 147(1):69–77. https://doi.org/10.1053/j.gastro.2014.03.013

Crespi R, Pughe L, Dowd A (2022) Care of the child with a pediatric endocrine disorder. In: Pediatric diagnostic labs for primary care: an evidence-based approach. Springer International Publishing, pp 413–460. https://doi.org/10.1007/978-3-030-90642-9_11

Cross AJ, Sinha R (2016) Meat consumption and cancer. In: International encyclopedia of public health. https://doi.org/10.1016/B978-0-12-803678-5.00273-3

Crowe W, Elliott CT, Green BD (2019) A review of the in vivo evidence investigating the role of nitrite exposure from processed meat consumption in the development of colorectal cancer. Nutrients 11(11). https://doi.org/10.3390/nu11112673

Dai Y, Zhang M, Liu X, Sun T, Qi W, Ding W, Chen Z, Zhang P, Liu R, Chen H, Chen S, Wang Y, Yue Y, Song N, Wang W, Jia H, Ma Z, Li C, Chen Q, Li B (2024) Salmonella manipulates macrophage migration via SteC-mediated myosin light chain activation to penetrate the gut-vascular barrier. EMBO J 43(8):1499. https://doi.org/10.1038/S44318-024-00076-7

Damotharan K, Sudhakaran G, Ramu M, Krishnan M, Karthick Raja Namasivayam S, Arockiaraj J (2024) Biochemical processes mediating neurotoxicity induced by synthetic food dyes: a review of current evidence. Chemosphere 364. https://doi.org/10.1016/j.chemosphere.2024.143295

Darshan TG, Monicashree KV, Bhat B, Bathinapatla A, Kanchi S (2024) Metallic silver and copper oxide nanoparticles: uses in food preservation and impacts on the environment, pp 183–230. https://doi.org/10.1016/bs.coac.2024.08.015

Dasti JI, Tareen AM, Lugert R, Zautner AE, Groß U (2010) Campylobacter jejuni: a brief overview on pathogenicity-associated factors and disease-mediating mechanisms. Int J Med Microbiol: IJMM 300(4):205–211. https://doi.org/10.1016/J.IJMM.2009.07.002

de Carvalho JGR, Augusto HC, Ferraz R, Delerue-Matos C, Fernandes VC (2024) Micro(nano) plastic and related chemicals: emerging contaminants in environment, food and health impacts. Toxics 12(10):762. https://doi.org/10.3390/toxics12100762

De la Mora ZV, Macías-Rodríguez ME, Arratia-Quijada J, Gonzalez-Torres YS, Nuño K, Villarruel-López A (2020) Clostridium perfringens as foodborne pathogen in broiler production: pathophysiology and potential strategies for controlling necrotic enteritis. Anim Open Access J MDPI 10(9):1718. https://doi.org/10.3390/ANI10091718

de Oliveira CAF, da Cruz AG, Tavolaro P, Corassin CH (2016) Food safety. In: Antimicrobial food packaging. Elsevier, pp 129–139. https://doi.org/10.1016/B978-0-12-800723-5.00010-3

De RK, Tomar N (2012) Modeling the optimal central carbon metabolic pathways under feedback inhibition using flux balance analysis. J Bioinform Comput Biol 10(6):1250019. https://doi.org/10.1142/S0219720012500199

Deon BC, Medeiros LB, Lúcia de Freitas Saccol A, Hecktheuer LH, Saccol S, Naissinger M (2014) Good food preparation practices in households: a review. Trends Food Sci Technol 39(1):40–46. https://doi.org/10.1016/j.tifs.2014.06.008

Dev N, Sankar J, Vinay MV (2016) Functions of thyroid hormones. In: Thyroid disorders: basic science and clinical practice. https://doi.org/10.1007/978-3-319-25871-3_2

Dhiman SK, Dureja H (2020) Significance of and challenges in regulating endocrine disruptors – how regulators and industry can conquer? Endocr Metab Immune Disord Drug Targets 20(10):1664–1681. https://doi.org/10.2174/1871530320666200606225104

Di Filippo M, Pescini D, Galuzzi BG, Bonanomi M, Gaglio D, Mangano E, Consolandi C, Alberghina L, Vanoni M, Damiani C (2022) INTEGRATE: model-based multi-omics data integration to characterize multi-level metabolic regulation. PLoS Comput Biol 18(2):e1009337. https://doi.org/10.1371/journal.pcbi.1009337

Dian Z, Sun Y, Zhang G, Xu Y, Fan X, Yang X, Pan Q, Peppelenbosch M, Miao Z (2021) Rotavirus-related systemic diseases: clinical manifestation, evidence and pathogenesis. Crit Rev Microbiol 47(5):580–595. https://doi.org/10.1080/1040841X.2021.1907738

Díaz-Villaseñor A, Burns AL, Hiriart M, Cebrián ME, Ostrosky-Wegman P (2007) Arsenic-induced alteration in the expression of genes related to type 2 diabetes mellitus. Toxicol Appl Pharmacol 225(2):123–133. https://doi.org/10.1016/j.taap.2007.08.019

Dimitrakopoulou ME, Kotsiri Z, Vantarakis A (2024) Food-borne pathogens and sources of contamination. In: Biosensors for foodborne pathogen detection: a rapid detection approach. Elsevier, pp 1–16. https://doi.org/10.1016/B978-0-323-95586-7.00001-0

Diviš P, Štávová E, Pořízka J, Drábiková J (2017) Determination of tin, chromium, cadmium and lead in canned fruits from the Czech market. Potravinarstvo Slovak J Food Sci 11(1):564–570. https://doi.org/10.5219/749

Ehlers CF, Malone DJ (2024) Endocrine diseases and disorders. In: Acute care physical therapy: a clinician's guide, 2nd edn. Routledge, pp 421–451. https://doi.org/10.4324/9781003522485-10

Ehling-Schulz M, Lereclus D, Koehler TM (2019) The Bacillus cereus group: Bacillus species with pathogenic potential. Microbiol Spectr 7(3). https://doi.org/10.1128/MICROBIOLSPEC.GPP3-0032-2018/ASSET/652B9375-99B6-4A8E-9127-ACBA975A0A48/ASSETS/GRAPHIC/GPP3-0032-2018-FIG5.GIF

Elmassry MM, Zayed A, Farag MA (2022) Gut homeostasis and microbiota under attack: impact of the different types of food contaminants on gut health. Crit Rev Food Sci Nutr 62(3):738–763. https://doi.org/10.1080/10408398.2020.1828263

Encarnação T, Pais AACC, Campos MG, Burrows HD (2019) Endocrine disrupting chemicals: impact on human health, wildlife and the environment. Sci Prog 102(1):3–42. https://doi.org/10.1177/0036850419826802

Evans RD, Heather LC (2016) Metabolic pathways and abnormalities. Surgery (United Kingdom) 34(6):266–272. https://doi.org/10.1016/j.mpsur.2016.03.010

Evans RD, Heather LC (2022) Human metabolism: pathways and clinical aspects. Surgery (United Kingdom) 40(4):219–226. https://doi.org/10.1016/j.mpsur.2022.01.004

Fanaselle W, Pouillot R, Papafragkou E, Liggins G, Williams L, van Doren JM (2022) Evaluation of the impact of compliance with mitigation strategies and frequency of restaurant surface cleaning and sanitizing on control of norovirus transmission from ill food employees using an existing quantitative risk assessment model. J Food Prot 85(8):1177–1191. https://doi.org/10.4315/JFP-21-423

Feng P, Ye Z, Kakade A, Virk AK, Li X, Liu P (2019) A review on gut remediation of selected environmental contaminants: possible roles of probiotics and gut microbiota. Nutrients 11(1). https://doi.org/10.3390/nu11010022

Fernandez-Tresguerres Hernández JA (2005) Effect of monosodium glutamate given orally on appetite control (a new theory for the obesity epidemic) | Efecto del glutamato monosódico por

vía oral sobre el control del apetito (una nueva teoría para la epidemia de la obesidad). An R Acad Nac Med 122(2)
Ferreira RG, Cardoso MV, de Souza Furtado KM, Espíndola KMM, Amorim RP, Monteiro MC (2019) Epigenetic alterations caused by aflatoxin b1: a public health risk in the induction of hepatocellular carcinoma. Transl Res 204:51–71. https://doi.org/10.1016/j.trsl.2018.09.001
Fink G (2007) Endocrine systems. In: Encyclopedia of stress. Elsevier Inc, pp 926–927. https://doi.org/10.1016/B978-012373947-6.00143-4
Foster-Nyarko E, Pallen MJ (2022) The microbial ecology of Escherichia coli in the vertebrate gut. FEMS Microbiol Rev 46(3):fuac008. https://doi.org/10.1093/FEMSRE/FUAC008
Fraser AM, Miller C (2014) Educating for food safety. In: Practical food safety. Wiley, pp 31–48. https://doi.org/10.1002/9781118474563.ch3
Fung F, Wang HS, Menon S (2018) Food safety in the 21st century. Biomed J 41(2):88–95. Elsevier B.V. https://doi.org/10.1016/J.BJ.2018.03.003
Gałkowska D, Pycia K, Kapuśniak K, Juszczak L (2023) Food contamination – characteristics and legal regulations. Part I – plant toxins | Zanieczyszczenia żywności – charakterystyka i regulacje prawne. Czesc I – Toksyny roślinne. Zywnosc Nauka Technol Jakosc/Food Sci Technol Qual 30(4):245–263. https://doi.org/10.15193/zntj/2023/137/482
Galligan TM, Temkin AM, Hale MD (2023) Endocrine system. In: Environmental contaminants and endocrine health. Elsevier, pp 3–23. https://doi.org/10.1016/B978-0-12-824464-7.00010-6
Ganakumar V, Sruthi K, Ghatnatti VB, Goroshi M (2024) Endocrine disruptors and the heart: unraveling the cardiovascular impact. Indian J Cardiovasc Dis Women – WINCARS 9(4):230–240. https://doi.org/10.25259/IJCDW_68_2024
Gatto A, Pulitanò SM, Conti G, Soave PM, Di Sarno L, Chiaretti A (2021) Botulism during SARS-CoV-2 pandemic: the importance of differential diagnoses. Anaerobe 70. https://doi.org/10.1016/j.anaerobe.2021.102389
Gatzke-Kopp LM, Willoughby M, Kress AM, McArthur K, Wychgram C, Folch DC, Brunswasser S, Dabelea D, Elliott AJ, Hartert T, Wright RO, Wright RJ (2024) Airborne lead exposure and childhood cognition: the Environmental Influences on Child Health Outcomes (ECHO) cohort (2003–2022). Am J Public Health 114(3):309–318. https://doi.org/10.2105/AJPH.2023.307519
Gerardi A (2023) Global Food Safety Initiative (GFSI): underpinning the safety of the global food chain, facilitating regulatory compliance, trade, and consumer trust. In: Present knowledge in food safety. Elsevier, pp 1089–1098. https://doi.org/10.1016/B978-0-12-819470-6.00058-5
Ghazanfar S, Abid R, Ajmal W, Haider Z, Khan B, Rani S, Basuli D, Seal S, Das T, Sinha D, Jyoti SU (2024) An introduction to food safety and its impact on human health. Food Saf Contam Risk Assess:1–22. https://doi.org/10.1201/9781003333913-1
Ghosh T, Raj GVSB, Dash KK (2022) A comprehensive review on nanotechnology based sensors for monitoring quality and shelf life of food products. Meas Food 7:100049. https://doi.org/10.1016/j.meafoo.2022.100049
Ghosh N, Sinha K, Sil PC (2024) Pesticides and the gut microbiota: implications for Parkinson's disease. Chem Res Toxicol 37(7):1071–1085. https://doi.org/10.1021/acs.chemrestox.4c00057
Giri B, Dey S, Das T, Sarkar M, Banerjee J, Dash SK (2018) Chronic hyperglycemia mediated physiological alteration and metabolic distortion leads to organ dysfunction, infection, cancer progression and other pathophysiological consequences: an update on glucose toxicity. Biomed Pharmacother 107:306–328. https://doi.org/10.1016/j.biopha.2018.07.157
Gizaw Z (2019) Public health risks related to food safety issues in the food market: a systematic literature review. Environm Health Prev Med 24(1):1–21. https://doi.org/10.1186/S12199-019-0825-5
Gonya S, Kallmerten P, Dinapoli P (2024) Are infants and children at risk of adverse health effects from dietary deoxynivalenol exposure? An integrative review. Int J Environ Res Public Health 21(6):808. https://doi.org/10.3390/ijerph21060808
Gorman LS, Chiasera JM (2013) Endocrinology review – adrenal and thyroid disorders. Clin Lab Sci: J Am Soc Med Technol 26(2):107–111. https://doi.org/10.29074/ascls.26.2.107
Gourama H (2020) Foodborne pathogens, pp 25–49. https://doi.org/10.1007/978-3-030-42660-6_2

Grace D (2023) Burden of foodborne disease in low-income and middle-income countries and opportunities for scaling food safety interventions. Food Secur 15(6):1475–1488. https://doi.org/10.1007/s12571-023-01391-3

Griffin AJ, McSorley SJ (2011) Development of protective immunity to Salmonella, a mucosal pathogen with a systemic agenda. Mucosal Immunol 4(4):371. https://doi.org/10.1038/MI.2011.2

Griffin M, Smith C, Chauret C (2008) Microarray analysis of gene expression of escherichia coli O157: H7 in drinking water. In: Water resources research progress. Nova Science Publishers, Inc, pp 225–240

Groopman JD, Kensler TW, Wild CP (2008) Protective interventions to prevent aflatoxin-induced carcinogenesis in developing countries. Annu Rev Public Health 29. https://doi.org/10.1146/annurev.publhealth.29.020907.090859

Habib AA (2014) Botulism. In: Encyclopedia of the neurological sciences. https://doi.org/10.1016/B978-0-12-385157-4.00609-6

Hachimura S (2022) Immunomodulation by food: novel collaborations between food components and microbiota. J Nutr Sci Vitaminol 68:S126–S127. https://doi.org/10.3177/jnsv.68.S126

Hadi J, Brightwell G (2021) Safety of alternative proteins: technological, environmental and regulatory aspects of cultured meat, plant-based meat, insect protein and single-cell protein. Foods 10(6):1226. https://doi.org/10.3390/foods10061226

Hadian Z, Eslamizad S, Yazdanpanah H (2019) Pesticide residues analysis in Iranian fruits and vegetables by gas chromatography-mass spectrometry. Iran J Pharm Res 18(1):275–285

Hafez M, Gourlie R, Telfer M, Schatz N, Turkington TK, Beres B, Aboukhaddour R (2022) Diversity of Fusarium spp. associated with wheat node and grain in representative sites across the Western Canadian prairies. Phytopathology 112(5):1003–1015. https://doi.org/10.1094/PHYTO-06-21-0241-R

Halimeh FB, Rafei R, Osman M, Kassem II, Diene SM, Dabboussi F, Rolain JM, Hamze M (2021) Historical, current, and emerging tools for identification and serotyping of Shigella. Braz J Microbiol 52(4):2043. https://doi.org/10.1007/S42770-021-00573-5

Hall M, Lanphear B, Chevrier J, Hornung R, Green R, Goodman C, Ayotte P, Martinez-Mier EA, Zoeller RT, Till C (2023) Fluoride exposure and hypothyroidism in a Canadian pregnancy cohort. Sci Total Environ 869. https://doi.org/10.1016/j.scitotenv.2022.161149

Hamlin HJ (2013) Endocrine disruption. In: The physiology of fishes, 4th edn. CRC Press, pp 235–256. https://doi.org/10.1201/b17948-8

Haneishi Y, Watanabe S, Okada A, Takemae H, Bastoni D, Treppiccione L, Saggese A, Mizutani T, Rossi M, Miyamoto J (2024) The intake of dietary lipids improves glucose tolerance via modulating gut microbiota. J Nutr Sci Vitaminol 70(4):336–343. https://doi.org/10.3177/jnsv.70.336

Hassan EM, Örmeci B, DeRosa MC, Dixon BR, Sattar SA, Iqbal A (2021) A review of Cryptosporidium spp. and their detection in water. Water Sci Technol 83(1):1–25. https://doi.org/10.2166/WST.2020.515

Havira MS, Ta A, Kumari P, Wang C, Russo AJ, Ruan J, Rathinam VA, Vanaja SK (2020) Shiga toxin suppresses noncanonical inflammasome responses to cytosolic LPS. Sci Immunol 5(53). https://doi.org/10.1126/SCIIMMUNOL.ABC0217

Heilandt T, Mulholland C, Younes M (2014) Institutions involved in food safety: FAO/WHO Codex Alimentarius Commission (CAC). In: Encyclopedia of food safety. Elsevier, pp 345–353. https://doi.org/10.1016/B978-0-12-378612-8.00006-8

Heimesaat MM, Backert S, Alter T, Bereswill S (2021) Human campylobacteriosis-a serious infectious threat in a One Health perspective. Curr Top Microbiol Immunol 431:1–23. https://doi.org/10.1007/978-3-030-65481-8_1

Hill V (2014) Government regulation of food quality: international and in France and the U.S, pp 53–82. https://doi.org/10.1007/978-3-319-04250-3_2

Hof H, Schrecker J (2024) Fusarium spp.: infections and intoxications. GMS Infect Dis 12:Doc04. https://doi.org/10.3205/ID000089

Houbraken J, Frisvad JC, Seifert KA, Overy DP, Tuthill DM, Valdez JG, Samson RA (2012) New penicillin-producing Penicillium species and an overview of section Chrysogena. Persoonia 29:78–100. https://doi.org/10.3767/003158512x660571

Howard EJ, Lam TKT, Duca FA (2022) The gut microbiome: connecting diet, glucose homeostasis, and disease. Annu Rev Med 73. https://doi.org/10.1146/annurev-med-042220-012821

Hu SF, Yu YG, Li R, Zhuang P, Xia XZ, Xiao XL, Yang HN (2016) Simultaneous detection of nine foodborne pathogenic bacteria using high-resolution melting analysis. Mod Food Sci Technol 32(3):271–280. https://doi.org/10.13982/j.mfst.1673-9078.2016.3.043

Humphrey J (2012) Convergence of US and EU production practices under the new FDA Food Safety Modernization Act. World Econ 35(8):994–1005. https://doi.org/10.1111/j.1467-9701.2012.01458.x

Ispas S, Tuta LA, Botnarciuc M, Ispas V, Staicovici S, Ali S, Nelson-Twakor A, Cojocaru C, Herlo A, Petcu A (2023) Metabolic disorders, the microbiome as an endocrine organ, and their relations with obesity: a literature review. J Pers Med 13(11). https://doi.org/10.3390/jpm13111602

Jackson LA, Jansen M (2010) Risk assessment in the international food safety policy arena. Can the multilateral institutions encourage unbiased outcomes? Food Policy 35(6):538–547. https://doi.org/10.1016/j.foodpol.2010.07.004

Jalili C, Kazemi M, Cheng H, Mohammadi H, Babaei A, Taheri E, Moradi S (2021) Associations between exposure to heavy metals and the risk of chronic kidney disease: a systematic review and meta-analysis. In: Critical reviews in toxicology, vol 51, Issue 2. Taylor and Francis Ltd, pp 165–182. https://doi.org/10.1080/10408444.2021.1891196

Janda R, Ukić Š, Mikulec N, Vitale K (2021) Bisphenol A – an environmental and human threat. Agric Conspec Sci 86(4):295–304

Jiang Q-Y (2006) Court delay and law enforcement in China. Gabler. https://doi.org/10.1007/978-3-8350-9012-5

Jorgensen F, Rodgers J, Duncan D, Lawes J, Byrne C, Swift C (2022) Levels and trends of antimicrobial resistance in Campylobacter spp. from chicken in the UK. https://doi.org/10.46756/SCI.FSA.DUD728

Joyce EE, Chavarro JE, Rando J, Song AY, Croen LA, Fallin MD, Hertz-Picciotto I, Schmidt RJ, Volk H, Newschaffer CJ, Newschaffer CJ, Lyall K (2022) Prenatal exposure to pesticide residues in the diet in association with child autism-related traits: results from the EARLI study. Autism Res 15(5):957–970. https://doi.org/10.1002/aur.2698

Kamil Lakho M (2021) Social reasons for and repercussions of judicial delays in Subordinate Courts of Karachi: the need for sociological analysis. Sri Lanka J Soc Sci 44(2):181. https://doi.org/10.4038/sljss.v44i2.7928

Kanaan MHG, Tarek AM (2020) Clostridium botulinum, a foodborne pathogen and its impact on public health. Ann Trop Med Public Health 23(5):346–359. https://doi.org/10.36295/ASRO.2020.2357

Kaur N, Agarwal A, Sabharwal M, Jaiswal N (2021) Natural food toxins as anti-nutritional factors in plants and their reduction strategies. In: Food chemistry: the role of additives, preservatives and adulteration. Wiley, pp 217–266. https://doi.org/10.1002/9781119792130.ch8

Kendall H, Clark B, Rhymer C, Kuznesof S, Hajslova J, Tomaniova M, Brereton P, Frewer L (2019) A systematic review of consumer perceptions of food fraud and authenticity: a European perspective. Trends Food Sci Technol 94:79–90. https://doi.org/10.1016/j.tifs.2019.10.005

Khalaf EM, Taherian M, Almalki SG, Asban P, Kareem AK, Alhachami FR, Almulla AF, Romero-Parra RM, Jawhar ZH, Kiani F, Noroozi Manesh I, Mohammadi MJ (2024) Relationship between exposure to heavy metals on the increased health risk and carcinogenicity of urinary tract (kidney and bladder). Rev Environ Health 39(3):539–549. https://doi.org/10.1515/reveh-2022-0245

Khan A, Qadeer A, Wajid A, Ullah Q, Rahman SU, Ullah K, Safi SZ, Ticha L, Skalickova S, Chilala P, Bernatova S, Samek O, Horky P (2024) Microplastics in animal nutrition: occurrence, spread, and hazard in animals. J Agric Food Res 17:101258. https://doi.org/10.1016/j.jafr.2024.101258

Khoshnoud MJ, Siavashpour A, Bakhshizadeh M, Rashedinia M (2018) Effects of sodium benzoate, a commonly used food preservative, on learning, memory, and oxidative stress in brain of mice. J Biochem Mol Toxicol 32(2). https://doi.org/10.1002/jbt.22022

Kim K, Argos M, Persky VW, Freels S, Sargis RM, Turyk ME (2022) Associations of exposure to metal and metal mixtures with thyroid hormones: results from the NHANES 2007–2012. Environ Res 212. https://doi.org/10.1016/j.envres.2022.113413

Kinsey J (2005) Will food safety jeopardize food security? Agric Econ 32(SUPPL):149–158. https://doi.org/10.1111/j.0169-5150.2004.00020.x

Kirchsteiger-Meier E, Baumgartner T (eds) (2014) Global food legislation. Wiley. https://doi.org/10.1002/9783527680795

Koilparambil D, Varghese S, Shaikmoideen JM (2020) Chitosan nanoparticles. In: Nanobiotechnology in diagnosis, drug delivery, and treatment. Wiley, pp 197–215. https://doi.org/10.1002/9781119671732.ch10

Kumar A, Mangla SK, Kumar P, Kayikci Y (2021) Investigating enablers to improve transparency in sustainable food supply chain using F-BWM. In: Advances in intelligent systems and computing, vol 1197. AISC. https://doi.org/10.1007/978-3-030-51156-2_65

Kumar V, Sharma N, Sharma P, Pasrija R, Kaur K, Umesh M, Thazeem B (2023) Toxicity analysis of endocrine disrupting pesticides on non-target organisms: A critical analysis on toxicity mechanisms. Toxicol Appl Pharmacol 474. https://doi.org/10.1016/j.taap.2023.116623

Kunanitthaworn N, Mueangmo O, Saheng J, Wongjak W, Lertsiriladakul T, Chaito T, Nantarat P, Sudjaritruk T (2023) Seroprevalence of hepatitis A virus antibodies among children and adolescents living in Northern Thailand: an implication for hepatitis A immunization. Sci Rep 13(1):1–10. https://doi.org/10.1038/s41598-023-44643-0

Kureljušić J, Žutić J, Kureljušić B, Rokvić N, Tasić A, Ljubojević Pelić D, Vesković Moračanin S (2021) Salmonella – foodborne pathogen and antimicrobial resistance. IOP Conf Ser: Earth Environ Sci 854(1):012049. https://doi.org/10.1088/1755-1315/854/1/012049

Kuswandi B, Futra D, Heng LY (2017) Nanosensors for the detection of food contaminants. In: Nanotechnology applications in food. Elsevier, pp 307–333. https://doi.org/10.1016/B978-0-12-811942-6.00015-7

Kwon H, Lim DJ, Choi C (2025) Prevention of foodborne viruses and pathogens in fresh produce and root vegetables, pp 219–285. https://doi.org/10.1016/bs.afnr.2024.09.015

Lamas B, Martins Breyner N, Houdeau E (2020) Impacts of foodborne inorganic nanoparticles on the gut microbiota-immune axis: potential consequences for host health. Part Fibre Toxicol 17(1):19. https://doi.org/10.1186/s12989-020-00349-z

Latgé JP, Chamilos G (2020) Aspergillus fumigatus and aspergillosis in 2019. Clin Microbiol Rev 33(1). https://doi.org/10.1128/CMR.00140-18/ASSET/6C0F8C25-6E25-4B2F-B0D3-6A74510D0FC7/ASSETS/GRAPHIC/CMR.00140-18-F0020.JPEG

Lavrek D, Lava SAG, Milani GP, Simonetti GD, Bianchetti MG, Giannini O (2018) Hemolytic-uremic syndrome after Escherichia coli urinary tract infection in humans: systematic review of the literature. J Nephrol 31(6):919–924. https://doi.org/10.1007/s40620-018-0543-x

Leitch GJ, He Q (2011) Cryptosporidiosis-an overview. J Biomed Res 25(1):1. https://doi.org/10.1016/S1674-8301(11)60001-8

Lemyre L, Boutette P, Karyakina N, Markon MPL, Brazeau I, Krewski D (2009) International case studies of psychosocial ripple effects of bovine spongiform encephalopathy (BSE) in European countries. J Toxic Environ Health A 72(17–18):1092–1095. https://doi.org/10.1080/15287390903084611

Li Q (2022) Mechanisms for the invasion and dissemination of Salmonella. Can J Infect Dis Med Microbiol 2022(1):2655801. https://doi.org/10.1155/2022/2655801

Li J, Li Y, Sha R, Zheng L, Xu L, Xie HQ, Zhao B (2022) Effects of perinatal TCDD exposure on colonic microbiota and metabolism in offspring and mother mice. Sci Total Environ 832:154762. https://doi.org/10.1016/j.scitotenv.2022.154762

Liang S, Sutham P, Wu K, Mallikarjunan K, Wang J-P (2022) Giant magnetoresistance biosensors for food safety applications. Sensors 22(15):5663. https://doi.org/10.3390/s22155663

Liao X, Guo Y, He Y, Xiao Y, Li J, Liu R (2021) Metabolic enzymes function as epigenetic modulators: A Trojan Horse for chromatin regulation and gene expression. Pharmacol Res 173:105834. https://doi.org/10.1016/j.phrs.2021.105834

Lin CF (2014a) Public-private interactions in global food safety governance. Food Drug Law J 69(2)

Lin M (2014b) Nanotechnology and its applications to improve the detection of chemical hazards in foods. In: Food chemical hazard detection. Wiley, pp 249–261. https://doi.org/10.1002/9781118488553.ch8

Lin C-J, Shih H-M, Wu P-C, Pan C-F, Lin YH, Wu C-J (2023) Plasma selenium and zinc alter associations between nephrotoxic metals and chronic kidney disease: results from NHANES database 2011–2018. Ann Acad Med Singapore 52(8):398–410. https://doi.org/10.47102/annals-acadmedsg.202357

Lindner LF (2008) Regulating food safety: the power of alignment and drive towards convergence. Innov Eur J Soc Sci Res 21(2):133–143. https://doi.org/10.1080/13511610802214487

Liu H, Whitehouse CA, Li B (2018) Presence and persistence of Salmonella in water: the impact on microbial quality of water and food safety. Front Public Health 6:366505. https://doi.org/10.3389/FPUBH.2018.00159/PDF

Liu X, Zhu Y, Yang L, Liao H, Gan P, Yi Y (2023) A case of type A botulinum toxin injection poisoning treated with antitoxin | 抗毒素治疗 A 型肉毒毒素注射中毒 1 例. Chin J Plast Surg 39(2):187–189. https://doi.org/10.3760/cma.j.cn114453-20220315-00066

Liu Z, Yu X, Liu N, Liu C, Jiang A, Chen L (2025) Integrating AI with detection methods, IoT, and blockchain to achieve food authenticity and traceability from farm-to-table. Trends Food Sci Technol 158:104925. https://doi.org/10.1016/j.tifs.2025.104925

Llop S, Ibarlucea J, Sunyer J, Ballester F (2013) Current dietary exposure to mercury during pregnancy and childhood, and public health recommendations | Estado actual sobre la exposición alimentaria al mercurio durante el embarazo y la infancia, y recomendaciones en salud pública. Gac Sanit 27(3):273–278. https://doi.org/10.1016/j.gaceta.2012.09.002

Lu X, Lin Z (2021) COVID-19, economic impact, mental health, and coping behaviors: a conceptual framework and future research directions. Front Psychol 12. https://doi.org/10.3389/fpsyg.2021.759974

Lu T-H, Su C-C, Chen Y-W, Yang C-Y, Wu C-C, Hung D-Z, Chen C-H, Cheng P-W, Liu S-H, Huang C-F (2011) Arsenic induces pancreatic β-cell apoptosis via the oxidative stress-regulated mitochondria-dependent and endoplasmic reticulum stress-triggered signaling pathways. Toxicol Lett 201(1):15–26. https://doi.org/10.1016/j.toxlet.2010.11.019

Lucero Y, Matson DO, Ashkenazi S, George S, O'ryan M (2021) Norovirus: facts and reflections from past, present, and future. Viruses 13(12):2399. https://doi.org/10.3390/V13122399

Madilo FK, Kunadu AP, Tano-Debrah K (2024) Challenges with food safety adoption: a review. J Food Saf 44(1). https://doi.org/10.1111/jfs.13099

Malin AJ, Riddell J, McCague H, Till C (2018) Fluoride exposure and thyroid function among adults living in Canada: effect modification by iodine status. Environ Int 121:667–674. https://doi.org/10.1016/j.envint.2018.09.026

Manibusan MK, Touart LW (2018) Toxicology assessment of endocrine-active substances. In: Comprehensive toxicology, vol 9, 3rd edn. Elsevier, pp 136–171. https://doi.org/10.1016/B978-0-12-801238-3.64226-5

Mantovani A (2016) Endocrine disrupters and the safety of food chains. Horm Res Paediatr 86(4):279–288. https://doi.org/10.1159/000441496

Marchetti G, Serraino A, Giacometti F, Bonfante E, Rosmini R (2013) Review on aflatoxins in food and feed: Sources of human exposure and control | Rassegna sulle aflatossine negli alimenti e nei mangimi: Fonti di contaminazione per l'uomo e controllo. Ind Aliment 52(535):7–19

Marques PH, Prado LCDS, Felice AG, Rodrigues TCV, Pereira UDP, Jaiswal AK, Azevedo V, Oliveira CJF, Soares S (2022) Insights into the Vibrio genus: a One Health perspective from host adaptability and antibiotic resistance to in silico identification of drug targets. Antibiotics 11(10):1399. https://doi.org/10.3390/ANTIBIOTICS11101399/S1

Martínez-Ibarra A, Martínez-Razo LD, MacDonald-Ramos K, Morales-Pacheco M, Vázquez-Martínez ER, López-López M, Rodríguez Dorantes M, Cerbón M (2021) Multisystemic alterations in humans induced by bisphenol A and phthalates: experimental, epidemiological and clinical studies reveal the need to change health policies. Environ Pollut 271. https://doi.org/10.1016/j.envpol.2020.116380
Martinson JNV, Walk ST (2020) Escherichia coli residency in the gut of healthy human adults. https://doi.org/10.1128/ecosalplus.ESP-0003-2020
Massari S, Principato L, Antonelli M, Pratesi CA (2022) Learning from and designing after pandemics. CEASE: a design thinking approach to maintaining food consumer behaviour and achieving zero waste. Socio Econ Plan Sci 82:101143. https://doi.org/10.1016/j.seps.2021.101143
Mattock E, Blocker AJ (2017) How do the virulence factors of shigella work together to cause disease? Front Cell Infect Microbiol 7(MAR):249163. https://doi.org/10.3389/FCIMB.2017.00064/PDF
May SG (2010) Importing a change in diet: the proposed food safety law of 2010 and the possible impact on importers and international trade. Food Drug Law J 65(1):1–35, i
Medzhitov R (2007) Recognition of microorganisms and activation of the immune response. Nature 449(7164):819–826. https://doi.org/10.1038/nature06246
Mesa-Garcia MD, Plaza-Diaz J, Gomez-Llorente C (2018) Molecular basis of oxidative stress and inflammation. In: Obesity: oxidative stress and dietary antioxidants. https://doi.org/10.1016/B978-0-12-812504-5.00003-9
Miano T, D'orazio V, Zaccone C (2014) Trace elements and food safety. In: PHEs, environment and human health: potentially harmful elements in the environment and the impact on human health. Springer Netherlands, pp 339–370. https://doi.org/10.1007/978-94-017-8965-3_9
Miao Y, Rong M, Li M, He H, Zhang L, Zhang S, Liu C, Zhu Y, Deng Y-L, Chen P-P, Miao X-P, Zeng Q (2021) Serum concentrations of organochlorine pesticides, biomarkers of oxidative stress, and risk of breast cancer. Environ Pollut 286. https://doi.org/10.1016/j.envpol.2021.117386
Miclotte L, Van de Wiele T (2022) 3.12 – Microbiota interactions with processed foods, food additives and metabolic disorders. In: Comprehensive gut microbiota, vol 3. https://doi.org/10.1016/B978-0-12-819265-8.00092-9
Migliarini B, Piccinetti CC, Martella A, Maradonna F, Gioacchini G, Carnevali O (2011) Perspectives on endocrine disruptor effects on metabolic sensors. Gen Comp Endocrinol 170(3):416–423. https://doi.org/10.1016/j.ygcen.2010.11.025
Milton A, Mullan B (2010) Consumer food safety education for the domestic environment: a systematic review. Br Food J 112(9):1003–1022. https://doi.org/10.1108/00070701011074363
Mobeen N, Kumar P (2017) Defluoridation techniques – a critical review. Asian J Pharm Clin Res 10(6):64–71. https://doi.org/10.22159/ajpcr.2017.v10i6.13942
Mohammed SB, Elseory AMA (2024) Animal health and food security in Saudi Arabia. In: Food and nutrition security in the Kingdom of Saudi Arabia, Vol. 1: National analysis of agricultural and food security. Springer International Publishing, pp 207–227. https://doi.org/10.1007/978-3-031-46716-5_9
Moszak M, Szulińska M, Bogdański P (2020) You are what you eat—the relationship between diet, microbiota, and metabolic disorders—a review. Nutrients 12(4). https://doi.org/10.3390/nu12041096
Motarjemi Y, Smithers GW (2024) Encyclopedia of food safety. 2500. http://www.sciencedirect.com:5070/referencework/9780128225202/encyclopedia-of-food-safety
Mountzouris KC (2007) Nutritional strategies targeting the beneficial modulation of the intestinal microflora with relevance to food safety: the role of probiotics and prebiotics. In: Food safety: a practical and case study approach. Springer US, pp 133–152. https://doi.org/10.1007/978-0-387-33957-3_7
Moy GG, Godefroy S (2024) International standards and harmonization of food safety legislation. In: Encyclopedia of food safety. Elsevier, pp 465–473. https://doi.org/10.1016/B978-0-12-822521-9.00153-2

Mukherjee J, Das PK, Banerjee D, Mukherjee A (2023) Immune system. In: Textbook of veterinary physiology. Springer Nature, pp 89–110. https://doi.org/10.1007/978-981-19-9410-4_5
Mukhopadhyay S, Mondal SA, Goswami S, Dutta D (2020) Dietary fat, salt, and sugar: a clinical perspective of the social catastrophe. In: Dietary sugar, salt and fat in human health. https://doi.org/10.1016/B978-0-12-816918-6.00003-2
Munekata PES, Pateiro M, Rodríguez-Lázaro D, Domínguez R, Zhong J, Lorenzo JM (2020) The role of essential oils against pathogenic Escherichia coli in food products. In: Microorganisms, vol 8, Issue 6. MDPI AG, pp 1–16. https://doi.org/10.3390/microorganisms8060924
Muñoz NC, Arango-Ruiz Á (2020) Mercury as a pollutant and risk factor for human health [El mercurio como contaminante y factor de riesgo para la salud humana]. Rev Lasallista Investig 17(2):280–296. https://doi.org/10.22507/rli.v17n2a21
Murray R, Glass-Kaastra S, Gardhouse C, Marshall B, Ciampa N, Franklin K, Hurst M, Thomas MK, Nesbitt A (2017) Canadian consumer food safety practices and knowledge: foodbook study. J Food Prot 80(10):1711–1718. https://doi.org/10.4315/0362-028X.JFP-17-108
Mussagulova A (2021) The twenty-first-century public servant: a developing country perspective. In: The Palgrave handbook of the public servant. Springer International Publishing, pp 1597–1622. https://doi.org/10.1007/978-3-030-29980-4_31
Muthusamy G, Karthikeyan S, Arun Giridhari V, Alhimaidi AR, Balachandar D, Ammari AA, Paranidharan V, Maruthamuthu T (2024) Identification of potential biomarkers and spectral fingerprinting for detection of foodborne pathogens in raw chicken meat matrix using GCMS and FTIR. Foods 13(21):3416. https://doi.org/10.3390/foods13213416
Muzzaffar S, Nazir T, Bhat MM, Wani IA, Masoodi FA (2022) Goitrogens. In: Handbook of plant and animal toxins in food: occurrence, toxicity, and prevention. https://doi.org/10.1201/9781003178446-6
Nagashima S, Ko K, Yamamoto C, Bunthen E, Ouoba S, Chuon C, Ohisa M, Sugiyama A, Akita T, Hossain MS, Ork V, Mao B, Tanaka J (2021) Prevalence of total hepatitis A antibody among 5 to 7 years old children and their mothers in Cambodia. Sci Rep 11(1):1–7. https://doi.org/10.1038/s41598-021-83710-2
Nakazato G, Kobayashi RKT, Seabra AB, Duran N (2017) Use of nanoparticles as a potential antimicrobial for food packaging. In: Food preservation. Elsevier, pp 413–447. https://doi.org/10.1016/B978-0-12-804303-5.00012-2
Nath PC, Nandi NB, Tiwari A, Das J, Roy B (2023) Applications of nanotechnology in food sensing and food packaging. In: Nanotechnology applications for food safety and quality monitoring. Elsevier, pp 321–340. https://doi.org/10.1016/B978-0-323-85791-8.00006-9
Naville D, Labaronne E, Vega N, Pinteur C, Canet-Soulas E, Vidal H, Le Magueresse-Battistoni B (2015) Metabolic outcome of female mice exposed to a mixture of low-dose pollutants in a diet-induced obesity model. PLoS One 10(4):e0124015. https://doi.org/10.1371/journal.pone.0124015
Neeti K, Singh R (2023) Groundwater quality assessment and health risks from fluoride in Jamui, Bihar. Eng Technol Appl Sci Res 13(1):10204–10208. https://doi.org/10.48084/etasr.5576
Nerín C, Aznar M, Carrizo D (2016) Food contamination during food process. Trends Food Sci Technol 48:63–68. https://doi.org/10.1016/j.tifs.2015.12.004
Nettleton JE, Reimer RA, Shearer J (2016) Reshaping the gut microbiota: impact of low calorie sweeteners and the link to insulin resistance? Physiol Behav 164:488–493. https://doi.org/10.1016/j.physbeh.2016.04.029
Nichols KT, McConnell DT (2012) Food safety for the consumer. J Consum Health Internet 16(1):85–92. https://doi.org/10.1080/15398285.2012.647563
Nie J (2023) Exposure to aluminum in daily life and Alzheimer's disease. In: Neurotoxicity of aluminum, 2nd edn. https://doi.org/10.1007/978-981-99-1592-7_6
Nollet LML, Ali F (2024) Food products regulations in the United States. In: Global regulations of medicinal, pharmaceutical, and food products. https://doi.org/10.1201/9781003296492-21
O'Neill A, Quinn MR (2013) Hypothyroidism: diagnosis and screening, genetic influences and treatment options. In: Hypothyroidism: diagnosis and screening, genetic influences and treatment options. Nova Science Publishers

O'Neill C, McCarthy MB, O'Reilly S, Alfnes F (2023) Food interests, preferences and behaviours: a profile of the sustainable food consumer. Br Food J 125(13):352–374. https://doi.org/10.1108/BFJ-09-2022-0762

Obadina AO (2023) Foodborne diseases and vulnerable groups. In: Encyclopedia of food safety, vol 1–4, 2nd edn. Elsevier, pp V1-80–V1-86. https://doi.org/10.1016/B978-0-12-822521-9.00002-2

Oberoi S, Devleesschauwer B, Gibb HJ, Barchowsky A (2019) Global burden of cancer and coronary heart disease resulting from dietary exposure to arsenic, 2015. Environ Res 171:185–192. https://doi.org/10.1016/j.envres.2019.01.025

Odegaard AO, Pereira MA (2006) Trans fatty acids, insulin resistance, and type 2 diabetes. Nutr Rev 64(8):364–372. https://doi.org/10.1301/nr.2006.aug.364-372

Ogunsakin RE, Ebenezer O, Ginindza TG (2022) A bibliometric analysis of the literature on norovirus disease from 1991–2021. Int J Environ Res Public Health 19(5):2508. https://doi.org/10.3390/IJERPH19052508

Onohuean H, Agwu E, Nwodo UU (2022) A global perspective of vibrio species and associated diseases: three-decade meta-synthesis of research advancement. Environ Health Insights 16:11786302221099406. https://doi.org/10.1177/11786302221099406

Osek J, Lachtara B, Wieczorek K (2022) Listeria monocytogenesr – how this pathogen survives in food-production environments? Front Microbiol 13:866462. https://doi.org/10.3389/FMICB.2022.866462

Otairu A, Umar AA, Zawawi NAA, Pakir AHK (2013) Public sector perception of barriers to the use of public-private partnerships (PPP) in Nigeria-findings from a questionnaire survey. Res J Appl Sci Eng Technol 5(9):2809–2820. https://doi.org/10.19026/rjaset.5.4810

Pakbin B, Brück WM, Brück TB (2023) Molecular mechanisms of Shigella pathogenesis; recent advances. Int J Mol Sci 24(3):2448. https://doi.org/10.3390/IJMS24032448

Pakdel M, Olsen A, Bar EMS (2023) A review of food contaminants and their pathways within food processing facilities using open food processing equipment. J Food Prot 86(12):100184. https://doi.org/10.1016/j.jfp.2023.100184

Pal A, Kant K (2019) Using blockchain for provenance and traceability in Internet of Things-integrated food logistics. Computer 52(12):94–98. https://doi.org/10.1109/MC.2019.2942111

Parashar UD, Nelson EAS, Kang G (2013) Diagnosis, management, and prevention of rotavirus gastroenteritis in children. BMJ (Clinical Research Ed) 347:f7204. https://doi.org/10.1136/BMJ.F7204

Patel S, McCormick BA (2014) Mucosal inflammatory response to Salmonella typhimurium infection. Front Immunol 5(JUL):87909. https://doi.org/10.3389/FIMMU.2014.00311/BIBTEX

Pearlman M, Obert J, Casey L (2017) The association between artificial sweeteners and obesity. Curr Gastroenterol Rep 19(12). https://doi.org/10.1007/s11894-017-0602-9

Pepito CPC, Aleluya ERM, Alagon FJA, Clar SE, Abayan JJA, Salaan CJO, Bahinting MFP (2023) Mask, hairnet, and handwash detection system for food manufacturing health and safety monitoring. In: 2023 IEEE 15th international conference on humanoid, nanotechnology, information technology, communication and control, environment, and management (HNICEM), pp 1–6. https://doi.org/10.1109/HNICEM60674.2023.10589183

Perera RA, Perera RT, Liyanage UP, Premaratne J, Liyanage JA (2024) Chemometric assessment of bioaccumulation and contamination pathways for toxic metals in diet and environment: implications for chronic kidney disease of unknown etiology (CKDu) in Sri Lankan agricultural regions. Environ Monit Assess 196(12). https://doi.org/10.1007/s10661-024-13316-4

Petrović T, D'Agostino M (2016) Viral contamination of food. Antimicrob Food Packag 65. https://doi.org/10.1016/B978-0-12-800723-5.00005-X

Pierre C (2014) Foodborne outbreaks. In: Handbook of fermented meat and poultry, 2nd edn. https://doi.org/10.1002/9781118522653.ch49

Pires SM, Devleesschauwer B (2021) Estimates of global disease burden associated with foodborne pathogens. In: Foodborne infections and intoxications. https://doi.org/10.1016/B978-0-12-819519-2.00020-7

Pizarro-Cerdá J, Cossart P (2018) Listeria monocytogenes: cell biology of invasion and intracellular growth. Microbiol Spectr 6(6). https://doi.org/10.1128/MICROBIOLSPEC.GPP3-0013-2018/ASSET/9CFAEF32-8296-4919-B4D1-152CF9E0DBFE/ASSETS/GRAPHIC/GPP3-0013-2018-FIG2.GIF

Pleadin J, Samardžija M (2019) Hormonally active substances in the food chain from farm animals to consumers. Vet Stanica 50(6):501–512

Poladia N, Peters RW, Angus RA (2007) Investigation of the effectiveness of various treatment methods to remove endocrine disrupting compounds from domestic wastewater. In: AIChE annual meeting, conference proceedings

Pomajzl RJ, Varman M, Holst A, Chen A (2009) Hemolytic uremic syndrome (HUS)-incidence and etiologies at a regional children's hospital in 2001–2006. Eur J Clin Microbiol Infect Dis 28(12):1431–1435. https://doi.org/10.1007/s10096-009-0800-y

Pray L, Yaktine A (2009) Managing food safety practices from farm to table. In: Managing food safety practices from farm to table: workshop summary, pp 1–116. https://doi.org/10.17226/12594

Pyatha S, Kim H, Lee D, Kim K (2022) Association between heavy metal exposure and Parkinson's disease: a review of the mechanisms related to oxidative stress. Antioxidants 11(12). https://doi.org/10.3390/antiox11122467

Qian C, Liu Y, Barnett-Neefs C, Salgia S, Serbetci O, Adalja A, Acharya J, Zhao Q, Ivanek R, Wiedmann M (2023) A perspective on data sharing in digital food safety systems. Crit Rev Food Sci Nutr 63(33):12513–12529. https://doi.org/10.1080/10408398.2022.2103086

Qiu J, Lu Y, He D, Lee YW, Ji F, Xu J, Shi J (2020) Fusarium fujikuroi species complex associated with rice, maize, and soybean from Jiangsu Province, China: phylogenetic, pathogenic, and toxigenic analysis. Plant Dis 104(8):2193–2201. https://doi.org/10.1094/PDIS-09-19-1909-RE

Quereda JJ, Morón-García A, Palacios-Gorba C, Dessaux C, García-del Portillo F, Pucciarelli MG, Ortega AD (2021) Pathogenicity and virulence of Listeria monocytogenes: a trip from environmental to medical microbiology. Virulence 12(1):2509. https://doi.org/10.1080/21505594.2021.1975526

Radford S (2019) Sources of contamination in food. In: Encyclopedia of food security and sustainability. Elsevier, pp 518–522. https://doi.org/10.1016/B978-0-08-100596-5.22264-6

Rah DK (2000) Art of replacing craniofacial bone defects. Yonsei Med J 41(6):756–765. https://doi.org/10.3349/ymj.2000.41.6.756

Rai VR, Bai JA (eds) (2014) Microbial food safety and preservation techniques. CRC Press. https://doi.org/10.1201/b17465

Ranjan Sahoo S, Deb M (2022) Gas gangrene: a deadly lethal disease due to Clostridium perfringens. J Res Med Dent Sci 10(10). www.jrmds.in

Rasooly A, Herold KE (2006) Biosensors for the analysis of food- and waterborne pathogens and their toxins. J AOAC Int 89(3):873–883. https://doi.org/10.1093/jaoac/89.3.873

Redmond EC, Griffith CJ (2006) Assessment of consumer food safety education provided by local authorities in the UK. Br Food J 108(9):732–752. https://doi.org/10.1108/00070700610688377

Rehman K, Fatima F, Waheed I, Akash MSH (2018) Prevalence of exposure of heavy metals and their impact on health consequences. J Cell Biochem 119(1):157–184. https://doi.org/10.1002/jcb.26234

Reitano M, Pappalardo G, Selvaggi R, Zarbà C, Chinnici G (2024) Factors influencing consumer perceptions of food tracked with blockchain technology. A systematic literature review. Appl Food Res 4(2):100455. https://doi.org/10.1016/j.afres.2024.100455

Reshma R, Preetha R, Thangadurai D (2023) Microbial pathogens and mycotoxins in food and control measures. In: Food microbial and molecular biology: from fundamentals to applications. Apple Academic Press, pp 311–343. https://doi.org/10.1201/9781003300601-12

Robilotti E, Deresinski S, Pinsky BA (2015) Norovirus. Clin Microbiol Rev 28(1):134–164. https://doi.org/10.1128/CMR.00075-14/ASSET/229E800D-B839-4C2D-86AD-7D07A9DABBD1/ASSETS/GRAPHIC/ZCM9990924950002.JPEG

Rocourt J (2014) Foodborne diseases: foodborne diseases and vulnerable groups. In: Encyclopedia of food safety, vol 1. Elsevier, pp 323–331. https://doi.org/10.1016/B978-0-12-378612-8.00080-9
Rodríguez-Melcón C, Alonso-Calleja C, Capita R (2024) The One Health approach in food safety: challenges and opportunities. Food Front 5(5):1837–1865. https://doi.org/10.1002/fft2.458
Rossi ML, Paiva A, Tornese M, Chianelli S, Troncoso AR (2008) Listeria monocytogenes outbreaks: a review of the routes that favor bacterial presence | Brotes de infección por Listeria monocytogenes: Una revisión de las vías que llevan a su aparición. Rev Chil Infectol 25(5):328–335
Rotavirus (for Parents) | Nemours KidsHealth (n.d.) Retrieved March 28, 2025, from https://kidshealth.org/en/parents/rotavirus.html
Roya AQ, Akram A (2016) Impact of toxic heavy metals in food systems: a systemic review. Int J ChemTech Res 9(8):422–429
Rozno AI (1959) Large-frame fluorokymography of the heart. Klin Med 37(4):127–132
Ruiz N, Silhavy TJ (2022) How Escherichia coli became the flagship bacterium of molecular biology. J Bacteriol 204(9):e00230–e00222. https://doi.org/10.1128/JB.00230-22
Rusconi G (2016) Food safety and policy in the European Union. In: International food law and policy. Springer International Publishing, pp 451–483. https://doi.org/10.1007/978-3-319-07542-6_20
Sadrieh F, Annavarjula M (2005) Firm-specific determinants of corporate lobbying participation and intensity. Int J Public Adm 28(1–2):179–202. https://doi.org/10.1081/PAD-200044552
Sahl JW, Morris CR, Emberger J, Fraser CM, Ochieng JB, Juma J, Fields B, Breiman RF, Gilmour M, Nataro JP, Rasko DA (2015) Defining the phylogenomics of Shigella species: a pathway to diagnostics. J Clin Microbiol 53(3):951. https://doi.org/10.1128/JCM.03527-14
Sanitation and HACCP (2013) Surimi and surimi seafood. CRC Press, pp 375–388. https://doi.org/10.1201/b16009-22
Santa Maria MP, Hill BD, Kline J (2019) Lead (Pb) neurotoxicology and cognition. Appl Neuropsychol Child 8(3):272–293. https://doi.org/10.1080/21622965.2018.1428803
Saxena AK, Kumar A (2020) Effect of arsenic exposure in reproductive health. In: Springerbriefs in applied sciences and technology. https://doi.org/10.1007/978-981-15-4700-3_5
Schmidt RH, Rodrick GE (2003) Food safety handbook. https://muhammadsubchi.wordpress.com/wp-content/uploads/2010/04/food-safety-handbook.pdf
Schnupf P, Sansonetti PJ (2019) Shigella pathogenesis: new insights through advanced methodologies. Microbiol Spectr 7(2). https://doi.org/10.1128/MICROBIOLSPEC.BAI-0023-2019/ASSET/23F12BFC-F3BD-46B8-8C92-6E0548C1B69F/ASSETS/GRAPHIC/BAI-0023-2019-FIG4.GIF
Schwendimann L, Sivaprakasam I, Buvaneshwari S, Gurumurthy GP, Mishra S, Ruiz L, Sekhar M, Fleiss B, Riotte J, Mani S, Mani S, Gressens P (2021) Agricultural groundwater with high nitrates and dissolved salts given to pregnant mice alters brain development in the offspring. Ecotoxicol Environ Saf 224. https://doi.org/10.1016/j.ecoenv.2021.112635
See WZC, Naidu R, Tang KS (2024) Paraquat and Parkinson's disease: the molecular crosstalk of upstream signal transduction pathways leading to apoptosis. Curr Neuropharmacol 22(1):140–151. https://doi.org/10.2174/1570159X21666230126161524
Segal BH, Walsh TJ (2012) Current approaches to diagnosis and treatment of invasive Aspergillosis. Am J Respir Crit Care Med 173(7):707–717. https://doi.org/10.1164/RCCM.200505-727SO
Seibold F (2005) Food-induced immune responses as origin of bowel disease? Digestion 71(4):251–260. https://doi.org/10.1159/000087051
Selvasubramaniam S, Cyrus I, Jayasundar S (2008) Toxicodynamics of dimethoate on central dopaminergic system. Indian Vet J 85(3):259–261
Senay TE, Ferrell JL, Garrett FG, Albrecht TM, Cho J, Alexander KL, Myers-Morales T, Grothaus OF, D'Orazio SEF (2020) Neurotropic lineage III strains of Listeria monocytogenes disseminate to the brain without reaching high titer in the blood. MSphere 5(5). https://doi.org/10.1128/MSPHERE.00871-20

Sensoy I (2021) A review on the food digestion in the digestive tract and the used in vitro models. Curr Res Food Sci 4:308–319. https://doi.org/10.1016/J.CRFS.2021.04.004
Seymour IJ, Appleton H (2008) Foodborne viruses and fresh produce. J Appl Microbiol 91(5):759. https://doi.org/10.1046/J.1365-2672.2001.01427.X
Shahid M, Dumat C, Khalid S, Niazi NK, Antunes PMC (2017) Cadmium bioavailability, uptake, toxicity and detoxification in soil-plant system. In: Reviews of environmental contamination and toxicology, vol 241. Springer New York LLC, pp 73–137. https://doi.org/10.1007/398_2016_8
Shapiro K, Bahia-Oliveira L, Dixon B, Dumètre A, de Wit LA, VanWormer E, Villena I (2019) Environmental transmission of toxoplasma gondii: oocysts in water, soil and food. Food Waterborne Parasitol 15:e00049. https://doi.org/10.1016/J.FAWPAR.2019.E00049
Sharma H, Mutharasan R (2013) Review of biosensors for foodborne pathogens and toxins. Sensors Actuators B Chem 183:535–549. https://doi.org/10.1016/j.snb.2013.03.137
Sheptulina AF, Golubeva JA, Drapkina OM (2023) Fructose consumption as a risk factor for metabolic syndrome and nonalcoholic fatty liver disease. Russ J Evid-Based Gastroenterol 12(1):85–92. https://doi.org/10.17116/dokgastro20231201185
Shoaib H, Naz S, Panhwar A, Khaskheli AR, Sidhu AR, Siyal FJ, Mahesar SA, Kara H, Ayyildiz HF, Kandhro AA, Raza J, Almani S (2024) Review on health concern: trans fatty acids and hydrogenation process. Pak J Sci Ind Res Ser B: Biol Sci 67B(1):91–102
Sibanda T, Buys EM (2022) Listeria monocytogenes pathogenesis: the role of stress adaptation. Microorganisms 10(8):1522. https://doi.org/10.3390/MICROORGANISMS10081522
Silva C, Calva E, Maloy S (2014) One Health and food-borne disease: Salmonella transmission between humans, animals, and plants. Microbiol Spectr 2(1):10.1128/MICROBIOLSPEC.OH-0020-2013/ASSET/98179CA4-C3EE-4A9E-B607-06246DE562DC/ASSETS/GRAPHIC/OH-0020-2013-FIG1.GIF
Silvestre C, Cimmino S (eds) (2013) Ecosustainable polymer nanomaterials for food packaging. CRC Press. https://doi.org/10.1201/b13754
Singh R, Puniya AK (2024) Role of food safety regulations in protecting public health. Indian J Microbiol 64(3):1376–1378. https://doi.org/10.1007/s12088-024-01240-7
Singh PK, Singh RP, Singh P, Singh RL (2019) Food hazards: physical, chemical, and biological. Food Saf Hum Health:15–65. https://doi.org/10.1016/B978-0-12-816333-7.00002-3
Singh H, Bhardwaj SK, Khatri M, Kim K-H, Bhardwaj N (2021) UVC radiation for food safety: an emerging technology for the microbial disinfection of food products. Chem Eng J 417:128084. https://doi.org/10.1016/j.cej.2020.128084
Singh H, Kumar D, Deep A, Puri S, Khatri M, Bhardwaj N (2024) Fluorescent nanosensors for detection of microbial toxins in food matrices: a review. J Food Meas Charact 18(9):7669–7699. https://doi.org/10.1007/s11694-024-02757-7
Skaper SD, Facci L, Zusso M, Giusti P (2018) An inflammation-centric view of neurological disease: beyond the neuron. Front Cell Neurosci 12. https://doi.org/10.3389/fncel.2018.00072
Smith TJ, Hill KK, Raphael BH (2014) Historical and current perspectives on Clostridium botulinum diversity. Res Microbiol 166(4):290. https://doi.org/10.1016/J.RESMIC.2014.09.007
Song YH, Yu HQ, Tan Y, Lv W, Fang DH, Liu D (2020) Similarity matching of food safety incidents in China: aspects of rapid emergency response and food safety. Food Control 115:107275. https://doi.org/10.1016/J.FOODCONT.2020.107275
Sonia, Seth R (2019) L-cysteine functionalized gold nanoparticles as a colorimetric sensor for ultrasensitive detection of toxic metal ion cadmium. Mater Today: Proc 24:2375–2382. https://doi.org/10.1016/j.matpr.2020.03.767
Sreevathsan S, Bhavana BK, Debnath S, Mudliar SN (2023) Advanced computational tools for enhanced food quality and safety, pp 207–247. https://doi.org/10.1007/978-3-031-30683-9_8
Stambas J, Lu C, Tripp RA (2020) Innate and adaptive immune responses in respiratory virus infection: implications for the clinic. Expert Rev Respir Med 14(11):1141–1147. https://doi.org/10.1080/17476348.2020.1807945
Stankovic I (2016) Codex alimentarius. In: Encyclopedia of food and health. Elsevier, pp 191–196. https://doi.org/10.1016/B978-0-12-384947-2.00180-X

Steffi PF, Thirumalaiyammal B, Priya NS, Manivel T (2024) A modern focus on foodborne microbes. In: Antimicrobials in environment. CRC Press, pp 144–162. https://doi.org/10.1201/9781003268925-10

Stephen Inbaraj B, Chen BH (2016) Nanomaterial-based sensors for detection of foodborne bacterial pathogens and toxins as well as pork adulteration in meat products. J Food Drug Anal 24(1):15–28. https://doi.org/10.1016/j.jfda.2015.05.001

Stroustrup A, Swan SH (2024) Endocrine disruptors. In: Textbook of children's environmental health, 2nd edn. https://doi.org/10.1093/oso/9780197662526.003.0040

Sultana R, Mahmud A, Koli SM, Nayema J, Ghosh A, Sushree SB, Shom P, Siddiqui TA, Das KK, Acharjee M (2024) Isolation and identification of Vibrio species from different types of water sources along with their drug susceptible pattern. Biomed Biotechnol Res J 8(2):207–212. https://doi.org/10.4103/BBRJ.BBRJ_138_24

Sun L, Yang S, Deng Q, Dong K, Li Y, Wu S, Huang R (2020) Salmonella effector SpvB disrupts intestinal epithelial barrier integrity for bacterial translocation. Front Cell Infect Microbiol 10:606541. https://doi.org/10.3389/FCIMB.2020.606541/BIBTEX

Suresh S, Singh SA, Rushendran R, Vellapandian C, Prajapati B (2023) Alzheimer's disease: the role of extrinsic factors in its development, an investigation of the environmental enigma. Front Neurol 14. https://doi.org/10.3389/fneur.2023.1303111

Switaj TL, Winter KJ, Christensen SR (2015) Diagnosis and management of foodborne illness. Am Fam Physician 92(5):358–365

Takahashi T, Shimohata T (2019) Vascular dysfunction induced by mercury exposure. Int J Mol Sci 20(10). https://doi.org/10.3390/ijms20102435

Tayour C, Ritz B, Langholz B, Mills PK, Wu A, Wilson JP, Shahabi K, Cockburn M (2019) A case-control study of breast cancer risk and ambient exposure to pesticides. Environ Epidemiol 3(5). https://doi.org/10.1097/EE9.0000000000000070

Thomas MK, Vriezen R, Farber JM, Currie A, Schlech W, Fazil A (2015) Economic cost of a *Listeria monocytogenes* outbreak in Canada, 2008. Foodborne Pathog Dis 12(12):966–971. https://doi.org/10.1089/fpd.2015.1965

Thomer L, Schneewind O, Missiakas D (2016) Pathogenesis of Staphylococcus aureus bloodstream infections. Annu Rev Pathol 11:343. https://doi.org/10.1146/ANNUREV-PATHOL-012615-044351

Thompson FL, Iida T, Swings J (2004) Biodiversity of Vibrios. Microbiol Mol Biol Rev 68(3):403. https://doi.org/10.1128/MMBR.68.3.403-431.2004

Tian J, Hu J, Liu D, Yin J, Chen M, Zhou L, Yin H (2021) Cadmium chloride-induced transgenerational neurotoxicity in zebrafish development. Environ Toxicol Pharmacol 81. https://doi.org/10.1016/j.etap.2020.103545

Tiwari M (2019) Arsenic in rice grain: role of transporters in arsenic accumulation. In: Energy, environment, and sustainability. https://doi.org/10.1007/978-981-15-0690-1_10

Tiwari A, Barbhuiya RI, Dash KK (2023) Nanotechnology-based sensors for shelf-life determination of food materials. In: Nanotechnology applications for food safety and quality monitoring. Elsevier, pp 289–300. https://doi.org/10.1016/B978-0-323-85791-8.00014-8

Tong SYC, Davis JS, Eichenberger E, Holland TL, Fowler VG (2015) Staphylococcus aureus infections: epidemiology, pathophysiology, clinical manifestations, and management. Clin Microbiol Rev 28(3):603–661. https://doi.org/10.1128/CMR.00134-14/ASSET/CCDDEE5F-6E8B-4E7E-81D4-A80716493F86/ASSETS/GRAPHIC/ZCM0031525140001.JPEG

Trachtman H, Austin C, Lewinski M, Stahl RAK (2012) Renal and neurological involvement in typical Shiga toxin-associated HUS. Nat Rev Nephrol 8(11):658–669. https://doi.org/10.1038/nrneph.2012.196

Turatti A (2007) Sanitation design of fresh-cut processing and plant processing. Acta Hortic 746:147–158. https://doi.org/10.17660/ActaHortic.2007.746.17

Turesky RJ (2018) Mechanistic evidence for red meat and processed meat intake and cancer risk: a follow-up on the international agency for research on cancer evaluation of 2015. Chimia 72(10):718–724. https://doi.org/10.2533/chimia.2018.718

Tyler CR, Allan AM (2014) The effects of arsenic exposure on neurological and cognitive dysfunction in human and rodent studies: a review. Curr Environ Health Rep 1(2):132–147. https://doi.org/10.1007/s40572-014-0012-1

Udom GJ, Abdulyekeen BR, Osakwe MO, Ezejiofor AN, Orish CN, Orish FC, Frazzoli C, Orisakwe OE (2025) Reconsideration of the health effects of monosodium glutamate: from bench to bedside evidence. J Environ Sci Health C Toxicol Carcinog 43(1):51–81. https://doi.org/10.1080/26896583.2024.2415202

Ummulkhair Salamah I, Faridah Laraba T, Jeremiah David B, Mercy N (2024) International regulations in food contaminants. In: Emerging contaminants in food and food products. CRC Press, pp 29–43. https://doi.org/10.1201/9781003438236-4

Uzal FA, Freedman JC, Shrestha A, Theoret JR, Garcia J, Awad MM, Adams V, Moore RJ, Rood JI, Mcclane BA (2014) Towards an understanding of the role of Clostridium perfringens toxins in human and animal disease. Future Microbiol 9(3):361. https://doi.org/10.2217/FMB.13.168

Valerón PF, Pestano JJ, Luzardo OP, Zumbado ML, Almeida M, Boada LD (2009) Differential effects exerted on human mammary epithelial cells by environmentally relevant organochlorine pesticides either individually or in combination. Chem Biol Interact 180(3):485–491. https://doi.org/10.1016/j.cbi.2009.04.010

Vandenberg LN (2014) Hazards of food contact material: bisphenol A and endocrine disruption. In: Encyclopedia of food safety, vol 2. https://doi.org/10.1016/B978-0-12-378612-8.00428-5

Varzakas T (2016) HACCP and ISO22000: risk assessment in conjunction with other food safety tools such as FMEA, Ishikawa Diagrams and Pareto. In: Encyclopedia of food and health. Elsevier, pp 295–302. https://doi.org/10.1016/B978-0-12-384947-2.00320-2

Vasanthi S, Bhat RV (2018) Management of food safety risks in India. Proc Indian Natl Sci Acad 84(4):937–943. https://doi.org/10.16943/ptinsa/2018/49439

Walia A, Mehra R, Kumar N, Singh TP, Kumar H (2023) Good manufacturing practices and safety issues in functional foods and nutraceuticals. In: Bioactive components. Springer Nature, Singapore, pp 167–182. https://doi.org/10.1007/978-981-19-2366-1_11

Wang XC, Chen K, Zeng ZQ, Zhuang WY (2017) Phylogeny and morphological analyses of Penicillium section Sclerotiora (Fungi) lead to the discovery of five new species. Sci Rep 7(1):1–14. https://doi.org/10.1038/s41598-017-08697-1

Wang G, Chen H, Yang J, Wang J, Yan B, Chen M, Wu S (2024a) Harmful effects of metal deposits in the brain. In: Waste clearance in the brain. https://doi.org/10.2174/97898153130861240100 07

Wang S, Wang S, Chen T, Yu J, Shi Y, Chen G, Xu J, Qiu J, Zhu F, Ouyang G (2024b) Detection and health implications of phthalates in tea beverages in market: application of novel solid-phase microextraction fibers. Sci Total Environ 953:176031. https://doi.org/10.1016/j.scitotenv.2024.176031

Webb M (2015) Overview of food safety standards. In: Food safety, market organization, trade and development. Springer International Publishing, pp 45–58. https://doi.org/10.1007/978-3-319-15227-1_3

WHO (2025) Food safety. https://www.who.int/news-room/fact-sheets/detail/food-safety

Winder N, Gohar S, Muthana M (2022) Norovirus: an overview of virology and preventative measures. Viruses 14(12):2811. https://doi.org/10.3390/V14122811

Witczak A, Pokorska-Niewiada K (2023) Food safety. In: Chemical and functional properties of food components, 4th edn, pp 419–432. https://doi.org/10.1201/9781003265955-15/FOOD-SAFETY-AGATA-WITCZAK-KAMILA-POKORSKA-NIEWIADA

Worley MJ (2023) Salmonella bloodstream infections. Trop Med Infect Dis 8(11). https://doi.org/10.3390/TROPICALMED8110487

Wu J, Zou Y, Xu Y, Xie M, Zheng Z, Geng J (2021) Neurological disease. In: Clinical molecular diagnostics. Springer Nature, pp 717–749. https://doi.org/10.1007/978-981-16-1037-0_41

Xiao Y, Huang Z, Yu K, Wang M, Gao H, Bai X, Jiang M, Wang D (2022) Distribution and molecular characteristics of Vibrio species isolated from aquatic environments in China, 2020. Microorganisms 10(10):2007. https://doi.org/10.3390/MICROORGANISMS10102007

Xu P (2022) Effects of xenobiotics on nutrition. In: Nutritional toxicology. Springer Nature, pp 29–41. https://doi.org/10.1007/978-981-19-0872-9_2

Xu L, Ren H, Gao G, Zhou L, Malik MA, Li P (2016) The progress and challenges in metabolic research in China. IUBMB Life 68(11):847–853. https://doi.org/10.1002/iub.1563

Yaliman A, Şen EI (2011) Parkinson's disease and rehabilitation | Parkinson hastaliği ve rehabilitasyonu. Turkiye Fiziksel Tip ve Rehabilitasyon Dergisi 57(1):38–44. https://doi.org/10.4274/tftr.57.07

Yin H, Nair MS, Liu Y, Amalaradjou MA, Venkitanarayanan K (2017) Controlling foodborne pathogens using antimicrobial metals. In: Food safety and consumption: assessment, practices and current issues. Nova Science Publishers

Yoo A, Lin M, Mustapha A (2021) Zinc oxide and silver nanoparticle effects on intestinal bacteria. Materials 14(10):2489. https://doi.org/10.3390/ma14102489

Yuan H, Zhao J, Xie E, Yi L, Zheng Z, Geng J (2021) Endocrine and metabolic diseases. In: Clinical molecular diagnostics. Springer Nature, pp 665–716. https://doi.org/10.1007/978-981-16-1037-0_40

Yunker AG, Patel R, Page KA (2020) Effects of non-nutritive sweeteners on sweet taste processing and neuroendocrine regulation of eating behavior. Curr Nutr Rep 9(3):278–289. https://doi.org/10.1007/s13668-020-00323-3

Zarazúa S, Jiménez-Capdeville ME, Schliebs R, Ríos R (2013) Central nervous system targets of chronic arsenic exposure. In: Arsenic: sources, environmental impact, toxicity and human health – a medical geology perspective. Nova Science Publishers, Incorporated

Zaujan NAM, Ali A, Osman M, Chee HY, Ithnin NR, Misni N, Sukeri S, Chin CP-Y (2021) Effect of smartphone app's intervention on consumers' knowledge, attitude, practice, and perception of food poisoning prevention when dining out at selected rural areas in Terengganu. Int J Environ Res Public Health 18(19):10294. https://doi.org/10.3390/ijerph181910294

Zhao X, Shen C, Zhu H, Wang C, Liu X, Sun X, Han S, Wang P, Dong Z, Ma X, Sun A, Ge J (2016) Trans-fatty acids aggravate obesity, insulin resistance and hepatic steatosis in C57BL/6 mice, possibly by suppressing the IRS1 dependent pathway. Molecules 21(6). https://doi.org/10.3390/molecules21060705

Zia AI, Mukhopadhyay SC (2016) Human endocrine system and hormonal measurement. In: Smart sensors, measurement and instrumentation, vol 20. Springer International Publishing, pp 1–20. https://doi.org/10.1007/978-3-319-32655-9_1

Zietek T, Giesbertz P, Ewers M, Reichart F, Weinmüller M, Urbauer E, Haller D, Demir IE, Ceyhan GO, Kessler H, Rath E (2020) Organoids to study intestinal nutrient transport, drug uptake and metabolism – update to the human model and expansion of applications. Front Bioeng Biotechnol 8:577656. https://doi.org/10.3389/fbioe.2020.577656

Zochodne DW (2020) Our wired nerves: the human nerve connectome. In: Our wired nerves: the human nerve connectome. Elsevier. https://doi.org/10.1016/B978-0-12-821487-9.09995-5

Chapter 2
Chrono Biological Considerations in Food Safety

Aniruddha Sarker, Isha Datta, Anirbandeep Bose, Sanjay Dey, Dibya Das, Uttam Budathoki, and Rajdip Goswami

2.1 Introduction

2.1.1 Defining Chronobiology

Chronobiology is the formal study of biological rhythms that control the physiological functions of living organisms in accordance with time cues, primarily the light and dark cycles. Chronobiology examines how physiological processes are regulated in organisms by the time-associated signals of light and dark cycles. These rhythms are involved in immune functioning, digestion, hormone secretion, metabolism, and sleep. Three major categories of biological rhythms constitute chronobiology (Park et al. 2017). **Circadian Rhythms** Happen on a 24-hour cycle and control sleep-wake patterns, feeding behaviour, and metabolic processes. **Ultradian Rhythms** Happen repeatedly during a 24-hour period, including heart rate, hormonal rhythms, and digestion cycles. **Infradian Rhythms** Take place over a duration greater than 24 hours, e.g., menstrual cycles or changes in metabolism during the seasons (Sonnier et al. 2014). From a food safety point of view, chronobiology determines the fact that the body behaves differently on different days concerning digestion, metabolism, and excretion of food. All these functions are under the control of the body's master clock, called the Superchiasmatic Nucleus (SCN), located in the hypothalamus, which synchronizes the processes of numerous peripheral

A. Sarker · I. Datta · A. Bose · S. Dey · R. Goswami (✉)
Department of Pharmaceutical Technology, School of Health and Medical Sciences, Adamas University, Kolkata, West Bengal, India

D. Das
Department of Pharmaceutical Technology, JIS University, Kolkata, India

U. Budathoki
Department of Pharmacy, School of Science, Kathmandu University, Kathmandu, Nepal

T. Sarkar et al. (eds.), *Physiological Perspectives on Food Safety: Exploring the Intersection of Health and Nutrition*, https://doi.org/10.1007/978-3-031-94582-3_2

clocks found in organs like the liver, pancreas, and intestinal tract (Spezani et al. 2020). The SCN is controlled by external signals, primarily light, to regulate the activities of the body, while peripheral clocks are controlled by feeding schedules and food consumption. However, it should be noted that from the perspective of food safety, chronobiology tells the reason for differences in the occurrence as well as functioning of the body's digestions, metabolisms, and excretions of food based on the day of the week (Clawson et al. 2019). Chronobiology plays an important role in food safety because it determines how toxins, foodborne pathogens, and metabolism are incurred in the body. Nutrients and harmful substances utilized and excreted by the body depend on what and when one consumes them. Therefore, with this knowledge-based approach, one can formulate and develop better approaches to improving food safety and nutrition status (Kumar et al. 2023).

2.1.2 Biological Rhythms and Human Physiology

Biological rhythms are internal periodic mechanisms that determine basic physiological processes of the human organism. Their function includes stimulation of hormone release, participation in digestion, metabolism, sleep, and immune responses. The circadian rhythms of the body are 24-h biological clocks regulated by the superchiasmatic nucleus (SCN), the master biological clocks of an organism that ensure synchronization of peripheral clocks in others organs like the liver, intestines, and pancreas. This section discusses biological rhythms' role in digestion, metabolism, and detoxification and how biological rhythms are involved in food safety (Teong and Heilbronn 2021).

2.1.2.1 Circadian Influence on Digestion and Metabolism

It is worthy to be noted that the body functions with variances in digestion efficiency at different times of the day; increased enzyme activity during daytime gives way to decreased enzyme activity during nighttime. Maximizing insulin sensitivity and glucose metabolism in this active phase would allow for maximal energy use and storage with food intake in early hours. Any food intake that varies from this timing, for example, midnight munching, disrupts these natural rhythms further to induce metabolic derangements like insulin resistance and obesity (Castro-de-Paiva et al. 2022).

2.1.2.2 Hormonal Regulation and Appetite Control

The release of hunger-controlling hormones such as ghrelin (hunger hormone) and leptin (satiety hormone) is regulated in a circadian rhythm. Alterations in such rhythms lead to inappropriate triggering of hunger, causing overconsumption of foods and poor diet (Varkaneh Kord et al. 2021).

2.1.2.3 Immune Function and Food Safety

This system participates in the circadian rhythm of the immune system with heightened offense during active and restorative functions by night. An optimally functioning immune system can protect better against any food pathogens when the timings of the meal are in sync with the natural rhythm. The disruption of the circadian rhythm, for example, shift work and irregular sleep cycle, leads to the impairment of pituitary production. These two factors consequently affect immune defence and increase vulnerabilities against infections through the food route (Spencer 1968).

2.1.2.4 Gut Microbiota and Chronobiology

Circadian rhythms have a role in the regulation of activities and the composition of gut microbiota. Regularly scheduled meal times were "beneficial bacteria's" hours, while non-regularly scheduled meals or eat times "messed up" the balance of microbes, making the host more vulnerable to being infected by various intestinal diseases and infections (Arora and Dey 2014).

2.1.2.5 Detoxification and Liver Function

Detoxifying harmful substances is a function of the liver and its enzymatic activity is circadian, meaning that detoxification occurs mainly in the relaxing hours to help the body rid itself of toxins absorbed from ingested food items or environmental exposure. Any alteration of this cycle via irregular meal schedules can sabotage the efficacy of detoxification, thereby exposing the body to more toxic accumulation and metabolic-related stress. Understanding biological rhythms in human physiology is crucial to maximizing food safety measures. Synchronizing meal times with circadian rhythms maximizes digestion, boosts the immune system, and minimizes health risks from foodborne toxins and pathogens. Food safety chronobiological awareness offers a promising way to realise maximum public health benefits (Kord-Varkaneh et al. 2023).

2.1.3 Importance of Chronobiology in Food Safety

Chronobiology is at the heart of food safety, which, through its influence, affects the body's management of contaminants and pathogens. Food safety becomes all the more necessary since metabolic competence and immune processes are diurnally programmed. The efficient functioning of metabolic enzymes, homeostatic maintenance of the gut microbiota, and immune defence mechanisms work obliquely and thus affect body detoxification of noxious agents and body response to food agents, depending on the time of day (Yanishi et al. 2017). Studies reveal that food poisoning might be enhanced in severity upon meal consumption in the evening hours when digestion and immune responses are depressed. Besides, food preparation, storage, and consumption involve interaction with the body's rhythm. Inefficiency in eating times increases disease risks by promoting the growth of bacteria in the gastrointestinal tract, modifying gut microbiota, and degrading body detoxification competence. Synchronizing eating timing with biological rhythm not only advances food safety but also minimizes disease susceptibility, enhances nutritional absorption, and boosts the overall health of the metabolisms. Education in chronobiology in the context of food safety may drive more efficient guidelines for public health, tailored to nutritional approaches for individualized guidance and efficacious dietary strategies to minimize dangers related to foodborne disease. This section assesses how foodborne illness vulnerability is affected by eating timing and the benefits of synchronizing eating habits and circadian rhythms for maximizing the safety of the food and, in general, an individual's well-being (Hatori et al. 2012). The Importance of Chronobiology in Food Safety is described pictorially in Fig. 2.1.

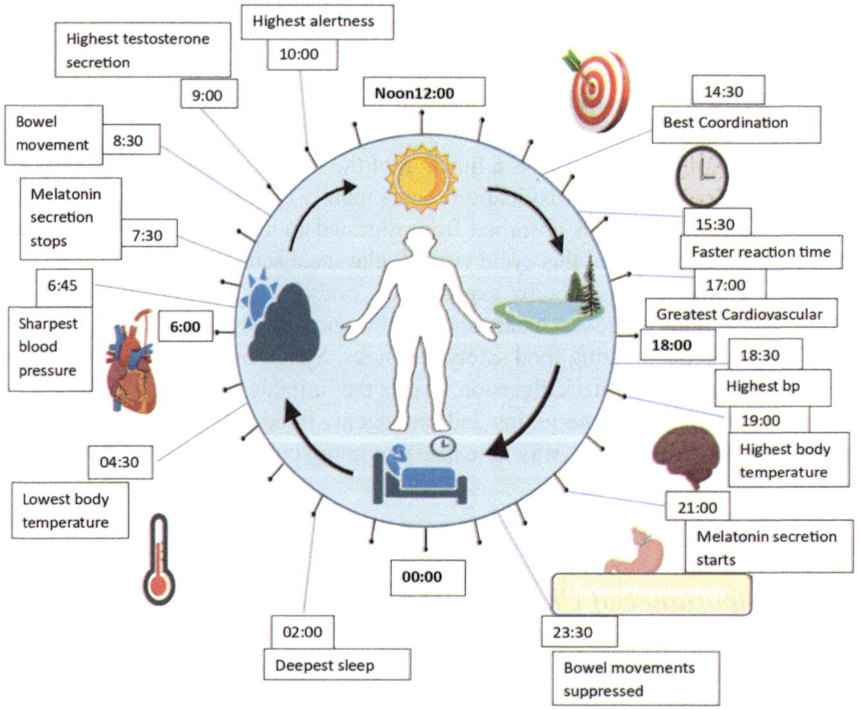

Fig. 2.1 The importance of chronobiology in food safety (Flanagan et al. 2021)

2.2 Physiological Foundations of Circadian Rhythms

2.2.1 *The Role of the Internal Clock in Human Health*

The human biological clock is governed by the suprachiasmatic nucleus (SCN). It controls the body's rhythms. The hypothalamic master clock serves the purpose of coordinating several external stimuli, such as light and darkness, with physiological processes controlling metabolism, digestion, immunity, and general health. The next section describes how circadian disruption can undermine food safety by compromising the body's ability to metabolize nutrients and repel disease-causing microorganisms (Santos et al. 2020).

2.2.1.1 Neural and Hormonal Regulation of the Internal Clock

The SCN signals to peripheral clocks within organs such as the liver, pancreas, and gut through neural and hormonal mechanisms. Melatonin is the hormone released by the pineal gland for controlling the sleep-wake cycle and adjusting the metabolic efficiency during the hours of the day. The hormone cortisol, which is elevated in the morning, induces alertness and activates metabolism; it is then decreased during the night to allow recovery and detoxification (Berg et al. 2002).

2.2.1.2 Impact of Light on Circadian Rhythms

The morning sun encourages the strongest circadian rhythms for the best possible digestion and metabolic action. Artificial lights at night will interfere with circadian rhythms, thus creating metabolic derangements and increasing the risk of foodborne illness (Carter et al. 2016).

2.2.1.3 The Internal Clock and Nutrient Metabolism

The functionality of the organism concerning the digestion of macronutrients changes with time, enhancing carbohydrate metabolism in the morning and fat oxidation in the evening. Hence, there will be much less efficient energy expenditure and a higher risk of metabolic disorders from eating at night when digestion does not work at its best (de Pureza et al. 2021).

2.2.1.4 Immune System and Circadian Influence

Immune system functions are characterized by a circadian rhythm that elicits immune activation during waking hours and promotes repair mechanisms during sleep. Circadian misalignment will thus debilitate immunity and impair the body's ability to ward off foodborne infection (Coutinho et al. 2018).

2.2.2 *Key Circadian Mechanisms Governing Digestion and Metabolism*

Circadian rhythms have a major role to play in maintaining human digestion and metabolism, for their timing purposes allow for maximum nutrient uptake, energy expenditure, and detoxification with reference to the body's internal clock. Some of the main circadian mechanisms influencing digestion and metabolism. Here, we discuss the metabolic and hormonal paths regulating food digestion and their ramifications for food safety (Meng et al. 2020).

2.2.2.1 Gastrointestinal Function and Enzyme Secretion

The digestive enzymes are produced on a circadian rhythm primarily when the body is active, which would be during the daytime. Digestion, during this day phase, is positively facilitated by gastric acid secretion, bile production, and the release of pancreatic enzymes. Digestion during the night is relatively poor (Chen et al. 2019).

2.2.2.2 Glucose and Insulin Regulation

Insulin sensitivity is greatest in the morning and early afternoon, allowing for effective glucose uptake and metabolism. Eating late at night disturbs this rhythm, resulting in impaired glucose metabolism and a heightened risk of insulin resistance and type 2 diabetes (Scholtens et al. 2020).

2.2.2.3 Lipid Metabolism and Energy Storage

The capacity of the body to oxidize and store fat depends on circadian signals. Featuring high-fat meals in the late evening at night adds to fat storage and metabolic strain (Philippens et al. 1977).

2.2.2.4 Gut Microbiota and Circadian Influence

The gut microbiome has a daily rhythm, with some populations of bacteria having peak activity at certain times. Changes in eating habits result in dysbiosis, making individuals more susceptible to gastrointestinal disease and foodborne infection (Garaulet et al. 2020).

2.2.2.5 Detoxification and Liver Function

The detoxification functions of the liver are circadian, with maximal activity during resting periods. Consuming food at night overloads liver function and diminishes its capacity to eliminate toxins effectively (Davis et al. 2020).

2.2.3 Detoxification and Immune Function: A Chronobiological Perspective

The circadian nature by which immune function and liver function regulate detoxification and resistance to pathogens is being surveyed. Disturbed rhythmicity, therefore, reciprocally impairs detoxification in respect to food-borne toxins and pathogens. Thus, one might consider that detoxification and immune-competence in the body are strictly tuned to circadian rhythms for optimum protection against toxins and pathogens (Regmi and Heilbronn 2020).

2.2.3.1 Liver Detoxification Rhythms

It is the active phase of the enzymatic detoxification system of the liver from night time until early morning which breaks down food toxins as well as environmental toxins. Disturbances such as late-night meals would disrupt such rhythms somehow interfering with detoxification causing oxidative stress (Gu et al. 2020).

2.2.3.2 Circadian Rhythms Affecting Immune Responses

Most immune defense mechanisms and inflammatory activities occur during the day when activities are at their peak, thus the immune system works optimally also during the daytime. Sleep during the night hours facilitates the restoration and repair of tissues; any alteration that distorts restoration and repair may lead to altered immune response and great susceptibility to infection (Genario et al. 2021).

2.2.3.3 Circadian Asynchrony

The long-term outcome of foodborne illness onset among those with chronic circadian rhythm disruption, as with shiftwork or jet lag, is modified. Long exposures to circadian rhythm disruption were shown to be associated with increased risk for metabolic disorders, inflammatory conditions, or gastrointestinal diseases (Genario et al. 2019).

2.3 Chrono-biological Influence on Food Safety

2.3.1 Timing of Foodborne Hazard Exposure

The hormone is at its peak activity in the early hours of morning and early afternoon. While waiting for it to enter an optimum state for glucose uptake and subsequent metabolism, anything ingested after midnight tends to disturb this rhythm, which consequently reduces glucose uptake and eventually gives way to resistance against insulin and type 2 diabetes (Santos and Macedo 2018).

2.3.1.1 Gastric Acid Secretion and Pathogen Resilience

Gastric acid production in our body shows a corresponding circadian rhythm, with maximum production during working hours and with minimum production at night. The period of low acid production following the ingestion of contaminated food, during which pathogens are likely to persist, significantly increases the risk of infection (Varady et al. 2009).

2.3.1.2 Enzymatic Activity and Toxin Resistance

The time of day exercised, their usage and the availability of digestive enzymes differ in different periods of 24 hours, and those variations might affect the breakdown and absorption of nutrients as well as detoxification reactions. At night, reduced enzymatic activity could prevent the organism's neutralization of toxins from food sources and thus render the organism more vulnerable to those poisons (Persynaki et al. 2017).

2.3.1.3 Immune System Efficiency and Foodborne Pathogens

Immune function is most potent in the active phase, with more vigorous inflammatory responses and pathogen defence mechanisms. Foodborne illnesses that are contracted during the night may be more intense due to compromised immune defences in the rest phase (Poggiogalle et al. 2018).

2.3.1.4 Gastrointestinal Motility and Clearance of Contaminants

Motility patterns regulated by circadian rhythms have implications for the elimination rate of ingested pathogens from the body. One possible effect of reduced gastrointestinal motility at night would be a greater exposure to toxins and their by-products (Qian et al. 2018).

2.3.1.5 Practical Implications for Food Safety

The time of proper aging relative to most optimum digestive and immune capacity would reduce an individual's vulnerability to foodborne diseases. Renouncing midnight meals, especially for high-risk foods, may enhance the capability of the body to neutralize probable contaminants (Grandner 2017).

2.3.2 Circadian Rhythms and the Body's Response to Contaminants

The circadian rhythms maintain a great influence over the detoxification and immune response mechanisms toward food contaminants and toxins. The circadian system regulates major physiological functions relevant to defence against foodborne threats, such as immune responses, liver detoxification, and gut barrier integrity (Roenneberg and Merrow 2016).

2.3.2.1 Circadian Regulation of Immune Responses to Contaminants

The functioning of the immune system-dependent time varies; an immune system in action during the active phase (daytime in diurnal *Homo sapiens*) is least active during the rest phase. Immune cells, including macrophages and neutrophils, also demonstrate intrinsic circadian variations in their abilities to recognize and destroy pathogens. Such misalignment of the circadian rhythm, such as working night shifts, greatly compromises immune surveillance and adds to the risk of acquiring infection via contaminated food (Kord Varkaneh et al. 2022).

2.3.2.2 Liver Detoxification and Circadian Timing

The liver detoxifies and removes foodborne toxins, and detoxification activity in the liver is controlled in a circadian manner. The detoxifying enzymes, including cytochrome P450, exhibit circadian (daily) rhythm, with peak activity at certain times of the day. The ingestion of contaminated food during peak detoxification times could maintain the activity of the toxin in the body for a long time and create a high body burden of the toxin (Fernández-Rodríguez et al. 2022).

2.3.2.3 Gut Barrier Function and Circadian Influence

The gastrointestinal barrier is the initial line of defence against pathogens and toxins ingested, with circadian rhythms modulating its integrity. Disturbances in circadian synchronization, including abnormal meal timing or sleep deprivation, compromise gut barrier function, allowing easier entry of contaminants into the bloodstream (Liu et al. 2022).

2.3.2.4 Metabolic Implications of Circadian Disruptions on Contaminant Processing

The metabolic breakdown of foodborne chemicals, such as preservatives, additives, and naturally occurring toxins, depends on the timing of intake. Research indicates that ingestion of certain contaminants when the body is in its resting phase leads to greater levels of toxicity due to decreased metabolic clearance rates. Circadian Rhythms and the Body's Response to Contaminants are described in Table 2.1 (Chowdhury 2022).

2.4 Impact of Meal Timing on Health

2.4.1 Synchronization of Meal Times with Circadian Rhythms

Syncing meal times with circadian rhythms is essential in maximizing digestion, metabolism, and overall well-being. The internal body clock-sequencing system regulates the activity of certain enzymes and hormones and metabolic functions,

Table 2.1 Circadian rhythms and the body's response to contaminants (Stenger et al. 2023; Santos and Tinsley 2024; Bo et al. 2014, 2015; Madjd et al. 2016; Richter et al. 2020)

Aspect	Description	Circadian influence
Gastrointestinal function	The ability of the digestive system to process and eliminate contaminants	Peaks during the active phase, ensuring efficient digestion and pathogen clearance
Gut microbiota	The composition and activity of microbial populations in the gut that interact with contaminants	Fluctuates with feeding and fasting cycles; dysbiosis during circadian disruption increases vulnerability to toxins
Detoxification pathways	Enzymatic processes in the liver and other organs neutralize and eliminate toxins	Cyclical expression of detoxification enzymes, with higher activity during the active phase
Immune response	The body's ability to detect and respond to pathogens or harmful contaminants	Immunological functions, such as cytokine production, are rhythmically regulated and more robust during certain times
Metabolic processing of toxins	Breakdown and removal of foodborne contaminants via metabolic pathways	Enhanced metabolic activity during the day (or active phase), with reduced efficacy during rest periods
Oxidative stress management	Cellular mechanisms to mitigate damage from contaminants	Antioxidant defenses show circadian peaks, reducing toxin-induced oxidative stress during optimal times
Circadian disruption	Effects of misalignment on the body's contaminant response	Leads to impaired detoxification, weakened gut barrier, and heightened susceptibility to contaminants
Pathogen susceptibility	Likelihood of infection or adverse effects from contaminants	Increased during circadian misalignment or rest phases when immune defences are downregulated

ensuring that food timing is of utmost importance for the physiological balance. Ideal meal timing would enhance digestion, metabolism, and nutrient absorption. This part summarizes the studies that support the benefits of eating meals synchronized with biological rhythms (Howell and Kones 2017).

2.4.1.1 Energy Utilization and Metabolic Efficiency

Overall during the day, macronutrient digestion efficiency varies; highest metabolic effect occurs during the day. Advancing meal timing enhances glucose metabolism and insulin sensitivity, whereas delaying meal timing could disturb glucose metabolism and thus trigger the development of metabolic diseases-e.g., diabetes and obesity. Late-night eating disturbances would disrupt regular metabolic cycles leading to dysfunction in energy production, fat deposition, and metabolic stress (Tinsley et al. 2019).

2.4.1.2 Coordination of Enzymes and Hormones

Digestive enzymes, amylases, and proteases have circadian-driven oscillations, being most active during the day. In that sense, hormones such as insulin, glucagon, and ghrelin have also shown rhythmic behavior, suggesting that consumption of food during this high activity period would be associated with untimely nutrient absorption and storage (Ho 2018).

2.4.1.3 Impact on Gastrointestinal Function

Daytime maximal peristalsis and gastric emptying rates embody the gut circadian rhythm. Nighttime food intake is an affront to these natural activities and will lead to increased occurrences of acid reflux, bloating, and a slowed process of digestion. Meal timing synchronization aids in gut microbiota balance due to the healthy proliferation of good bacteria when food intake coincides with circadian rhythms (Bouassida et al. 2010).

2.4.1.4 Role in Weight Management and Appetite Control

Eating food at abnormal times disrupts the release of leptin and ghrelin; hence, hunger and cravings are brought about. The study shows that late eaters consume excessive calories and have an increase in BMI as compared to early eaters (Hall 2008).

2.4.1.5 Circadian Synchronization and Food Safety

The body's detoxifying processes and immune responses against foodborne pathogens are more efficient if the timing of eating corresponds to circadian rhythms. Having proper mealtimes will minimize toxin buildup, maximize the immune response, and improve overall food safety performance (Corley et al. 2018).

2.4.2 *Effects of Irregular Eating Patterns*

An irregular meal schedule often favours the body's natural circadian rhythms and mechanisms, resulting in a cascade of ill health effects that may include metabolic derangements, disturbances in hormone levels, impaired immune function, and susceptibility to foodborne diseases. Irregular meal timing disturbs metabolic homeostasis and heightens the risk of foodborne pathogens; this section then discusses health consequences of eating late and having meals at unpredictable times (Cho et al. 2019).

2.4.2.1 Metabolic Consequences

Such feed patterns change glucose metabolism and have a possible effect on the risk of developing insulin resistance and type-2 diabetes. Night eating is supposed to be implicated in obesity and weight gain for its effect on energy utilization and poor utilization of other nutrients (Carter et al. 2018).

2.4.2.2 Hormonal Disruptions

A person would have to be a ghrelin (hunger hormone) and leptin (satiety hormone) oscillate with a circadian rhythm, but erratic eating patterns disturb both hormones and lead to an urge to overeat or unhealthy eat. Their levels get disturbed by disrupted meal timing since cortisol is responsible for modulating metabolism and stress, promoting stress-induced eating behaviours (Carter et al. 2018).

2.4.2.3 Gut Microbiota Disturbances

What happens is that some irregular eating habits disturb the gut microbiome's circadian rhythm, leading to gastrointestinal disturbances, bloating, and inflammation. An improper microbiota composition makes a person more susceptible to foodborne pathogens and also impairs digestion (Guo et al. 2021).

2.4.2.4 Weakened Immune System

It has been observed that the time accept eaten noun phrase is directly affects the immune response. Eating time irregularities affect the production of immune cells, thus affecting the ability of the body to fight infections. The individuals will have more chances of suffering from bacterial and viral infectious diseases such as food contamination diseases (Li et al. 2020).

2.4.2.5 Food Safety Risks

Poor meal timings inhibit the normal functioning of the digestive system and detoxification mechanisms and lead to an increased possibility of food poisoning due to improper breakdown and detoxification of the toxins. Night shifters or individuals whose eating habits change while travelling are particularly likely to have compromised immunity and impaired digestion, making such individuals more susceptible to foodborne diseases. Abiding by a consistent meal time helps an individual to perfect his digestion and metabolism effects, and therefore reduces immune problems related to a person's health caused by irregular eating schedules (Mangaraj et al. 2016).

2.4.3 Gut Microbiota Composition and Immune Response Modulation

Gut microbiota, which are complex microbial community organisms residing in the gastrointestinal tract, play an important role in digestion, metabolism, and the immune system. The circadian rhythm shows an influence on the structure of gut bacteria, their function, and the health of an individual in general, specifically, food safety. This section will study the effects of meal timing on the diversity of microbiota, immunity and food safety (Antoni et al. 2018).

2.4.3.1 Circadian Regulation of Gut Microbiota

Specifically, they show the daily rhythms of the gut microbiome, where certain bacteria flourish at different times of day. A regular meal routine promotes microbial diversity and stability, while an irregular meal and feeding regime disturbs the microbial balance and brings about dysbiosis (literally meaning 'lack of balance' in gut bacteria) (Lowe et al. 2020).

2.4.3.2 Impact on Digestive Efficiency

Following a synchronized gut microbiome, there are better processes of digestion and nutrient reuptake. Disturbances in the rhythm of microbes can damage digestion and lead to symptoms of abdominal discomfort, bloating, and heightened sensitivities to food (Pilz et al. 2007).

2.4.3.3 Immune System Modulation

The gut microbiota regulates immunity by mediating inflammation and the ability to resist infection. Dysbiosis causes damage to the gut barrier and increases susceptibility to infections and food-related diseases (Trepanowski et al. 2017).

2.4.3.4 Food Safety Implications

Microbiota implicated in disease will therefore lead to over-colonisation of the gut with potentially pathogenic bacteria such as Salmonella and E. coli, thereby increasing the risk of foodborne illness. Probiotics and prebiotics, consumed following biological rhythms, could thus be beneficial for gut health and improving immunity. Gut microbiota are stable and varied due to a normal meal frequency pattern, which improves digestive health, boosts the immune system, and makes one less vulnerable to foodborne illnesses (Kunduraci and Ozbek 2020).

2.5 Circadian Disruptions and Food Safety Risks

2.5.1 Effects of Shift Work and Time Zone Travel on Food Safety

The body clock's natural rhythm is distorted by work timings and constantly traveling in two directions. This leads to metabolic changes and increases one's susceptibility to foodborne illness. Every above affects the digestive system and immunity, and also detoxification capacity against toxins (Kul et al. 2014).

2.5.1.1 Circadian Misalignment in Shift Workers

People working night duty always fall victim to reversed biological clocks, diminished digesting efficiencies, and dysfunctional metabolism. These changes predispose night-shift workers to insulin resistance and abnormal glucose metabolism, which are high-risk factors for obesity and diabetes. House irregularities in the gastrointestinal microbiota resulting from improper meal timings also bring out inflammatory changes in addition to impaired immunity (Koutsari et al. 2011).

2.5.1.2 Food Safety Challenges for Shift Workers

Because digestive enzymes and the intestines are less effective during the night, eating at night increases the risk for indigestion and the lack of absorption of nutrients. Workers at night tend to utilize processed or convenience foods, many of which add preservatives and contaminants to make it even more difficult for the body's detoxification processes. Immune function is also compromised during nocturnal activity, leading to high exposure to food pathogens (Morgan et al. 1999).

2.5.1.3 Effects of Time Zone Travel on Foodborne Illness Susceptibility

This causes the desynchronization of our circadian rhythms induced by frequent travel across time zones, causing digestive dysfunction, increased bloating, and changes in bowel regimens. Jet lag is thus an immune-deficient status that exposes a traveller to succumbing to foodborne illness, particularly in places where eating habits and food hygiene practices differ from their own. Time meal desynchronization, along with internal body clock exposure, influences the gut microbiome, rendering it more receptive to pathogenic strains. Attention to and management of shift work and time-zone travel risks will lead to enhanced food safety or broader health outcomes. Meal timing adjustment, exposure to natural daylight, and diet modification are among the approaches available for the synchronization of circadian rhythms and the minimization of adverse effects (Yoshino et al. 2014).

2.5.2 Weakened Defence Mechanisms and Pathogen Susceptibility

Chrono disruption compromises immune behaviour, thereby making one more prone to foodborne diseases. This section highlights immune impairment mechanisms due to circadian disharmonies. Disruption of the circadian rhythm diminishes the body's frontline defence against foodborne pathogens and infection. There are several physiological phenomena for this enhanced vulnerability (Santos 2022).

2.5.2.1 Impaired Immune Function

It is also widely appreciated that the immune system functions on a daily basis, with a well-functioning immune response during the day hours and a focus on restorations at night. However, working at nonstandard times and having inconsistent sleep patterns may interfere with normal cytotoxic immune sync-rony and probably contribute to decreased immune responses against infectious agents. The lower counts of cells like T-cells and natural killer cells compromise the immunity against infections that might result from food being contaminated (Jakubowicz et al. 2013).

2.5.2.2 Disruptions in Gut Barrier Integrity

The barrier for pathogenic bacteria and toxic material in the gastrointestinal tract is formed by the gastrointestinal mucosa. Circadian desynchronization disrupts gut permeability and bacterial translocation systemically; thus, the chances of infection are increased. A dysregulated gut microbiota composition also negatively impacts the gut barrier-supporting functions, leaving pathways open for pathogens such as Salmonella and Listeria (Kutsuma et al. 2014).

2.5.2.3 Reduced Detoxification Efficiency

The detoxifying functions of the liver, especially those pertaining to the function of the cytochrome P450 enzymes, are governed by the circadian rhythm. Such disruptions diminish the liver's capacity to detoxify ingested toxins, thereby extending the exposure to such toxic agents (Beraldo et al. 2005).

2.5.2.4 Metabolic and Hormonal Alterations

Cortisol and insulin are hormonal secretions that alter metabolism as disruption med circadian rhythms, while changes in their levels impair immune responses so that infections might host self. Metabolic disorders cause the responses to foodborne diseases, which are due to insulin resistance. Mediate under these vulnerabilities in appropriate meal timing, improve sleep hygiene, and intervene strategically in dietary use to recover immune resilience and achieve better food safety outcomes (Carroll and Nestel 1973).

2.5.3 Challenges in Effective Dietary Interventions

Indeed, there is a considerable barrier against circadian disruption for dietary therapies supporting food safety and health. Misaligning work schedules with circadian rhythms, irregularly timed food intake, and jet lag are the interfering factors that really hinder the success of the dietary intervention. These aspects are key to blocking the effective execution of dietary recommendations under those circumstances. This section emphasizes how food safety is the most difficult area with regard to application of chrononutrition (Garaulet et al. 2013).

2.5.3.1 Disrupted Nutrient Absorption and Metabolism

Absorption and utilization of the nutrients depend on the circadian rhythm and are optimally active during specific parts of the day. Poor absorption would certainly put one at risk for several essential vitamins and minerals if the hours of food intake were erratic. Irregular meal times are also another cause of metabolic stress, heightening the risk of becoming obese or suffering from diabetes and cardiovascular conditions (Xiao et al. 2019).

2.5.3.2 Difficulties in Meal Timing Compliance

Bringing off a meal routine is tough, if not impossible, for shift or pilot crews. Their meal routines shift according to shifts and compel alternate schedules. Then there are social and cultural factors, from family dining traditions to those of the workplace, that are elements in compliance with the routine mealtimes (Saad et al. 2012).

2.5.3.3 Increased Susceptibility to Foodborne Illnesses

Dysregulation of circadian rhythms compromises immune defense mechanisms within the organism, thus increasing the susceptibility of the individual toward foodborne pathogens. Disregulation of digestion and alteration of gut microbiota composition can, in turn, increase risks associated with the ingestion of contaminated food (Dashti et al. 2021).

2.5.3.4 Compromised Detoxification Processes

Circadian rhythms dictate the performance of detoxification enzymes in the liver, best functioning at rest. Disruption of eating and sleeping, therefore, takes a toll on the elimination capacity and adds to the toxin load and metabolic stress (Lindgren et al. 2009).

2.5.3.5 Behavioural and Psychological Barriers

Chronic mismatching of circadian rhythms has been tied to an increased craving for high-calorie and unhealthy foods, which in turn would reduce one's ability to adhere to dietary recommendations. Sleep loss and stress further impair decision-making toward convenience rather than nutritional content. It shall take multifaceted approaches in personalised meal planning, workplace nutritional policies and public health activities in fostering alignments of diet-biological rhythms to overcome these hurdles (Ha and Song 2019).

2.6 Nutritional Implications of Circadian Alignment

2.6.1 Enhanced Nutrient Absorption and Metabolism

Nutrient absorption and metabolism are closely linked to circadian rhythms emanating from biological clocks, reflecting how efficiently the body processes and uses various foods. Synchronizing diet with biological clocks optimizes digestion, absorption, and energy utilization. Multiple mechanisms represent how circadian rhythms affect the metabolism of nutrients. Meals would be timed at those moments when the absorption and efficiency of metabolism for nutrients are optimal while the risk from food would be minimized (Tosini et al. 2014).

2.6.1.1 Enzymatic Activity and Digestive Efficiency

Assimilation of food occurs when digestive enzymes are released in response to circadian rhythms, with peak activity around possible feeding times. Primarily, enzymes involved in carbohydrates, lipids, and proteins perform better or more efficiently during the day because breakdown and assimilation of nutrients always takes place. A person is actually active, and digestion works well when a meal is taken during this active phase than consuming it at late evening when such a process would be lessened, and nutrient assimilation would be below its optimal level (Paoli et al. 2019).

2.6.1.2 Hormonal Regulation of Metabolism

Insulin sensitivity varies according to circadian rhythms, being more sensitive in the morning and less efficient in the evening. Ghrelin, the appetite hormone, and leptin, the satiety hormone, have a circadian rhythm, regulating appetite control and energy balance. Disrupted meal times may cause hormonal disturbances, promoting obesity and metabolic diseases (Emerenziani and Sifrim 2005).

2.6.1.3 Gastrointestinal Motility and Nutrient Absorption

The gut is not always actively moving nutrients at varying speeds throughout the day, as it is at higher rates during the active levels. The timing of food ingestion can affect gut permeability with subsequent alterations in the absorption of micronutrients such as vitamins and minerals. When such an alteration in motility, rhythmically synchronized with the clock, occurs, it leads to various gastrointestinal disorders, such as bloating and malabsorption and inflammation of the gut (Stockman et al. 2018).

2.6.1.4 Mitochondrial Function and Energy Production

Mitochondria as the energy production power plants of the cell, display circadian rhythmic fluctuations in energy output and efficiency. Eating during efficient metabolism maximizes ATP yield and reduces oxidative stress while maximizing energy balance. Food intake during the late night desynchronizes mitochondrial rhythmicity, causing inefficient energy expenditure and increased stress on metabolism. Through chronobiological dietary practices, people might strengthen nutrition uptake, optimize metabolism functions, and avert metabolic disorders. The chronobiological approach to nutrition is therefore a very important intervention for food safety and health enhancement. The circadian clock governs powerhouses of the cells showing variations in energy production and efficiency. The consumption of food is therefore best aligned within the windows of greatest metabolic activity when ATP production can occur efficiently, thus preventing oxidative stress and maintaining the energy balance of the body. Late-night eating disrupts mitochondrial rhythms, resulting in inefficient energy metabolism and increasing metabolic stress (Santos et al. 2021).

2.6.2 Toxin Accumulation and Metabolic Stress Reduction

Circadian rhythm plays an important role in modulating the elimination of toxins and reducing their metabolic stress within the body. The body seems to be very organized about detoxification since it is best at night, under the minimum energy demand from digestion. Eating outside the circadian peaks may disrupt such processes and cause toxin accumulation and higher metabolic stress between food intake and biological clocks (Genta et al. 2021).

2.6.2.1 Liver Detoxification and Chronobiology

Since the biotransformation done by the liver is primarily through enzymes activated in a circadian rhythm, like those involving cytochrome P450, eating late at night interferes with the hepatic detoxification processes that render toxins and food metabolic by-products poorly metabolized. This long term interference heightens the chances of such accumulation leading to metabolic disorders and oxidative stress (Vetter 2020).

2.6.2.2 Gastrointestinal Barrier and Toxin Defence

The gut barrier is responsible for the control of nutrient flow into the bloodstream and keeps out unwelcome entities from intruding into circulation. Circadian rhythm modifications weaken gut integrity and enhance intestinal permeability (leaky gut syndrome) by allowing toxic bacteria and toxins to be absorbed into the bloodstream and thus provoking systemic inflammation and metabolic stress (Flanagan et al. 2021).

2.6.2.3 Oxidative Stress and Mitochondrial Function

Mitochondria that serve a function for cellular energy production show a circadian rhythm. The disturbance in the feeding pattern leads to a disturbance in mitochondrial function, increased oxidative stress, and cellular damage. Timely meal intake synchronizes mitochondrial efficiency with energy demand, thus lessening metabolic stress and optimizing cellular function (Santos et al. 2022).

2.6.2.4 Elimination of Foodborne Contaminants

The circadian rhythm governs the ability of the human body to detoxify harmful food chemical preservatives and pesticides. Food intake synchronized with detoxification windows will enhance their elimination and diminish long-term health risks. In contrast, conflicting eating schedules delay the detoxification mechanisms and thus prolong the exposure time of these chemicals. By assimilating natural detoxification patterns to eating habits, one can minimize toxin accumulation and metabolic stress and eventually attain better health and possibly food security. Just as manipulable detoxifying diet schedule interventions maximize toxin removal, they avert inflammation and keep metabolic fitness on terms of time (Rivera et al. 2020).

2.6.3 Risks of Misaligned Feeding Times

Food timing misaligned with circadian physiology might severely affect digestion, metabolism, and health. Meals delayed may contribute to the onset of metabolic syndromes and foodborne diseases. The body's inner clock system is programmed to work optimally for digestion and nutrient absorption at specific suitable times during the day, typically on a cycle with daylight. Food intake that is offset from the natural circadian cycle can disturb various physiological functions and lead to metabolic disorders, disturbed digestion, and increased susceptibility to food-borne diseases. Thus, various studies have begun to illuminate the effects of eating at different time of day (Oike et al. 2014).

2.6.3.1 Metabolic Dysfunction

Late-night food intake has been associated with an increase in insulin resistance and glucose intolerance that predisposes individuals to type 2 diabetes. The disturbances in hormones, such as insulin and glucagon, regulated by circadian rhythms, may meddle in optimal glucose metabolism and result in the wastage of energy (Manfredini et al. 2013).

2.6.3.2 Impaired Digestive Efficiency

The enzymatic activity of the gut is also circadian-dependent, maximally digestively effective in daylight. Eating at night reduces gastric motility and enzyme secretions, such that indigestion, bloating, and acid reflux may occur (De Giorgi et al. 2013).

2.6.3.3 Disruption of Gut Microbiota

Feeding times alter the gut microbiota composition. Erratic and unpredictable eating habits can drive microbial diversity and consequently accelerate the growth of pathogenic bacteria diagnosed with inflammation and gastro-intestinal disorders (Raboso Moreno et al. 2023).

2.6.3.4 Increased Susceptibility to Foodborne Illnesses

At night, the immune system becomes weak and less able to combat foodborne pathogens. Eating during metabolic rest periods can significantly compromise immune function and can leave the body susceptible to infection by dirty or contaminated food. Eating in synchrony with biological rhythms maximizes metabolic well-being and aids in digestion and food safety. Using chronobiological principles for eating minimizes the adverse effects of irregular feeding schedules and thus bolsters long-term health (Portaluppi et al. 2012).

2.7 Integrating Chronobiology into Food Safety Practices

2.7.1 Developing Temporal Food Safety Guidelines

Chrono biological principles have to be incorporated into the food safety regulations to ensure improvement in public health and mitigation of risks from foodborne illnesses. Traditional food safety measures mainly focus on avoiding contamination, storage conditions, and hygienic practices. However, available recent studies have shown that timing has become an important factor in food safety as well as general health outcomes (Chokroverty 2009).

2.7.1.1 Synchronizing Meal Times with Biological Rhythms

Establishing guidelines for eating behaviour by natural circadian rhythms can aid digestion, metabolism, and immune function. Encouraging earlier eating and the avoidance of late-night eating decreases metabolic stress and enhances food safety outcomes (Pears et al. 2018).

2.7.1.2 Food Storage and Preparation Timing

Chronobiologic studies show that rather than being constant throughout the day, the activities of digestive-system enzymes and the effectiveness of detoxification processes in the liver change at different times of the day. Hence, food storage conditions will have to consider biological rhythms, too. The regulation could allow for the establishment of appropriate ingestion times for food, depending on type and relevant storage duration, to reduce the risk of microbial activity and spoilage (Abreu and Bragança 2015).

2.7.1.3 Consideration for Shift Workers and Time Zone Travelers

In order to counteract the adverse effects of circadian desynchronization on food safety, special protocols should be developed for employees who work nontraditional shifts. Planned meal breaks can then be put back into the working environment to mitigate health risks (Sundararaj et al. 1982).

2.7.1.4 Public Health Policies and Education

The insertion of chronobiology into the public health policy and dietary advice should concern government organizations and health institutions alike. Health education campaigns will make people interested in meal timing such that individuals will be more compliant with food safety protocols and, therefore, more likely to reap the benefits of health outcomes. The society would develop an integrated approach towards food safety by integrating such considerations into food safety measures, not only assuring the quality of food but also ensuring the consumption of food along with biological cycles to improve total health outcomes (Some Aspects of the Chronobiology of Nutrition 2025).

2.7.2 *Public Health Strategies for Chrono-nutrition*

Spurring changes in public health activities will not necessarily include conventional sourcing of nutrition diets, but would rather incorporate the influence of chronobiology on dietary recommendations and consumption behaviour in light of a greater extent of wellness while minimizing risk under foodborne diseases. Chrono-nutrition is the study of how one aligns nutritional behaviours with biological rhythms to achieve maximal digestion, metabolism, and immune function. The next section outlines the evidence-based interventions and awareness concerning such (Walker et al. 1985).

2.7.2.1 Education and Awareness Campaigns

Governments and public health organizations must inform the public about the value of eating time. Information campaigns can emphasize how circadian synchrony impacts digestion, metabolic disease, and foodborne illness (Nelson and Halberg 1986a).

2.7.2.2 Policy Implementation

Public health policies must incorporate chrononutrition principles into dietary recommendations. School lunch programs and corporate cafeterias can implement structured meal times that correspond to natural biological cycles (Kim et al. 2017).

2.7.2.3 Healthcare Integration

Doctors and nutritionists should include chrono-nutrition in their dietary advice. Personalized nutrition plans can maximize food safety by recommending meal timing that is in sync with the individual's circadian rhythm. and dietitians should incorporate chrono-nutrition in dietary recommendations. Personalized nutrition plans can optimize food safety by advising meal timing that aligns with the individual's circadian rhythm (Nelson and Halberg 1986b).

2.7.2.4 Special Considerations for Vulnerable Populations

Shift workers, the aged, and metabolic disorder patients require personalized chrono-nutritional counselling. Timed meal consumption and light therapy can minimize the negative consequences of circadian disruption.

Through the integration of chrono-nutrition into public health, we are able to optimize food safety, maximize metabolic wellness, and reduce the incidence of foodborne diseases, leading to better health on a population scale (Nelson 1988).

2.7.3 *The Role of Education and Awareness*

Awareness and education form the cornerstone for implementation of chronobiology in food safety measures. Although research is done on the influence of biological rhythms on food safety and nutrition, awareness and implementation among the population are low. With increased awareness, individuals, policymakers, and healthcare practitioners make sound choices by coordinating food safety measures with biological rhythms. Education is key in advocating for safe consumption of food in the framework of biological rhythms. This section beckons efforts towards raising awareness of chronobiology in food safety (Sono et al. 2019).

2.7.3.1 The Need for Education on Chronobiology and Food Safety

Public awareness of food safety has traditionally been all about hygiene, storage, and avoiding contamination. Evidence shows that what, and when, is consumed is also critical. Mischanneled eating patterns-eating late at night, shift work eating times, and eating at nontraditional times-have adverse health effects. Educating individuals about circadian rhythms and their role in digestion, metabolism, and immune function can decrease foodborne illness risks and improve overall health (Jin et al. 2017).

Bridging the Knowledge Gap: This relationship between food metabolic rate and circadian rhythms is unknown to most people. This information must be added to school syllabi, colleges, and health facilities. Shifting Public Perception: Chronobiology is typically underrepresented in popular debate on nutrition and food safety. Campaigns must demonstrate how the timing of food consumption impacts digestion, toxin removal, and the immune system. Addressing Misinformation: More people are interested in meal timing and intermittent fasting dietary trends, but these can provide misleading information and lead to misinformed eating and lifestyle habits. Evidence-based guidelines can dispel dangerous misconceptions (Della Penna and Orefice 2018).

2.7.3.2 Strategies for Raising Awareness

To facilitate the integration of chronobiological principles into food safety and public health policy, special awareness programs must be designed. Community Engagement and Outreach: The government and health organizations should organize workshops, seminars, and campaigns to raise awareness of the importance of meal timings to health and food safety. Social Media and Digital Platforms: Online media can do wonders for information transfer. Such as infographics, short videos, and interacting content can be used to present complex scientific concepts in a more accessible way. Labelling and Consumer Guidance: This would be to put time-based guidelines on food packaging-and such as “optimal eating hours for efficient nutrient absorption”-to assist consumers in making better choices (Yang et al. 2022).

2.7.3.3 Healthcare- and Nutrition-Related Roles

Medical professionals, dietitians, and nutritionists play a central role in enhancing the knowledge of their patients with regard to chrononutrition and food safety. Incorporating Chronobiology into Dietary Advice: Nutritionists have to make independent timing of meals in accordance with the circadian rhythms. Training Programs for Healthcare Providers: The medical education must include issues concerning time-related eating and food safety so that health professionals could advise the patients well. Targeted interventions for at risk groups: Shift workers, metabolically disordered people, and elders are highly susceptible to negative outcomes of eating at the wrong time (Nelson and Halberg 1986a).

2.7.3.4 Educational Policies and School-Based Initiatives

Schools are instrumental in influencing lifelong eating patterns. The ingraining of chronobiology in school curricula is likely to help instil good eating habits in children (Kim et al. 2017).

Incorporating Chronobiology into Science and Health Classes: Educating students on circadian rhythms and meal timing can promote awareness and long-term compliance with chrono-nutritional principles. **Structured Meal Timing in Schools**: School lunch schedules should be timed during optimal metabolic functioning hours to ensure students' physical and cognitive health. **Parental Engagement**: Uniform nutrition is a goal that would be easy to achieve only with effective measures. This could be achieved by going about teaching parents the chrono-nutrition approach-to using basic tools and diverse materials so that they can completely understand how to effectively incorporate the approach into their families (Nelson and Halberg 1986b).

2.7.3.5 The Long-Term Impact of Education and Awareness

Behavioural interventions being applied persistently generate long-term effects on human behaviour, and at the end of the day, these changes have also served a good purpose in improving public health and alleviating the burden of foodborne diseases (Vetter 2020).

Improved Public Health Outcomes: Application of food timing by the circadian rhythm may dramatically accelerate metabolic well-being and immune defence power. **Economic Benefits**: The decrease in health care expenditure due to fewer diet-dependent diseases and illness from food-borne diseases could mean gains in the economy. **Enhanced Food Industry Practices**: Heightened consumer awareness will steer the food industry toward Chrono nutritionist practices, such as minder services predicated on designated time screens for food ingestion. Transforming food safety and nutrition to promote healthier societies and lower disease risks through awareness and education is the most important way people and institutions can tap the potential of chronobiology. Educating and providing awareness about themselves and institutions could harness the promise given by chronobiology to convert food safety and nutrition into healthier living and reduced incidence of diseases (Santos et al. 2022).

2.8 Future Directions in Chronobiological Research and Food Safety

2.8.1 Emerging Technologies in Chrono-nutrition Studies

New technology has made a way for new advances in chronobiology and nutrition in food safety. With new ways to understand how biological rhythm affects digestion, detoxification, immunity, and metabolism, research could go on for better innovations. These technologies further help researchers and health providers in the formulation of individualized nutritional interventions that are scheduled according to the circadian rhythm of the person, which would ultimately support public health and food safety (Rivera et al. 2020).

2.8.1.1 Wearable and Smart Devices for Circadian Monitoring

Wearable technology today plays a significant role in monitoring physiological markers of circadian rhythms. Wearables include smartwatches and biosensors that track sleep-wake patterns, heart rate variability, glucose, and metabolic responses; all data are transmitted back in real time for chrono-nutritional evaluation. **Continuous Glucose Monitoring (CGM)**: The devices monitor how glucose levels change throughout the day in order to determine the best times to eat for maximum metabolic efficiency. **Circadian Rhythm Tracking Apps**: Applications based on AI monitor sleeping and eating patterns and offer personalized dietary advice based on an individual's body clock. **Metabolic Rate Sensors**: Wearable technology can track metabolic changes at different times of the day to maximize food intake to prevent metabolic stress (Oike et al. 2014; De Giorgi et al. 2013; Chronobiology 2025).

2.8.1.2 Artificial Intelligence and Big Data in Chrono-nutrition

The studies of chrono-nutrition have been revitalized in the glare of artificial intelligence and machine learning, which allow the correlation of the timing of eating with health outcomes on the basis of large data evaluation. **AI-Based Nutritional Algorithms**: Machine learning programs predict how individuals respond to different meal schedules depending on their circadian rhythm, metabolic rate, and genes. **Big Data Analysis in Public Health**: The big data analysis predicts circadian desynchronization patterns and foodborne disease risk using artificial intelligence and provides policy recommendations concerning meal timing policy. **Predictive Modelling**: Computer simulations using AI allow researchers to decipher how the timing of meals influences gut microbiota, immunity, and detoxification (Manfredini et al. 2013).

2.8.1.3 Chrono Pharmacology and Nutrigenomics Integration

Chrono-pharmacology, the pharmacology of drug-biological rhythm interaction, is being applied more and more to nutrition. Likewise, nutrigenomics investigates the ways genes affect the metabolism of nutrients, providing a personalized modality for chrononutrition (Ha and Song 2019; Tosini et al. 2014; Paoli et al. 2019).

Time-specific nutritional Supplementation: Personalized supplement regimens take into consideration a person's circadian rhythm to enhance nutrient uptake and reduce metabolic stress. **Gene Expression Analysis for Meal Timing**: With the help of technological advances in nutrigenomics, researchers are beginning to achieve a better understanding of how genes encode metabolic responses to diet throughout the day. **Chrono-targeted Functional Foods**: Technological advancements have enabled nutrigenomics to be used for revealing patterns of gene regulation in metabolic response to food over the 24 hours (Flanagan et al. 2021; Emerenziani and Sifrim 2005; Santos et al. 2021; Genta et al. 2021; Vetter 2020).

2.8.1.4 Lab-on-a-Chip and Biomarker Analysis

Microfluidic technology, lab-on-chip devices, enable real-time tracking of chrononutrition biomarkers. The small diagnostic devices are hormone secreted at different times of the day, monitor the activity of digestive enzymes, and also measure any changes in the function of the immune system during the entire day. Salivary Cortisol and Melatonin Tests: These help assess circadian disruptions and their implications on metabolic health. Gut Microbiome Analysis: New sequencing technologies reveal day-to-day fluctuations in gut bacteria, informing personalized dietary recommendations. Blood-Based Biomarker Detection: Portable devices monitor inflammation markers and oxidative stress related to circadian misalignment (Raboso Moreno et al. 2023).

2.8.1.5 Food Safety Innovations Guided by Chronobiology

The galactic immune system in circumspection modulates the microbiomes and cell inhibition pathogenic guess immune systems of individuals in the gastric tract every day, which alters food safety technology protocols. Smart Food Labelling: Smart packaging has provisions to indicate to the consumers on when to consume it, to get the fullest digestion and absorption of nutriments. Food Safety AI Models: AI food surveillance systems assess the contamination risks posed by environmental and circadian variables so as to provide a safer in taking of food. Personalized Meal Delivery Services: These are programmes that create personalized meals to their consumers using the biological cycles of the individual for optimum metabolism and digestion. Therefore, the novel and upcoming technologies have a lot to contribute especially in keeping studies related to chrono-nutrition ahead towards food safety protocols, reducing metabolic disease afflictions, and giving long-term pro-health guarantees. They would also bring fine changes and innovative developments to the next generation of personalized precision nutrition and synchronizing it with biological rhythms. Personalized Nutrition and Food Safety Potential Applications (Portaluppi et al. 2012).

2.8.2 Chrono Biological Applications Regarding Personalized Nutrition and Food Safety

Chrono-bioapplications regarding personalized nutrition and food safety have much potential to achieve enhanced health benefits. With burgeoning technological innovation and determination of circadian differences at the individual level, personalized dietary interventions can aid maximum nutrient absorption while alleviating the health risks posed by foodborne pathogens and metabolic diseases (Pears et al. 2018).

2.8.2.1 Personalized Meal Timing Plans

Such Personal Nutritional Algorithms built with the help of AI will thus generate custom diets according to the internal clock of the individual. This could also harness smart applications and wearables that will measure physiologic parameters to help determine meal timing for optimal digestion and metabolic activity (Abreu and Bragança 2015).

2.8.2.2 Genetic and Metabolic Profiling

Nutrigenomics studies tell how genetic variations affect the way a person responds to different foods at different times of the day. Metabolomics investigation can identify optimal eating windows that might enhance nutrient bioavailability and reduce metabolic stress (Sundararaj et al. 1982).

2.8.2.3 Targeted Food Safety Measures

Guidelines for food safety can be attuned to circadian rhythms of immunity as a way to curb instances of foodborne diseases. Dietary advice on strengthening immune defence at certain times of the day can benefit high-risk groups- for example, shift workers- since they may be more susceptible to these diseases than average healthy populations (Walker et al. 1985).

2.8.2.4 Customized Dietary Interventions for Chronic Disease Prevention

With circadian-nutrient-oriented dietary recommendations, chronic diseases such as diabetes, obesity, and cardiovascular illnesses may be prevented and/or treated. Synchronizing food intake with biological rhythms will indeed strengthen metabolic and digestive disorder treatments. The implementation of chronobiology-based personalized nutrition and food safety habits has tremendous potential for optimizing health and minimizing risks of disease. This will require major endorsement through future efforts of large clinical trials and technology in securing the feasibility of such approaches (Nelson and Halberg 1986a).

2.8.3 Challenges and Opportunities for Implementation

The significance of integrating chronobiology into food safety standards is that such integration can be challenging and helpful at the same time. Knowledge of these factors is vital for effective implementation. Integration of chrono-nutrition and food safety policy is faced with many challenges; therefore, more research and regulatory adjustments should be allowed (Ekmekcioglu and Touitou 2011).

2.8.3.1 Challenges

Lack of Awareness and Education: Many times, healthcare practitioners or food safety experts would fail to understand the role of chronobiology in the framework of nutrition or food safety. **Individual Variability**: Individual variation in circadian rhythms makes the setting of universal dietary recommendations difficult. **Logistical Barriers**: Extensive infrastructure change would be needed to implement time-based food safety regulations in large-scale food industries and public health programs. **Scientific Complexity**: Additional investigation is necessary to form standard guidelines that relate circadian biology to food safety protocols (Almoosawi et al. 2016).

2.8.3.2 Opportunities

Advances in Wearable Technology: It is common knowledge that wrist sensors are normally the only types of wearables accurate enough for purposes of estimating circadian markers. **Policy Development**: Governments and health agencies can tailor chrononutrition principles to dietary recommendations and public health programs. **Food Industry Innovation**: Food manufacturers can produce products that are circadian biology-aligned to optimize digestion and nutrient absorption. **Research and Collaboration**: More inter-disciplinary research can bridge the gaps between chronobiology, nutrition, and food safety. Through these challenges and available opportunities, chronobiology can be successfully incorporated into food safety protocols to result in better health gains and lower risks related to inappropriate eating habits (Halberg 1989).

2.9 Conclusion

A chronobiological approach to food safety sheds new light on optimizing health. Synchronization of food intake with biological rhythms optimizes metabolic efficiency, reduces the risk of foodborne disease, and optimizes well-being. Subsequent work will have to consider how to integrate the principles of chrononutrition into food safety policy to optimize additional public health. The use of chronobiology to ensure food safety is a paradigm change in the appreciation of how nutrition, metabolism, and disease are related to each other. We stressed throughout the chapter the way biological rhythms control digestion, metabolism, immunity, and detoxification and how food consumption should be synchronized with circadian rhythms. Eating timing is crucial in determining if the body can metabolize nutrients in an effective way, defend itself against foodborne pathogens, and maintain a healthy balance of gut microbes. Circadian disruption caused by shift work, travel, or non-traditional meal times can compromise these built-in mechanisms, leaving an individual more susceptible to illness and less sensitive to dietary intervention. Implementation of the principles of chrono-nutrition to food safety protocols is faced with opportunities and challenges. There are both logistical and education-based hurdles, but technical innovation in wearables, personalized nutrition, and policy-making provide avenues for likely solutions. Future studies must continue to determine the significance of biological rhythms in food behaviour, with an eye to public health interventions that optimize the synchronization of meals for the greatest health benefit. An integrated food safety system based on chronobiological principles will not only promote individual health and well-being but will also strengthen the overall general public health system. By promoting awareness, education, and research in this area, we can progress toward a more holistic approach to nutrition and health based on the science of biological rhythms.

References

Abreu T, Bragança M (2015) The bipolarity of light and dark: a review on bipolar disorder and circadian cycles. J Affect Disord 185:219–229
Almoosawi S, Vingeliene S, Karagounis LG, Pot GK (2016) Chrono-nutrition: a review of current evidence from observational studies on global trends in time-of-day of energy intake and its association with obesity. Proc Nutr Soc 75(4):487–500
Antoni R, Johnston KL, Collins AL, Robertson MD (2018) Intermittent v. continuous energy restriction: differential effects on postprandial glucose and lipid metabolism following matched weight loss in overweight/obese participants. Br J Nutr 119(5):507–516
Arora A, Dey CS (2014) SIRT2 negatively regulates insulin resistance in C2C12 skeletal muscle cells. Biochim Biophys Acta Mol basis Dis 1842(9):1372–1378
Beraldo FH, Almeida FM, Da Silva AM, Garcia CRS (2005) Cyclic AMP and calcium interplay as second messengers in melatonin-dependent regulation of Plasmodium falciparum cell cycle. J Cell Biol 170(4):551–557
Berg AH, Combs TP, Scherer PE (2002) ACRP30/adiponectin: An adipokine regulating glucose and lipid metabolism. Trends Endocrinol Metab 13(2):84–89
Bo S, Musso G, Beccuti G, Fadda M, Fedele D, Gambino R et al (2014) Consuming more of daily caloric intake at dinner predisposes to obesity. A 6-year population-based prospective cohort study. PLoS One 9(9):e108467
Bo S, Fadda M, Castiglione A, Ciccone G, De Francesco A, Fedele D et al (2015) Is the timing of caloric intake associated with variation in diet-induced thermogenesis and in the metabolic pattern? A randomized cross-over study. Int J Obes 39(12):1689–1695
Bouassida A, Chamari K, Zaouali M, Feki Y, Zbidi A, Tabka Z (2010 Jul) Review on leptin and adiponectin responses and adaptations to acute and chronic exercise. Br J Sports Med 44(9):620–630
Carroll KF, Nestel PJ (1973) Diurnal variation in glucose tolerance and in insulin secretion in man. Diabetes 22(5):333–348
Carter S, Clifton PM, Keogh JB (2016) The effects of intermittent compared to continuous energy restriction on glycaemic control in type 2 diabetes; a pragmatic pilot trial. Diabetes Res Clin Pract 122:106–112
Carter S, Clifton PM, Keogh JB (2018) Effect of intermittent compared with continuous energy restricted diet on glycemic control in patients with type 2 diabetes: a randomized noninferiority trial. JAMA Netw Open 1(3):e180756
Chen HJ, Chuang SY, Chang HY, Pan WH (2019) Energy intake at different times of the day: its association with elevated total and LDL cholesterol levels. Nutr Metab Cardiovasc Dis 29(4):390–397
Cho Y, Hong N, Kim KW, Cho SJ, Lee M, Lee YH et al (2019) The effectiveness of intermittent fasting to reduce body mass index and glucose metabolism: a systematic review and meta-analysis. J Clin Med 8(10)
Chokroverty S (2009) An overview of normal sleep. Sleep disord med basic sci tech considerations. Clin Asp Third Ed:5–21
Chowdhury R (2022) Maternal body size at birth, childhood, and adolescence: a window of opportunity to reduce preterm birth. Am J Clin Nutr 115(4):989–990
Chronobiology – an overview | ScienceDirect Topics [Internet]. [cited 2025 Mar 30]. Available from: https://www.sciencedirect.com/topics/medicine-and-dentistry/chronobiology
Clawson RC, de la Cruz LN, Allen S, Wolgemuth T, Maner A, Dorsett A et al (2019) Continuous access to snacks from weaning onwards in female rats causes weight gain, insulin insensitivity, and sustained leptin resistance in adulthood. Physiol Behav 201:165–174
Corley BT, Carroll RW, Hall RM, Weatherall M, Parry-Strong A, Krebs JD (2018) Intermittent fasting in type 2 diabetes mellitus and the risk of hypoglycaemia: a randomized controlled trial. Diabet Med 35(5):588–594

Coutinho SR, Halset EH, Gåsbakk S, Rehfeld JF, Kulseng B, Truby H et al (2018) Compensatory mechanisms activated with intermittent energy restriction: a randomized control trial. Clin Nutr 37(3):815–823

Dashti HS, Gómez-Abellán P, Qian J, Esteban A, Morales E, Scheer FAJL et al (2021) Late eating is associated with cardiometabolic risk traits, obesogenic behaviors, and impaired weight loss. Am J Clin Nutr 113(1):154–161

Davis R, Bonham MP, Nguo K, Huggins CE (2020) Glycaemic response at night is improved after eating a high protein meal compared with a standard meal: a cross-over study. Clin Nutr 39(5):1510–1516

de Castro-de-Paiva P, de Marinho TS, Mandarim-de-Lacerda CA, Aguila MB (2022) Intermittent fasting, high-intensity interval training, or a combination of both have beneficial effects in obese mice with nonalcoholic fatty liver disease. J Nutr Biochem 104:104

De Giorgi A, Mallozzi Menegatti A, Fabbian F, Portaluppi F, Manfredini R (2013 Dec) Circadian rhythms and medical diseases: does it matter when drugs are taken? Eur J Intern Med 24(8):698–706

de Pureza IROM, de Macena ML, da Silva Junior AE, Praxedes DRS, Vasconcelos LGL, Bueno NB (2021) Effect of early time-restricted feeding on the metabolic profile of adults with excess weight: a systematic review with meta-analysis. Clin Nutr 40(4):1788–1799

Della Penna G, Orefice S (2018) Qualitative representation of spatio-temporal knowledge. J Vis Lang Comput 49:1–16

Ekmekcioglu C, Touitou Y (2011) Chronobiological aspects of food intake and metabolism and their relevance on energy balance and weight regulation. Obes Rev 12(1):14–25

Emerenziani S, Sifrim D (2005) Gastroesophageal reflux and gastric emptying, revisited. Curr Gastroenterol Rep 7(3):190–195

Fernández-Rodríguez R, Martínez-Vizcaíno V, Mesas AE, Notario-Pacheco B, Medrano M, Heilbronn LK (2022) Does intermittent fasting impact mental disorders? A systematic review with meta-analysis. Crit Rev Food Sci Nutr 63(32):11169–11184

Flanagan A, Bechtold DA, Pot GK, Johnston JD (2021) Chrono-nutrition: from molecular and neuronal mechanisms to human epidemiology and timed feeding patterns. J Neurochem 157(1):53–72

Garaulet M, Gómez-Abellán P, Alburquerque-Béjar JJ, Lee YC, Ordovás JM, Scheer FAJL (2013) Timing of food intake predicts weight loss effectiveness. Int J Obes 37(4):604–611

Garaulet M, Qian J, Florez JC, Arendt J, Saxena R, Scheer FAJL (2020) Melatonin effects on glucose metabolism: time to unlock the controversy. Trends Endocrinol Metab 31(3):192–204

Genario R, Morello E, Bueno AA, Santos HO (2019) The usefulness of melatonin in the field of obstetrics and gynecology. Pharmacol Res 147:104337

Genario R, Cipolla-Neto J, Bueno AA, Santos HO (2021) Melatonin supplementation in the management of obesity and obesity-associated disorders: a review of physiological mechanisms and clinical applications. Pharmacol Res 163

Genta FD, Neto GBR, Sunfeld JPV, Porto JF, Xavier AD, Moreno CRC et al (2021) COVID-19 pandemic impact on sleep habits, chronotype, and health-related quality of life among high school students: a longitudinal study. J Clin Sleep Med 17(7):1371–1377

Grandner MA (2017) Sleep, health, and society. Sleep Med Clin 12(1):1–22

Gu C, Brereton N, Schweitzer A, Cotter M, Duan D, Børsheim E et al (2020) Metabolic effects of late dinner in healthy volunteers – a randomized crossover clinical trial. J Clin Endocrinol Metab 105(8):2789–2802

Guo Y, Luo S, Ye Y, Yin S, Fan J, Xia M (2021) Intermittent fasting improves cardiometabolic risk factors and alters gut microbiota in metabolic syndrome patients. J Clin Endocrinol Metab 106(1):64–79

Ha K, Song Y (2019) Associations of meal timing and frequency with obesity and metabolic syndrome among Korean adults. Nutrients 11(10)

Halberg F (1989) Some aspects of the chronobiology of nutrition: more work is needed on "when to eat". J Nutr 119(3):333–343

Hall KD (2008) What is the required energy deficit per unit weight loss? Int J Obes 32(3):573–576
Hatori M, Vollmers C, Zarrinpar A, DiTacchio L, Bushong EA, Gill S et al (2012) Time-restricted feeding without reducing caloric intake prevents metabolic diseases in mice fed a high-fat diet. Cell Metab 15(6):848–860
Ho KKY (2018) Diet-induced thermogenesis: fake friend or foe? J Endocrinol 238(3):R185–R191
Howell S, Kones R (2017) "Calories in, calories out" and macronutrient intake: the hope, hype, and science of calories. Am J Physiol Endocrinol Metab 313(5):E608–E612
Jakubowicz D, Barnea M, Wainstein J, Froy O (2013) High caloric intake at breakfast vs. dinner differentially influences weight loss of overweight and obese women. Obesity 21(12):2504–2512
Jin JL, Guo YL, Li JJ (2017) Plasma free fatty acids in relation with the severity of coronary artery disease in non-diabetics: a Gensini score assessment. IJC Metab Endocr 14:48–52
Kim M, Farrell MJ, Steinberg DR, Burdick JA, Mauck RL (2017) Enhanced nutrient transport improves the depth-dependent properties of tri-layered engineered cartilage constructs with zonal co-culture of chondrocytes and MSCs. Acta Biomater 58:1–11
Kord Varkaneh H, Salehi Sahlabadi A, Găman MA, Rajabnia M, Sedanur Macit-Çelebi M, Santos HO et al (2022) Effects of the 5:2 intermittent fasting diet on non-alcoholic fatty liver disease: a randomized controlled trial. Front Nutr 9
Kord-Varkaneh H, Salehi-sahlabadi A, Tinsley GM, Santos HO, Hekmatdoost A (2023) Effects of time-restricted feeding (16/8) combined with a low-sugar diet on the management of non-alcoholic fatty liver disease: a randomized controlled trial. Nutrition 105:111847
Koutsari C, Ali AH, Mundi MS, Jensen MD (2011 Aug) Storage of circulating free fatty acid in adipose tissue of postabsorptive humans: quantitative measures and implications for body fat distribution. Diabetes 60(8):2032–2040
Kul S, Savaş E, Öztürk ZA, Karadağ G (2014) Does Ramadan fasting Alter body weight and blood lipids and fasting blood glucose in a healthy population? A meta-analysis. J Relig Health 53(3):929–942
Kumar M, Orkaby A, Tighe C, Villareal DT, Billingsley H, Nanna MG et al (2023) Life's essential 8: optimizing health in older adults. JACC Adv 2(7):100560
Kunduraci YE, Ozbek H (2020) Does the energy restriction intermittent fasting diet alleviate metabolic syndrome biomarkers? A randomized controlled trial. Nutrients 12(10):1–13
Kutsuma A, Nakajima K, Suwa K (2014) Potential association between breakfast skipping and concomitant late-night-dinner eating with metabolic syndrome and proteinuria in the Japanese population. Scientifica (Cairo) 2014:1–9
Li W, Wu M, Zhang Y, Wei X, Zang J, Liu Y et al (2020) Intermittent fasting promotes adult hippocampal neuronal differentiation by activating GSK-3β in 3xTg-AD mice. J Neurochem 155(6):697–713
Lindgren O, Mari A, Deacon CF, Carr RD, Winzell MS, Vikman J et al (2009) Differential islet and incretin hormone responses in morning versus afternoon after standardized meal in healthy men. J Clin Endocrinol Metab 94(8):2887–2892
Liu M, He P, Zhou C, Zhang Z, Zhang Y, Li H et al (2022) Association of waist-calf circumference ratio with incident cognitive impairment in older adults. Am J Clin Nutr 115(4):1005–1012
Lowe DA, Wu N, Rohdin-Bibby L, Moore AH, Kelly N, Liu YE et al (2020) Effects of time-restricted eating on weight loss and other metabolic parameters in women and men with overweight and obesity: the TREAT randomized clinical trial. JAMA Intern Med 180(11):1491–1499
Madjd A, Taylor MA, Delavari A, Malekzadeh R, MacDonald IA, Farshchi HR (2016) Beneficial effect of high energy intake at lunch rather than dinner on weight loss in healthy obese women in a weight-loss program: a randomized clinical trial. Am J Clin Nutr 104(4):982–989
Manfredini R, Salmi R, Fabbian F, Manfredini F, Gallerani M, Bossone E (2013) Breaking heart: chronobiologic insights into Takotsubo cardiomyopathy. Heart Fail Clin 9(2):147–156
Mangaraj M, Nanda R, Panda S (2016) Apolipoprotein A-I: a molecule of diverse function. Indian J Clin Biochem 31(3):253–259

Meng H, Zhu L, Kord-Varkaneh H, Santos OH, Tinsley GM, Fu P (2020) Effects of intermittent fasting and energy-restricted diets on lipid profile: a systematic review and meta-analysis. Nutrition 77

Morgan LM, Aspostolakou F, Wright J, Gama R (1999) Diurnal variations in peripheral insulin resistance and plasma non-esterified fatty acid concentrations: a possible link? Ann Clin Biochem 36(4):447–450

Nelson W (1988) Food restriction, circadian disorder and longevity of rats and mice. J Nutr 118(3):284–289

Nelson W, Halberg F (1986a) Schedule-shifts, circadian rhythms and lifespan of freely-feeding and meal-fed mice. Physiol Behav 38(6):781–788

Nelson W, Halberg F (1986b) Meal-timing, circadian rhythms and life span of mice. J Nutr 116(11):2244–2253

Oike H, Oishi K, Kobori M (2014) Nutrients, clock genes, and chrononutrition. Curr Nutr Rep 3(3):204–212

Paoli A, Tinsley G, Bianco A, Moro T (2019) The influence of meal frequency and timing on health in humans: the role of fasting. Nutrients 11(4)

Park S, Yoo KM, Hyun JS, Kang S (2017) Intermittent fasting reduces body fat but exacerbates hepatic insulin resistance in young rats regardless of high protein and fat diets. J Nutr Biochem 40:14–22

Pears S, Makris A, Hennessy A (2018) The chronobiology of blood pressure in pregnancy. Pregnancy Hypertens 12:104–109

Persynaki A, Karras S, Pichard C (2017) Unraveling the metabolic health benefits of fasting related to religious beliefs: a narrative review. Nutrition 35:14–20

Philippens KMH, Mayersbach V, Scheving HLE (1977) Effects of the scheduling of meal feeding at different phases of the circadian system in rats. J Nutr 107(2):176–193

Pilz S, Scharnagl H, Tiran B, Wellnitz B, Seelhorst U, Boehm BO et al (2007 Nov) Elevated plasma free fatty acids predict sudden cardiac death: a 6.85-year follow-up of 3315 patients after coronary angiography. Eur Heart J 28(22):2763–2769

Poggiogalle E, Jamshed H, Peterson CM (2018) Circadian regulation of glucose, lipid, and energy metabolism in humans. Metabolism 84:11–27

Portaluppi F, Smolensky MH, Haus E, Fabbian F (2012) Relevance of chronobiology to the research and clinical practice of neurorehabilitation. Sleep Med Clin 7(4):655–666

Qian J, Dalla Man C, Morris CJ, Cobelli C, Scheer FAJL (2018) Differential effects of the circadian system and circadian misalignment on insulin sensitivity and insulin secretion in humans. Diabetes Obes Metab 20(10):2481–2485

Raboso Moreno B, López Riolobos C, Díaz-García JM, Matesanz López C, Abad FA (2023) Influence of Chronotypes in respiratory pathology. Open Respir Arch 5(1):100228

Regmi P, Heilbronn LK (2020) Time-restricted eating: benefits, mechanisms, and challenges in translation. iScience 23(6):101161

Richter J, Herzog N, Janka S, Baumann T, Kistenmacher A, Oltmanns KM (2020) Twice as high diet-induced thermogenesis after breakfast vs dinner on high-calorie as well as low-calorie meals. J Clin Endocrinol Metab 105(3):e3047–e3048

Rivera AS, Akanbi M, O'Dwyer LC, McHugh M (2020) Shift work and long work hours and their association with chronic health conditions: a systematic review of systematic reviews with meta-analyses. PLoS One 15(4):e0231037

Roenneberg T, Merrow M (2016) The circadian clock and human health. Curr Biol 26(10):R432–R443

Saad A, Man CD, Nandy DK, Levine JA, Bharucha AE, Rizza RA et al (2012) Diurnal pattern to insulin secretion and insulin action in healthy individuals. Diabetes 61(11):2691–2700

Santos HO (2022) Intermittent fasting and fat mass: what is the clinical magnitude? Obesities 2(1):1–7

Santos HO, Macedo RCO (2018) Impact of intermittent fasting on the lipid profile: assessment associated with diet and weight loss. Clin Nutr ESPEN 24:14–21

Santos HO, Tinsley GM (2024) Is breakfast consumption detrimental, unnecessary, or an opportunity for health promotion? A review of cardiometabolic outcomes and functional food choices. Diabetes Metab Res Rev 40(2):e3684

Santos HO, Earnest CP, Tinsley GM, Izidoro LFM, Macedo RCO (2020) Small dense low-density lipoprotein-cholesterol (sdLDL-C): analysis, effects on cardiovascular endpoints and dietary strategies. Prog Cardiovasc Dis 63(4):503–509

Santos HO, Genario R, Macedo RCO, Pareek M, Tinsley GM (2021) Association of breakfast skipping with cardiovascular outcomes and cardiometabolic risk factors: an updated review of clinical evidence. Crit Rev Food Sci Nutr 62(2):466–474

Santos HO, Genario R, Tinsley GM, Ribeiro P, Carteri RB, Coelho-Ravagnani CDF et al (2022) A scoping review of intermittent fasting, chronobiology, and metabolism. Am J Clin Nutr 115(4):991–1004

Scholtens EL, Krebs JD, Corley BT, Hall RM (2020) Intermittent fasting 5:2 diet: what is the macronutrient and micronutrient intake and composition? Clin Nutr 39(11):3354–3360

Some Aspects of the Chronobiology of Nutrition: More Work is Needed on "When to Eat" – ScienceDirect [Internet]. [cited 2025 Mar 30]. Available from: https://www.sciencedirect.com/science/article/abs/pii/S0022316622171816?via%3Dihub

Sonnier T, Rood J, Gimble JM, Peterson CM (2014) Glycemic control is impaired in the evening in prediabetes through multiple diurnal rhythms. J Diabetes Complicat 28(6):836–843

Sono T, Meyers CA, Miller D, Ding C, McCarthy EF, James AW (2019) Overlapping features of rapidly progressive osteoarthrosis and Charcot arthropathy. J Orthop 16(3):260–264

Spencer IO (1968) Death during therapeutic starvation for obesity. Lancet 1(7555):1288–1290

Spezani R, da Silva RR, Martins FF, de Souza MT, Aguila MB, Mandarim-de-Lacerda CA (2020) Intermittent fasting, adipokines, insulin sensitivity, and hypothalamic neuropeptides in a dietary overload with high-fat or high-fructose diet in mice. J Nutr Biochem 83:108419

Stenger S, Grasshoff H, Hundt JE, Lange T (2023) Potential effects of shift work on skin autoimmune diseases. Front Immunol 13

Stockman MC, Thomas D, Burke J, Apovian CM (2018) Intermittent fasting: is the wait worth the weight? Curr Obes Rep 7(2):172–185

Sundararaj BI, Nath P, Halberg F (1982) Circadian meal timing in relation to lighting schedule optimizes catfish body weight gain. J Nutr 112(6):1085–1097

Teong XT, Heilbronn LK (2021) Evidence gaps and potential roles of intermittent fasting in the prevention of chronic diseases. Exp Gerontol 153:111506

Tinsley GM, Moore ML, Graybeal AJ, Paoli A, Kim Y, Gonzales JU et al (2019) Time-restricted feeding plus resistance training in active females: a randomized trial. Am J Clin Nutr 110(3):628–640

Tosini G, Owino S, Guillaume JL, Jockers R (2014) Understanding melatonin receptor pharmacology: latest insights from mouse models, and their relevance to human disease. BioEssays 36(8):778–787

Trepanowski JF, Kroeger CM, Barnosky A, Klempel MC, Bhutani S, Hoddy KK et al (2017) Effect of alternate-day fasting on weight loss, weight maintenance, and cardioprotection among metabolically healthy obese adults: a randomized clinical trial. JAMA Intern Med 177(7):930–938

Varady KA, Hudak CS, Hellerstein MK (2009 Jun) Modified alternate-day fasting and cardioprotection: relation to adipose tissue dynamics and dietary fat intake. Metabolism 58(6):803–811

Varkaneh Kord HM, Tinsley GO, Santos H, Zand H, Nazary A, Fatahi S et al (2021) The influence of fasting and energy-restricted diets on leptin and adiponectin levels in humans: a systematic review and meta-analysis. Clin Nutr 40(4):1811–1821

Vetter C (2020) Circadian disruption: what do we actually mean? Eur J Neurosci 51(1):531–550

Walker WV, Russell JE, Simmons DJ, Scheving LE, Cornelissen G, Halberg F (1985) Effect of an adrenocorticotropin analogue, ACTH 1-17, on DNA synthesis in murine metaphyseal bone. Biochem Pharmacol 34(8):1191–1196

Xiao Q, Garaulet M, Scheer FAJL (2019) Meal timing and obesity: interactions with macronutrient intake and chronotype. Int J Obes 43(9):1701–1711

Yang C, Xue L, Wu Y, Li S, Zhou S, Yang J et al (2022) PPARβ down-regulation is involved in high glucose-induced endothelial injury via acceleration of nitrative stress. Microvasc Res 139:104272

Yanishi M, Kimura Y, Tsukaguchi H, Koito Y, Taniguchi H, Mishima T et al (2017) Factors associated with the development of sarcopenia in kidney transplant recipients. Transplant Proc 49(2):288–292

Yoshino J, Almeda-Valdes P, Patterson BW, Okunade AL, Imai SI, Mittendorfer B et al (2014) Diurnal variation in insulin sensitivity of glucose metabolism is associated with diurnal variations in whole-body and cellular fatty acid metabolism in metabolically normal women. J Clin Endocrinol Metab 99(9):E1666–E1670

Chapter 3
Nutrient Absorption Dynamics and Food Contamination

Nizwa Itrat, Beenish Israr, Sana Arif, Maryam Narjis, Sobia Asghar, and Akhtar Ali

3.1 Introduction

Food safety is critical to public health, ensuring dietary intake supports proper growth, development, and overall well-being. However, food contamination has emerged as a growing concern due to industrialization, environmental degradation, and the extensive use of agrochemicals (Haque et al. 2025). Contaminants in food can be categorized into biological, chemical, and physical agents, all of which have detrimental effects on human health. Chemical contaminants, including heavy metals, pesticides, mycotoxins, and food additives, significantly impact nutrient absorption and metabolism (Anwar et al. 2025).

Nutrient absorption is a complex process influenced by various physiological and environmental factors. When food is contaminated with harmful substances, the bioavailability of essential nutrients is compromised, leading to deficiencies and associated health disorders. For example, heavy metals such as lead and cadmium compete with essential minerals like calcium and iron for absorption sites in the intestine, reducing their bioavailability (Ishra et al. 2025). Pesticide residues disrupt gut microbiota and enzyme function, hindering the digestion and absorption of macronutrients and micronutrients (Subedi et al. 2025). Similarly, fungi produce mycotoxins in improperly stored food, damaging the intestinal lining and impairing nutrient uptake (Hoxha et al. 2025).

N. Itrat (✉) · B. Israr · S. Arif · M. Narjis · S. Asghar
Institute of Home Science, Faculty of Food Nutrition and Home Sciences, University of Agriculture, Faisalabad, Pakistan
e-mail: nizwa.itrat@uaf.edu.pk

A. Ali
Department of Health, Nutrition, and Food Sciences, Florida State University, Tallahassee, FL, USA

T. Sarkar et al. (eds.), *Physiological Perspectives on Food Safety: Exploring the Intersection of Health and Nutrition*, https://doi.org/10.1007/978-3-031-94582-3_3

The long-term effects of food contaminants on nutrient absorption can lead to serious health issues such as anemia, weakened immune function, and developmental disorders in children. In adults, chronic exposure can contribute to metabolic diseases, cognitive impairments, and increased susceptibility to infections (Jha and Singh 2025). Given the widespread occurrence of food contamination, it is crucial to implement stringent food safety regulations, promote consumer awareness, and encourage research into safer agricultural and food processing practices.

3.2 Nutrient Absorption Dynamics

Nutrient absorption involves various mechanisms that facilitate the uptake of essential nutrients into the body (Basile et al. 2023). Passive transport, including diffusion and facilitated diffusion, allows essential minerals to move along concentration gradients without requiring energy. This process enables the uptake of ions such as nitrates, potassium, and calcium, ensuring a steady supply of nutrients for plant growth. Additionally, osmosis is key in water and solute absorption, helping maintain cellular hydration and nutrient balance (Stumpff and Manneck 2025). Since hydroponic systems provide a continuous flow of nutrients, passive transport mechanisms contribute significantly to nutrient uptake, especially in the early growth stages (Deng et al. 2025).

3.3 Passive Diffusion

Passive diffusion enables small, non-polar molecules, such as fatty acids and fat-soluble vitamins (A, D, E, and K), to move across cell membranes without energy expenditure. This process relies on a concentration gradient, allowing nutrients to move from areas of higher concentration in the intestinal lumen to lower concentration within enterocytes. The efficiency of passive diffusion is influenced by factors such as membrane fluidity and lipid composition. Recent studies have highlighted the role of membrane lipid rafts in facilitating the absorption of certain lipophilic nutrients (Vieira et al. 2025).

3.4 Facilitated Diffusion

Facilitated diffusion involves specific transport proteins that assist in moving nutrients across cell membranes without using cellular energy. This mechanism is crucial for absorbing nutrients like fructose. Transport proteins, such as GLUT5, bind to these nutrients and undergo conformational changes to shuttle them across the membrane. Recent research has elucidated the regulation of these transporters by dietary components and gut microbiota metabolites, suggesting that diet-induced modulation of transporter expression can significantly impact nutrient absorption efficiency (Lavin et al. 2007).

3.5 Active Transport

Active transport is an energy-dependent process that moves nutrients against their concentration gradients. This mechanism is vital for absorbing essential nutrients like glucose and amino acids. Transporters such as SGLT1 (sodium-glucose-linked transporter 1) utilize ATP to function. Recent studies have shown that ion channels and transporters play significant roles in regulating nutrient uptake in the gastrointestinal tract, highlighting their importance in maintaining nutrient homeostasis (Batterman et al. 2002).

3.6 Endocytosis

Endocytosis enables the uptake of large molecules, such as specific vitamins and lipoproteins, through vesicle formation and engulfment by enterocytes. This process allows the cell membrane to invaginate, encasing extracellular substances in vesicles internalized. Recent research has shed light on the role of specific receptors and signaling pathways in mediating endocytosis, emphasizing its importance in nutrient uptake and cellular regulation (Schiller 2014).

3.6.1 Role of Digestive Enzymes

The increasing prevalence of complex diseases linked to genetic factors, lifestyle choices, and metabolic disorders has driven a shift from precision-targeted therapies to multitarget drug development. Chalcones, a class of bioactive compounds with an aromatic system conjugated to a Michael acceptor, have gained attention for their diverse pharmacological properties, including antioxidant, anti-inflammatory, anticancer, and antimicrobial effects. Their ability to modulate digestive enzymes, such as amylase, lipase, and protease, (Fig. 3.1) highlights their potential to regulate nutrient digestion and absorption, which is crucial for managing metabolic disorders, obesity, and gastrointestinal conditions (Table 3.1) (Jackson and McLaughlin 2009).

Incorporating a thiophene pharmacophore in chalcone derivatives further enhances their therapeutic potential, as thiophene-based drugs are already established in pharmaceuticals (Goyal et al. 2025).

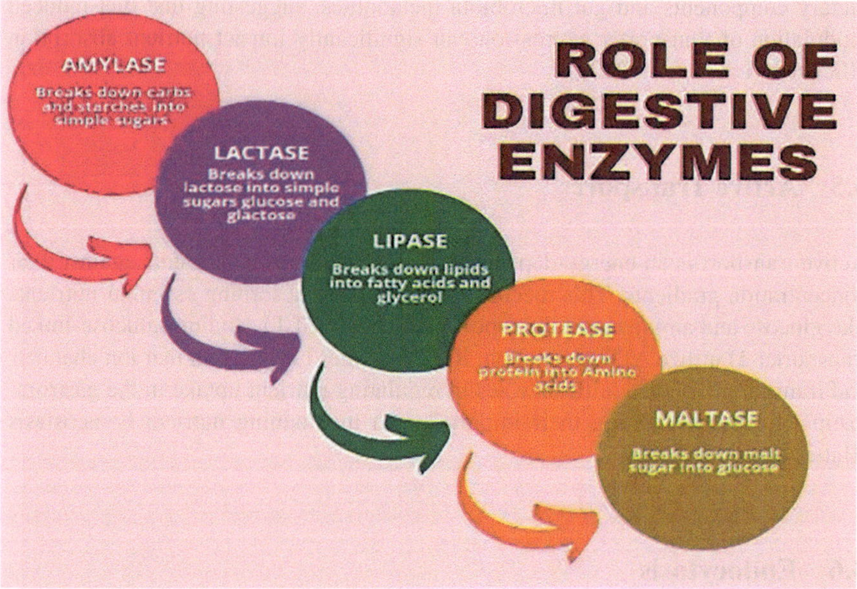

Fig. 3.1 Role of digestive enzymes in absorption

Table 3.1 Role of digestive enzymes

Enzyme	Source	Substrate	End products	Function	References
Amylase	Salivary glands, pancreas	Starch (polysaccharides)	Maltose (disaccharide)	Breaks down carbohydrates	Holtmann et al. (1997)
Maltase	Small intestine	Maltose	Glucose	Converts maltose into absorbable glucose	Nichols et al. (2018)
Lactase	Small intestine	Lactose	Glucose, galactose	Breaks down lactose (milk sugar)	Felicilda-Reynaldo and Kenneally (2016)
Sucrase	Small intestine	Sucrose	Glucose, fructose	Breaks down table sugar	Dahlqvist (1972)
Pepsin	Stomach	Proteins	Peptides	Begins protein digestion in an acidic condition	Schade and Schilling (1967)
Trypsin	Pancreas	Peptides	Smaller peptides	Further protein digestion	Torrissen and Male (2000)
Peptidases	Small intestine	Peptides	Amino acids	Final breakdown of proteins for absorption	Das (1959)

3.6.2 *Impact of Gut Microbiota*

The gut microbiota (GM) plays a crucial role in nutrient absorption by influencing the integrity of the intestinal barrier, enzyme activity, and metabolic processes. A balanced GM composition enhances the breakdown and assimilation of essential nutrients, such as vitamins, minerals, and amino acids, by producing key metabolites like short-chain fatty acids (SCFAs) and specific enzymes. Conversely, dysbiosis an imbalance in GM can compromise intestinal permeability, leading to malabsorption, inflammation, and disrupted nutrient uptake. This imbalance may hinder the bioavailability of critical nutrients like calcium, vitamin D, and iron, affecting overall health. Thus, maintaining a healthy GM is vital for optimal nutrient absorption and metabolic homeostasis (Krajmalnik-Brown et al. 2012).

The gut microbiota maintains overall health by regulating metabolic functions, immune responses, and neural signaling pathways. It comprises a complex community of microorganisms that interact to sustain intestinal homeostasis. When this microbial balance is disrupted, known as dysbiosis, it can contribute to the development and progression of various diseases, including metabolic disorders, inflammatory conditions, and neurological impairments. Research has shown that an imbalance in gut microbiota composition is linked to inflammatory bowel disease, type 2 diabetes, colorectal cancer, and cerebrovascular diseases. This dysregulation affects the body's immune responses, increases systemic inflammation, and alters the gut-brain axis, a bidirectional communication network between the gut and the central nervous system. In cerebrovascular diseases, particularly ischemic stroke, changes in gut microbiota composition have been associated with increased inflammation and vascular dysfunction, which may worsen disease outcomes. The gut microbiota also plays a role in modulating neurotransmitter production and influencing cognitive and emotional health, making its impact extend beyond digestion and metabolism. Given its broad influence, maintaining a balanced gut microbiome is increasingly recognized as a potential strategy for preventing and managing various health conditions, including those affecting the brain and cardiovascular system (Chen et al. 2025a).

3.7 Factors Affecting Nutrient Absorption

3.7.1 *Microbial Community Dynamics*

Microbial succession during composting is significantly essential for nutrient absorption. Different microbial species contribute to the breakdown and transformation of organic matter into bioavailable nutrients. The richness and diversity of bacterial and fungal communities enhance the decomposition process and improve nutrient retention (Carrero-Colón et al. 2006).

3.7.2 Temperature and Composting Phases

The thermophilic phase (high-temperature stage) accelerates microbial activity and nutrient release. Peanut straw entered the thermophilic phase earlier than wheat bran and poplar leaf litter, resulting in better nutrient retention. Proper temperature regulation is essential for the optimal conversion of organic matter into absorbable nutrients.

3.7.3 Humification and Organic Matter Transformation

The formation of humus (humic substances) enhances nutrient stability and bioavailability. The humic acid to fulvic acid (HA/FA) ratio is a key indicator of humification, influencing how well plants can absorb nutrients.

3.7.4 pH and Electrical Conductivity (EC) Regulation

Stable pH conditions promote efficient microbial activity and nutrient solubility. Low electrical conductivity (EC < 4 mS/cm) ensures the absence of excess salts that could hinder nutrient uptake. Mature compost with a balanced pH prevents phytotoxicity and enhances plant nutrient absorption.

3.7.5 Heavy Metal Content and Agricultural Safety

Low heavy metal content in compost ensures that toxic elements do not hinder nutrients. The study confirmed that peanut straw compost had the lowest heavy metal accumulation, making nutrients more available for plant absorption (Chen et al. 2025b).

3.8 Nutrient Bioavailability

Bioavailability refers to the extent and rate at which bioactive compounds, such as antioxidants and phytochemicals, are absorbed and utilized in the body. These compounds found abundantly in fruits and vegetables, play a crucial role in promoting health by reducing oxidative stress and preventing metabolic disorders. However, their effectiveness is often limited due to factors such as poor solubility, instability in the gastrointestinal environment, and undesirable taste profiles. Enhancing the bioavailability of these nutrients is essential to maximize their physiological benefits and therapeutic potential.

Nanotechnology has emerged as an innovative approach to improve the bioavailability of bioactive food components. Techniques such as nano-encapsulation, nano-emulsions, and nanoparticle delivery systems enhance the solubility, stability, and targeted release of antioxidants, ensuring their efficient absorption in the body. However, safety concerns, including potential toxicity and bioaccumulation, necessitate the use of biodegradable, food-grade materials to ensure their safe application in functional foods. The integration of nanotechnology in food science offers promising solutions to overcome bioavailability challenges, contributing to better health outcomes and disease prevention.

Nutrient bioavailability refers to the proportion of a nutrient absorbed and utilized by the body after digestion. Several factors influence it, including the food matrix, nutrient form, digestive enzymes, gut microbiota, and interactions with other dietary components (Fig. 3.2). For instance, minerals like iron and zinc may have reduced bioavailability in the presence of phytates found in whole grains, while vitamin C enhances non-heme iron absorption. Similarly, polyphenols in cereals and legumes can act as antioxidants but may also hinder the absorption of certain minerals. Processing techniques such as fermentation, germination, and enzymatic treatments can enhance bioavailability by breaking down antinutrients

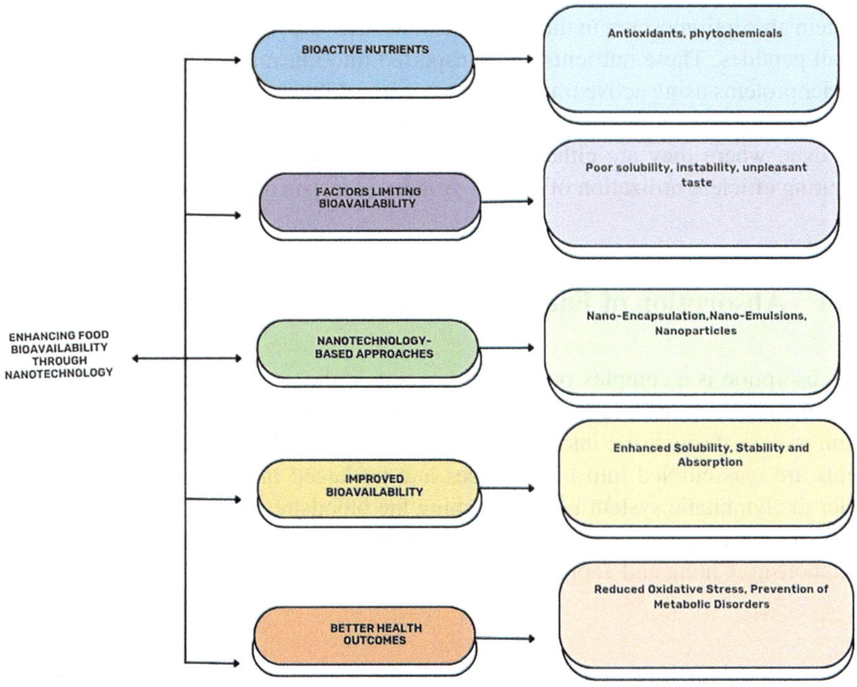

Fig. 3.2 Enhancing nutrient bioavailability through nanotechnology

and improving nutrient release. Additionally, digestion efficiency and individual physiological factors, such as gut health and genetic variations, play crucial roles in determining the extent to which nutrients are absorbed and utilized by tissues. Understanding these factors is essential for optimizing dietary intake and developing nutrient-dense functional foods (Żyła and Duda 2025).

3.9 Absorption of Carbohydrates

Carbohydrate absorption occurs mainly in the small intestine, where monosaccharides like glucose and galactose are actively transported through sodium-dependent glucose transporters, while fructose is absorbed via facilitated diffusion (Table 3.2). These sugars enter the intestinal cells and are released into the bloodstream, where they travel to the liver for metabolism. The rapid absorption of glucose ensures a steady supply of energy, while unabsorbed carbohydrates, such as fiber, pass into the colon for fermentation by gut bacteria (Jackson and McLaughlin 2009).

3.10 Absorption of Proteins

Protein absorption occurs in the small intestine after digestion into amino acids and small peptides. These nutrients are transported into intestinal cells via specialized carrier proteins using active transport mechanisms (Table 3.2). Inside the cells, peptides into amino acids before entering the bloodstream. They are then transported to the liver, where they are either used for protein synthesis or energy production, ensuring efficient utilization of dietary proteins (Jackson and McLaughlin 2009).

3.11 Absorption of Fats

Fat absorption is a complex process that occurs in the small intestine. After digestion, monoglycerides and free fatty acids form micelles with bile salts, allowing them to pass through the intestinal cell membrane. Inside the cells, these components are reassembled into triglycerides and packaged into chylomicrons, which enter the lymphatic system before reaching the bloodstream (Table 3.2). This process ensures efficient transport of dietary fats to tissues for energy storage and metabolism (Chung and Tappenden 2024).

Table 3.2 Absorption of macronutrients

Macronutrient	Site of absorption	The end product of digestion	Absorption mechanism	Transport pathway	References
Carbohydrates	Small intestine (deudenum, jejunum)	Monosaccharides (glucose, galactose, fructose)	Active transport (glucose, galactose)	Portal vein to the liver	Pereira et al. (2018)
			Facilitated diffusion (fructose)		
Proteins	Small intestine (deudenum,jejunum)	Amino acids dipeptides, tripeptides	Active transport (sodium-dependent)	Portal vein to the liver	Wapnir (2024)
Fats (lipids)	Small intestine (jejunum,ileum)	Fatty acids, monoglycerides	Passive diffusion (micelle formation)	The lymphatic system (via chylomicrons)	Lichtenstein and Jones (2012)

3.12 Absorption of Vitamins

3.12.1 Water-Soluble Vitamins

Water-soluble vitamins dissolve quickly in the intestinal lumen, allowing direct absorption through the gut lining (Table 3.3). Their absorption efficiency is influenced by factors such as molecular weight and ionization. Typically, smaller molecules and weak acids or bases are absorbed more readily, while strong electrolytes face challenges crossing the intestinal membrane. Vitamins such as B3, B6, and C, along with biotin, have relatively low molecular weights (ranging from 176 to 244), enabling them to diffuse more efficiently into the bloodstream (Halsted 2003).

3.12.2 Fat-Soluble Vitamins

In contrast, fat-soluble vitamins require a different mechanism for absorption due to their hydrophobic nature (Table 3.3). These vitamins dissolve in dietary fats and form emulsions in the upper gastrointestinal tract during digestion. The emulsification process, aided by mastication and gastric churning, enhances their solubility. However, since the fat droplets in emulsions are relatively large, they cannot directly interact with the intestinal lining for absorption. The presence of pancreatic lipase

Table 3.3 Absorption of micronutrients

Category	Micronutrient	Sources	Absorption	Storage	References
Fat-soluble	Vitamin A	Liver, eggs, dairy, carrots	Absorbed with dietary fats in the small intestine using bile and pancreatic enzymes	Stored in liver and adipose tissue	Kuksis (2018)
	Vitamin D	Sunlight, fish, fortified foods	Absorbed with fats, bile salts are required for emulsification.	Stored in liver and fat tissue	Kuksis (2018)
	Vitamin E	Nuts, seeds, vegetable oils	Absorbed in the small intestine wqith fats	Stored in adipose tissues	Drevon (1991)
	Vitamin K	Leafy green broccoli	Absorbed in the small intestine with fat	Stored in liver	Andrès et al. (2024)
Water-soluble	Vitamin B complex	Whole grains, meat, dairy	Directly absorbed into the bloodstream into the small intestine	Not stored excess excreted in urine	Schellack et al. (2019)
	Vitamin C	Citrus fruits, pepper	Absorbed in the small intestine	Not stored excess excreted in urine	Doseděl et al. (2021)

in the intestinal lumen plays a critical role in breaking down these fat droplets, facilitating the absorption of fat-soluble vitamins in the small intestine (Allen 2025).

3.13 Absorption of Minerals

The absorption of essential minerals like zinc, iron, calcium, magnesium, and copper involves specific transport mechanisms in the small intestine. Zinc uptake is regulated by ZIP4, which increases absorption during deficiency, while ZnT1 facilitates its excretion when levels are sufficient. Similarly, DMT1 controls iron absorption, transporting ferrous ions into enterocytes, while ferroportin regulates their release into the circulation. Calcium is absorbed via TRPV6 channels and is stored in vesicles or bound to proteins like calbindin. Magnesium uptake occurs through TRPM6, while copper transport involves CTR1 and ATP7A. These minerals interact within the body, influencing each other's absorption efficiency (Duan et al. 2025).

3.14 Influence of Dietary Fiber in Food Waste Valorization

The increasing global focus on food waste valorization highlights the need for sustainable strategies to maximize resource utilization. Dietary fiber, a crucial component in plant-based food waste, plays a significant role in both nutrition and industrial applications. Food processing byproducts, such as fruit and vegetable peels, cereal bran, and legume residues, are rich in fiber but often remain underutilized due to processing limitations. Typically, dietary fiber consists of insoluble (IDF) and soluble (SDF) fractions, with SDF being more beneficial due to its functional properties and fermentation potential in the gut. However, most food waste-derived fibers exist in insoluble forms, reducing their effectiveness in value-added applications. Various treatment methods, including enzymatic hydrolysis, chemical modification, and fermentation, have been explored to enhance SDF content, making fiber-rich food waste more suitable for human consumption and industrial use. This transformation not only improves the nutritional quality of fiber but also increases its commercial viability, supporting the global movement toward sustainable food systems.

Valorization strategies incorporating dietary fiber into food and non-food products have been widely studied. Research highlights its potential use in bakery items, dairy alternatives, animal feed, biodegradable materials, and even as an absorbent for environmental contaminants. In many cases, fermentation precedes fiber modification, enhancing its digestibility and bioavailability. Additionally, functional improvements such as increased water-holding capacity, viscosity, and gel-forming ability make fiber-enriched byproducts valuable in food formulation. Despite extensive studies on fiber modification techniques, a systematic comparison of their efficiency in enhancing SDF content remains limited (Table 3.4) (Azevedo et al. 2025).

Table 3.4 Effect of food processing on nutrient absorption

Form of food	Effects on nutrient absorption
Raw food (natural state)	Contains natural vitamins, minerals, and enzymes
Processing (cooking, preserving, refining, etc.)	This can lead to nutrient loss
Nutrient loss (heat, oxidation, leaching)	Reduces vitamins and minerals(e.g vitamin C loss in heat)
Nutrient enhancement (fortification, bioavailability)	Increase availability of some nutrients (e.g cooking tomatoes increases lycopene
Altered digestibility	Processing can make some nutrients easier to absorb (e.g. breaking down fiber)
Digestion and absorption	The body's ability to absorb nutrients easier to absorb
Health impact	This can lead to nutrient deficiencies

3.14.1 Pesticide and Veterinary Drug Residues

Pesticides and veterinary drugs are widely used in agriculture and animal husbandry to enhance yield and prevent diseases. However, their residues can persist in food products, leading to potential health risks. Long-term exposure has been associated with neurological disorders, hormonal imbalances, and increased cancer risks (Fig. 3.3).

3.14.2 Heavy Metal Ions

Due to environmental pollution, heavy metals such as lead (Pb), mercury (Hg), cadmium (Cd), and arsenic (As) are major food contaminants. These metals accumulate in the food chain and can cause severe health issues, including kidney damage, neurological disorders, and developmental impairments in children (Pettersson 1976).

3.14.3 Foodborne Pathogens

Bacteria, viruses, and parasites that contaminate food can lead to serious foodborne illnesses. Common pathogenic microorganisms include:

- Salmonella spp.—Associated with poultry, eggs, and unpasteurized dairy.
- Escherichia coli (E. coli)—Found in contaminated meat and vegetables.
- Listeria monocytogenes—Poses risks in dairy, processed meats, and refrigerated foods.
- Norovirus—A leading cause of viral foodborne outbreaks.

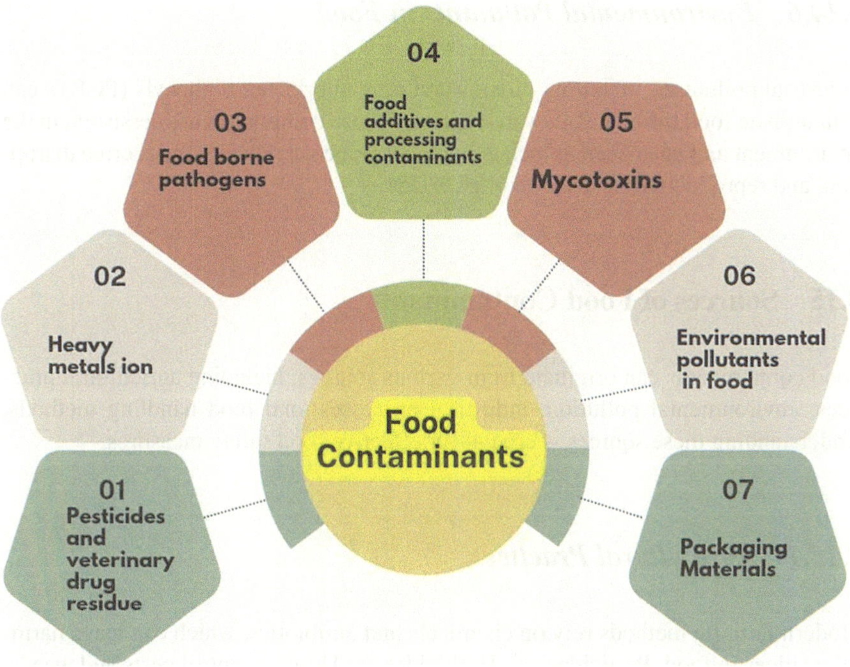

Fig. 3.3 Common food contaminants

3.14.4 *Mycotoxins*

Mycotoxins are toxic compounds produced by certain molds (fungi) that grow on crops like cereals, nuts, and dried fruits. Common types include:

- Aflatoxins—Produced by Aspergillus species, highly carcinogenic.
- Ochratoxin A—Affects cereals, coffee, and grapes, linked to kidney toxicity.
- Fumonisins—Found in corn, associated with esophageal cancer (Staib and Fuchs 2014).

3.14.5 *Food Additives and Processing Contaminants*

While food additives enhance taste, color, and preservation, some can have adverse effects. Artificial sweeteners (Aspartame)Linked to metabolic disorders. Preservatives (sodium nitrite) can increase the risk of carcinogenic nitrosamines. Acrylamide, is a byproduct of high-temperature cooking frying, and baking, with potential neurotoxic and carcinogenic properties (Roca-Saavedra et al. 2018).

3.14.6 Environmental Pollutants in Food

Industrial pollutants, including dioxins and polychlorinated biphenyls (PCBs), can contaminate food through soil, water, and air. These compounds are persistent in the environment and have been associated with immune suppression, endocrine disruption, and reproductive issues (Fan et al. 2025).

3.15 Sources of Food Contaminants

Food contaminants can originate from various sources, including agricultural practices, environmental pollution, industrial processes, and food handling methods. Understanding these sources is crucial for effective food safety measures.

3.15.1 Agricultural Practices

Modern farming methods rely on chemicals and antibiotics, which can leave harmful residues in food. Pesticides and Herbicides are Used to control pests and weeds, these chemicals can remain in fruits, vegetables, and grains. Veterinary Drugs are Antibiotics and growth hormones administered to livestock that can persist in meat, milk, and eggs. Fertilizers Excessive use of chemical fertilizers may lead to contamination with heavy metals such as cadmium and lead.

3.15.2 Environmental Pollution

Food can be contaminated by pollutants present in soil, water, and air. Heavy Metals found in Industrial waste and mining activities introduce toxic metals (lead, mercury, arsenic) into soil and water, affecting crops and seafood. Airborne Pollutants found in Factories release harmful substances like dioxins and polychlorinated biphenyls (PCBs) that settle on food crops and animal feed. Contaminated Water Sources present in Polluted irrigation water or contaminated seafood habitats can introduce pathogens and chemical toxins into the food chain.

3.15.3 *Food Processing and Packaging*

Contamination can occur during food processing, storage, and packaging. Food Additives and Preservatives present in Excessive use of artificial colours, sweeteners, and preservatives can pose health risks. Processing Byproducts in High-temperature cooking methods (e.g., frying, grilling) can produce harmful compounds like acrylamide and polycyclic aromatic hydrocarbons (PAHs). Packaging Materials found in plastic cans and storage containers may leach chemicals (e.g., bisphenol A or BPA) into food over time.

3.15.4 *Microbial Contamination*

Pathogenic bacteria, viruses, and fungi can contaminate food at different stages of production and handling. Key sources include:

Poor Hygiene in Food Handling—Cross-contamination due to unwashed hands, dirtyutensils, or unclean food preparation areas.
Improper Storage Conditions—Failure to maintain proper temperature and humidity levels can lead to mold growth and bacterial proliferation.
Raw or Undercooked Foods—Foods like meat, seafood, and dairy can harbor bacteria such as Salmonella, E. coli, and Listeria if not handled or cooked properly.

3.15.5 *Natural Toxins*

Some contaminants occur naturally in food due to biological processes. Mycotoxins are Produced by molds that grow on grains, nuts, and dried fruits, leading to health risks. Bacterial Toxins Produced by bacteria in food, water, and the human body cause severe illness, food poisoning, and fatal paralysis. Plant Toxins These are the naturally occurring chemicals in plants and can be toxic. They are found in cassava, almonds, cherry pits, and green potatoes, which may release cyanide, causing nausea and neurological symptoms. Algal Toxins are Found in seafood due to harmful algal blooms, causing conditions like shellfish poisoning. Toxic Plant Compounds Some plants produce natural toxins (e.g., cyanogenic glycosides in cassava) that require proper processing to eliminate risks.

3.15.6 Accidental and Intentional Contamination

Food contamination can sometimes result from accidents or deliberate actions. Industrial Accidents from Chemical spills or leaks from factories can introduce contaminants into food supply chains. Food Fraud and Adulteration, Intentional contamination, such as adding unauthorized substances to increase product weight or appearance, can pose serious health hazards. Household Contaminants found in Poor storage, use of expired ingredients, or exposure to household chemicals can also contribute to food contamination (Fan et al. 2025).

3.16 Chemical Contaminants in Foods

Chemical contaminants in food originate from various sources, posing risks to human health and food safety (Fig. 3.4). These contaminants can enter the food chain through environmental pollution, agricultural practices, industrial processes, and food processing methods. Heavy metals, pesticide residues, industrial chemicals, and persistent organic pollutants are common sources of contamination in food products. Additionally, the misuse of veterinary drugs in livestock and aquaculture contributes to chemical residues in food. Understanding these sources is essential for developing effective food safety strategies and minimizing health hazards associated with chemical contaminants (Liu et al. 2025).

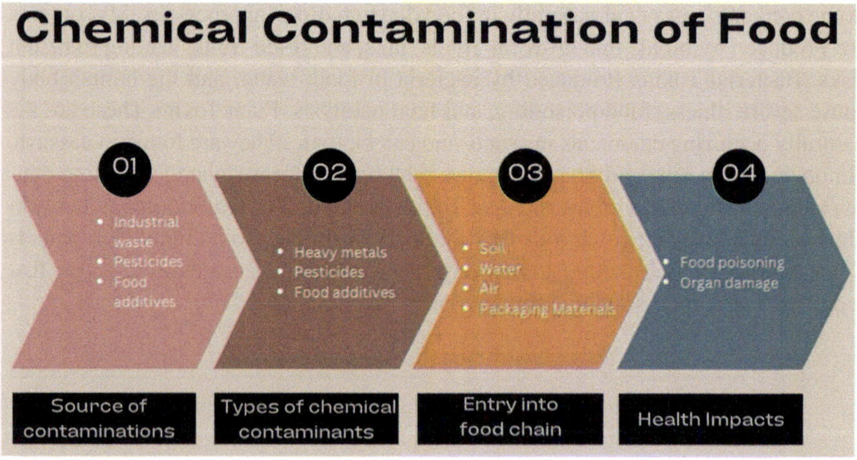

Fig. 3.4 Chemical contaminants in food

3.17 Biological Contaminants of Food

3.17.1 Microalgae Contaminants

Certain microalgae species, if unintentionally introduced, can compete with cultivated microalgae, reducing nutrient yield by up to 10 times. Some unapproved microalgae may produce toxins, making food and feed unsafe for consumption.

3.17.2 Predatory Microorganisms (Grazers & Rotifers)

Predators such as rotifers and protozoa can consume microalgae, reducing their concentration by up to 90% within 2–5 days. This affects biomass quality and production efficiency.

3.17.3 Heterotrophic Bacteria

These bacteria compete for nutrients, reducing the efficiency of microalgae production. They can alter carbon, nitrogen, and phosphorus balance, impacting the final nutritional quality of food products.

3.17.4 Toxin-Producing Microorganisms

Some microorganisms in contaminated cultures release toxins, which can be harmful if consumed. These contaminants must be quantified to assess their impact on food safety and quality (Flores-Tinoco et al. 2025) (Fig. 3.5).

3.18 Impact of Food Contaminants on Nutrient Absorption

Food contaminants, such as potentially toxic elements (PTEs) like arsenic, can significantly impact nutrient absorption by altering soil quality and plant uptake. When present in contaminated soils, these elements can compete with essential nutrients, reducing their bioavailability to plants like Lactuca sativa (lettuce). This disruption can lead to nutrient deficiencies in crops, ultimately affecting human nutrition when consumed. The study highlights how organic and inorganic amendments, such as compost, manure, and iron oxides, influence the mobility and uptake of these contaminants, which can either mitigate or exacerbate their impact on nutrient

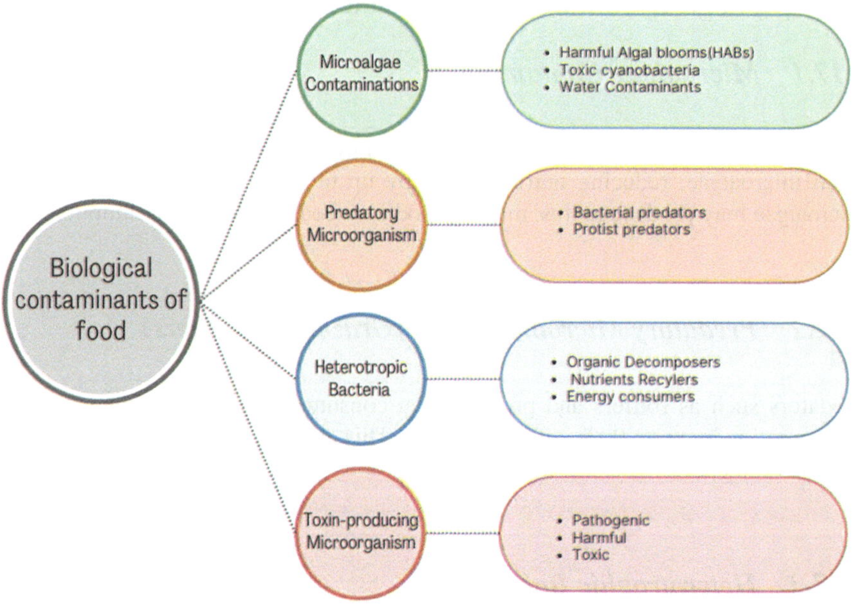

Fig. 3.5 Biological contaminants in food

absorption. Additionally, the interaction between contaminants and soil amendments plays a crucial role in determining how effectively plants can absorb necessary nutrients. For instance, while biochar and iron oxides may reduce the availability of certain toxic elements, they can also influence essential nutrient retention. If amendments are not carefully selected, they may unintentionally facilitate the release of harmful substances, further hindering nutrient absorption. Thus, proper soil management strategies are essential to ensure that food grown in contaminated environments maintains its nutritional value while minimizing health risks associated with toxic element exposure (Mensah et al. 2025).

3.19 Detection and Measurement of Food Contaminatets

Ensuring food safety requires effective detection and measurement of contaminants, including biological, chemical, and physical hazards. Modern methods like spectroscopy, chromatography, and AI-driven detection enhance food safety management.

3.19.1 Types of Food Contaminants (Fig. 3.6)

Biological:

Biological contaminants (Fig. 3.5) are the living organisms or the substance they produce.They are the leading cause of foodborne illlness and also include

- Bacteria- (Salmonella, E. coli),These bacteria can contaminate a wide range of food including meat, poultry,dairy products and vegetables.
- Viruses- Viruses like Norovirus and Hepatitis A can easily spread through contaminated food and are highly infectious.They are often transmitted trough improper handling and poor hygiene.
- Fungi-Molds and yeast can grow on food especially in warm and humid conditions
- Parasites—Parasatic organisms including Giardia and Trichnella can contaminate food especially undercooked meat.

Chemical:

- Pesticides-Residues of pesticides can remain on fruits,vegetables and grains.
- Heavy metals-Such as (lead, mercury), can contaminate food through water and soil.
- Industrial toxins- Contaminants like dioxins and polychlorinated biphenyls (PCBs) can enter the food chain through environmental pollution.

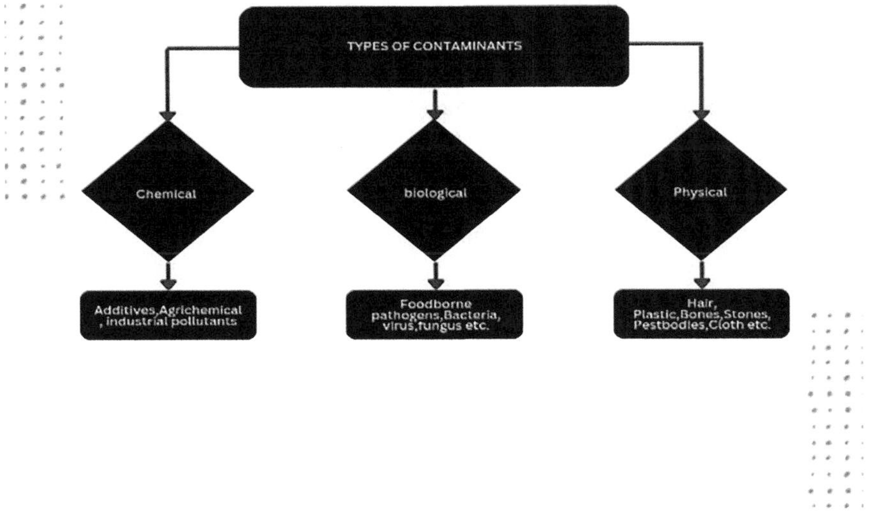

Fig. 3.6 Types of contaminants

Physical:
They are the foreign objects that enter in the food accidentally

- Metal fragments-Machinery or types of equipment which are used in food processing can end up in food products.
- Plastic particles- pieces of plastic from packaging material can contaminate food
- Glass-Broken pieces from container can cause serious risk if ingested.

3.19.2 Detection Methods

- **Microbiological Testing**: PCR, ELISA, and Next-Generation Sequencing (NGS) detect pathogens.
- **Spectroscopic & Imaging**: Infrared, Raman, and Hyperspectral Imaging identify contaminants.
- **Chromatography**: GC-MS and HPLC analyze chemical residues.
- Biosensors: Electrochemical and optical sensors provide real-time detection.
- AI & Deep Learning: Image recognition and machine learning improve accuracy (Fan et al. 2025).

3.20 Regulatory Standards for Food Security

3.20.1 International Food Safety Standards

Global organizations like the FAO and WHO establish food safety standards through Codex Alimentarius, ensuring food quality, hygiene, and safety. These guidelines help nations develop food policies to protect consumer health and facilitate fair trade. Food safety is a major public health concern that affects countless households with global economic integration and evolving consumer health preferences, the food industry continues to grow in both diversity and complexities (Abd El-Hack et al. 2022).

3.20.2 National Food Security Laws

Countries implement legal frameworks to ensure stable food production, distribution, and access. These laws regulate food safety, agricultural practices, and emergency food reserves to address shortages. The universal human right to food must be recognized this means securing both food security, the basic necessity of food access for survival, and food safety which protects people from the health risks associated with food and drinking water (Kirchsteiger-Meier and Baumgartner 2014).

3.20.3 Land Use and Agricultural Policies

Governments regulate land use to maintain food security by preventing excessive urbanization of farmland. Policies promote sustainable land management, soil fertility conservation, and efficient agricultural resource allocation. The environmental benefits are expected to be at least as significant as the reduced financial burden on consumers and taxpayers. This highlights the importance of environmental quality in driving policy reforms and offers a more effective approach to achieving the environmental goals of land retirement programs (Ye et al. 2025).

3.20.4 Food Import and Export Regulations

Strict import and export policies control food quality, preventing contaminated or substandard products from entering markets. These regulations ensure compliance with national and international food safety standards. After reviewing scientific literature researchers identified eight key factors related to importing agricultural products supply risk, demand risk, production risk, management and operational risk, logistical and infra-structural risk, and political and regulatory risk. To evaluate these risks, experts used three multicriteria decision support methods (MCDM), SAW, TOPSIS, and geometrical means (Aday and Aday 2020).

3.20.5 Food Quality Control and Testing

Food safety agencies enforce regular inspections, quality checks, and certifications to detect contaminants and ensure compliance with safety standards. Laboratories test for pesticides, heavy metals, and microbial hazards. Ensuring food safety and high standards presents major challenges for the food industry. In recent years, artificial intelligence|(AI) and machine learning(ML) have emerged as comprising solutions to address these issues (Chhetri 2024).

3.20.6 Environmental and Sustainability Standards

Regulations promote eco-friendly agricultural practices, reducing pollution and soil degradation. Sustainable farming techniques, biodiversity conservation, and water management are key components. The commission has published the reporting standards, but keeping up with them can be challenging. The mandate and standards are closely connected and evolving quickly making it harder to stay on top of the changes (Baboukardos et al. 2023).

3.20.7 Emergency Food Security Policies

Governments establish contingency plans for food crises, including food stockpiling, price stabilization mechanisms, and disaster relief programs to ensure continuous food availability. Effective intervention is necessary for tackling the insecurity of food.However, research on food security interventions on different levels remains limited.additionally, past studies have rarely explored how these interventions align with the four key factors of food security availability, access, stability, and utilization (Ammar et al. 2024).

3.20.8 Consumer Protection and Food Labeling

Strict labeling laws require accurate nutritional information, ingredient lists, and allergen warnings to help consumers make informed choices. Misleading claims and unsafe food marketing are prohibited. When faced with complex choices consume depend upon labeling to make informed decisions. At the core of this struggle is the fight for visibility, as each group vies for limited space to convey important information (Franca et al. 2024).

3.20.9 Biotechnology and GMO Regulations

Laws govern the use of genetically modified organisms (GMOs) in food production, ensuring they meet safety standards before entering the market. Transparent labeling informs consumers about GMO content. Regulations play a crucial role in protecting public health and safety,ensuring that products with harmful ingrediets never make it to the market (Wray-Cahen et al. 2024).

3.20.10 Food Waste Management Regulations

Policies aim to reduce food wastage by promoting redistribution, composting, and responsible disposal practices. Businesses and consumers are encouraged to minimize food loss throughout the supply chains (Yerkinbayeva et al. 2025). Throwing food waste into landfills is significantly contributes into the emmission of green house gas as it breaks down with oxygen and releases methane—a potent green house gas (Jacob-John et al. 2023).

3.21 Strategies for Reducing Food Contaminants

To ensure food safety and minimize health risks associated with hazardous chemicals in food processing, a comprehensive approach is essential. The following strategies can be implemented to reduce the presence of food contaminants:

3.21.1 Good Agricultural Practices (GAPs)

Use organic farming techniques to reduce pesticide and heavy metal contamination.

Implement crop rotation and soil testing to prevent excessive accumulation of harmful substances.

Avoid the use of untreated sewage water for irrigation to minimize microbiological and chemical contamination.

In developing economies,food safety indicators are important for ensuring public health and protecting that food which remain for consumption (Leong et al. 2020).

3.21.2 Good Manufacturing Practices (GMPs)

Maintain strict hygiene protocols in food processing facilities to prevent cross-contamination.

Regularly inspect and clean machinery, storage units, and production lines to reduce chemical residues.

Use food-grade equipment and packaging materials to prevent the leaching of harmful substances.

3.21.3 Minimization of Process-Induced Contaminants

Adjust cooking temperatures to prevent the formation of carcinogenic compounds like acrylamide and polycyclic aromatic hydrocarbons (PAHs).

Use alternative processing methods such as vacuum frying and steaming to reduce harmful by-products.

Implement enzyme-based processing to minimize the need for chemical additives.

3.21.4 Control of Pesticide and Antibiotic Residues

Enforce stringent regulations on the use of pesticides and veterinary drugs in food production.

Encourage the use of biopesticides and integrated pest management (IPM) techniques.

Implement residue monitoring programs to detect and control excessive chemical levels.

3.21.5 Water and Food Quality Monitoring

Regularly test raw materials and food products for the presence of contaminants.

Implement advanced water filtration systems to remove heavy metals, toxins, and pathogens.

Conduct hazard analysis and critical control points (HACCP) assessments to identify contamination risks.

3.21.6 Food Packaging Safety

Use biodegradable and non-toxic food packaging materials to prevent chemical leaching.

Avoid plastic packaging containing bisphenol A (BPA) and phthalates.

Store food in temperature-controlled environments to prevent the migration of harmful substances.

3.21.7 Public Awareness and Policy Implementation

Conduct consumer education programs on safe food handling and storage practices.

Strengthen food safety regulations and implement stricter enforcement of food quality standards.

Encourage research and development of alternative food preservation techniques to reduce the need for chemical preservatives.

By integrating these strategies, food manufacturers, policymakers, and consumers can contribute to reducing food contaminants and ensuring safer food for public consumption (Vignesh et al. 2024).

3.22 Innovative Approaches in Food Waste Volarization

Recent advancements in sustainable food waste management focus on biorefinery processes that integrate multiple eco-friendly techniques to extract bioactive compounds from fruit and vegetable by-products. Novel approaches, such as enzymatic hydrolysis, supercritical fluid extraction, and membrane-based separation, offer greener alternatives to traditional chemical-intensive methods. These techniques not only reduce environmental impact but also enhance the recovery of high-value compounds like polyphenols, dietary fibers, and natural colorants for clean-label food applications (Zhang and Zhang 2024).

Additionally, emerging technologies, including fermentation and microbial bioconversion, are gaining attention for their ability to transform food waste into biofuels, bioplastics, and functional ingredients. By adopting circular economy principles, these strategies help minimize waste, generate new revenue streams, and improve food system sustainability. Integrating these innovative approaches into existing food processing industries can lead to cost-effective and energy-efficient waste recycling, ultimately contributing to a more sustainable food chain (Vignesh et al. 2024). A company's technological experties is the to drive the innovation,making it one of the most impotrtant skill needed to br compitetive in long run.

3.23 Flowchart: Sustainable Biorefinery Process for Food Waste Valorization

3.24 Future Directions in Nutrient Absorption Research

Nutrient absorption is a critical aspect of human nutrition, influencing overall health, disease prevention, and therapeutic applications (Fig. 3.7). Based on the existing research trends and gaps, future studies should focus on the following directions to advance our understanding and application of nutrient absorption mechanisms.

3.24.1 Advancements in Nutrient Bioavailability Studies

Future research should explore the bioavailability of various nutrients in different food matrices, emphasizing how processing methods, storage conditions, and cooking techniques influence nutrient retention and absorption. Additionally, personalized nutrition approaches, including genetic predisposition and gut microbiota variations, must be studied to understand individual differences in nutrient metabolism.The body's ability to absorb minerals and vitamins from fortified food can be influenced by various factors (Coelho et al. 2021).

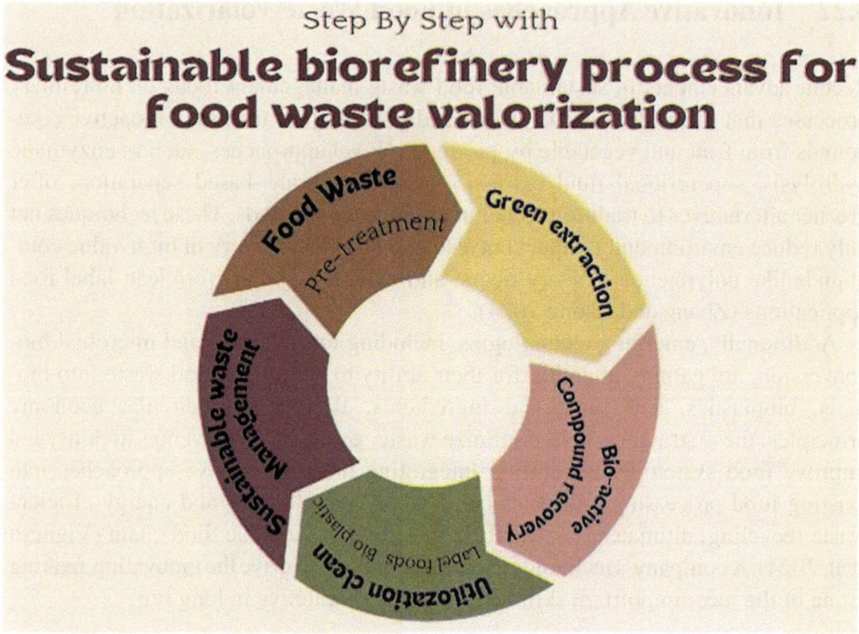

Fig. 3.7 Sustainable biorefinery process for food waste valorization

3.24.2 *Role of Food Contaminants and Additives in Nutrient Absorption*

The impact of food contaminants, such as heavy metals, pesticide residues, and mycotoxins, on nutrient absorption remains an underexplored area. Future studies should investigate how these contaminants interact with nutrients at the molecular level, potentially leading to deficiencies or toxic effects. Similarly, the role of food additives, preservatives, and artificial sweeteners in altering gut permeability and nutrient transport requires further examination.Eating food that have antibiotics may triger allergies and also contributes in the antibiotic resistant bacteria in human body (Dong et al. 2024).

3.24.3 *Gut Microbiota and Nutrient Absorption Interactions*

Emerging evidence suggests that gut microbiota play a crucial role in the breakdown and assimilation of nutrients. Future research should focus on the modulation of gut bacteria through probiotics, prebiotics, and synbiotics to enhance nutrient absorption. Moreover they are also responsible for diabetes \insulin resistance,fatty liver and cardiovascular disease.The impact of gut microbuta had less attention and the impact of minor foods.Additionally, metagenomics and metabolomics techniques can provide deeper insights into how microbial metabolites influence nutrient bioavailability.

3.24.4 Nutrient-Nutrient and Drug-Nutrient Interactions

Understanding interactions between nutrients (e.g., iron and calcium competition) and between nutrients and medications (e.g., metformin affecting vitamin B12 absorption) is essential for optimizing dietary recommendations. Future research should employ advanced imaging and biochemical techniques to unravel these interactions at the cellular level, leading to improved dietary guidelines and clinical interventions.Encapsulating vitamin,mineral and antioxidants in nanoparticles offers a groundbreaking way to enhance stability and solubility making nutrient more effective in supporting human health (Vieira et al. 2023).

3.24.5 Nanotechnology and Targeted Nutrient Delivery Systems

Nanotechnology-based approaches, such as nanoencapsulation, offer promising solutions to improve nutrient stability and absorption. Future studies should focus on developing smart delivery systems that enhance the bioavailability of vitamins, minerals, and bioactive compounds, particularly for individuals with malabsorption disorders or chronic diseases.Nanotechnology is set to revolutionized the field with breaking through like advanced disease diagnostics and better nutrient absorption. (Abdallah et al. 2023)

3.24.6 Sustainable and Alternative Food Sources for Enhanced Nutrient Absorption

With the increasing demand for sustainable food sources, future research should investigate how novel foods, including edible insects, algae, and plant-based proteins, affect nutrient absorption. Special attention should be given to anti-nutritional factors, such as phytates and oxalates, and their mitigation strategies to improve bioavailability.The thermal and biochemical treartment has also various application in food (Sharma et al. 2023).

3.24.7 *Clinical and Epidemiological Studies on Nutrient Deficiency Disorders*

Longitudinal studies assessing nutrient absorption efficiency across different populations, age groups, and disease conditions are necessary. Research should focus on developing non-invasive diagnostic tools for assessing nutrient absorption status, leading to better public health strategies for combating deficiencies.Adequate study designs under strict methodolgies and careful interpretation of resultsare of atmost progressing in nutrition field.Following them provdes greater credibility and trust worthiness to clinical trials,SRs and results of meta analysis (Wimalawansa 2025).

3.25 Conclusion

Future research in nutrient absorption should integrate multidisciplinary approaches, including molecular biology, gut microbiome studies, advanced imaging techniques, and bioinformatics (Baynes et al. 2016). Addressing these key areas will contribute to optimizing dietary strategies, improving therapeutic interventions, and enhancing overall human health (Wang et al. 2025).

References

Abd El-Hack ME, El-Saadony MT, Salem HM, El-Tahan AM, Soliman MM, Youssef GB et al (2022) Alternatives to antibiotics for organic poultry production: types, modes of action and impacts on bird's health and production. Poult Sci 101(4):101696

Abdallah Y, Nehela Y, Ogunyemi SO, Ijaz M, Ahmed T, Elashmony R et al (2023) Bio-functionalized nickel-silica nanoparticles suppress bacterial leaf blight disease in rice (Oryza sativa L.). Front Plant Sci 14:1216782

Aday S, Aday MS (2020) Impact of COVID-19 on the food supply chain. Food Qual Saf 4(4):167–180

Allen LH (2025) Micronutrients – assessment, requirements, deficiencies, and interventions. N Engl J Med 392(10):1006–1016

Ammar KA, Kheir AMS, Ali BM, Sundarakani B, Manikas I (2024) Developing an analytical framework for estimating food security indicators in The United Arab Emirates: a review. Environ Dev Sustain 26(3):5689–5708

Andrès E, Lorenzo-Villalba N, Terrade J-E, Méndez-Bailon M (2024) Fat-soluble vitamins A, D, E, and K: review of the literature and points of interest for the clinician. J Clin Med 13(13):3641

Anwar RS, Ahmed RR, Streimikiene D, Streimikis J, Zamek D (2025) Securing food futures: the interplay of safety governance, hygiene, supplier beliefs and consumer engagement. Br Food J 127(2):758–778

Azevedo P, Jin S, Wu Y, Derakhshani H, Xu H, Lei H et al (2025) Effects of dietary fiber on fecal microbiota of grower–finisher pig offspring from parents with divergent estimated breeding value for feed conversion ratio. Can J Anim Sci 105:1–11

Baboukardos D, Gaia S, Lassou P, Soobaroyen T (eds) (2023) The multiverse of non-financial reporting regulation. Accounting Forum. Taylor & Francis

Basile EJ, Launico MV, Sheer AJ (2023) Physiology, nutrient absorption. StatPearls [Internet]. StatPearls Publishing
Batterman S, Metts T, Kalliokoski P (2002) Diffusive uptake in passive and active adsorbent sampling using thermal desorption tubes. J Environ Monit 4(6):870–878
Baynes RE, Dedonder K, Kissell L, Mzyk D, Marmulak T, Smith G et al (2016) Health concerns and management of select veterinary drug residues. Food Chem Toxicol 88:112–122
Carrero-Colón M, Nakatsu CH, Konopka A (2006) Effect of nutrient periodicity on microbial community dynamics. Appl Environ Microbiol 72(5):3175–3183
Chen L, Wang X, Wang S, Liu W, Song Z, Liao H (2025a) The impact of gut microbiota on the occurrence, treatment, and prognosis of ischemic stroke. Neurobiol Dis 207:106836
Chen H, Sheng J, Ye Q, Li J, Yu X, Wu H et al (2025b) Efficient resource recovery from food waste digestate via hydrothermal treatment and its application as organic fertilizer. Bioresour Technol 416:131742
Chhetri KB (2024) Applications of artificial intelligence and machine learning in food quality control and safety assessment. Food Eng Rev 16(1):1–21
Chung BM, Tappenden KA (2024) Macronutrient digestion, absorption, and metabolism. Clinical nutrition in gastrointestinal disease. CRC Press, pp 77–96
Coelho RC, Barsotti RCF, Maltez HF, Júnior CAL, de Sousa BH (2021) Expanding information on the bioaccessibility and bioavailability of iron and zinc in biofortified cowpea seeds. Food Chem 347:129027
Dahlqvist A (1972) Intestinal absorption of sucrose. Acta Med Scand 192(S542):13–18
Das M (1959) Role of peptidases and peptide transport in the intestinal absorption of proteins. World Rev Nutr Diet 24:58
Deng R, Nau F, Lucas T, Maillard M-B, Leduc A, Ossemond J et al (2025) First assessments of nutrient bioaccessibility with an INFOGEST semi-dynamic gastric digestion in vitro protocol adapted to model proton pump inhibitor use. Food Res Int:115833
Dong S, Li L, Hao F, Fang Z, Zhong R, Wu J et al (2024) Improving quality of poultry and its meat products with probiotics, prebiotics, and phytoextracts. Poult Sci 103(2):103287
Doseděl M, Jirkovský E, Macáková K, Krčmová LK, Javorská L, Pourová J et al (2021) Vitamin C – sources, physiological role, kinetics, deficiency, use, toxicity, and determination. Nutrients 13(2):615
Drevon CA (1991) Absorption, transport and metabolism of vitamin E. Free Radic Res Commun 14(4):229–246
Duan H, Guo Y, Ren C, Wei R, Yan W (2025) Mechanisms of vitamin and mineral absorption and their potential for eye fatigue relief. J Fut Foods 5(3):219–228
Fan C, Zhang R, Jin M (2025) The materials synthesized in deep eutectic solvents for the detection of food contaminants. Food Qual Saf 9:fyaf006
Felicilda-Reynaldo RFD, Kenneally M (2016) Digestive enzyme replacement therapy: pancreatic enzymes and lactase. Medsurg Nurs 25(3):182–185
Flores-Tinoco V, Brito-Bello AA, Le VP, Lopez-Arredondo D (2025) Control of biological contamination in microalgae cultures. Algal bioreactors. Elsevier, pp 239–262
Franca P, Pierucci AP, Boukid F (2024) Analysis of ingredient list and nutrient composition of plant-based burgers available in the global market. Int J Food Sci Nutr 75(2):159–172
Goyal MR, Boldhane R, Gonzalez D, Gupta P, Kadam B (2025) Biofluid dynamics of the human digestive system. Apple Academic Press, Biofluid dynamics of human body systems, pp 457–487
Halsted CH (2003) Absorption of water-soluble vitamins. Curr Opin Gastroenterol 19(2):113–117
Haque ML, Oosterveer P, Reynolds C, Vignola R (2025) Does food safety governance in Bangladesh include the urban poor? An analysis of government strategies and policies for the retail food environment in Bangladesh, 2013–2022. Habitat Int 159:103372
Holtmann G, Kelly DG, Sternby B, DiMagno EP (1997) Survival of human pancreatic enzymes during small bowel transit: effect of nutrients, bile acids, and enzymes. Am J Physiol Gastrointest Liver Physiol 273(2):G553–G5G8

Hoxha M, Abbasciano M, Avantaggiato G, Zappacosta B, Tricarico D (2025) Mycotoxins and neuropsychiatric symptoms: possible role in special refugee populations. Front Pharmacol 16:1524152

Ishra R, Sharif S, Soar J, Khanam R (2025) Food safety knowledge among adult consumers in Bangladesh: a COVID-19 comparative study. Br Food J 127:96–113

Jackson AD, McLaughlin J (2009) Digestion and absorption. Surgery (Oxford) 27(6):231–236

Jacob-John J, D'Souza C, Marjoribanks T, Singaraju S (2023) Sustainable Development Goals: a review of SDG 12.3 in food supply chain literature. BIJ 30(9):3465–3481

Jha P, Singh AK (2025) Regulatory compliance and food safety standards. Engineering solutions for sustainable food and dairy production. In: Innovations and techniques in food processing and dairy engineering. Springer, pp 463–487

Kirchsteiger-Meier E, Baumgartner T (2014) Global food legislation: an overview. Wiley

Krajmalnik-Brown R, Ilhan ZE, Kang DW, DiBaise JK (2012) Effects of gut microbes on nutrient absorption and energy regulation. Nutr Clin Pract 27(2):201–214

Kuksis A (2018) Absorption of fat-soluble vitamins. CRC Press, Fat absorption, pp 65–86

Lavin SR, McWhorter TJ, Karasov WH (2007) Mechanistic bases for differences in passive absorption. J Exp Biol 210(15):2754–2764

Leong W-H, Teh S-Y, Hossain MM, Nadarajaw T, Zabidi-Hussin Z, Chin S-Y et al (2020) Application, monitoring and adverse effects in pesticide use: the importance of reinforcement of Good Agricultural Practices (GAPs). J Environ Manag 260:109987

Lichtenstein AH, Jones PJ (2012) Lipids: absorption and transport. Present Knowl Nutr:118–131

Liu Q, Ou Y, Florence M, Li Y, Wang Y, Zhang Y et al (2025) Exploration of the development of a food safety risk index system for chemical contamination of aquatic products. Food Control:111189

Mensah AK, Sekyi-Annan E, Addai P, Ulzen OO, Salifu M, Adams S (2025) Manure and iron oxide show potential for reducing uptake of arsenic and mercury in lettuce grown in a contaminated mining site. J Hazard Mater Adv 17:100545

Nichols BL, Baker SS, Quezada-Calvillo R (2018) Metabolic impacts of maltase deficiencies. J Pediatr Gastroenterol Nutr 66:S24–SS9

Pereira S, Kildegaard HF, Andersen MR (2018) Impact of CHO metabolism on cell growth and protein production: an overview of toxic and inhibiting metabolites and nutrients. Biotechnol J 13(3):1700499

Pettersson O (1976) Heavy-metal ion uptake by plants from nutrient solutions with metal ion, plant species and growth period variations. Plant Soil 45:445–459

Roca-Saavedra P, Mendez-Vilabrille V, Miranda JM, Nebot C, Cardelle-Cobas A, Franco CM et al (2018) Food additives, contaminants and other minor components: effects on human gut microbiota – a review. J Physiol Biochem 74:69–83

Schade SG, Schilling RF (1967) Effect of pepsin on the absorption of food vitamin B12 and iron. Am J Clin Nutr 20(6):636–640

Schellack G, Harirari P, Schellack N (2019) Vitamin B-complex deficiency, supplementation and management. SA Pharm J 86(3):23–29

Schiller LR (2014) Absorption of nutrients. Gastrointestinal anatomy and physiology: the essentials, pp 108–128

Sharma M, Bains A, Sridhar K, Chawla P, Sharma M (2023) Process optimization for spray dried Aegle marmelos fruit nanomucilage: characterization, functional properties, and in vitro antibiofilm activity against food pathogenic microorganisms. Int J Biol Macromol 249:126050

Staib L, Fuchs TM (2014) From food to cell: nutrient exploitation strategies of enteropathogens. Microbiology 160(6):1020–1039

Stumpff F, Manneck D (2025) Prebiotics as modulators of colonic calcium and magnesium uptake. Acta Physiol 241(2):e14262

Subedi D, Paudel M, Poudel S, Koirala N (2025) Food safety in developing countries: common foodborne and waterborne illnesses, regulations, organizational structure, and challenges of food safety in the context of Nepal. Food Front 6(1):86–123

Torrissen KR, Male R (2000) Trypsin isozymes: development, digestion, and structure. Food science and technology. Marcel Dekker, New York, pp 215–270
Vieira IRS, Tessaro L, Lima AKO, Velloso IPS, Conte-Junior CA (2023) Recent progress in nanotechnology improving the therapeutic potential of polyphenols for cancer. Nutrients 15(14):3136
Vieira BM, de Mello e Silva GN, Silva MI (2025) Nutritional elements II: vitamins and minerals. In: Fundamentals of drug and non-drug interactions: Physiopathological perspectives and clinical approaches. Springer, pp 57–86
Vignesh A, Amal TC, Vasanth K (2024) Food contaminants: impact of food processing, challenges and mitigation strategies for food security. Food Res Int:114739
Wang Y, Wei W, Wang Y, Yu L, Xing Z, Zhang J et al (2025) Innovative applications of medium- and long-chain triacylglycerol in nutritional support: current perspectives and future directions. Compr Rev Food Sci Food Saf 24(2):e70116
Wapnir RA (2024) Protein nutrition and mineral absorption. CRC Press
Wimalawansa SJ (2025) Enhancing the design of Nutrient Clinical trials for disease prevention – a focus on vitamin D: a systematic review. Nutr Rev nuae164
Wray-Cahen D, Hallerman E, Tizard M (2024) Global regulatory policies for animal biotechnology: overview, opportunities and challenges. Front Genome Edit 6:1467080
Ye Z, Du H, Zhang Y, Tan S (2025) Exploring the impacts of land titling on eco-environment: a case of livestock farming system in Inner Mongolia, China. Sci Tot Environ 958:177988
Yerkinbayeva L, Myrzakarim A, Moldagaliyeva A, Kalymbek B, Yerezhepkyzy R, Nurmukhankyzy D (2025) The role of legal regulation in promoting organic agriculture to achieve sustainable food security. J Lifestyle SDGs Rev 5(1):e04197-e
Zhang Q, Zhang H (2024) Assessing agri-food waste valorization challenges and solutions considering smart technologies: an integrated Fermatean fuzzy multi-criteria decision-making approach. Sustain For 16(14):6169
Żyła K, Duda A (2025) Towards improved bioavailability of cereal inositol phosphates, myo-inositol and phenolic acids. Molecules 30(3)

Chapter 4
Neural Control of Appetite and Food Choices in the Context of Food Safety

Nida Kanwal, Mahwish Tanveer, Aamir Shehzad, and Imran Pasha

4.1 Introduction

An Overview of Appetite and Food Choices

The psychological need for food or drink is known as appetite. Sensual reactions to the sights, sounds, smells, and tastes of food are among the many variables that affect hunger. Social and behavioral problems are additional elements that could influence these sensory reactions. For instance, even if someone is not hungry, the sight of food may encourage him/her to eat. Alternatively, dinner may be persuaded to eat because others are eating even though they are not hungry. In each of these cases, the brain receives the visual cue of food **through** neurons and registers a desire to eat. It might take precedence over the bodily sensations of contentment from previous meals (Whitney et al. 2019).

Chemical senses begin to learn and navigate the world around us in the third trimester of pregnancy by detecting stimuli in our immediate surroundings. We learn what is safe, pleasurable, and fulfilling to consume during our early years because of our initial exposure to the fundamental flavors and scents of our food environment. Food selection and consumption are significantly acquired behaviors

N. Kanwal (✉)
Department of Food Science and Technology, Riphah International University, Faisalabad Campus, Faisalabad, Pakistan
e-mail: nidakanwal26@gmail.com

M. Tanveer · I. Pasha
National Institute of Food Science and Technology, Faculty of Food, Nutrition and Home Sciences, University of Agriculture, Faisalabad, Pakistan

A. Shehzad
UniLaSalle, Univ. Artois, ULR7519 – Transformations & Agro-ressources, Normandie Université, Mont-Saint-Aignan, France

T. Sarkar et al. (eds.), *Physiological Perspectives on Food Safety: Exploring the Intersection of Health and Nutrition*, https://doi.org/10.1007/978-3-031-94582-3_4

based on the connection between sensory data and the feelings of fullness and satisfaction experienced after eating. Food has no nutritional value until it is selected and eaten, and further consumption will rely on how well this experience goes (Forde 2018).

Together with vision and hearing, taste, smell, and texture are crucial in determining how palatable the first foods are from infancy onward, and they continue to pique our appetite. A food's flavor and texture are important indicators of the possible after-ingestive effects of ingestion, and as the food environment grows more complicated, these indicators can be useful in daily energy choices and hunger control. For instance, while food texture might decrease energy intake by influencing eating rate, food odors can increase hunger and affect our food choices. Subtle variations in the sensory quality and intensity of foods that are equally loved can affect eating habits and the amount of energy needed to feel full (Boesveldt et al. 2018).

The development of food-based strategies to save people from diseases like type-2 diabetes and obesity throughout the lifetime will depend heavily on our ability to comprehend how these food sensations influence eating behaviors associated with overconsumption (Forde 2018). The relative contribution of each sense to the perceptual response to food and beverages can be affected as we age because older consumers' sensory systems start to lose acuity. In terms of their dietary experiences and perceptual capacities, older consumers are the most diverse population. These changes in perception take place at a time when proper nutrition is crucial for preserving immunological and general health. Although we now have a better knowledge of these sensory alterations, more work must be done to increase food acceptance and encourage food intake in a population that is becoming more and more sensitive. Age-related increases in the prevalence of chronic illness and disease cause significant alterations in sensory perception. Perception and appetite can also be affected by the pathophysiology and treatment of diseases like cancer and neurological disorders like Alzheimer's. Early diagnosis of these disorders can also be made using changes in the sensory systems, and during treatment, malnutrition can be decreased by using sensory-based methods to increase food intake and appetite (Boesveldt et al. 2018).

Understanding the various elements that affect a person's food choices and how culture shapes them requires an understanding of the personal food system. People make complicated food choices based on a variety of biological, psychological, economic, social, cultural, physical, and political aspects. Food choice is the study of why people eat the foods they do. Food choice behaviors are a significant determinant of nutritional status and health and are linked to the social and economic expression of identities, preferences, and cultural meanings (Enriquez and Archila-Godinez 2022).

Individual food choices involve a variety of actions, such as purchasing, cooking, storing, distributing, serving, consuming, and cleaning up. Numerous factors, including what to eat, when, where, and with whom to participate in eating behavior, are considered while making these difficult decisions. During history and during a person's life, decisions are made in both short- and long-term periods. Three main viewpoints are used to analyze eating choices (Chen and Antonelli 2020). According to the rationalist viewpoint, people make choices to reduce expenses and maximize benefits. According to the structuralist viewpoint, societal structures and external circumstances influence (i.e., strengthen or limit) people's decisions. According to the constructionist viewpoint, people experience, define, interpret, negotiate, manage, and symbolize the reality around them when making decisions. In this section, we employ a structuralist approach to comprehend the factors that influence personal food preferences and a constructionist approach to comprehend the mechanisms that underlie food selection (Chen and Antonelli 2020).

Attitudes and ideas about food are influenced by the interaction of several multiple factors. Dietary practices and food preferences are influenced by an individual's interactions with their social and physical surroundings. The innermost circle represents how biology affects the foods we choose to eat. Genes, physiology, and metabolism all affect the basic tastes—sweet, sour, salty, bitter, and umami—as well as smell, which all play a part in how food is perceived overall (Grzymisławska et al. 2020). Strong sensory pleasure reactions are evoked by some foods. Particularly when sugar is present, the oral feeling of fat is very rewarding. The combination of sugar and fat is also associated with the activation of the brain's pleasure receptors. The food business has come under fire for producing foods that are heavy in fat, salt, and sweetness to gain profit from natural biological tendencies (Paarlberg 2023). When food was limited or unpredictable, the pursuit of sweet and high-energy meals might have served an adaptive purpose. Because the current food supply offers inexpensive, appealing, energy-dense meals with poor nutritious value and great reward potential, the biological processes that control food intake might not be able to keep up. Learning, experience, and illness conditions can all modify the physiological processes that regulate food intake (Watts et al. 2022).

The second circle addresses how a person interacts with their food choices and is related to conditioned food preferences, conditioned satiety, and familiarity and learnt safety. People learn most of their food preferences and acceptance patterns through physiological conditioning that is gained through exposure to various foods. With frequent exposure, new foods become more popular. Additionally, physiological conditioning and emotional reactions to the social setting in which eating takes place might alter feelings of satiety. By giving, modeling, encouraging, limiting, and rewarding with food, parents significantly influence the environment in which children interact with food (Anzman-Frasca et al. 2018).

The third circle consists of interpersonal determinants (like family and social networks), societal and cultural norms, and intrapersonal determinants (such attitudes, beliefs, motives and values, personal meanings, knowledge, and abilities). As people progress through life and are exposed to social and cultural norms, their attitudes, beliefs, motivations, values, knowledge, and abilities all change. The foods that people purchase and prepare for eating are influenced by these factors. People adopt food-related identities, which in turn affects their eating preferences. Family and social networks also influence food choices by observing what other people eat (Higgs and Ruddock 2020).

The fourth circle includes more general environmental and social factors that affect food choices, many of which have to do with marketing, the food supply, and social food and nutrition policies and initiatives. In the food choice process model, "context" refers to the social and environmental (economic, informational, cultural, and physical) factors (Ziegler et al. 2021). As time passes advancement in neuroscience reveals different regions of brain and different neural circuits which are coordinated during different phases of eating these. Several hypothalamic nuclei are also involved in feeding. Beside this there are some pathways like mesolimbic pathways and nodes in the hind brain which interconnect to modulate the process of appetite. We also look at how feeding circuits are rewired and how motivational systems react to food after prolonged exposure to a high-calorie diet. The development of medical solutions that assist people in maintaining healthy body weight will be strengthened by an understanding of how the neural system controls eating behavior (Yoo et al. 2021).

4.2 Neural Mechanisms of Appetite Regulation

Survival depends on the exact control of fluid and energy equilibrium. It is commonly known that both central appetite signals and peripheral sensory inputs tightly govern ingestive behaviors. Our knowledge of fundamental taste characteristics, the molecular and/or cellular underpinnings of taste perception, and the brain circuits for hunger and thirst have advanced significantly with the advent of modern neurogenetic technologies.

4.2.1 Basic Neuroanatomy and Physiology

An essential part of the brain that controls homeostasis throughout the body is the hypothalamus. By producing and releasing neurohormones into the bloodstream, hypothalamic neuroendocrine cells regulate homeostasis1. Some hypothalamus neuroendocrine cells emit end hormones, such as vasopressin [VP] and oxytocin [OXT], which are transported to their target organs and have actions there, just like in other endocrine organs. Hormones which help to release corticotrophin (CRH) and other hormones are secreted by different kinds of hypothalamus neuroendocrine cells. In anterior pituitary there is also a second population of neuroendocrine cells which release many stimulating hormones when they're stimulated, these hormones include adrenocorticotropic hormone (ACTH). These are then released into hypophyseal portal system. The HP (hypothalamic-pituitary) axis, a key element of the hypothalamic-neuroendocrine system, is the name given to this kind of relationship of hypothalamus and pituitary neuroendocrine cells due to their functionality (Goel et al. 2023).

Important humoral reactions to stressors like cold and stress are provided by the HP axis. The HP-thyroid (HPT) axis releases thyroid hormones from the thyroid gland in response to cold, while the HP-adrenal (HPA) axis releases glucocorticoids from the adrenal cortex in response to stress (Yoo et al. 2021). It has also been demonstrated that hormones secreted by the HPA and HPT axes control metabolism and appetite. Thyroid hormones have been demonstrated to enhance appetite, increase heat output, and decrease body weight, whereas glucocorticoids have been demonstrated to increase appetite and impact glucose homeostasis (Goel et al. 2023).

The effects of these hormones are confirmed by the fact that altered feeding behavior and/or disturbed metabolism are hallmarks of human disorders linked to malfunction of the HPA axis (e.g., Cushing syndrome) and the HPT axis (e.g., Graves' disease). Furthermore, it has been observed that CRH and thyrotropin-releasing hormone (TRH) regulate metabolism and appetite (Yoo et al. 2021). These results demonstrate the possible role of the HPA and HPT axes, as well as hypothalamic CRH and TRH neurons, in controlling energy balance and metabolism.

Certain neurons that express neurohormones, including CRH and TRH neurons, do not project to the median eminence these neurons are not all "neuroendocrine" cells. Thus, neurons that project to the median eminence and release neurohormones are the only neurons that can be referred to as "neuroendocrine cells." Non-neuroendocrine cells and neuroendocrine cells, both come under the umbrella of neurohormone expressing cells (Yoo et al. 2021).

The hypothalamus's neurohormone-expressing cells which are non-neuroendocrinic make up a neural circuitry that makes use of traditional neurotransmitters. For instance, prior research has shown that orexigenic (appetite-promoting) agouti-related peptide (AgRP) neurons in the arcuate nucleus of the hypothalamus (ARH) make glutamatergic connections with TRH neurons in the PVH, and that activation of these neurons improves appetite. Additionally, neurons of the LHA, which is said to be implicated in stress responses, receive glutamatergic fibers from PVH CRH neurons. To control fluid intake, it has also been demonstrated that OXT neurons in the PVH send glutamatergic projections to the lateral parabrachial nucleus (PBN). To fully comprehend the function of hypothalamic neurohormone-expressing cells, it is crucial to consider both the neuroendocrine axis and the neuronal (non-neuroendocrine) circuitry (Yoo et al. 2021).

As a response to micronutrients in the meal and the byproducts of their digestion the cell processes lead to the release of gut hormones, with an emphasis on the "incretin" hormones, glucose-dependent insulinotropic polypeptide (GIP) and glucagon-like peptide-1 (GLP-1). It is now understood that a variety of processes are involved in the production of gut hormones triggered by nutrients. New targets for the treatment of disordered eating are probably going to be found because of a better knowledge of these systems. Vagus nerves can transmit signals related to meal and signals from gastrointestinal luminal to the brain for further processing. Evidence suggests that endocannabinoids interact directly or indirectly with vagal afferents along the gut-brain axis for transmitting the signals and play its modulatory role by this (Fig. 4.1) (Page and Li 2018).

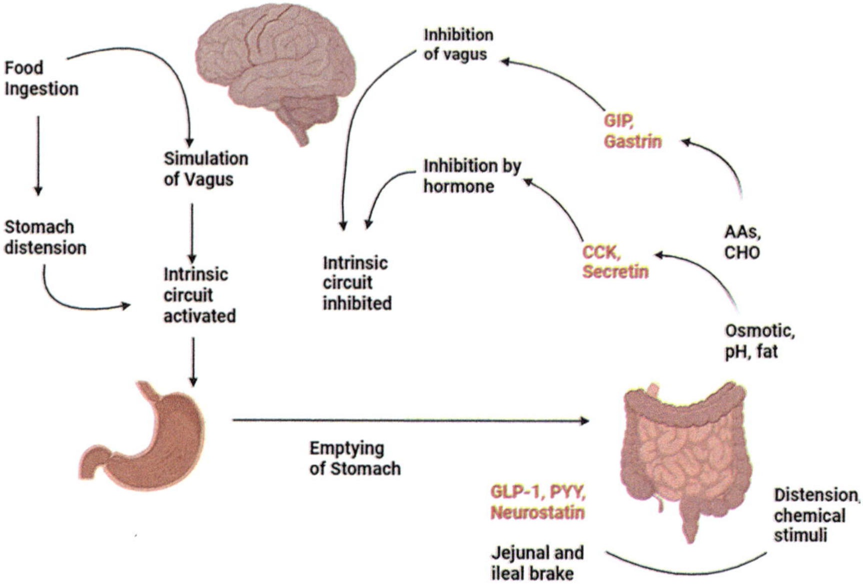

Fig. 4.1 Neural mechanism of appetite regulation

Endocannabinoids mediate nutrient-induced secretion of gut hormones from enteroendocrine cells and further activate, directly or indirectly, the cannabinoid receptors on vagal afferents. Evidence suggests that disruptions in the gut microbiota can lead to various dysfunctions, including obesity and type 2 diabetes. Furthermore, growing evidence indicates that changes in the composition of microbiota, resulting from dietary changes, can impact body weight and eating habits. This process, influenced by altered microbiota, is mediated by gut hormone release and vagus nerve signaling, which are key pathways through which microbiota exert their effects (Longo et al. 2023). Rather than metabolic needs, making them potentially significant for controlling food intake there are ways in which the gut hormones ghrelin and GLP-1 influence behaviors that are driven by food rewards, with an emphasis on how the two hormones have conflicting impacts on several behavioral variables associated with food reward and reinforcement. Through the nucleus of the solitary tract, sensory data from the GI tract is sent to the central nervous system, where it is combined with inputs from the midbrain, forebrain, and other brainstem nuclei. The dorsal motor nucleus of the vagus receives this information and coordinates feedback signals to the periphery (Decarie-Spain and Kanoski 2021).

Food's appearance, taste, texture, and smell are all subjective sensory experiences that are known to have a significant impact on how much energy is consumed. The impact of orosensory elements on appetite and energy intake during food ingestion and mastication, such as food texture and palatability, as well as the primary neurophysiological processes that mediate reactions to orosensory exposure. Food intake is driven by the hedonic experience, which encompasses mood changes and feelings of satisfaction. These factors are fundamental to a positive postprandial experience (Livovsky and Azpiroz 2021).

In the mouth cavity and throughout the gastrointestinal system lumen, the composition of a meal is sensed. Fat, umami, kokumi, and carbohydrates are among the other tastes that have been shown to exist in addition to the four fundamental ones of sweet, bitter, salty, and sour. The stomach is an important organ for acute management of food intake because it stores and receptacles food, typically for several hours after consumption. A meal causes the stomach to expand, sending a signal of fullness. After emptying of the stomach, the nutrients flow to the small intestine which lead to the absorption of these nutrients and release of different gut hormones for further digestion (Goyal et al. 2019). Dietary nutrients appear to have region-specific effects once they are in the small intestinal lumen. One of the research elaborate the process known as 'ileal brake' which is actually the slow stomach emptying when nutrients are delivered into the ileum. This concept of intestinal brake was found to affect the calorie intake and to lower hunger (Goyal et al. 2019). Dysregulations in the systems governing energy intake and balance are hallmarks of obesity. Obesity alters gut hormone secretion, which reflects altered gastrointestinal nutrient receptor function and the impact of various weight loss techniques. "Healthy" aging is linked to a decrease in desire to eat and hunger, which makes people more vulnerable to malnutrition than obesity. The gastrointestinal processes that control appetite and energy intake alter with age. Chapman and colleagues summarize these changes and discuss the consequences for using dietary supplements, especially those high in protein, to meet nutritional needs and prevent age-related muscle loss (Feinle-Bisset and Horowitz 2021).

4.2.2 Neurotransmitters and Neuropeptides

The brain's hypothalamus is highly integrated with feeding control and fuel expenditure signals. The brain's hypothalamus produces coordinated feedback for body signals, like metabolic and hormonal, nutritional input, peripheral circulation input, and central neural input. It also coordinates the ARC (arcuate nucleus) activity which involves in suppressing appetite via transcripts like cocaine- and amphetamine-regulated transcript (CART) and proopiomelanocortin (POMC), and also stimulates appetite via neuropeptide Y (NPY) and agouti-related protein (AgRP), and arcuate nucleus lesions lead to obesity and overweight (Slutzki 2023). Energy expenditure and food intake is controlled by five hypothalamic nuclei which are the dorsomedial nucleus, paraventricular nucleus, ventromedial nucleus, lateral hypothalamic area, and arcuate nucleus. The arcuate nucleus regulates eating activity by activating oxytocin and corticotropin-releasing hormone after receiving neuronal input from the paraventricular nucleus and lateral hypothalamic area. After activated by ghrelin, the AgRP and NPY are further involved in the start of the eating process. While hormones that produce α-MSH promote fullness, which is brought on by POMC. The 241 amino acid residues that make up pro-opiomelanocortin (POMC) work through

MC1 R through MC5 R receptors to reduce food intake by activating melanocortin peptides such as α, β, and γ-melanocyte-stimulating hormone (MSH) and ACTH. The pathogenic action of POMCs and MC4 receptor mutation may contribute to obesity (Slutzki 2023).

The details of neuropeptides described in Table 4.1 are detailed below:

Neurotensin (NT)

NT is a peptide of 13 amino acid which is found in GIT and the brain. It activates neurotensin receptors and helps in suppression of food intake. It also plays a significant role in different processes of pituitary hormones secretion, thermoregulation, and dopaminergic transmission modulation (Barchetta et al. 2022).

Corticotropin-releasing hormone (CRH)

Corticotropin-releasing hormone (CRH) is produced by the paraventricular nucleus and suppresses food intake. Corticotropin Releasing Hormone administration reduces stomach acid secretion and gastric emptying while increasing colonic motility, which causes the bowels to empty. In reaction to stress, the hypothalamic medial parvocellular paraventricular nucleus, which controls acute hunger, produces CRH. Information from type 1 and type 2 CRH to the hypothalamic ARC suppresses the production of NPY/AgRP neurons, which in turn stimulates appetite (Huang et al. 2024).

NPY, or neuropeptide Y

In the paraventricular nucleus, the central nervous system (CNS), and outside the CNS, primarily in the gastrointestinal tract (GIT), NPY, a 36-amino acid peptide, is extremely prevalent. NPY shares 50% homology with the pancreatic polypeptide and 70% with peptide YY (PYY). A crucial function in regulating appetite and

Table 4.1 Neuropeptides involved in regulating appetite

Neuropeptide/Hormone	Function	Reference
Neurotensin (NT)	Activates neurotensin receptors to suppress food intake	Barchetta et al. (2022)
Corticotropin-Releasing Hormone (CRH)	Help in reducing appetite	Huang et al. (2024)
Neuropeptide Y (NPY)	Regulates appetite and energy balance by enhancing food intake	Asuku et al. (2024)
Agouti-Related Peptide (AgRP)	Increases appetite while reducing metabolism and energy expenditure	Bruschetta et al. (2018)
Orexins (Hypocretins)	Promoting appetite	Mogavero et al. (2023)
Galanin (Gal)	Promotes eating	Marcos and Coveñas (2021)
Melanin-Concentrating Hormone (MCH)	Regulate appetite, obesity, and insulin resistance	Prida et al. (2022)
Endocannabinoids (ECS)	Activate CB1 receptors to promote food intake and energy balance	Kurtov et al. (2024)

energy balance, NPY is activated by neuropeptide 1 and neuropeptide 5 receptors to modify food intake, enhance appetite, and promote eating (Asuku et al. 2024).

Agouti-related peptides (AgRP)

Peptides associated with Agouti (AgRP) AgRP/NPY neurons in the brain generate the neuropeptide AgRP. The gene produces 132 agouti-like amino acid peptides. It is a crucial regulator of melanocortin signaling. Neuropeptide Y and the neurotransmitter gammaaminobutyric acid are both co-exhibited by AgRP neurons, which are in the hypothalamus. When injected centrally, this strong orexigenic peptide enhances hunger. AgRP increases appetite while suppressing metabolism and energy expenditure (Bruschetta et al. 2018).

Orexins

Orexins are peptide hormones, namely Orexin A/hypocretin 1 and Orexin B/hypocretin 2. (OX B/Hcrtr2) is made up of 28 amino acids and is activated by both Orexin/hypocretin-1 and Orexin/hypocretin-2 receptors. OX A/Hcrtr1 constitute of 33 amino acids have two chains linked via di-sulfide bond. These hormones were initially shown to regulate hunger centers, but they are now also engaged in modulating the sleep/wakefulness cycle and feeding behavior. The Orixin 2 receptor has identical affinity for both neuropeptides, however the Orexin/hypocretin-1 receptor greatly increases an activity. To maintain a lengthy, cohesive waking time, orexins excite orexin neurons, which function in the central nervous system as monoamine and Ach neurotransmitters. Additionally, orexins are involved in reward systems and drug addiction pathways. Research has shown that orexin-A decreases the behaviorally normal satiety region, which in turn promotes appetite (Mogavero et al. 2023).

Galanin (Gal)

A 29-amino acid peptide is released via the stomach, spinal cord, dorsal raphe nucleus, and locus coeruleus, known as galanin. In satiated rats, galanin, either given intraventricularly or through the hypothalamus, promotes eating. Galanin regulates pituitary hormone secretion, affects energy balance, neuropathic pain, alcohol intake, and decreased insulin activity. Molecular cloning approaches have so far identified three distinct types of galanin receptors (GalR1, GalR2, and GalR3). Research has indicated that galanin is involved in the regulation of body weight and feeding (Marcos and Coveñas 2021).

Level of Melanin Concentrating Hormone (MCH)

This peptide constitutes of 19 amino acid. It interacts with MCH-1, and MCH-2 receptor types and it is produced by the lateral hypothalamus. MCH neurons are directly linked to orexin neurons through their cannabinoid CB1 and OX1-R receptors. In contrast to MCH-R1, which is widely distributed throughout the human brain, MCH-R2 is expressed in the hippocampus and amygdala. While central MCH injection boosts appetite, its overexpression led to obesity and insulin resistance. MCH is less effective than NPY, however it is just as potent as orexins. MCH is impacted by dietary limitations. Despite not being a glucose-sensitive hormone, leptin inhibits the release of MCH, and fasting affects its secretion (Prida et al. 2022).

ECS, or endocannabinoids

Endocannabinoids are eicosanoid molecules that are produced by arachidonic acid. By interacting with the brain's cannabinoid1 receptors, it promotes eating and digestive habits. After delivery, CB1 modulators have a role in the motor regulation of breastfeeding behavior, indicating the presence of endocannabinoids in the developing newborn brain. Consequently, it plays a role in controlling hunger and energy expenditure (Kurtov et al. 2024).

4.3 Cognitive and Emotional Influences on Food Choices

Throughout their lives, people encounter and enjoy a wide variety of foods and drinks, which are referred to as "foods". Food-related behavior, including food choice and affective responses to foods, has been thought to be influenced by certain emotions. Food-evoked emotions are a more accurate predictor of a person's food preference than liking scores.

A food choice model based on hedonic scales gains predictive value from the emotions of the consumer. Beyond conventional sensory and acceptability assessments, research suggests that measuring emotional reactions to foods may uncover previously unknown product features, which could be a valuable source of data for product development and marketing. Consequently, it is critical to acquire accurate and trustworthy assessments of food-evoked emotions (Prinyawiwatkul 2023).

Emotion is defined differently by various researchers, despite its significance. Some define emotions as a mental state of readiness resulting from cognitive appraisals of events or thoughts, while others describe them as short-term affective responses to stimuli. Another perspective considers emotions as brief, intense physiological and mental reactions, whereas others propose that emotions are any mental experiences with high intensity and hedonic content (pleasure/displeasure). These definitions incorporate internal and external, top-down and bottom-up, physiological and cognitive components, emphasizing that no single metric can fully capture all relevant factors (Prinyawiwatkul 2023).

Researchers have suggested the development of implicit measurement tools and integrating them with explicit measurements when practical. Examples of such methods include physiological indicators that track autonomic nervous system activity and facial expression recognition for emotions like happiness, sadness, anger, surprise, fear, and disgust. However, there is no universally accepted standard for evaluating food-evoked emotions (Prinyawiwatkul 2023).

Emotional eating refers to the tendency to eat in response to emotional cues rather than physiological hunger. It is a concern for individuals across different weight categories but is particularly problematic for those who are overweight. A significant portion of overweight or obese individuals report engaging in excessive emotional eating. Research indicates that emotional eating is associated with increased body fat, waist circumference, and body mass index (BMI). Individuals with higher levels of emotional eating are significantly more likely to be overweight or obese than those with lower levels. Additionally, emotional eating has been linked to difficulty in losing weight and long-term weight gain (Frayn and Knäuper 2018).

Emotional eaters tend to eat more frequently, consume more sweet and high-fat foods, and eat in response to stress. Research on emotional eating has often focused on general emotions rather than specific ones, despite its negative effects. Moreover, most studies do not consider positive emotions as potential triggers for eating. Understanding the relationships between different forms of emotional eating and health outcomes can enhance the conceptualization of emotional eating behavior (Frayn and Knäuper 2018).

A study by Braden et al. (2018) investigated the connections between physical health variables and psychological factors related to emotional eating. These include eating in response to anxiety/anger (EE-A), depression (EE-D), boredom (EE-B), and positive emotions (EE-P). Correlational analyses indicated that EE-D, EE-A, and EE-B were linked to higher eating disorder symptoms, greater emotional regulation difficulties, and lower psychological well-being. However, EE-P did not show a significant correlation with these outcomes.

Regression analyses revealed that EE-D was the strongest predictor of eating disorder symptoms, emotional regulation issues, and psychological well-being. Further exploratory analyses indicated that aspects of emotion regulation were linked to EE-D, EE-A, and EE-B, as well as specific disordered eating symptoms. The findings suggest that different forms of emotional eating exhibit unique patterns in relation to psychological outcomes (Braden et al. 2018).

4.4 Food Safety and Its Neural Impact

The amygdala and prefrontal cortex come together to evaluate many food safety issues, and in this process, they integrate decision-making skills, learned experiences, and emotional reactions to come on a final decision. These two parts of the brain, the amygdala and prefrontal cortex, also decide whether a meal should be eaten or not. Different emotions like disgust, fear, and aversion also play a crucial role in evaluating food risk, and these emotions are primarily handled by the amygdala, specifically its limbic system. Certain sensory clues, such as a bitter taste, odd color, or unpleasant odor of food, can help identify potential risks of contamination or toxicity in food. When all these processes come together, they can make a person refrain from eating that food. This process further assists in developing associative memories linked to foodborne illnesses. For instance, if a particular type of food has made you ill in the past, you tend to memorize it and are compelled to avoid that food in the future to keep yourself safe (Šimić et al. 2021).

However, the prefrontal cortex uses logical thinking, prior experiences, and societal standards for risk assessment, logical decision-making, and sense of self-control when it comes to deciding whether to eat a particular food or not. Sometimes, it does a good job than amygdala and enables a person to identify strange or unfamiliar foods by using learned or cultural cues (Küçükay 2018). So, we can say both amygdala and prefrontal cortex work together when question comes about food safety. On one side PFC assesses the situation by confirming the risk while amygdala help to give reaction to any such hazard. Other than these two, brain stem also play its role and lead to some physiological reactions like vomiting or nausea which help to flush out any kind of hazardous substance or toxic compound if diet seems poisonous. So, we can say these portions of the brain play its role in assessing food risks and give better reaction of it (Küçükay 2018).

4.4.1 Food Safety Warnings and Decision-Making

The olfactory brain, which detects smells, and the gustatory cortex, which perceives taste, are the first organs involved in sensory processing and risk detection; these organs help in spotting possible contamination or spoiling; the visual cortex also plays a role by interpreting visual clues like mildew, discoloration, or food labels with expiration dates; and even before conscious decision-making takes place, the brain may immediately initiate avoidance behavior in response to these sensory cues that signify danger. Neural systems that integrate memory, emotional reactions, cognitive evaluation, and sensory processing are integral to the decision-making process regarding food safety (De Korte 2022).

Cognitive assessment and logical decision-making (PFC) are primarily handled by the prefrontal cortex, which weighs the benefits and risks of various external food safety warnings, such as government advisories, food labels, and expiration dates, to determine whether a food is dangerous or not. In some cases, if rational information suggests that the food is still safe to eat, the PFC takes precedence over automatic responses. Strong and trustworthy food safety information can help counteract learned behaviors or unconscious biases, resulting in more informed decisions (Steward et al. 2019).

The amygdala plays a critical role in evoking feelings of disgust or dread when there is a perceived risk to food safety, such as salmonella infection. The insular cortex also processes disgust, which increases food aversion and discourages people from eating potentially dangerous foods. These emotional responses result in strong avoidance behaviors that protect against foodborne diseases (Breslin 2019).

Long-term food safety procedures also depend on memory and learning. The hippocampus stores information about previous foodborne illness experiences, which promotes cautious decision-making when comparable risks arise in the future. People who have experienced food illness in the past are more likely to be cautious going forward, often avoiding similar items even when the danger is minimal. The brain effectively regulates food safety decisions by fusing memory, emotional reactions, logical analysis, and sensory detection, helping people to avoid possible health risks. The striatum influences habitual actions, such as routinely checking expiration dates before buying food (Jiao et al. 2024).

4.4.2 *Cognitive Bias in Risk Perception and Avoidance Behavior*

A person's cognitive biases also help him out to decide whether to eat a potentially dangerous food or not. But there is a problem with it. It can overestimate or underestimate the hazard. One of the biases is optimism bias, in which a person keeps a belief that food-related illness 'won't happen to me', even after clear warnings regarding that food. On the other hand, one of the kinds is negativity bias, in which a person only prioritizes their bad encounters without regarding good ones. Like if a person got food poisoning due to a specific meal, then he or she may permanently avoid that food, even if the danger is minimal. This can lead to needless limitations on diet (EFSA et al. 2021).

Some people take food safety decisions based on past events regarding a particular food. For example, the case of *E. coli* contamination in lettuce led people to become unreasonably afraid of lettuce. Just like that, confirmation bias also pushes people to ignore contradicting data about a particular food, selectively focusing on that information that supports their preconceived notions. For example, if someone believes organic food is always safer, then even if its contamination occurs, he or she would ignore that news. Even the status quo bias also causes resistance in people to a particular change. For example, if some community eats undercooked meat due to their cultural practices, then despite many health warnings, they will continue to eat that undercooked meat. So, due to optimism bias or status quo bias, these people indulge in unhealthy or unsafe behavior, despite having numerous warnings (EFSA et al. 2021).

There are many ways which can help in taking decisions about food safety by lessening the influence of these biases. We can raise awareness and educate people and give scientifically supported information to them to deal with all the misinformation. Furthermore, behavioral nudges like labeling the food with some readable content or some visual cues can help to overcome cognitive biases. All these steps can help people to make more logical decisions about food safety (Hellstrøm 2021).

4.5 Neuroscience of Food Choices in the Context of Public Health

4.5.1 *Public Health Campaigns and Neural Impacts*

We can go for public health campaigns to raise awareness among people, and in these campaigns, health workers use both psychological and neural mechanisms to make it successful. They use attention-grabbing cues, habit-building, emotional engagement of people, and cognitive reinforcements. The effectiveness of these campaigns will be dependent on neural mechanisms, like making your message clear by using vivid colors, warning labels or bold fonts (Table 4.2). There is an old

Table 4.2 Application of neuroscience principles in food safety practices

Neuroscience principle	Application in food safety practices
Attention & Engagement (Prefrontal Cortex)	Use bold warning labels and attention-grabbing visuals to highlight food risks.
Memory & Retention (Hippocampus)	Implement repetitive messaging, mnemonics, and step-by-step training to reinforce food safety habits.
Emotional Processing (Amygdala)	Utilize fear-based messaging (moderate levels) and real-life foodborne illness stories to increase risk perception.
Risk & Reward Processing (Striatum)	Pair food safety behaviors with positive reinforcement (e.g., rewards, incentives) to encourage compliance.
Habit Formation (Basal Ganglia)	Embed food safety habits into daily routines through visual reminders, checklists, and nudges

"Clean, Cook, Chill, Separate" campaign. In this campaign, the auditory cortex helps in transmitting the message via memorable slogans, social media videos or radio commercials. One of the main factors in this is emotional involvement, where our amygdala is involved. You can go for visual or imagery messages in it, like for elaborating or triggering disgust emotions, you can use images of bacteria or uncleaned hands (Kasza et al. 2022).

Health workers can also build up habits or memories in people via these campaigns, like putting up many hygiene posters in kitchens or supermarkets, which give signals to the hippocampus and improve long-term memory regarding food safety information among them. Giving reminders of hand washing or checking the expiration date of food, all of this can help to make long-term behavioral changes due to these campaigns, which show the same message repetitively at different venues (Gardner and Rebar 2019).

Social influences and behavior norms can also play a role in shaping food safety practices. Like in this era of social media, when people observe others following specific eating habits, they are more inclined to follow them. But there are some communities in which their culture has some norms that are contradicted with food safety guidelines. So, to target that kind of population during a campaign, you must align your message with their cultural values. For maximum effectiveness, wisely moderating fear or disgust, like emotional triggers, can be proved effective, but there is also an opposite case, if you spread excessive negativity, it may backfire. You should choose that kind of message which may seem relatable or personal to them. You can seek help from community leaders, influencers, or celebrities to influence such a kind of population during campaigning. Reputation and reinforcement should be top priority if you want long-term behavioral change in them. In reminders, you can also go for putting signs beside things, like encouraging people to wash their hands, as a powerful reinforcement for food safety practice. All this can lessen the incidence of food-borne illnesses (Bass et al. 2022).

4.5.2 Neuroscientific Principles for Food Safety Education

Neuroscience-based techniques can be used to improve education regarding food safety, and this can help in making memory, emotional engagement, and habit development in the population. This kind of training program can help to produce long-lasting behavioral changes. This section explains the techniques for food safety education with special reference to neuroscientific principles. To gauge people's attention and involvement, you can take help from virtual reality simulations and/or videos to engage the population. By keeping the stories of real-life food-borne illnesses instances as an example to emotionally engage the brain and seek attention towards that topic (Li et al. 2020).

You should avoid information overloading and deliver brief or targeted instructions for more effectiveness. Like, you can also incorporate gamified training modules instead of lengthy lectures. Retention or recalling of memory are the key elements for long-term learning. Memory occurs in the hippocampus, which can be strengthened via repetition of that message. You can use acronyms, such as "clean, separate, cook, chill" as a helpful tool for improving memory. You can also incorporate a kind of training that provides weekly text reminders on food hygiene, which can improve long-term retention (Varman et al. 2021).

Emotional involvement and risk awareness in the population are two main elements in food safety education. Our amygdala plays a significant role in feelings like fear and disgust, which can enhance the memory of emotion-dependant events. You can also use visually elaborated narratives or pictures to raise awareness about improper food handling hazards. The brain's reward system can also work in this kind of education, like complimenting and encouraging appropriate behavior (Varman et al. 2021).

Habit development and behavioral modification include some components like dopamine, which is necessary to reinforce the behavior, while the striatum and basal ganglia govern habits. We can put visual cues near the sink regarding hand-washing practices where food is prepared (Clark 2019).

4.5.3 Interventions to Influence Safe Food Choices

People's resistance to change and how humans make decisions are crucial topics studied in consumer research, particularly regarding healthy and ecologically sound food choices. To prevent metabolic diseases like obesity or type 2 diabetes, it's essential to understand why people choose unhealthy food and what encourages them to make healthier choices. A hypothesis, formed through numerous disciplines, suggests that dietary choices are influenced by environmental, social, and personal factors (Peeters 2018).

Policies should be made that encourage healthy food choices. Target audience variations in age, lifestyle choices, health values and socioeconomic standards should be considered while making policies. And check how these elements interact, and which would be the best strategy, both politically and socially acceptable, should be used as foundation while making policies. Recently different studies have spotted behavior-influencing policy instruments or "nudges" which help the consumers to make better food choices (Capoferri 2018).

You can improve policy efficacy via three primary ways, firstly making consumer information more accessible, relevant, or visible. Secondly, present adoptable policy instruments like healthy defaults in public canteens, and thirdly, support and adopt an empirically grounded approach to policy making that uses pilot studies and experiments to test, learn, and modify policies. Involvement of organizations and government can be helpful to develop more successful plans, which further help in promoting better eating habits by incorporating behavioural insights into the creation of policies (Capoferri 2018).

Behavioral therapies should be given to people to make them more conscious of their hunger and satiety cues. Both behavioral and neurological interventions should be combined to create and promote safe eating habits. This can lower the risk associated with unhealthy eating habits. CBT, cognitive behavioral therapy, can be helpful for the people who have a problem of emotional eating or binge eating disorder. Because this therapy will help in identifying and changing this problem. Nutritional literacy could be promoted by education and awareness initiatives at schools and workplaces (Silva et al. 2023).

Neurofeedback training and transcranial magnetic stimulation (TMS) can also be found helpful in enhancing self-control in overeating habits and help in managing the cravings of a person. Suppressants and drugs can also be found helpful by affecting the dopamine and serotonin pathways, these will further help in weight management. A new field of study, gut-brain axis, can also be modulated via gut microbes, so we can use probiotics and prebiotics in this case (Noël et al. 2019).

To lessen triggers of unhealthy eating and improve self-regulation, two digital tools can be used for it, one is AI-assisted brain training and other is virtual reality exposure treatment. These tools will be helpful in food-related health problems, eating disorders, and can also help people to lower the risk of obesity by integrating behavioral and neurological therapies. To can gain effective and long-lasting results, the best approach is to use a multi-faceted approach that incorporates psychological assistance, legislative nudges, education, awareness, and developments in neuroscientific research (Noël et al. 2019).

4.6 Future of Food Safety and Neural Research

4.6.1 Emerging Technologies and Tools

Deep learning is a significant branch of artificial intelligence that has demonstrated enormous promise in handling complex data, enabling automated analysis, and generating precise forecasts. It is therefore gradually becoming a crucial tool for determining the safety of food. Data mining is a component of deep learning technology that increases efficiency and speed. Additionally, it automatically extracts unique and desirable properties from vast datasets, enabling more accurate risk prediction and food safety issue detection than older methods. Among deep learning architectures, convolutional neural networks (CNNs) and recurrent neural networks (RNNs) are two popular models that excel at processing a variety of input types (Banerjee et al. 2019). CNNs' remarkable capacity to process image data and extract image attributes makes them very useful for identifying concerns linked to food safety in the food industry. Because CNNs can efficiently extract multi-level data, such as color, shape, and texture, from food images, they are helpful in scenarios like adulteration analysis, surface defect detection, and food appearance inspection. For instance, by evaluating the texture of fruit surfaces, CNNs can rapidly identify sections that are deteriorating or automatically identify defective goods on food packing lines. Additionally, when combined with super-resolution imaging technology, CNNs can precisely evaluate minute contaminants, significantly improving the detection accuracy of food safety (Banerjee et al. 2019).

RNN is helpful in handling time series data for risk evaluation in food production and distribution, it analyses temporal changes like temperature and humidity. It will be helpful in monitoring transportation conditions in cold chain logistics to prevent food spoilage, so it can predict the risk of contamination in food production. During food production, one of the variants of RNN is LSTM, Long Short-Term Memory network, which captures long-term dependencies, this makes it ideal for complex food safety data prediction tasks (Banerjee et al. 2019).

By combining their technological advantages in picture identification and time series prediction, CNNs and RNNs can provide a complementary deep learning framework for food safety. CNNs' multi-feature recognition capabilities enable efficient food component and appearance identification, including raw material sorting, spoiling inspection, and fault screening. Meanwhile, the time series prediction capabilities of RNNs significantly improve dynamic risk warning during food processing. CNNs and RNNs collaborate in the food supply chain to improve operational efficiency and ensure its safety by utilizing their individual strengths (Ding et al. 2025).

Neuroimaging and electrophysiology techniques

Scientists may learn more about how the brain interprets food choices, risk perception, and food safety issues by doing neuroscientific research. These techniques offer important insights that enhance consumer behavior and food safety education by looking at decision-making, emotional reactions, and sensory processing. Because they provide insight into how individuals evaluate nutritional value, identify dietary hazards, and respond to food safety signals, a variety of neuroimaging techniques are essential in food choice research. Researchers can detect cerebral activity during food-related decision-making by using Functional Magnetic Resonance Imaging (fMRI), which detects changes in blood flow in the brain (Mostafa 2025).

Although studies have shown that the prefrontal cortex (PFC) plays a part in determining the nutritional worth and safety of food, stimuli associated with harmful or disgusting meals cause the insula and amygdala to become activated.

Moreover, the striatum, an important reward area, responds to meals that are visually appealing or high in calories, even when they are harmful. Electroencephalography (EEG), which uses electrical impulses, provides real-time information on how the brain functions. According to EEG studies, alpha suppression is more noticeable when people watch images of contaminated food, whereas theta waves increase when people evaluate food hazards. Hazards to food safety are discovered more immediately when P300 responses are quicker. Other techniques like magnetoencephalography (MEG) and near-infrared spectroscopy (fNIRS) help us better understand how the brain reacts to food safety cues in real-world scenarios (Peixoto 2024).

These tactics significantly advance food safety research by determining how the brain processes risks, enhancing consumer education, refining public health message, and enabling customized treatments based on individual risk sensitivity. However, as neuroscience advances, the ability to influence food choices through brain stimulation, behavioral nudging, and neuromarketing also raises ethical concerns of fairness, autonomy, consent, and even abuse (Goncalves et al. 2024). Food interventions should be open, voluntary, and transparent so that people can make wise decisions, according to ethical principles. Since many brain-based food therapies work subconsciously, informed permission and transparency are also essential. When neural approaches are used in food marketing or public health initiatives, consumers should be properly informed and given the opportunity to opt out. Autonomy and free will are important ethical considerations. If brain interventions have a subconscious impact, people may lose control over their eating choices. For example, suppressing appetites for unhealthy meals with transcranial magnetic stimulation (TMS) may alter decisions without the user's knowledge (Goncalves et al. 2024). Fairness and accessibility provide more challenges. If only wealthy individuals or companies have access to nutritional interventions based on neuroscience, health inequities may arise. For instance, food companies with neurotechnology resources may create delicious, processed meals, while public health groups lack the tools to promote better choices (Grasty 2023).

References

Anzman-Frasca S, Ventura AK, Ehrenberg S, Myers KP (2018) Promoting healthy food preferences from the start: a narrative review of food preference learning from the prenatal period through early childhood. Obes Rev 19(4):576–604

Asuku AO, Ayinla MT, Ajibare AJ (2024) Central nervous system regulation of eating and brain functions. In: Nutrition and psychiatric disorders: an evidence-based approach to understanding the diet-brain connection. Springer Nature Singapore, Singapore, pp 69–87

Banerjee I, Ling Y, Chen MC, Hasan SA, Langlotz CP, Moradzadeh N et al (2019) Comparative effectiveness of convolutional neural network (CNN) and recurrent neural network (RNN) architectures for radiology text report classification. Artif Intell Med 97:79–88

Barchetta I, Baroni MG, Melander O, Cavallo MG (2022) New insights in the control of fat homeostasis: the role of neurotensin. Int J Mol Sci 23(4):2209

Bass SB, Brajuha J, Kelly PJ, D'Avanzo P, Lambertini E, Nordhagen S, Monterrosa EC (2022) Changing behavior, attitudes, and beliefs about food safety: a scoping review of interventions across the world and implications for empowering consumers. Foodborne Pathog Dis 19(1):19–30
Boesveldt S, Bobowski N, McCrickerd K, Maître I, Sulmont-Rossé C, Forde CG (2018) The changing role of the senses in food choice and food intake across the lifespan. Food Qual Prefer 68:80–89
Braden A, Musher-Eizenman D, Watford T, Emley E (2018) Eating when depressed, anxious, bored, or happy: are emotional eating types associated with unique psychological and physical health correlates? Appetite 125:410–417
Breslin PA (2019) Chemical senses in feeding, belonging, and surviving: or, are you going to eat that? Cambridge University Press
Bruschetta G, Jin S, Kim JD, Diano S (2018) Prolyl carboxypeptidase in Agouti-related Peptide neurons modulates food intake and body weight. Mol Metab 10:28–38
Capoferri VL (2018) Automaticity in decision-making and food consumption: a nudge intervention in a hospital
Chen PJ, Antonelli M (2020) Conceptual models of food choice: influential factors related to foods, individual differences, and society. Foods 9(12):1898
Clark J (2019) Reducing the risk: psychological and technological approaches for improving handwashing practices in the foodservice industry. University of Arkansas
De Korte EM (2022) Time will tell: material surface cues for the visual perception of material ageing insights from psychophysics, online experiments, image processing and a science festival. Doctoral dissertation. University of Bradford
Decarie-Spain L, Kanoski SE (2021) Ghrelin and glucagon-like peptide-1: a gut-brain axis battle for food reward. Nutrients 13(3):977
Ding H, Hou H, Wang L, Cui X, Yu W, Wilson DI (2025) Application of convolutional neural networks and recurrent neural networks in food safety. Foods 14(2):247
Enriquez JP, Archila-Godinez JC (2022) Social and cultural influences on food choices: a review. Crit Rev Food Sci Nutr 62(13):3698–3704
European Food Safety Authority (EFSA), Maxim L, Mazzocchi M, Van den Broucke S, Zollo F, Robinson T et al (2021) Technical assistance in the field of risk communication. EFSA J 19(4):e06574
Feinle-Bisset C, Horowitz M (2021) Appetite and satiety control—contribution of gut mechanisms. Nutrients 13(10):3635
Forde CG (2018) From perception to ingestion; the role of sensory properties in energy selection, eating behaviour and food intake. Food Qual Prefer 66:171–177
Frayn M, Knäuper B (2018) Emotional eating and weight in adults: a review. Curr Psychol 37:924–933
Gardner B, Rebar AL (2019) Habit formation and behavior change. In: Oxford research encyclopedia of psychology. Oxford University Press
Goel M, Mittal A, Jain VR, Bharadwaj A, Modi S, Ahuja G et al (2023) Integrative functions of the hypothalamus: linking cognition, emotion and physiology for well-being and adaptability. Ann Neurosci:09727531241255492
Goncalves M, Hu Y, Aliagas I, Cerdá LM (2024) Neuromarketing algorithms' consumer privacy and ethical considerations: challenges and opportunities. Cogent Bus Manag 11(1):2333063
Goyal RK, Guo Y, Mashimo H (2019) Advances in the physiology of gastric emptying. Neurogastroenterol Motil 31(4):e13546
Grasty K (2023) Edible learning: school food service programming and experiential education that promote sustainable agriculture and better nutritional health. Doctoral dissertation. San Jose State University
Grzymisławska M, Puch EA, Zawada A, Grzymisławski M (2020) Do nutritional behaviors depend on biological sex and cultural gender? Adv Clin Exp Med 29(1):165–172
Hellstrøm MCL (2021) Nudging: strategies and sustainability-label signifiers to influence potential sustainable food consumer behavior. Master's thesis. NTNU

Higgs S, Ruddock H (2020) Social influences on eating. In: Handbook of eating and drinking: interdisciplinary perspectives. Springer, pp 277–291

Huang IH, Schol J, Calder S, Gharibans AA, Van den Houte K, Verheyden A et al (2024) Effects of corticotropin-releasing hormone on gastric electrical activity and sensorimotor function in healthy volunteers: a double-blinded crossover study. Am J Physiol Gastrointest Liver Physiol 326(5):G622–G630

Jiao L, Ma M, He P, Geng X, Liu X, Liu F et al (2024) Brain-inspired learning, perception, and cognition: a comprehensive review. IEEE Trans Neural Netw Learn Syst 36(4):5921–5941

Kasza G, Csenki E, Szakos D, Izsó T (2022) The evolution of food safety risk communication: models and trends in the past and the future. Food Control 138:109025

Küçükay A (2018) Psychological and judicial factors influencing on decision making. Law Justice Rev 9(16)

Kurtov M, Rubinić I, Likić R (2024) The endocannabinoid system in appetite regulation and treatment of obesity. Pharmacol Res Perspect 12(5):e70009

Li L, Gow ADI, Zhou J (2020) The role of positive emotions in education: a neuroscience perspective. Mind Brain Educ 14(3):220–234

Livovsky DM, Azpiroz F (2021) Gastrointestinal contributions to the postprandial experience. Nutrients 13(3):893

Longo S, Rizza S, Federici M (2023) Microbiota-gut-brain axis: relationships among the vagus nerve, gut microbiota, obesity, and diabetes. Acta Diabetol 60(8):1007–1017

Marcos P, Coveñas R (2021) Neuropeptidergic control of feeding: focus on the galanin family of peptides. Int J Mol Sci 22(5):2544

Mogavero MP, Godos J, Grosso G, Caraci F, Ferri R (2023) Rethinking the role of orexin in the regulation of REM sleep and appetite. Nutrients 15(17):3679

Mostafa MM (2025) Functional Magnetic Resonance Imaging (fMRI) in food research: a three-decade retrospective bibliometric network analysis. J Culin Sci Technol 23(2):246–276

Noël X, Bechara A, Saeremans M, Kornreich C, Dousset C, Campanella S et al (2019) Addiction: brain and cognitive stimulation for better cognitive control and far beyond. In: Inhibitory control training-a multidisciplinary approach. IntechOpen

Paarlberg RL (2023) Food politics: what everyone needs to know. Oxford University Press

Page AJ, Li H (2018) Meal-sensing signaling pathways in functional dyspepsia. Front Syst Neurosci 12:10

Peeters A (2018) Obesity and the future of food policies that promote healthy diets. Nat Rev Endocrinol 14(7):430–437

Peixoto IDS (2024) Integrating neurofeedback and virtual reality to enhance exposure therapy for anxiety and food aversion. Doctoral dissertation

Prida E, Fernandez-Gonzalez S, Pena-Leon V, Perez-Lois R, Fernø J, Seoane LM et al (2022) Crosstalk between melanin concentrating hormone and endocrine factors: implications for obesity. Int J Mol Sci 23(5):2436

Prinyawiwatkul W (2023) Food-evoked emotion, product acceptance, food preference, food choice and consumption: some new perspectives. Foods 12(11):2095

Silva P, Araújo R, Lopes F, Ray S (2023) Nutrition and food literacy: framing the challenges to health communication. Nutrients 15(22):4708

Šimić G, Tkalčić M, Vukić V, Mulc D, Španić E, Šagud M et al (2021) Understanding emotions: origins and roles of the amygdala. Biomolecules 11(6):823

Slutzki T (2023) Cocaine-and amphetamine-regulated transcript neurons of the arcuate nucleus in the regulation of feeding behaviour. McGill University (Canada)

Steward T, Miranda-Olivos R, Soriano-Mas C, Fernández-Aranda F (2019) Neuroendocrinological mechanisms underlying impulsive and compulsive behaviors in obesity: a narrative review of fMRI studies. Rev Endocr Metab Disord 20:263–272

Varman SD, Cliff DP, Jones RA, Hammersley ML, Zhang Z, Charlton K, Kelly B (2021) Experiential learning interventions and healthy eating outcomes in children: a systematic literature review. Int J Environ Res Public Health 18(20):10824

Watts AG, Kanoski SE, Sanchez-Watts G, Langhans W (2022) The physiological control of eating: signals, neurons, and networks. Physiol Rev 102(2):689–813

Whitney EN, Rolfes SR, Crowe T, Walsh A (2019) Understanding nutrition. Cengage AU

Yoo ES, Yu J, Sohn JW (2021) Neuroendocrine control of appetite and metabolism. Exp Mol Med 53(4):505–516

Ziegler AM, Kasprzak CM, Mansouri TH, Gregory AM, Barich RA, Hatzinger LA et al (2021) An ecological perspective of food choice and eating autonomy among adolescents. Front Psychol 12:654139

Chapter 5
Neurophysiological Implications of Food Safety: Food Born Pathogens and Its Neurological Impact

Priti Nandi, Joysurya Kanthal, Surajit Hansda, and Semanti Ghosh

Abbreviations

15-AcDON	15-acetyldeoxynivalenol
3-AcDON	3-acetyldeoxynivalenol
BBB	Blood Brain Barrier
BEA	Beauvercin
BoNT	Botulinum Neurotoxin
CER	Cereulide
CNS	Central Nervous System
CSF	Cerebrospinal Fluid
DON	Deoxynivalenol
ENN	Enniatins
IBD	Inflammatory Bowel Disease
IL-1β	Interleukin-1 beta
IL-6	Interleukin-6
JAMs	Junctional Adhesion Molecules
LLO	Listeriolysin
lnc	Long non-coding RNA
LPS	Lipopolysaccharides
MON	Moniliformin

P. Nandi · J. Kanthal · S. Ghosh (✉)
Department of Biotechnology, School of Life Sciences, Swami Vivekananda University, Kolkata, India
e-mail: semantig@svu.ac.in

S. Hansda
Department of Pharmaceutical Sciences, Texas Tech University Health Sciences Centre, Lubbock, TX, USA

T. Sarkar et al. (eds.), *Physiological Perspectives on Food Safety: Exploring the Intersection of Health and Nutrition*, https://doi.org/10.1007/978-3-031-94582-3_5

ncRNA	Non-coding RNA
OMVs	Outer Membrane Vesicles
PI-IBS	Post-infectious irritable bowel syndrome
ROS	Reactive Oxygen Species
SAE	Sepsis-associated encephalopathy
SCFAs	Short Chain Fatty Acids
TJ	Tight Junction
TLRs	Toll Like Receptors
TNF- α	Tumor Necrosis Factor-alpha
VacA	Vacuolating cytotoxin A
ZO-1	Zonula occludend-1

5.1 Introduction

Foodborne infections are a dedicated worldwide public health risks which are caused by parasitic organism, bacteria, and viral pathogens (Tauxe et al. 2010). Recent research shows that viral infections may cause neuroinflammatory responses, which might result in long-term neurological disorders, though their immediate impacts are frequently restricted to the gastrointestinal system (Wouk et al. 2021). The primary method by which food-borne pathogens impact neuroinflammation, impacting immunological equilibrium and brain function, is the gut-brain axis (Nakhal et al. 2024). A variety of human diseases, including colon cancer, the metabolic syndrome, obesity, type 2 diabetes, asthma, inflammatory bowel disease, or IBD, cardiovascular disease, and neurodegenerative diseases, have been linked to the loss of balance caused on by modifications in the gut microbiota's structure restored on by dietary changes, antibiotics exposure, and an illness. A new study indicates that diseases cause modifications in the structure of the gut microbiota as well as inflammation changes in various kinds of organs and tissues, including brain cells (Zhu et al. 2020). A significant mechanism that occurs when food-borne pathogens influence neuroinflammation and impact immunological barriers regulation and function inside the brain is the gut-brain axis (Delaney and Hornig 2018). Increasing permeability in the gut (also referred to as "leaky gut") may occur from disruption carried on by food-borne pathogens (Camilleri et al. 2019), thereby rendering it probable for harmful bacteria and mediators of inflammation to enter the circulatory system and travel through to the brain. The metabolites of various foodborne bacteria lead to accumulation of inflammatory cytokines in brain tissue, causing Blood-Brain Barrier (BBB) disruption. Thus, these neurotransmitters can easily pass through the BBB (Haroon et al. 2017). Studies revealed that foodborne pathogens also influence the long-term illnesses. Hence, these pathogens toxic substances cause Brain abscesses, Meningoencephalitis, Alzheimer disease, Meningitis, Parkinson disease, Rhombencephalitis by raising the level of pro-inflammatory cytokines (Boyd et al. 2022; Disson and Lecuit 2012). The toxic methods that link foodborne infections to neuroinflammation and their potential association with neurodegenerative diseases are explored in this chapter (Shahba et al. 2021). In recent studies showed that foodborne pathogens derived metabolites such as neurotoxin and mycotoxin has been modifying BBB integrity and their implication in chronic illnesses. The prevention and limitation about foodborne pathogen mediated disease, has been discussed to get a better lifestyle through food intake.

5.2 Potential Mechanism of Pathogen Associated Neurotoxicity

A number of human illnesses may damage brain tissue, and infections caused by bacteria can lead to the central nervous system (CNS) in mortality and morbidity along with recurrent neurological disorders, even when antibiotics are employed (Taha 2020). The condition known as encephalopathy, or meningoencephalitis triggered by infectious agents have left survivors with permanent neurological consequences (Ludlow et al. 2016). These signs occur in disorders which spread through multiple routes, including gastrointestinal infections. Several streptococcal species have the ability of infecting the central nervous system directly resulting in encephalitis or meningitis. Toxins secreted by foodborne pathogens like *Clostridium botulinum* induced encephalopathy (Liu 2024).

5.2.1 Foodborne Pathogens and Their Neurological Impact

5.2.1.1 *Escherichia coli* and Neuroinflammation

The Shiga toxin is generated by certain *Escherichia Coli* (*E. Coli*) strains, particularly those that are enterohemorrhagic (EHEC), and it may travel through the intestinal wall and enter into the circulatory system (Detzner et al. 2022). Shiga toxins may lead to inflammatory and neurodegeneration by entering the blood-brain barrier (Goldstein et al. 2021). According to research experiments, inflammatory carried on by pathogens activates the nervous system's immune system cells known as microglia, which leads to chronic oxidative stress and damage to the neurons (Vida et al. 2014).

5.2.1.2 *Salmonella enterica* and Brain Inflammation

Since Salmonella infections primarily impact the intestinal tract, it may spread over the body and causing meningitis, especially to individuals who have weakened immune systems. Toll-like receptors (TLRs), in the nervous system have been stimulated by bacterial lipopolysaccharides (LPS) (Sima et al. 2024), which promotes the generation of cytokines that promote inflammation such IL-1β, TNF-α, and IL-6 (Li et al. 2021). Both of neurodegenerative disorders and cognitive decline can arise from this inflammation cascade.

5.2.1.3 *Listeria monocytogenes* and Meningitis

Among the more significant diseases caused by food that affects the central nervous system (CNS) is listeriosis, or which is brought via *L. monocytogenes*. For the purpose to trigger encephalitis and meningitis, the bacteria can penetrate the epithelium of the intestine, reach the circulation into the body, and cross the blood-brain barrier. Neurological degeneration and chronic neuroinflammation may result from *L. monocytogenes'* potent immune-mediated response, that involves activation of microglial cells and increased production of cytokines (Sam 2020).

5.2.1.4 *Clostridium botulinum* and Neurotoxicity

One of the strongest identified neurotoxins is botulinum neurotoxic (BoNT), that can be generated by the *Clostridium botulinum*. While BoNT primarily immobilizes the muscles, recent study evidence demonstrates that this substance could also cause problems with neuronal communication systems. In the backdrop of chronic neurodegenerative conditions, low-dose administration of botulinum toxins can trigger mild neuroinflammatory symptoms (McDew-White et al. 2023).

5.3 Neurotoxic Effects of Foodborne Toxins

A reciprocal interaction pathway between gut microbiota and the central nervous system, the gut-brain axis has a significant impact on neurological inflammation (Ma et al. 2019). Toxic substances or neurotoxin metabolites generated by numerous food-borne microbes may damage the effectiveness of the blood-brain barrier. Even inside the cerebral cortex, these toxins have a tendency to immediately stimulate cells known as glial cells, that can lead to neuroinflammation and disruption of neurotransmitter (Haroon et al. 2017). Therefore, the risk of neurological disease influenced the presence of microbial toxin, which is estimated by toxin mean response rate and concentration (Rajkovic 2014).

5.3.1 **Bacillus cereus** *and Its Enterotoxin*

Enterotoxin of *Bacillus cereus*, Cereulide (CER) foodborne toxin act as a highly selective ionophore, has a receptor mediated mechanism that have adverse effect on human health after ingestion of 6 hours (Rajkovic 2014).

5.3.2 **Fusarium spp.** *and Its Toxin*

Fusarium spp. derived two ionophore toxin such as Beauvericin (BEA) and enniatins (ENN) that has cytotoxic effect on the cell. These secondary metabolites act as a selective cation channel that led to fragmenting the internucleosomal DNA to pose the cytotoxic effect (Rajkovic 2014).

5.3.3 **Listeria monocytogenes** *and Its Neurotoxin*

Listeria monocytogenes released Listeriolysin (LLO) reach the blood-brain barrier through circulating cells and may able to recognize the barrier surface receptor such as InlA, InlB or Vip, that lead to neuroinflammation including meningitis, meningoencephalitis, brain abscesses and rhombencephalitis (Disson and Lecuit 2012).

5.3.4 **Clostridium botulinum** *and Its Neurotoxin*

Clostridium botulinum secretes the neurotoxin, botulinum toxin can easily penetrate the barrier of BBB, and alter the gut-brain axis to immobilize the neural muscles (McDew-White et al. 2023).

5.3.5 **Escherichia coli** *and Its Cytotoxin*

Escherichia coli produced Shiga toxin influence the neuroinflammation by raising the level of pro-inflammatory cytokines and interact with toll like receptor also damage the neurons by permitting into blood-brain barrier (Vida et al. 2014).

5.4 Chemicals and Contaminants in the Food Supply

Food is a pivotal contributor in human beings and the rate of food contamination getting increased day by day. The contamination is stimulated by chemical and biological substances. For our developing scenario, foods are contaminated during processing and packaging. To stabilize the processed food various chemicals such as additives are mixed to increased its shelf-life. Followingly, foodborne pathogens are main reason to spoilage the food. Hence, it possesses various impact on human health through entering in the blood brain barrier. Subsequently, environmental contamination such as pesticides, agrochemicals occur in raw materials, that has an adverse side effect on human health. These agricultural contaminates are mainly neurotoxin.

5.4.1 Agricultural Contaminates

Organophosphorus groups pesticides inhibit the acetylcholinesterase enzyme level that led to accumulate acetylcholine in the synaptic cleft. Similarly, carbamate cause acetylcholinesterase enzyme reverse carbamylating to accumulate acetylcholine in central nervous system. Organochlorine insecticides are bioaccumulate in the food chain by acting as a potent neurotoxin and estrogenic properties (Stefano and Avellone 2014).

5.4.2 Food Packaging Contaminates

Food packing related contaminates such as bisphenol A, arsenic, cadmium, polychlorinated biphenyl (PCBs) and lead cause neurological impairment in children as well as adult through mimicking the cell-cell signaling after ingestion via package contamination. High level of exposure can lead kidney and neural disruption, reproductive disorders, congenital disabilities, neurodevelopmental disorders such as cerebral palsy, autism, mental retardation, attention deficit disorder (Rather et al. 2017).

5.4.3 Food Processing Related Contaminates

During meat storage, carbonyl group of protein get oxidated and produce p-cresol metabolites which may raised the risk of ulcerative colitis and colon cancer, that are produced from colonic bacteria (Li et al. 2021).

5.5 Gut-Brain Axis and Foodborne Pathogens

The gut-brain axis (GBA) is a multi-pathway, intricate communication mechanism between the gut and the brain, including the immune system, hormones, the nervous system and the microbiota. This bidirectional communication is essential for controlling a number of physiological processes, such as biological breakdown, immune responses, and even behavior. When it comes to foodborne pathogens, these are microorganisms (bacteria, viruses, or parasites) that can be ingested through contaminated food and can cause gastrointestinal illness, such as food poisoning. The gut-brain axis may be involved in various ways when the body is infected by foodborne pathogens:

5.5.1 Immune Response

Foodborne pathogens often provoke an immune response in the gut, releasing cytokines and other inflammatory molecules. These molecules can signal the brain, influencing mood, cognition, and stress responses. Chronic inflammation associated with persistent infections can lead to conditions like irritable bowel syndrome (IBS), where gut inflammation has been linked to altered brain activity.

5.5.2 Microbiota Disruption

Pathogens can modify the composition and diversity of the gut microbiota, leading to dysbiosis (imbalance in the gut microbiome). This disruption can trigger signals transmitted to the brain from the gut, contributing to symptoms like anxiety, depression, or gastrointestinal discomfort. For instance, Salmonella or Clostridium difficile infections can cause alteration in gut microbiota that influence brain function and behavior (Marano et al. 2023).

5.5.3 Neurotransmitter Production

The gut microbiota is implicated in the synthesis of neurotransmitters, including serotonin, which contributes to regulating mood and gastrointestinal motility. When disrupted by a foodborne pathogen, the production of these neurotransmitters could be affected, leading to symptoms like nausea, anxiety, or even long-term cognitive effects (Mhanna et al. 2024).

5.5.4 Stress Responses

The gut-brain axis is essential for controlling the body's reaction to stress. Infections caused by foodborne pathogens can lead to psychological stress, which may worsen gut inflammation or contribute to conditions like post-infectious irritable bowel syndrome (PI-IBS) (Carabotti et al. 2015).

5.5.5 *Brain-Gut Signals and Pain*

Pathogens can increase gut sensitivity, leading to symptoms like abdominal pain and discomfort. Through the gut-brain axis, these pain signals are transmitted to the brain, which can lead to changes in behavior or mood. In severe cases, this could contribute to conditions such as functional gastrointestinal disorders (Chaudhry et al. 2023).

The gut-brain axis and foodborne pathogens have emerging relevance in understanding epigenetic regulation in chemotherapy resistance. The gut microbiota, a central player in the gut-brain axis, modulates systemic immunity and influences drug metabolism. Microbial metabolites, such as short-chain fatty acids (SCFAs), can regulate histone acetylation and methylation, impacting the level of genes expression associated with drug resistance. Additionally, gut-derived serotonin and other neurotransmitters influence cellular stress responses, potentially altering chemotherapy efficacy. Foodborne pathogens further complicate this interplay by inducing inflammation and epigenetic changes. Pathogens like Salmonella or *Helicobacter pylori* secrete virulence factors that interact with host chromatin, leading to aberrant DNA methylation or histone modifications. This can activate pathways like NF-κB or STAT3, which are implicated in both cancer progression and resistance to chemotherapeutic agents. Non-coding RNAs (ncRNAs) are critical mediators of these effects. MicroRNAs (miRNAs) and long non-coding RNAs (lncRNAs) are influenced by microbiota metabolites and pathogen infections, modulating gene networks that govern cell proliferation, apoptosis, and drug efflux. For instance, miR-21 and miR-155, both linked to inflammation, have been shown to drive chemoresistance by targeting apoptotic and drug transporter genes.

Therapeutically, modulating the gut microbiome or targeting pathogen-induced epigenetic alterations presents a promising avenue. Probiotics and prebiotics may restore microbiota balance, while inhibitors of specific epigenetic enzymes (e.g., HDACs or DNMTs) could counteract pathogen-driven changes. Exploring this axis could yield biomarkers and strategies to overcome chemotherapy resistance (Bultman 2017; Hamon and Cossart 2008).

5.6 Foodborne Pathogens and BBB Disruption

A highly specific barrier called the blood-brain barrier (BBB) shields the central nervous system (CNS) from infections and other dangerous chemicals. However, some foodborne pathogens have developed mechanisms to breach the BBB, leading to severe neurological conditions such as meningitis and encephalopathy. There are some key mechanisms of BBB disruption by foodborne pathogens.

5.6.1 Lipopolysaccharides (LPS) and Inflammation

Gram-negative bacteria, such as *Escherichia coli* and Salmonella, produce lipopolysaccharides (LPS), which can trigger systemic inflammation. LPS compromises BBB integrity by increasing permeability and disrupting tight junctions between endothelial cells. This allows neurotoxins and inflammatory mediators to enter the brain, leading to conditions like sepsis-associated encephalopathy (SAE) (Profaci et al. 2020).

5.6.2 Transcellular and Paracellular Invasion

Some bacteria cross the BBB by actively invading and passing through endothelial cells. This process can involve receptor-mediated transcytosis, where bacteria exploit host receptors to gain entry into the CNS. For instance, *E. coli* K1 utilizes receptor-ligand interactions to enter brain microvascular endothelial cells. Furthermore, pathogens like *Listeria monocytogenes* degrade tight junction proteins, allowing them to slip between endothelial cells and enter the CNS (Kim 2008).

5.6.3 Trojan Horse Mechanism

Some bacteria evade immune defenses by infecting immune cells (such as macrophages), which then cross the BBB, effectively transporting the pathogen into the brain. This mechanism has been observed with *Listeria and Salmonella* (Kim 2008).

5.6.4 Direct Interaction with Endothelial Cells

Certain bacteria, such as *Streptococcus suis*, produce surface proteins that bind to fibrinogen and other host factors, facilitating their penetration of the BBB. This mechanism has been linked to bacterial meningitis in humans and animals (Al-Obaidi and Desa 2018).

5.7 Neurological Consequences of BBB Disruption

Disruptions in the BBB can lead to severe neurological consequences, including neuroinflammation, neurodegenerative diseases, and cognitive impairment. BBB disruption occurs due to multiple factors, including infections, inflammation, oxidative stress, and metabolic disorders. Once compromised, harmful substances such as toxins, immune cells, and pathogens infiltrate the brain, causing neuronal damage. Some mechanisms which involved of BBB disruption.

5.7.1 *Loss of Tight Junction Integrity*

The BBB relies on tight junction proteins to maintain selective permeability. Disruptions allow harmful agents to pass into the brain, triggering neuroinflammation and neuronal death.

5.7.2 *Neuroinflammation*

Once the BBB is breached, pro-inflammatory molecules like cytokines (TNF-α, IL-6, IL-1β) activate glial cells, leading to chronic neuroinflammation associated with diseases like Alzheimer's and Parkinson's.

5.7.3 *Oxidative Stress and Mitochondrial Dysfunction*

Increased reactive oxygen species (ROS) cause disruption of endothelial cell, further weakening the BBB and exacerbating neurodegenerative processes.

Research highlights the role of BBB breakdown in multiple neurological disorders.

5.7.4 *Alzheimer's Disease (AD)*

One of the primary causes of Alzheimer's disease is neurological inflammation. BBB dysfunction allows amyloid-beta (Aβ) plaques to accumulate in the brain, contributing to cognitive decline. Additionally, impaired clearance of toxic proteins worsens neurodegeneration. Chronic foodborne infections caused by pathogens can exacerbate the symptoms of tau and beta-amyloid accumulation, accelerating cognitive decline (Blake et al., 2024).

5.7.5 *Parkinson's Disease (PD)*

BBB permeability increases exposure to neurotoxicants, leading to the deteriorating of substantia nigra's dopaminergic neurons. A signature of Parkinson's disease, α-synuclein accumulation may be promoted by long-term involvement with endotoxins from bacteria and inflammatory compounds (Tansey et al., 2022).

5.7.6 *Multiple Sclerosis (MS)*

BBB breakdown facilitates the arrival of immune cells (T-cells, macrophages) into the CNS, triggering demyelination and motor deficits. Some microbial and fungal infections transmitted through food trigger autoimmune symptoms since the structures of their molecules resemble those that make up the host's neurological enzymes. *Campylobacter jejuni*, which is a major source of food-borne illness, has been scientifically associated in triggering inflammation corresponding to Multiple Sclerosis and Guillain-Barré Syndrome (Brooks and Mansfield, 2017).

5.7.7 *Stroke and Ischemia*

Vascular damage compromises BBB function, leading to cerebral edema and excitotoxicity, which contribute to secondary brain injuries (Hansda and Das 2025).

5.7.8 *Amyotrophic Lateral Sclerosis (ALS)*

Studies show vascular dysfunction and BBB defects contribute to motor neuron degeneration in ALS patients.

Pathogens, including infectious foodborne bacteria and viruses, can breach the BBB and cause neurological complications. *Listeria monocytogenes* (foodborne pathogen) can cross the BBB, leading to meningoencephalitis. *Escherichia coli (E. coli)* produces toxins that disrupt endothelial integrity, increasing BBB permeability. Viruses (e.g., SARS-CoV-2, HIV, Zika) invade the CNS by disrupting the BBB, leading to long-term cognitive impairments (Profaci et al. 2020; Kim 2008; Xiao et al. 2020; Che et al. 2024; Hansda et al. 2024; FASEB 2015).

Foodborne pathogens have developed complex ways to penetrate the blood-brain barrier, leading to severe CNS infections. Understanding these mechanisms is essential for establish the therapeutic approaches to prevent and diminish neurological complications caused by foodborne diseases.

5.8 Foodborne Toxins and Blood Brain Barrier Integrity

Food contamination influenced by microorganisms (bacterium and fungi) through releasing their toxic substances which lead to cause several health issues including gastrointestinal disorder, neurological damage. Blood Brain Barrier (BBB) sever as a protective shield which permit the nutrients and protect the barrier from foreign molecules (Behrens et al. 2015). However, emerging research shows that certain foodborne toxin can pass or disrupt the integrity of BBB through modifying tight junctions between endothelial cells, enhancing permeability. As a result, it enters the central nervous system (CNS) and increase the oxidative stress, inflammation which lead to cause neurodegenerative disease (Gaigé et al. 2014).

Fusarium fungi are ubiquitous parasites that predominantly infect cereals, documented worldwide and especially among Asia, European and North American climates. Trichothecenes, metabolite of *Fusarium*, also consider as mycotoxin divided into two categories- type A and type B, poses stable occurrence in food during cooking and processing and have impact on human health and animals (Van Der Fels-Klerx et al. 2012). Depending on their distribution in feed and food, toxicological prospect, T-2 toxin has a negative impact on BBB consider as a most toxic type A trichothecenes, while on the other hand type B toxin such as Deoxynivalenol (DON) or nivalenol, 3-acetyldeoxynivalenol (3-AcDON) distributed frequently in cereals but their entrance in BBB still remained unclear. Thus, about the processes through T-2 toxin significantly alters feeding habits are still not well understood but its effects on central nervous system (CNS) in general have lack of sufficient attention.

DON reduced the level of Caco-2 transepithelial monolayer and produce inflammatory markers includes mitogen-activated protein kinase which decreased the distribution of tight-junction related proteins such as claudin-4 and claudin-3, as a result barrier integrity get disrupted. A study shows that DON can potentially be identified inside the cerebrospinal fluid (CSF) following intragastric injected to pigs where the emphasis is on both blood and CSF similar after achieved its maxima within 4.5 h (Šarkanj et al. 2013). Thus, it represents the sole indication present that DON may enter neural tissues. At 10 μM concentration, DON can cross the BBB, leading modify the cell membrane and barrier integrity, decrease the cell viability and influence permeability (Behrens et al. 2015). The significance of DON for human health is demonstrated by the fact that prior research analyzing urinal individual mycotoxin biomarkers discovered that the TDI of DON of 1 μg/kg body mass was frequently surpassed (Gerding et al. 2014; Warth et al. 2012). Compared to morphine, DON can permit the BBB which shows the coefficient of higher transfer rate at a four to five times, but doesn't reach faster in brain than active xenobiotics of CNS (Behrens et al. 2015). In Caco-2 cells, 3-AcDON has a lower capacity to raise cellular inflammation indicators like MAPK than DON and 15-AcDON (Pinton et al. 2010). At significant concentration, it doesn't alter the barrier integrity and have a weaker cytotoxicity (Weidner et al. 2013). Hence, 3-AcDON is probably capable of passing through the blood-brain barrier and enter the tissues of neurons. The acetylation process of 3-AcDON at C-3 significantly increases its lipophilicity in comparison to DON, which explains its rapid passage through the BBB. Since 3-AcDON may readily breach the blood-brain barrier and release the more poisonous DON at both ends, it cannot be considered out that it is having toxicological significance even if it has not yet been observed in mammalian bloodstream following oral administration (Behrens et al. 2015). Moniliformin (MON) is highly polar in nature and potent organic acid because of its conjugated structure, which permits a negative charge to delocalize. Due to its low molecular weight and smaller in size enable to permeate easily through tissue forming barrier. Also, it has been reported that α-ketoglutarate and pyruvate dehydrogenases both are inhibited by MON at

5 μM concentration (von Bargen et al. 2012). In addition to its considerable bioavailability in vivo and partial BBB penetration, MON may have an impact on neural tissues (Behrens et al. 2015).

When T-2 toxin is administered orally, the intestinal epithelium changes, enabling the toxin to reach the bloodstream. As a result, it alters the cellular membrane of various peripheral organ, induce cytokines and increased reactive oxygen species (ROS) production, which concludes the systematic inflammation (Van der Fels-Klerx 2010). Moreover, a number of brain nuclei such as arcuate nucleus (Arc), central amygdala (CeA), nucleus tractus solitarus (NTS), paraventricular nucleus (PVN), dorsal motor nucleus of the vagus (X) involved in the endocrine and autonomic regulation of physiology and behaviour are activated upon consumption of T-2 toxin. Mammalian homeostasis may be altered by peripheral organs secreted cytokines that remain into the brain through pathways of humoral and neural (Gaigé et al. 2014).

Listeria monocytogenes, a high fatal gram- positive bacteria which has low pathogenicity and highly found in contaminated food. After infection, it releases virulent toxin which is Listeriolysin O (LLO) inside the phagocytes, aiding in cytosolic reproduction and evasion of the phagosome (Al-Obaidi and Desa 2018). Another toxin, Hemolysin O induce cytokines through NF-kB pathway, which lead to accumulation of lymphocytes and inflammatory cells into the nerve tissues that can disrupt the structure of BBB (Goraya et al. 2015). Recent research article demonstrate that RhoA GTPases promote macrophage absorption of *Listeria monocytogenes* and produced inflammatory cytokines such as TNF α, IL-6 and IL-1β which assist the bacteria traverse the BBB via macrophage (Shahid et al. 2021). Two pro-inflammatory cytokines include TNF α and IL-1, occur during viral infection that damage the integrity of blood-brain barrier and provide an environment that allows lymphocytes and macrophages to penetrate the central nervous system (Shahid et al. 2021). Studies showed that an infection with *L. monocytogenes* resulted in enhanced permeability of the BBB in murine model, which corresponded to a reduction in occludin and claudin-5 transcription in endothelial cells of cerebral microvascular (Ren et al. 2020).

Shiga toxin (Stx) produced by *Escherichia coli* bacteria can pass the BBB through macrophages. It increases the expression of microglial cells in nervous system which regulated the expression of surface marker Iba 1 that change the barrier integrity (Pradhan et al. 2016). *Streptococcus pneumoniae* release pneumolysin toxin which can breach endothelial cells, and cause cytotoxic impacts on cerebrovascular endothelial cells in the brain that can disrupt the blood-brain barrier (Al-Obaidi and Desa 2018). *Clostridium perfringens* producing epsilon toxin, ETX decreased the expression of claudin-5 that promote the permeability of BBB up to 800 Da molecules, as a result internalization and formation of caveolae increased. Mechanism of ETX on BBB permeability that interact with Myelin and Lymphocyte protein (MAL) receptor for formation of caveolae by recruiting CAV 1 on the apical side of the microvasculature central nervous system. Following their fusion, it releases content into the brain parenchyma (Linden et al. 2019). Another foodborne pathogen *Cronobacter sakazakii*, decreased the expression of Occludin and ZO-1 that considered as tight junction (TJ) proteins which facilitate the bacteria translocation in BBB. Also, studies shows that inflammatory factors and cytokines both regulated the permeability and integrity of the BBB structure (Jin et al. 2021). Additionally, *Bacillus anthracis* can interfere with tight junction proteins, like zonula occludens-1 (ZO-1), impairing the BBB's barrier function and allowing toxins and immune cells to enter the central nervous system, which can exacerbate inflammation and damage to neurons (Al-Obaidi and Desa 2018). Similarly, Streptococcus suis has been decreased the level of TJ proteins which sever as a vital role in breaching BBB (Zhao et al. 2024). With the respect of the above study revealed that both bacteria and fungi derived metabolites can permeable through BBB by altering its integrity.

5.9 Long-Term Neurological Implications of Foodborne Illnesses

Globally, millions of people die each year from foodborne illness, which is a widespread danger to human health. Therefore, it is crucial for the administration of public health to accurately anticipate the probability of foodborne diseases (Du et al. 2021). World Health Organization reported over 600 million individuals globally struggle with disorders brought on by food contamination each year, with 4.2 million people losing their lives for foodborne illness (Oliver 2019). Blood-Brain Barrier (BBB) preserves distinct milieu of the central nervous system (CNS) and facilitates interaction with the circulatory compartment. Zonula occludens (ZO-1), occludin, junctional adhesion molecules (JAMs) and claudin proteins considered among tight junctions (TJs) that connect the brain capillaries' endothelial cells and control the function and composition of the BBB. These proteins regulate whether chemicals cross the BBB. However, the integrity of the BBB can be changed by a number of CNS clinical events, including bacterial infections (Al-Obaidi and Desa 2018). The majority of neurological illnesses, including bacterial meningitis, sepsis, and the development of brain abscesses, are caused by bacterial infections. According to the research, microbial meningitis develops whenever blood-borne microbes penetrate the blood-brain barrier and invade the central nervous system (Iovino et al. 2013). Foodborne illnesses majorly brought on by gram-negative bacteria that coupled with chronic neurological disorder. However, *Clostridium botulinum, Campylobacter jejuni, Listeria monocytogenes, Helicobacter pylori, Salmonella typhi* has long- term effect on neurodegenerative diseases. Nonetheless, it is difficult to compare various health outcomes across a spectrum of foodborne diseases, each of which can lead to a diverse array of symptoms, complications, and long-term effects. Moreover, frequently occurring neurological disorder such as Parkinson disease (PD) and Alzheimer disease (AD) influenced by the foodborne pathogen colonization. Thus, mechanism of stimulating the chronic neurological disease by blood- borne bacteria remain unclear but few of studies revealed that targeting immune system to travel across the blood brain barrier and influenced to form the long-term neurological illnesses.

Helicobacter pylori, a gram-negative foodborne carcinogen produce virulent toxin VacA (vacuolating cytotoxin A). Emerging studies revealed that *H. pylori* not only cause gastrointestinal disorders as well as neurological complications (Kim et al. 2017). Thus, several investigations showed that *H. pylori* infection may play a role in Alzheimer's disease (AD) initiation and progression. The prospective mechanism of *H. pylori* to translocate inside brain through three pathways including circulating monocyte pathway, gastrointestinal retrograde neural pathway and oral-nasal olfactory pathway (Gravina et al. 2018; Xie et al. 2023). Similarly, its metabolite OMVs known as outer membrane vesicles can also be able to travel through biological obstacles and finally enter the inside of brain. Following this, *H. pylori* cause accumulation of amyloid protein by increasing the expression of transmembrane APP protein and AD- associated risk genes (Clu, CD33 and ABCA7) and modifying the level of cholesterol transporter ApoE lead effect on microglial homeostasis, glucose metabolism and lipid transport (Noori et al. 2023). Subsequently, it can cause neuroinflammation by upregulated the production of IL-8 and TNF-α proinflammatory cytokines and stimulate NF-κB signaling and Toll like receptor (TLR-4), which eventually results in behavioural and neurological abnormalities (Lin et al. 2016).

Furthermore, the pathogenic effect of *Campylobacter jejuni* colonization on gastroenteritis contaminated from consumption of unpasteurized milk, dairy product and contaminated poultry. An inflammatory chronic dementia disorder, Guillain-Barré Syndrome that effect on the peripheral nerve tissue which shows correlation with *C. jejuni (Brooks and Mansfield, 2017)*. The mechanism of correlation between GBS and *C. jejuni* remain unclear. Despite, lipopolysaccharide of *C. jejuni* has a structural analogue of gangliosides in nerve cell membrane. Therefore, it disrupts the peripheral nerve tissue and lead to the demyelination (Allos 1998).

Traditional foodborne gram-negative pathogen, *Clostridium botulinum* released neurotoxin like botulism which interfere the receptor of peripheral nerve cell by bind irreversibly on the ends of the nerve cell. Its neurotoxin can travel to neuromuscular junction through bloodstream (Bintsis 2017).

In the recent case study revealed that 11-year child with a severe fibrous tumor discovered in the skull has been connected to a brain abscess caused by *Salmonella typhi* (Shehbaz et al. 2023). Another finding shows that an outbreak with *Listeria monocytogenes* caused α-synuclein to oligomerize in the ileum and a strong pro-inflammatory activity both locally and systemically, which eventually led to mitochondrial dysfunction in neurons (Magalhães et al. 2023). Majorly, 86% brain abscesses caused by *Listeria monocytogenes* where three serotypes show huge impact on human health. The middle cerebral artery allows the brain parenchyma to penetrate the cerebral capillary endothelium, which leads to the formation of an abscess. In a case report, 4b serotype infection involved four virulence genes (hlyA, iap, actA and plcA) of *L. monocytogenes* in a sporadic brain abscesses formation (Khan et al. 2018). Subsequently, another case study summarized that a 64 years old female patient has been suffered with severe brain abscesses caused by *L. monocytogenes* toxin (Zhang et al. 2021).

Additionally, mycotoxin also example of foodborne pathogenic components that lead to impact on the neurological disorder. This mycotoxin interacts with IgE (Immunoglobulin E) to activate the mast cell to produce high level of histamine, proinflammatory cytokines, heparin and prostaglandin GD2 that lead to induce neurological impairment mainly autism spectrum disorder (ASD) as well as brain fog (Ehsanifar et al. 2023). The communication is regulated through bidirectional cross-talk namely gut-brain axis. Above information provide that foodborne gram- negative pathogens cause not only gastrointestinal disorders but also it induces chronic neurological illnesses with the distortion of blood brain barrier through inhibiting the interaction of receptors.

5.10 Prevention and Public Health Implications

Food represents one of the primary mechanisms that diseases are spread throughout the world because of contamination by microbes (Odeyemi 2016). According to the World Health Organization (WHO), 33 million years of excellent life are lost globally each year as a result of eating tainted food, though this number is likely understated (Noah 2024). In terms of leading healthy lives and preventing disease, consumer knowledge of the nutritional value of food and safety are crucial. Inappropriate domestic food management has been involved in a substantial proportion of cases of foodborne diseases. Consumers must be prepared to modify habits that are inconsistent with safe food preparation and preservation methods in an effort to reduce the chance of developing a foodborne illness. Changes in these behaviours are closely linked to consumers' awareness of safe food handling procedures (Losasso et al. 2012). This approach could be used to more effectively and affordably spread knowledge about nutritious foods on an enormous scale, earning substantial advantages in regards to early prevention and improving consumer-level management of food (Doyle et al. 2015). The prevention maintained properly to avoid the risk of contamination from foodborne pathogens are—avoiding the cross contamination, avoid food from unsafe sources, keep food at proper temperature to prevent the growth of the foodborne pathogen and adequate cooking (Hoelzer et al. 2018). Employing current methods like pasteurization and refrigeration, as well as taking a few easy precautions, like keeping both cooked and raw foods separate to prevent cross-contamination and practicing excellent hygiene, can lower the risk of illness caused by foodborne pathogens (Tauxe et al. 2010). Emerging studies shows various antibiotics and antimicrobial agents helps to treat the foodborne illnesses, but chronic disease couldn't treat very well. Hence, prevention led to proceed better lifestyle.

5.11 Conclusion and Future Perspective

Although recognized for its impact on digestive tract wellness, foodborne bacteria are now progressively associated with neurological disorders and neurological inflammation. These pathogenic microorganisms may accelerate disorders like Parkinson's and Alzheimer's by generating constant neuroinflammatory conditions via routes like blood-brain barrier destruction, systemic immune activation, and change of the gut microbiota. Additional studies are essential for understanding how diseases impact long-term neurologic health since the relationship connecting the brain and gut is a crucial pathway for microbiological relations within the central nervous system. Understanding the particular biological mechanisms that lead to infection from food cause neurological inflammation and support in the growth of disease ought to constitute the primary focus for further study. Determining connections between variables and developing targeted treatments requires the application of complex microbiome studies, biomarker verification, and neuroscience methodologies. Possible prevention and therapeutic possibilities involve probiotic-based medications, modifications to diet, and pharmacological approaches to regulate neurological inflammation. Preventing the exposure of neuroinflammatory microorganisms requires upgrading global food safety regulations and health promotion efforts. To diminish the adverse impacts of foodborne diseases on mental health and create strategies for preventing pathogen-induced neurological inflammation, an integrated strategy involving the disciplines of food microbiology, immunological research, and neurobiology is required.

References

Allos BM (1998) Campylobacter jejuni infection as a cause of the Guillain-Barré syndrome. Infect Dis Clin N Am 12(1):173–184. https://doi.org/10.1016/S0891-5520(05)70416-5

Al-Obaidi MMJ, Desa MNM (2018) Mechanisms of blood brain barrier disruption by different types of bacteria, and bacterial–host interactions facilitate the bacterial pathogen invading the brain. Cell Mol Neurobiol 38:1349–1368. https://doi.org/10.1007/s10571-018-0609-2

Behrens M, Hüwel S, Galla HJ, Humpf HU (2015) Blood-brain barrier effects of the Fusarium mycotoxins deoxynivalenol, 3 acetyldeoxynivalenol, and moniliformin and their transfer to the brain. PLoS One 10(11):e0143640. https://doi.org/10.1371/journal.pone.0143640

Bintsis T (2017) Foodborne pathogens. AIMS Microbiol 3(3):529–563. https://doi.org/10.3934/microbiol.2017.3.529

Blake S, Harding T, Baroni L, Harding M, Blake C, Piboolnurak P, Grant W (2024) Reducing dietary lipopolysaccharides to slow progression of cognitive impairment and Alzheimer's disease. J Brain Sci 7(1):10–18488. https://doi.org/10.18488/83.v7i1.3801

Boyd RJ, Avramopoulos D, Jantzie LL, McCallion AS (2022) Neuroinflammation represents a common theme amongst genetic and environmental risk factors for Alzheimer and Parkinson diseases. J Neuroinflammation 19(1):223. https://doi.org/10.1186/s12974-022-02584-x

Brooks PT, Mansfield LS (2017) Effects of antibiotic resistance (AR) and microbiota shifts on campylobacter jejuni-mediated diseases. Anim Health Res Rev 18(2):99–111. https://doi.org/10.1017/S1466252318000014

Bultman SJ (2017) Interplay between diet, gut microbiota, epigenetic events, and colorectal cancer. Mol Nutr Food Res 61(1):1500902. https://doi.org/10.1002/mnfr.201500902

Camilleri M, Lyle BJ, Madsen KL, Sonnenburg J, Verbeke K, Wu GD (2019) Role for diet in normal gut barrier function: developing guidance within the framework of food-labeling regulations. Am J Physiol Gastrointest Liver Physiol 317(1):G17–G39. https://doi.org/10.1152/ajpgi.00063.2019

Carabotti M, Scirocco A, Maselli MA, Severi C (2015) The gut-brain axis: interactions between enteric microbiota, central and enteric nervous systems. Ann Gastroenterol *28*(2):203. PMID: 25830558; PMCID: PMC4367209

Chaudhry TS, Senapati SG, Gadam S, Mannam HPSS, Voruganti HV, Abbasi Z, Abhinav T, Challa AB, Pallipamu N, Bheemishetty N, Arunachalam SP (2023) The impact of microbiota on the gut–brain axis: examining the complex interplay and implications. J Clin Med 12(16):5231. https://doi.org/10.3390/jcm12165231

Che J, Sun Y, Deng Y, Zhang J (2024) Blood-brain barrier disruption: a culprit of cognitive decline? Fluids Barriers CNS 21(1):63. https://doi.org/10.1186/s12987-024-00563-3

Delaney S, Hornig M (2018) Environmental exposures and neuropsychiatric disorders: what role does the gut–immune–brain axis play? Curr Environm Health Rep 5:158–169. https://doi.org/10.1007/s40572-018-0186-z

Detzner J, Pohlentz G, Müthing J (2022) Enterohemorrhagic Escherichia coli and a fresh view on Shiga toxin-binding glycosphingolipids of primary human kidney and colon epithelial cells and their toxin susceptibility. Int J Mol Sci 23(13):6884. https://doi.org/10.3390/ijms23136884

Stefano V Di, Avellone G (2014) Food contaminants. J Food Stud 3(1):88–103. https://doi.org/10.5296/jfs.v3i1.6192

Disson O, Lecuit M (2012) Targeting of the central nervous system by listeria monocytogenes. Virulence 3(2):213–221. https://doi.org/10.4161/viru.19586

Doyle MP, Erickson MC, Alali W, Cannon J, Deng X, Ortega Y, Smith MA, Zhao T (2015) The food industry's current and future role in preventing microbial foodborne illness within the United States. Clin Infect Dis 61(2):252–259. https://doi.org/10.1093/cid/civ253

Du Y, Wang H, Cui W, Zhu H, Guo Y, Dharejo FA, Zhou Y (2021) Foodborne disease risk prediction using multigraph structural long short-term memory networks: algorithm design and validation study. JMIR Med Inform 9(8):e29433. https://doi.org/10.2196/29433

Ehsanifar M, Rajati R, Gholami A, Reiss JP (2023) Mold and mycotoxin exposure and brain disorders. J Integr Neurosci 22(6):137. https://doi.org/10.31083/j.jin2206137

Federation of American Societies for Experimental Biology (FASEB) (2015, March 30) Neurological diseases share common blood-brain barrier defects. ScienceDaily. Retrieved March 22, 2025 from www.sciencedaily.com/releases/2015/03/150330173827.htm

Gaigé S, Djelloul M, Tardivel C, Airault C, Félix B, Jean A, Lebrun B, Toradec JD, Dallaporta M (2014) Modification of energy balance induced by the food contaminant T-2 toxin: a multimodal gut-to-brain connection. Brain Behav Immun 37:54–72. https://doi.org/10.1016/j.bbi.2013.12.008

Gerding J, Cramer B, Humpf HU (2014) Determination of mycotoxin exposure in Germany using an LC-MS/MS multibiomarker approach. Mol Nutr Food Res 58(12):2358–2368. https://doi.org/10.1002/mnfr.201400406

Goldstein J, Nuñez-Goluboay K, Pinto A (2021) Therapeutic strategies to protect the central nervous system against Shiga toxin from enterohemorrhagic Escherichia coli. Curr Neuropharmacol 19(1):24–44. https://doi.org/10.2174/1570159X18666200220143001

Goraya MU, Wang S, Munir M, Chen JL (2015) Induction of innate immunity and its perturbation by influenza viruses. Protein Cell 6(10):712–721. https://doi.org/10.1007/s13238-015-0191-z

Gravina AG, Zagari RM, De Musis C, Romano L, Loguercio C, Romano M (2018) Helicobacter pylori and extragastric diseases: a review. World J Gastroenterol 24(29):3204–3221. https://doi.org/10.3748/wjg.v24.i29.3204

Hamon MA, Cossart P (2008) Histone modifications and chromatin remodeling during bacterial infections. Cell Host Microbe 4(2):100–109. https://doi.org/10.1016/j.chom.2008.07.009

Hansda S, Das H (2025) Insights into cancer-associated thrombosis leading towards ischemic stroke. Biology 14(1):50. https://doi.org/10.3390/biology14010050
Hansda S, Prateeksha P, Das H (2024) Krüppel-like factor 2 (KLF2), a potential target for neuroregeneration. Neural Regen Res 19(11):2327–2328
Haroon E, Miller AH, Sanacora G (2017) Inflammation, glutamate, and glia: a trio of trouble in mood disorders. Neuropsychopharmacology 42(1):193–215. https://doi.org/10.1038/npp.2016.199
Hoelzer K, Switt AIM, Wiedmann M, Boor KJ (2018) Emerging needs and opportunities in foodborne disease detection and prevention: from tools to people. Food Microbiol 75:65–71. https://doi.org/10.1016/j.fm.2017.07.006
Iovino F, Orihuela CJ, Moorlag HE, Molema G, Bijlsma JJ (2013) Interactions between blood-borne Streptococcus pneumoniae and the blood-brain barrier preceding meningitis. PLoS One 8(7):e68408. https://doi.org/10.1371/journal.pone.0068408
Jin T, Guan N, Du Y, Zhang X, Li J, Xia X (2021) Cronobacter sakazakii ATCC 29544 translocated human brain microvascular endothelial cells via endocytosis, apoptosis induction, and disruption of tight junction. Front Microbiol 12:675020. https://doi.org/10.3389/fmicb.2021.675020
Khan S, Kumar A, Kale S, Kurkure N, Nair G, Dinesh K (2018) Multiple cortical brain abscesses due to Listeria monocytogenes in an immunocompetent patient. Trop Dr 48(2):160–163. https://doi.org/10.1177/0049475517728670
Kim KS (2008) Mechanisms of microbial traversal of the blood–brain barrier. Nat Rev Microbiol 6(8):625–634. https://doi.org/10.1038/nrmicro1952
Kim TJ, Sinn DH, Min YW, Son HJ, Kim JJ, Chang Y, Baek SY, Ahn SH, Lee H, Ryu S (2017) A cohort study on Helicobacter pylori infection associated with nonalcoholic fatty liver disease. J Gastroenterol 52:1201–1210. https://doi.org/10.1007/s00535-017-1337-y
Li L, Acioglu C, Heary RF, Elkabes S (2021) Role of astroglial toll-like receptors (TLRs) in central nervous system infections, injury and neurodegenerative diseases. Brain Behav Immun 91:740–755. https://doi.org/10.1016/j.bbi.2020.10.007
Lin HJ, Hsu FY, Chen WW, Lee CH, Lin YJ, Chen YYM, Chen CJ, Huang MZ, Kao MC, Chen YA, Lai HC, Lai CH (2016) Helicobacter pylori activates HMGB1 expression and recruits RAGE into lipid rafts to promote inflammation in gastric epithelial cells. Front Immunol 7:341. https://doi.org/10.3389/fimmu.2016.00341
Linden JR, Flores C, Schmidt EF, Uzal FA, Michel AO, Valenzuela M, Dobrow S, Vartanian T (2019) Clostridium perfringens epsilon toxin induces blood brain barrier permeability via caveolae-dependent transcytosis and requires expression of MAL. PLoS Pathog 15(11):e1008014. https://doi.org/10.1371/journal.ppat.1008014
Liu D (2024) Clostridium botulinum and associated neurotoxins. In: Molecular medical microbiology. Academic Press, pp 933–944. https://doi.org/10.1016/B978-0-12-818619-0.00058-7
Losasso C, Cibin V, Cappa V, Roccato A, Vanzo A, Andrighetto I, Ricci A (2012) Food safety and nutrition: improving consumer behaviour. Food Control 26(2):252–258. https://doi.org/10.1016/j.foodcont.2012.01.038
Ludlow M, Kortekaas J, Herden C, Hoffmann B, Tappe D, Trebst C, Griffin DE, Brindle HE, Solomon T, Brown AS, van Riel D, Wolthers KC, Pajkrt D, Wohlsein P, Osterhaus AD (2016) Neurotropic virus infections as the cause of immediate and delayed neuropathology. Acta Neuropathol 131:159–184. https://doi.org/10.1007/s00401-015-1511-3
Ma Q, Xing C, Long W, Wang HY, Liu Q, Wang RF (2019) Impact of microbiota on central nervous system and neurological diseases: the gut-brain axis. J Neuroinflammation 16:1–14. https://doi.org/10.1186/s12974-019-1434-3
Magalhães JD, Candeias E, Melo-Marques I, Silva DF, Esteves AR, Empadinhas N, Cardoso SM (2023) Intestinal infection triggers mitochondria-mediated α-synuclein pathology: relevance to Parkinson's disease. Cell Mol Life Sci 80(6):166. https://doi.org/10.1007/s00018-023-04819-3
Marano G, Mazza M, Lisci FM, Ciliberto M, Traversi G, Kotzalidis GD, Berardis DD, Laterza L, Sani G, Gasbarrini A, Gaetani E (2023) The microbiota–gut–brain axis: psychoneuroimmunological insights. Nutrients 15(6):1496. https://doi.org/10.3390/nu15061496

McDew-White M, Lee E, Premadasa LS, Alvarez X, Okeoma CM, Mohan M (2023) Cannabinoids modulate the microbiota–gut–brain axis in HIV/SIV infection by reducing neuroinflammation and dysbiosis while concurrently elevating endocannabinoid and indole-3-propionate levels. J Neuroinflammation 20(1):62. https://doi.org/10.1186/s12974-023-02729-6

Mhanna A, Martini N, Hmaydoosh G, Hamwi G, Jarjanazi M, Zaifah G, Kazzazo R, Haji AM, Alshehabi Z (2024) The correlation between gut microbiota and both neurotransmitters and mental disorders: a narrative review. Medicine 103(5):e37114. https://doi.org/10.1097/MD.0000000000037114

Nakhal MM, Yassin LK, Alyaqoubi R, Saeed S, Alderei A, Alhammadi A, Alshehhi M, Almehairbi A, Al Houqani S, BaniYas S, Qanadilo H, Ali BR, Shehab S, Statsenko Y, Meribout S, Sadek B, Akour A, Hamad MI (2024) The microbiota–gut–brain axis and neurological disorders: a comprehensive review. Life 14(10):1234. https://doi.org/10.3390/life14101234

Noah DL (2024) Incidence and impact of foodborne illness. https://www.msdvetmanual.com/public-health/food-safety/incidence-and-impact-of-foodborne-illness

Noori M, Mahboobi R, Nabavi-Rad A, Jamshidizadeh S, Fakharian F, Yadegar A, Zali MR (2023) Helicobacter pylori infection contributes to the expression of Alzheimer's disease-associated risk factors and neuroinflammation. Heliyon 9(9):e19607. https://doi.org/10.1016/j.heliyon.2023.e19607

Odeyemi OA (2016) Public health implications of microbial food safety and foodborne diseases in developing countries. Food Nutr Res 60(1):29819. https://doi.org/10.3402/fnr.v60.29819

Oliver SP (2019) Foodborne pathogens and disease special issue on the national and international PulseNet network. Foodborne Pathog Dis 16(7):439–440. https://doi.org/10.1089/fpd.2019.29012.int

Pinton P, Braicu C, Nougayrede JP, Laffitte J, Taranu I, Oswald IP (2010) Deoxynivalenol impairs porcine intestinal barrier function and decreases the protein expression of claudin-4 through a mitogen-activated protein kinase-dependent mechanism. J Nutr 140(11):1956–1962. https://doi.org/10.3945/jn.110.123919

Pradhan S, Pellino C, MacMaster K, Coyle D, Weiss AA (2016) Shiga toxin mediated neurologic changes in murine model of disease. Front Cell Infect Microbiol 6:114. https://doi.org/10.3389/fcimb.2016.00114

Profaci CP, Munji RN, Pulido RS, Daneman R (2020) The blood–brain barrier in health and disease: important unanswered questions. J Exp Med 217(4):e20190062. https://doi.org/10.1084/jem.20190062

Rajkovic A (2014) Microbial toxins and low level of foodborne exposure. Trends Food Sci Technol 38(2):149–157. https://doi.org/10.1016/j.tifs.2014.04.006

Rather IA, Koh WY, Paek WK, Lim J (2017) The sources of chemical contaminants in food and their health implications. Front Pharmacol 8:830. https://doi.org/10.3389/fphar.2017.00830

Ren JJ, Yang MW, Wang PY, Jiang JJ, Yan GQ (2020) Specific role of the tight junction proteins occludin and claudin-5 on the blood-brain barrier during Listeria monocytogenes infection. Medycyna Weterynaryjna 76(06):6416–2020. https://doi.org/10.21521/mw.6416

Sam L (2020) Impact of Parkinson's disease-linked-Lrrk2 mutation (Lrrk2G2019S) on the innate immune response during infection with Listeria Monocytogenes (Doctoral dissertation, Université d'Ottawa/University of Ottawa)

Šarkanj B, Warth B, Uhlig S, Abia WA, Sulyok M, Klapec T, Krska R, Banjari I (2013) Urinary analysis reveals high deoxynivalenol exposure in pregnant women from Croatia. Food Chem Toxicol 62:231–237. https://doi.org/10.1016/j.fct.2013.08.043

Shahba S, Mehrzad J, Malvandi AM (2021) Neuroimmune disruptions from naturally occurring levels of mycotoxins. Environ Sci Pollut Res 28(25):32156–32176

Shahid AD, Lu Y, Iqbal MA, Lin L, Huang S, Jiang X, Chen S (2021) Listeria monocytogenes crosses blood brain barrier through rho GTPases induced migration of macrophages and inflammatory interleukin expression. Microb Pathog 159:105143. https://doi.org/10.1016/j.micpath.2021.105143

Shehbaz M, Nizamuddin S, Parveen A, Abbas SM (2023) Salmonella enterica serovar typhi (salmonella typhi) brain abscess after brain surgery: a case report. Infect Dis J Pakistan 32(2):62–64. https://doi.org/10.61529/idjp.v32i2.138

Sima CM, Buzilă ER, Trofin F, Păduraru D, Luncă C, Duhaniuc A, Dorneanu OS, Nastase EV (2024) Emerging strategies against non-typhoidal salmonella: from pathogenesis to treatment. Curr Issues Mol Biol 46(7):7447–7472. https://doi.org/10.3390/cimb46070442

Taha AR (2020) Central nervous system infection. In: Highly infectious diseases in critical care: a comprehensive clinical guide. Springer International Publishing, Cham, pp 147–174

Tansey MG, Wallings RL, Houser MC, Herrick MK, Keating CE, Joers V (2022) Inflammation and immune dysfunction in Parkinson disease. Nat Rev Immunol 22(11):657–673

Tauxe RV, Doyle MP, Kuchenmüller T, Schlundt J, Stein CE (2010) Evolving public health approaches to the global challenge of foodborne infections. Int J Food Microbiol 139:S16–S28. https://doi.org/10.1016/j.ijfoodmicro.2009.10.014

Van der Fels-Klerx HJ (2010) Occurrence data of trichothecene mycotoxins T-2 toxin and HT-2 toxin in food and feed. EFSA Support Publicat 7(7):66E

Van Der Fels-Klerx HJ, Klemsdal S, Hietaniemi V, Lindblad M, Ioannou-Kakouri E, Van Asselt ED (2012) Mycotoxin contamination of cereal grain commodities in relation to climate in North West Europe. Food Addit Contamin Part A 29(10):1581–1592. https://doi.org/10.1080/19440049.2012.689996

Vida C, Gonzalez ME, De la Fuente M (2014) Increase of oxidation and inflammation in nervous and immune systems with aging and anxiety. Curr Pharm Des 20(29):4656–4678

von Bargen KW, Lohrey L, Cramer B, Humpf HU (2012) Analysis of the Fusarium mycotoxin moniliformin in cereal samples using 13C2-moniliformin and high-resolution mass spectrometry. J Agric Food Chem 60(14):3586–3591. https://doi.org/10.1021/jf300323d

Warth B, Sulyok M, Fruhmann P, Berthiller F, Schuhmacher R, Hametner C, Adam G, Fröhlich J, Krska R (2012) Assessment of human deoxynivalenol exposure using an LC–MS/MS based biomarker method. Toxicol Lett 211(1):85–90. https://doi.org/10.1016/j.toxlet.2012.02.023

Weidner M, Hüwel S, Ebert F, Schwerdtle T, Galla HJ, Humpf HU (2013) Influence of T-2 and HT-2 toxin on the blood-brain barrier in vitro: new experimental hints for neurotoxic effects. PLoS One 8(3):e60484. https://doi.org/10.1371/journal.pone.0060484

Wouk J, Rechenchoski DZ, Rodrigues BCD, Ribelato EV, Faccin-Galhardi LC (2021) Viral infections and their relationship to neurological disorders. Arch Virol 166:733–753

Xiao M, Xiao ZJ, Yang B, Lan Z, Fang F (2020) Blood-brain barrier: more contributor to disruption of central nervous system homeostasis than victim in neurological disorders. Front Neurosci 14:764. https://doi.org/10.3389/fnins.2020.00764

Xie J, Cools L, Van Imschoot G, Van Wonterghem E, Pauwels MJ, Vlaeminck I, Witte CD, Andaloussi S, Wierda K, Groef LD, Haesebrouck F, Hoecke LV, Vandenbroucke RE (2023) Helicobacter pylori-derived outer membrane vesicles contribute to Alzheimer's disease pathogenesis via C3-C3aR signalling. J Extracell Vesicles 12(2):12306. https://doi.org/10.1002/jev2.12306

Zhang J, Huang S, Xu L, Tao M, Zhao Y, Liang Z (2021) Brain abscess due to listeria monocytogenes: a case report and literature review. Medicine 100(31):e26839. https://doi.org/10.1097/MD.0000000000026839

Zhao G, Ying L, Shi Y, Dong Y, Fu M, Shen Z (2024) Potential mechanisms of Streptococcus suis virulence-related factors in blood–brain barrier disruption. One Health Adv 2(1):26. https://doi.org/10.1186/s44280-024-00059-7

Zhu S, Jiang Y, Xu K, Cui M, Ye W, Zhao G, Jin L, Chen X (2020) The progress of gut microbiome research related to brain disorders. J Neuroinflammation 17:1–20. https://doi.org/10.1186/s12974-020-1705-z

Chapter 6
Reproductive Physiology and Maternal-Fetal Food Safety

Sammra Maqsood, Muhammad Tayyab Arshad, Waseem Khalid, and Ayesha Bibi

6.1 Introduction

The biological mechanisms by which humans and other creatures reproduce are the subject of reproductive physiology (Ahmed et al. 2023). To pledge gamete production, fertilization, and a positive pregnancy, these processes involve the coordinated action of hormones, reproductive organs, and systems. In females, the hypothalamic-pituitary-gonadal axis controls ovulation, ovarian cycles, and uterine preparation for implantation (Al Rubaye et al. 2021). In males, it standardizes the relief of reproductive hormones and spermatogenesis (Apostolopoulou et al. 2024). The balance physiology required to reproduce is very well put by this exceedingly dynamic system (Plant and Zeleznik 2014; Das et al. 2023). The reproduction processes are highly physiological in many species and even in humans. As an example, reproduction's metabolic and energetic costs are clear in the context of increased demands on small mammals (Speakman 2008). Environmental and chemical factors that disrupt

S. Maqsood (✉)
National Institute of Food Science and Technology, University of Agriculture, Faisalabad, Pakistan

M. T. Arshad
University Institute of Food Science and Technology, The University of Lahore, Lahore, Pakistan

W. Khalid
Department of Organic Chemistry, Faculty of Chemical Sciences and Technologies, University of Castilla La Mancha, Ciudad Real, Spain
e-mail: waseem.khalid@uclm.es

A. Bibi
Deaprtment of Nutrition, Women University Mardan, KPK, Mardan, Pakistan
e-mail: drayeshabbb@wymardan.edu.pk

T. Sarkar et al. (eds.), *Physiological Perspectives on Food Safety: Exploring the Intersection of Health and Nutrition*, https://doi.org/10.1007/978-3-031-94582-3_6

these biological processes can affect reproductive health and fertility, such as exposure to phthalates (Hlisníková et al. 2020). Technological advances like organ-on-a-chip models have made it possible to research reproductive physiology in innovative ways that explain organ-to-organ connections and the pathways of diseases (Young and Huh 2021). Maternal health is important to healthy reproduction because it directly affects fetal growth and birth outcomes. Proper nutrition during pregnancy, which includes adequate intake of micronutrients such as iron and folic acid, reduces the risk of complications such as neural tube abnormalities and anemia (Marshall et al. 2022; Santander Ballestín et al. 2021). On the contrary, oxidative stress caused by aging or poor nutrition may degrade the quality of oocytes and the chances of reproduction success (Wang et al. 2021a). Pregnancy-related psychological stress alters hormone regulation, which in turn affects the outcomes for both the mother and the fetus. For example, the oxytocin levels that stress changes may affect the birthing process (Walter et al. 2021). The COVID-19 pandemic has worsened stress and healthcare access inequalities in healthcare systems worldwide, thus revealing vulnerabilities in maternal care (Meaney et al. 2022; Takemoto et al. 2022). The most significant issue about maternity care in the US compared to other developed countries is still maternal mortality, and better maternity support and care should be assured (Tikkanen et al. 2020; Declercq and Zephyrin 2020). Improving maternal health requires an integrated approach addressing stress levels, adequate health care availability, and proper nutrition. With such a significant impact on both the health of the mother and the fetus, food safety also assumes a great deal of importance during pregnancy. Pregnant women are more susceptible to foodborne pathogens and pollutants due to their altered immune response and increased metabolic needs (Plant and Zeleznik 2014). Since foodborne viruses, toxicants, and pollutants can lead to significant complications, such as miscarriage, preterm delivery, and developmental defects, safe food practices during this period are aimed at protecting the health of the developing fetus (Das et al. 2023). Preventive measures and awareness must reduce these risks. The most common foodborne pathogens, particularly virulent during pregnancy, include *Salmonella*, *Toxoplasma gondii*, and *Listeria monocytogenes*. As noted by Speakman (2008), these bacteria may penetrate the placental barrier to cause infections, which may lead to organ damage or neurological abnormalities that may even kill the fetus. For instance, *Toxoplasma gondii* causes prenatal toxoplasmosis, which impairs fetal sight and cognitive skills, whereas *Listeria monocytogenes* results in listeriosis, an illness associated with stillbirth and neonatal sepsis (Hlisníková et al. 2020; Mallott et al. 2020). Food pollutants and toxins also present serious risks. Pesticide residues interfere with endocrine activity, impacting fetal expansion and lifelong health impacts, while heavy metals such as lead and mercury may interfere with prenatal neural expansion (Zohar 2021; Wang et al. 2021a, b, c). Persistent organic pollutants and microplastics, now increasingly found in food systems, could exacerbate oxidative stress and impair nutritional absorption, thus exacerbating pregnancy consequences (Young and Huh 2021; Hlisníková et al. 2020). Further, it is shown that the mother's acquaintance with food pollutants early during pregnancy modifies the brain gut microbiota axis, which plays a noteworthy role in both mother and fetal health (Sarron et al. 2020). Mycotoxins, commonly found in grains and nuts, are dietary contaminants associated with delayed fetal development and gastrointestinal issues (Rizzo et al. 2021). Similarly, phthalates and bisphenols, which are often found in food packaging materials, have been linked to adverse developmental outcomes and impaired reproductive function, making it critical to minimize the disclosure of these toxicants during pregnancy (Yücel et al. 2021). Policymakers and health care providers believe that stringent regulations regarding food training and packaging can limit the ingestion of hazardous substances and avoid further harm (Nosek et al. 2022). Maternal nutrition cannot be overstated in importance for food safety. A diet rich in nutrients supports pregnancy's metabolic demands and protects against the adverse impacts of pollutants and foodborne pathogens. Nutritional deficiencies can weaken immune responses, making a person more susceptible to infections (Walter et al. 2021). For instance, antioxidants help mitigate oxidative stress caused by toxins, and proper intake of sufficient vitamins and minerals can enhance the immune response (Das et al. 2023; Areta et al. 2021). It is also advised that pregnant women should develop healthy eating habits to lower the risk of catching foodborne diseases. This means cooking meat thoroughly, avoiding eating raw or rare seafood, having good hygiene during food preparation, and ensuring fresh fruit is cleaned properly to eliminate pollutants and pesticides (Gallo et al. 2020). To improve healthier pregnancies and reduce the incidence of foodborne diseases, educating pregnant women on food safety practices and contamination risks is important (Aworh 2021; Gonya et al. 2024). Organ-on-a-chip models are one emerging technology that completely transforms our understanding of how foodborne toxicants affect the mother-fetal interaction. These emerging technologies open doors to better treatments for food safety by providing information on how pollutants interact with biological barriers during pregnancy (Young and Huh 2021). In addition, capabilities to monitor and evade foodborne risks are improved by the progression of pollutant detection, for example, using sensors based on noble metal nanoparticles (Putri et al. 2024). To put it together, maternal-fetal food safety encompasses several elements, such as reduction of toxin and infection exposure and education on dietary choices. Strong food safety procedures must be implemented to preserve the health of mothers and fetuses, while advocacy for stronger laws that regulate food production is also necessary (Onyeaka et al. 2024; Marshall et al. 2022). Ongoing study and technological innovation in this area will protect future generations from the negative impacts of foodborne illnesses and toxins (Pattabhiramaiah and Mallikarjunaiah 2023).

6.1.1 Objectives

1. To study the relationship between food safety during pregnancy and reproductive physiology.
2. To establish the primary physiological changes that influence food safety during pregnancy.
3. To examine the role of food safety and maternal diet in promoting fetal growth and development.

4. Evaluate the risk factors associated with contamination and foodborne pathogens when pregnant.
5. To provide means of ensuring pregnant women's food safety.
6. Determining further research and policy lines aimed at better food safety for mothers and their fetuses.

6.2 Reproductive Physiology

6.2.1 Hormonal Changes During Pregnancy

Several complex hormonal changes occur during pregnancy to enable the mother to make physiological adaptations and ensure the healthy development of the fetus. In the course of pregnancy, critical hormones such as progesterone, estrogen, and human chorionic gonadotropin (hCG) affect various systems to provide lodging for the emerging embryo (Abo et al. 2022).

6.2.1.1 Role of Progesterone

It requires progesterone to maintain the uterine lining and prevent contractions. While produced mainly by the corpus luteum early in pregnancy, this hormone is secreted by the placenta later in gestation. It inhibits immunological responses that would lead to the rejection of the developing fetus and also contributes to preparing the conditions that favor the growth of the fetus (Yaw et al. 2020). Moreover, progesterone influences gastrointestinal motility, often leading to constipation, which is one of the main issues in pregnant patients (Chen et al. 2020a).

6.2.1.2 Role of Estrogen

This means that the hormone estrogen has dramatically increased during pregnancy, promoting the development of breast and uterine tissue and the growth of the placenta (Mills et al. 2021). Besides, it helps supply fetal nutrients and promotes uterine blood flow. Furthermore, this hormone also significantly affects the mother's metabolism, encouraging lipogenesis and raising glucose and fats for the fetus to develop (Dunk and Serghides 2022).

6.2.1.3 Role of hCG

Shortly after implantation, trophoblast cells produce human chorionic gonadotropin (hCG), one of the earliest symptoms of pregnancy. Throughout the beginning stages of the pregnancy, the hormone sustains the corpus luteum, ensuring continued production of progesterone (Grant and Erickson 2022). Secondly, hCG aids the woman's immune system to accept the fetus by altering her immunological functions (Schjenken et al. 2021).

6.2.2 *Hormonal Impact on Digestion, Metabolism, and Immune Function*

These hormones'high levels are major influencers in metabolism and digestion. Although progesterone slows down the gastrointestinal transit time, it slows down nutrient absorption, can cause constipation, and induces heartburn, among other problems (Chen et al. 2020a, b). Although estrogen can sometimes expose moms to gestational diabetes, it also alters fat and glucose metabolism to ensure that the fetus receives a stable supply of energy (Zehravi et al. 2021). The immunological activity is modulated rather than reduced during pregnancy. To avoid infections of the mother and fetus, some components of the immune response from the mother need to be upregulated.

In contrast, others must be downregulated to prevent fetus rejection (Haggerty et al. 2021). Balance is crucial because altered immune responses can render people more vulnerable to infections, particularly foodborne illnesses (Chen et al. 2020a, b). Pregnancy-related hormonal changes are required to create a nurturing environment that supports both mother adaptation and fetal growth. Knowing what hCG, estrogen, and progesterone do explains how these hormones coordinate changes in many physiological systems. Although such adjustments are sometimes needed, they also suggest that monitoring maternal health is necessary to ensure effective management of problems when they occur.

6.3 Physiological Changes and Their Impact on Health

Pregnancy and all the other stages of life introduce physiological changes crucial for maintaining health but often modify the body's native immunological, cardiovascular, gastrointestinal, kidney, and liver physiology. These modifications influence how the body responds to immunological insults, detoxifies harmful chemicals, and processes nutrients.

6.3.1 *Immune System Alterations*

Throughout pregnancy and other physiological transitions, the immune system is drastically adapted. Immune suppression is one of these traits that renders the fetus less toxic to the mother but, in return, also upsurges the susceptibility of the mother to infections (Pipoyan et al. 2023). It may compromise systemic immunity and worsen illnesses, such as inflammatory bowel disease, by disturbing the gut microbiota, leading to infections and pathogen entry into the host's system (Chen et al. 2020a, b; Sarron et al. 2020). Food safety is significant because studies show that experience with pollutants at an early age distresses immunological programming (Sarron et al. 2020; Ventre et al. 2022).

6.3.2 Cardiovascular and Gastrointestinal Changes

Changes take place in the cardiovascular system with pregnancy to deal with the greater blood flow across the placenta and uterus; these changes have been known to worsen illnesses such as hypertension and preeclampsia, and among them are increased plasma volume and altered vascular resistance (Morton 2021). The efficiency of digestion and nutritional absorption is diminished because gastrointestinal motility slows with pregnancy (Bohn and Adeli 2022). Hormonal fluctuations and slowing of digestive processes can predispose to gastrointestinal upset and need to be monitored through diet carefully so that proper nutrition is achieved (Marshall et al. 2022).

6.3.3 Kidney and Liver Function Changes

The liver and kidneys adapt to the increased metabolic requirements. Typically, glomerular filtration rates are elevated, and this impacts the efficiency of metabolite and drug removal. This adaptation could put the mother's and fetus's health at risk because it may lead to inefficient detoxification or accumulation of toxic substances (Robinson and Phelan 2024; Chen et al. 2020a, b). Further, thorough dietary monitoring is essential because liver enzyme activity often varies, which in turn varies the metabolism of nutrients and xenobiotics (Bohn and Adeli 2022). Public health interventions rely on knowledge of these physiological changes. Research into food safety should focus on mitigating the risks of contaminants that exacerbate these vulnerabilities. For example, to safeguard populations in high-risk physiological states, it is necessary to limit acquaintance to toxins like microplastics, heavy metals, and foodborne pathogens (Zhu et al. 2023; Oladunjoye and Awani-Aguma 2023). Moreover, specific changes related to the trimester or other life-stage alterations could be measured in customized nutritional strategies to avoid difficulties and support optimal health outcomes (Marshall et al. 2022; Pannia et al. 2022).

6.4 Maternal-Fetal Nutrition and the Role of Food Safety

6.4.1 Nutritional Demands During Pregnancy

Physiological changes in the mother's body during pregnancy are so immense that there is a need for increased dietary intake to support the growth of both the mother and the fetus (Jeffs et al. 2020). The dietary requirements of the fetus support healthy growth and development of the fetus and the mother's overall health throughout pregnancy (Marshall et al. 2022). Dietary needs change with different stages of pregnancy, but some nutrients are essential for the growth of the fetus at all times. The developing fetus requires a range of nutrients for healthy development, organ development, and general well-being (Robinson and Phelan 2024). These nutrients are necessary for the expansion of the placenta, healthy brain growth, and the growth of the neurological and cardiovascular systems (Table 6.1). The nutritional intake of the mother directly prejudices the trajectory of fetal growth and long-term health consequences. For instance, nutrition can cause low birth weight, prematurity, or IUGR, all of which pose serious health risks to the fetus (Monfared 2024). Moreover, the mother's dietary shortages may accompany a greater risk of metabolic and developmental disorders later in life (Jain et al. 2022). In the second and third trimesters, a high need for metabolism must be satisfied with the consumption of a greater number of calories (Bordeleau et al. 2021). Although the quantity consumed is essential, the quality is also very crucial. A diet that satisfies both the mother's and the fetus's needs requires nutrient-rich intake of essential vitamins, minerals, proteins, and healthy fats (Jeffs et al. 2020).

6.4.2 Critical Nutrients and Their Sources

6.4.2.1 Folic Acid

Folic acid is one of the most vital nutrients for the embryo's development because it is a B-vitamin essential for DNA synthesis and cell division. Its use is mandatory in preventing neural tube defects such as spina bifida and anencephaly, especially when the neural tube is forming in the first trimester. Preconception and antenatal optimization of maternal folate levels is advisable. Citrus fruits, legumes, leafy green vegetables, and fortified cereals are all worthy dietary sources of folic acid (Pannia et al. 2022).

Table 6.1 Nutritional Demands, Maternal Diet Impact, and Food Safety Concerns During Pregnancy

Topic	Subtopic	Details	Reference
Nutritional demands	Folic acid	Essential for neural tube development, preventing defects. Sources: leafy greens, citrus fruits, fortified cereals.	Marshall et al. (2022)
	Iron	Vital for increasing blood volume and preventing anemia. Sources: red meat, beans, spinach.	Shukla et al. (2024)
	Calcium	It supports bone development in fetuses and prevents maternal bone loss. Sources are dairy products and leafy greens.	Bohn and Adeli (2022)
Impact of Maternal diet on Fetal development	Brain development	Adequate nutrition affects brain structure and cognitive outcomes. Omega-3 fatty acids and iron are crucial.	Bordeleau et al. (2021)
	Metabolic programming	Early nutrition influences fetal metabolic pathways, affecting obesity and disease risks later in life.	Marshall et al. (2022)
	Balanced diet	A well-rounded diet prevents pregnancy complications like gestational diabetes, preeclampsia, and low birth weight.	Jain et al. (2022)
	Gut microbiota	Maternal diet affects the fetal gut microbiome, which influences immune development and disease susceptibility.	Al Rubaye et al. (2021)
	Effects of micronutrients	Deficiencies in vitamin A, D, or folate can impair fetal development and immune function.	Ishaq et al. (2024)
	Maternal obesity	Obesity increases the risk of gestational diabetes, preeclampsia, and large-for-gestational-age infants.	Zehravi et al. (2021)
	Placental development	Nutrient-rich diets enhance placental development, improving nutrient transport to the fetus.	Abo et al. (2022)
	Epigenetic effects	Maternal nutrition can induce epigenetic changes affecting offspring's disease susceptibility.	Randunu and Bertolo (2020)

(continued)

Table 6.1 (continued)

Topic	Subtopic	Details	Reference
Food safety concerns	Neurodevelopmental effects	Maternal diet high in processed foods or low in omega-3 s may increase the risk of neurodevelopmental disorders.	Bordeleau et al. (2021)
	Salmonella	*Salmonella* infection may cause severe illness in pregnancy. Avoid raw or undercooked eggs, meat, and poultry.	Aworh (2021)
	Food handling	Proper food handling (e.g., washing hands, cooking to safe temperatures) is vital to prevent foodborne illnesses.	Chen et al. (2020a, b)
	Toxoplasma	Toxoplasmosis can lead to fetal brain damage or blindness. Avoid raw meats and cat litter.	Wastnedge et al. (2020)
	Risk of foodborne illness	Pregnant women have a weakened immune system, increasing susceptibility to foodborne pathogens.	Robinson and Phelan (2024)
	Role of food safety regulations	Food safety guidelines and regulations help mitigate risks from pathogens like *E. coli*, *Listeria*, and *Salmonella*.	Rizzo et al. (2021)
	Fish consumption	Avoid high-mercury fish (e.g., shark, swordfish) due to neurotoxic effects on the fetus.	Thakur et al. (2024)
	Antioxidants in diet	Antioxidants (e.g., vitamin C, E) can help prevent oxidative stress, supporting maternal and fetal health.	Obeagu and Obeagu (2024b)
	Dairy products	Avoid unpasteurized dairy products to reduce the risk of *Listeria* infection.	Watterson (2020)
	Foodborne illness prevention	Pregnant women should avoid cross-contamination by separating raw meats from ready-to-eat foods.	Curulli (2021)
	Environmental toxins	Pregnant women should avoid exposure to environmental toxins such as pesticides and plastics.	Chen et al. (2024a)

(continued)

Table 6.1 (continued)

Topic	Subtopic	Details	Reference
Nutritional demands During pregnancy	Vitamin D	Important for immune function and fetal skeletal development. Sources: sunlight, fortified milk, eggs.	Pannia et al. (2022)
	Folate supplementation	Supplementation helps prevent neural tube defects, especially in early pregnancy.	Marshall et al. (2022)
	Increased nutritional needs	Nutritional demands increase for fetal growth and development, with key nutrients like folic acid, iron, and calcium.	Robinson and Phelan (2024)
	Increased caloric needs	Caloric intake needs to increase by 300–500 kcal per day, especially in the second and third trimesters.	During (2022)
	Protein requirements	Protein needs increase, especially in the second and third trimesters. Sources: lean meats, beans, lentils.	During (2022)
	Zinc	Essential for fetal growth, immune function, and DNA synthesis. Sources: shellfish, nuts, seeds.	Thakur et al. (2024)

6.4.2.2 Iron

Since iron is essential to hemoglobin synthesis and oxygen transport, it is fundamental to the mother's and fetus's well-being. During pregnancy, increasing fetal demands and increased maternal blood volume raise the requirements for iron (Manta-Vogli et al. 2020). Insufficient iron supply may lead to iron deficiency anemia, thus increasing the chances of low birth weight, postpartum depression, and premature delivery. Foods rich in iron, such as red meat, chicken, lentils, beans, and fortified cereals, should be given importance, along with meals rich in vitamin C, which enhances the absorption of iron (Shukla et al. 2024).

6.4.2.3 Calcium

The fetal bones and teeth develop with calcium. It also plays a key role in neurological, blood circulation, and muscle functionality (Marshall et al. 2022). Calcium supplements are given to the fetus in the womb to support bone development for the fetus while the mother's body effectively absorbs the calcium. Great sources of calcium include leafy green vegetables, fortified plant-based milk, tofu, and dairy products such as milk, yogurt, and cheese (Hassan et al. 2020).

6.4.2.4 Vitamin D

Vitamin D is important in immunological activity and aids bone health and calcium absorption. Higher vitamin D levels reduce the risk of disorders like gestational diabetes and preeclampsia (Manta-Vogli et al. 2020). However, vitamin D is most widely obtained from sun exposure; a pregnant woman can also gain these nutrients by eating foods like egg yolks, fatty fish, and fortified dairy products if they are indoors or have less sunlight (Ishaq et al. 2024).

6.4.2.5 Magnesium

Most importantly, magnesium provides several physiological benefits: it keeps the blood sugar balanced, supports muscles and neurons functioning properly, and is essential to ensure a healthy pregnancy, thus producing DNA and bones besides supporting protein synthesis. A deficit in maternal magnesium causes problems associated with preterm labor, increased blood pressure, and diminished development of the baby. Good dietary sources of magnesium are leafy green vegetables, whole grains, nuts, and seeds (Fig. 6.1) (Shukla et al. 2024).

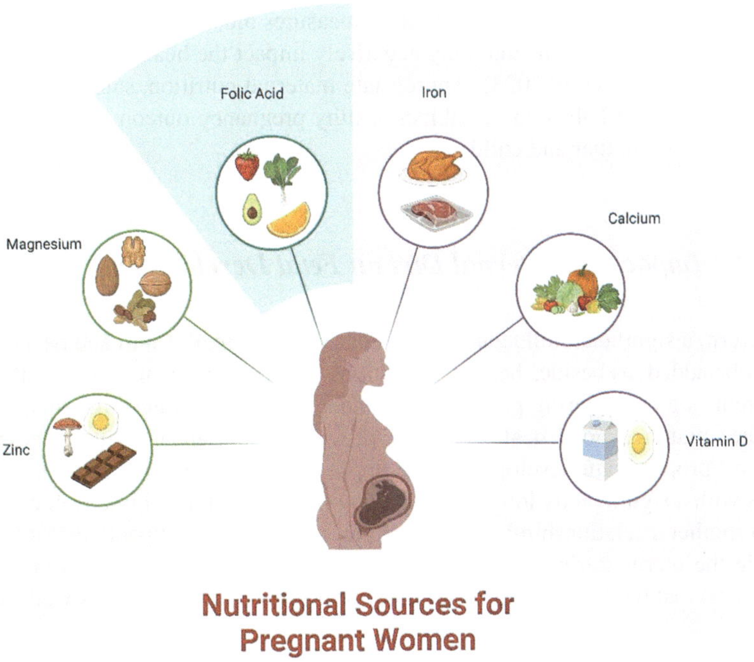

Fig. 6.1 Nutritional sources for pregnant women

6.5 Role of Food Safety in Maternal-Fetal Health

Food safety is an integral part of maternal nutrition because foodborne infections can have serious implications for both the mother and the fetus. Because the immune responses are modified, pregnant women are more susceptible to infections, and some viruses can even affect the development of the fetus by crossing the placenta. For instance, typical foodborne pathogens such as *Listeria monocytogenes* and *Toxoplasma gondii* can lead to miscarriage, premature delivery, or fetal neurological damage (Marshall et al. 2022). Pregnant women are advised to avoid raw eggs, undercooked meats, unpasteurized dairy products, and some kinds of shellfish due to their possible toxic or harmful germs to minimize health dangers (Hu et al. 2020). Thus, proper food handling procedures like properly washing fruits and vegetables, adequate cooking of meats, and proper cross-contamination prevention are required for safe consumption during pregnancy (Rajendran et al. 2021). More so, although some foods are healthy for the fetus, others, such as alcohol, high-mercury fish, and caffeine must be avoided since they can pose a risk to the development of the fetus. Apart from mitigating health risks, proper prenatal care will ensure that the mother's and the fetus's needs are met (Obeagu and Obeagu 2024a, b). This includes providing appropriate dietary recommendations and food safety knowledge. Maternal diet plays a major role in the health and development of the fetus. A healthy diet that includes all the necessary minerals, such as calcium, magnesium, iron, and folic acid, is essential for the growth and development of the fetus while ensuring the mother's health. Food safety measures must also be undertaken to prevent foodborne infections that may negatively impact the health of the mother and the fetus (Obeagu et al. 2023). Appropriate maternal nutrition, safe food handling, and consumption habits can establish healthy pregnancy outcomes and long-term health in both mother and child.

6.5.1 Impact of Maternal Diet on Fetal Development

In return, a significant influence of a mother's diet on fetal health and development must be added, as besides healthy conditions, intake of nutrients can directly affect her fetus's growth during pregnancy. An adequate maternal nutrition diet not only ensures maternal good health but also provides important nutrients necessary to support proper brain development and appropriate metabolic programming by the fetus with a significantly low risk for having pregnancy-related issues. It elaborates on a mother's relationship between her diet and the developments that take place inside the uterus during pregnancy; eating well or appropriately prevents certain problems that the fetus might suffer and could benefit in the future as well from it.

6.5.2 *Influence of Maternal Diet on Fetal Health*

The mother's nutrition has a big impact on the fetus's health, particularly in the development of critical organs such as the heart, the brain, and the immune system. The period from conception to age two of a child's life is believed to be the first 1000 days critical in determining long-term consequences for health (Jain et al. 2022). A healthy diet is recommended for both neurodevelopment and physical growth for pregnancy conditions. For example, embryonic brain development and cognitive function require omega-3 fatty acids, particularly DHA (docosahexaenoic acid) (Bowman et al. 2021). Children deficient in specific nutrients may develop cognitive impairments and developmental disabilities. In addition, folic acid and other micronutrients are necessary for reducing the incidence of neural tube defects (Pannia et al. 2022). The metabolic environment established by the mother's food also influences the growth and health of the fetus. For example, a higher rate of gestational diabetes has been associated with the mother's obesity, which may negatively impact fetal development. There is evidence that mothers' obese status correlates with overnutrition in the fetus, raising the chances of juvenile obesity and higher birth weights (Marshall et al. 2022). On the contrary, intrauterine growth restriction caused by undernutrition may lead to babies being born with low birth weight or even preterm (Gete et al. 2020).

6.5.3 *Maternal Diet and Brain Development*

The development of the embryonic brain is profoundly affected by a maternal diet, and it particularly requires the help of certain nutrients; these include iodine, iron, and folate. This nutrient is particularly implicated in folate deficiency, where associated cognitive deficits translate to an even higher chance of neurodevelopment problems (Pannia et al. 2022). For oxygen transport and neurodevelopment, iron generally performs a vital function. Maternal iron deficiency correlates with behavioral abnormalities and intellectual impairment in offspring (Thakur et al. 2024). Since fatty acids are a primary constituent of brain tissue, they are of greater importance for brain development. The embryonic brain grows extensively, particularly during the second and third trimesters of pregnancy, when omega-3 fatty acid DHA plays an especially important role (Fitzgerald et al. 2020). Two critical brain development processes are neurogenesis and synaptogenesis, and DHA supports both processes. Children may have neurodevelopmental disorders due to an insufficient accumulation of DHA during pregnancy (Bordeleau et al. 2021).

6.5.4 Metabolic Programming and Long-Term Health Implications

It refers to the metabolic programming process in which a pregnant mother's diet affects the subsequent susceptibility of the fetus to chronic conditions such as obesity, diabetes, and cardiovascular diseases. Epigenetic modifications cause the mother's diet to influence the gene expression changes in the offspring, affecting subsequent nutrition handling and stress response by the fetus (Randunu and Bertolo 2020). For example, the baby might be susceptible to asthma, obesity, and metabolic syndrome in adulthood if the mother consumes a diet rich in fats and sweets (Bordeleau et al. 2021). On the other hand, a high diet that incorporates fruits, vegetables, and lean meats may lead to a lower incidence of chronic diseases later in life, thus enabling a child to grow healthily (Mate et al. 2021). Studies show that the mother's nutrients might affect the development of the embryonic immune system. The mother's diet forms the fetus's immune system, and the immunological function of the fetus might be impaired by vitamin and mineral deficiencies (Obeagu and Obeagu 2024a, b). Antioxidants such as vitamins C and E, crucial for immunological function, can shield the fetus from oxidative stress that may cause low birth weight and preterm birth (Obeagu and Obeagu 2024a, b).

6.5.5 Preventing Pregnancy Complications Through a Balanced Diet

Some of the pregnancy problems that can be prevented through a balanced diet include gestational diabetes, hypertension, and preeclampsia. A diet rich in nutrients reduces the risk of gestational diabetes. This condition negatively affects the health of both the mother and the fetus and helps regulate blood sugar levels (Jain et al. 2022). A proper diet ensures proper blood pressure levels and helps reduce the risk of preeclampsia, which may lead to morbidity in both mother and fetus (Marshall et al. 2022). For example, folic acid is commonly provided to pregnant women and is key in preventing neural tube defects. Proper prenatal nutrition is emphasized by its role in preventing congenital disabilities (Pannia et al. 2022). Similarly, maintaining the bone strength of the mother with necessary calcium and vitamin D will benefit the fetus. Therefore, the mother's diet is essential for fetal development and health. A well-balanced diet of vital nutrients ensures proper fetal growth, brain development, and metabolic programming. Not only does it immediately impact the pregnancy outcome, but the mother's diet also has long-term effects on the health and wellbeing of the unborn child (Cipolla-Neto et al. 2022). Therefore, pregnant women must receive adequate nutrition to avoid pregnancy complications and create a healthy life for the mother and child (Fair et al. 2020). Health providers should stress the importance of maternal nutrition and encourage expectant mothers to adopt a balanced diet to promote the health of the developing fetus and the expectant mother's health.

6.5.6 *Food Safety Concerns for Pregnant Women*

This makes both the mother and the fetus more susceptible to foodborne infections due to the heavy physiological demand that pregnancy has on the body. Food safety is, therefore, crucial during pregnancy since foodborne infections can seriously threaten the health of both the emerging fetus and the mother. This section looks into the common foodborne pathogens, including *Salmonella*, *Toxoplasma*, and *Listeria*, with their imaginable hazards and the factors that inspiration the susceptibility of a mother to foodborne disease.

6.5.7 *Common Foodborne Pathogens and Their Potential Risks*

6.5.7.1 *Listeria monocytogenes*

Common sources of *Listeria* are raw vegetables, deli meats, and unpasteurized dairy products. It particularly targets those pregnant with them; if contracted, they cause listeriosis. Jeffs et al. (2020) describe the syndrome as causing severe newborn infections, premature labor, stillbirth, and miscarriage. Marshall et al. (2022) report that it can traverse the placenta, posing direct risks to the fetus due to sepsis and infections of the central nervous system.

6.5.7.2 *Salmonella*

The main concerns with the pregnancy stage are *Salmonella* infections, usually contracted through uncooked poultry, eggs, and contaminated fruits or vegetables. The susceptibility of pregnant women to *Salmonella* is heightened because of the immunological changes and changed gut activity. Salmonellosis signs include fever, abdominal cramps, and diarrhea, leading to dehydration and many other issues (Jain et al. 2022). The infection may rise the fetus's menace of low birth weight or initial delivery (Fig. 6.2) (Shukla et al. 2024).

6.5.7.3 *Toxoplasma gondii*

The protozoan parasite *Toxoplasma* is often spread by undercooked meat or contaminated water and soil (Fan et al. 2024). Congenital toxoplasmosis, caused by infection during pregnancy, can have serious implications for the fetus, including hydrocephalus, intellectual impairments, and vision problems. As cat litter can contain Toxoplasma oocysts, pregnant women are advised not to handle cat litter (Mate et al. 2021). Both mother and fetus may face challenges, so it's essential to identify this condition early on and begin treatment.

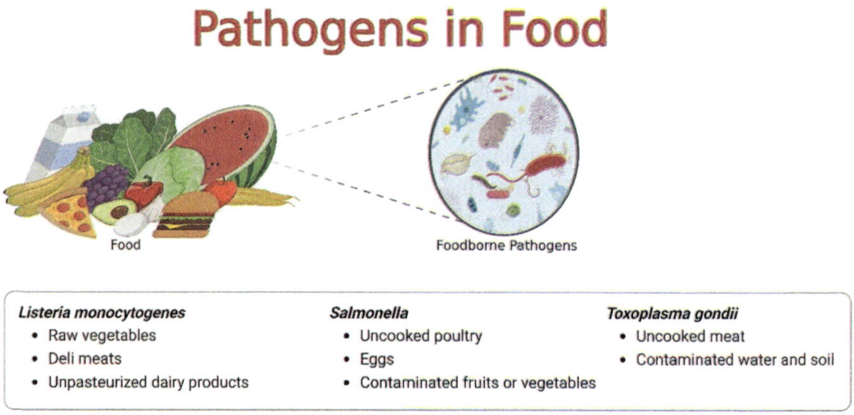

Fig. 6.2 Pathogens in Food

6.5.8 Risk Factors Influencing Maternal Susceptibility to Foodborne Illness

6.5.8.1 Altered Immune Function

The immune system experiences significant alterations during pregnancy as it tries to tolerate the fetus without losing its immunity to infection. These changes might, however, make people more susceptible to some infections, such as those caused by foodborne pathogens (Marshall et al. 2022). The decrease in cell-mediated immunity due to pregnancy may cause the body to have a harder time-fighting diseases like *Salmonella* and *Listeria*.

6.5.8.2 Hormonal Changes

Elevated progesterone and other pregnancy-related hormone levels affect the immune responses and gastrointestinal motility. Poorer digestion increases the time a person will be exposed to harmful microbes that can cause infection (Jain et al. 2022). Additionally, hormonal changes could affect how the body reacts to bacterial invasion by altering the microbiome (Mate et al. 2021).

6.5.8.3 Nutritional Deficiencies

Pregnant women are more prone to infections due to the weakening of the immune system caused by deficiencies in essential minerals like zinc, vitamin D, and folate (Pannia et al. 2022). Proper nutrition ensures maintaining immune function and preventing infections during pregnancy (Monfared 2024). Nutrient deficiencies may increase the risk of foodborne illnesses, compromising the body's ability to fight infections.

6.5.8.4 Age and Pre-existing Health Conditions

Younger, older, or pregnant women with pre-existing medical disorders (such as diabetes or hypertension) are more likely to be affected by foodborne diseases. Such women may have compromised immune systems or other conditions that exacerbate illnesses and complicate pregnancy (Obeagu and Obeagu 2024a, b). The best way to mitigate these risks is through proper prenatal care and monitoring.

6.5.8.5 Socioeconomic Factors

Socioeconomic status is another factor affecting wholesome food accessibility. Barriers in women's groups with low socio-economic status can be related to inappropriate handling of food, raw or unpasteurized foods, and access to medical services limited. As quoted by Hassan et al. (2020), this enhances the chances of exposure to foodborne pathogens and severity during pregnancy (Hassan et al. 2020).

Serious risks for foodborne infections in pregnant women and their developing fetuses include *Salmonella*, *Listeria*, and *Toxoplasma*. Some factors that raise these risks are pre-existing medical conditions, nutritional deficiencies, and immunocompromised states (Fan et al. 2024). Knowing these dangers and taking proper preventive measures in food handling and proper diet reduces food safety concerns during pregnancy. Food safety and nutritional necessities must be deliberated to guarantee mothers' health and well-being (Fig. 6.3) (Monfared 2024; Jain et al. 2022).

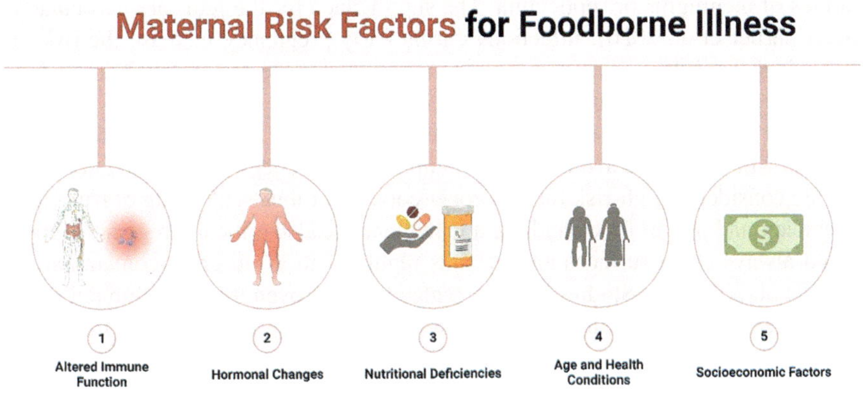

Fig. 6.3 Maternal risk factors for foodborne illness

6.6 Foodborne Pathogens and Maternal-Fetal Health

Foodborne diseases are especially a serious health threat to both mother and fetus due to changes in the immune system, impaired intestinal function, and increased susceptibility to infections that characteristically occur with pregnancy (Table 6.2). The four foodborne diseases associated with high-risk pathogens include *Salmonella* species, *Toxoplasma gondii*, *Escherichia coli*, and *Listeria monocytogenes*. This research describes the pathophysiology, maternal and fetal health consequences, and preventive efforts associated with the latter group of four foodborne illnesses during pregnancy.

6.6.1 Listeriosis

A gram-positive bacterium known as *Listeria monocytogenes* can cause listeriosis, a serious disease particularly dangerous for pregnant women. Both the mother and the fetus can suffer from extreme complications if the infection can cross the placental barrier (Huang et al. 2023). Due to the changes in immunological reactions during pregnancy, pregnant women are more susceptible to listeriosis (Neha 2021). Generally, listeriosis is caused by contaminated foods such as unpasteurized dairy products, deli meats, and ready-to-eat foods. *Listeria* continuously threatens food safety because it can survive under cold conditions, even in refrigerators. Some of the serious consequences of pregnancy from Listeriosis have been preterm birth, miscarriage, and stillbirth plus neonatal infection. Some signs and symptoms affecting the mother include fever and gastrointestinal upset through muscle pain that may occasionally feature more serious elements like sepsis (Dunphy et al. 2022). Vertical transmission through the bacterium may trigger very early labor, resulting in intra-uterine loss or neonate listeriosis, leading to listeriosis of neonates through features of meningitis or septicemia. The significance for the fetus are particularly dire (Charlier et al. 2020). Infections during early pregnancy increase the risk of miscarriage, but infections closer to term more often lead to preterm delivery and its complications for the infant (Kumar et al. 2022).

Several food safety measures are essential to evade listeriosis in pregnancy. Pregnant women should avoid raw milk, unpasteurized cheeses, and cold cuts since they are considered high-risk foods. Proper cooking of meats, cleaning of fruits and vegetables, and proper food handling and hygiene are also critical. Pregnant women should always clean surfaces and utensils regularly to avoid cross-contamination (Chen et al. 2024a, b). Studies in New Zealand have proven that infection rates can be abridged by public health campaigns and education on the danger of listeriosis (Jeffs et al. 2020).

Table 6.2 Foodborne pathogens and maternal-fetal health

Foodborne pathogen	Pathophysiology and risks	Effects on maternal and fetal health	Prevention strategies	References
Listeriosis	It is caused by *Listeria monocytogenes*, a pathogen that can cross the placenta and infect both mother and fetus.	It can cause miscarriage, stillbirth, preterm birth, and neonatal infection. Maternal sepsis is a risk (Dunphy et al. 2022).	Avoid unpasteurized dairy, raw meats, and soft cheeses. Ensure proper hygiene and food handling (Kraus Jr et al. 2024).	Jeffs et al. (2020); Dunphy et al. (2022); Wang et al. (2021a, b, c)
Salmonellosis	Caused by *Salmonella* species, often from contaminated poultry or eggs.	It can cause premature labor, sepsis, and fetal death, especially when there is bacteremia (Kumar et al. 2022).	Cook meats thoroughly, avoid unpasteurized dairy, and practice proper hygiene (Neha 2021).	Kumar et al. (2022)
Toxoplasmosis	Caused by *toxoplasma gondii*, a protozoan parasite that can cross the placenta and infect the fetus.	It can lead to miscarriage, stillbirth, or birth defects such as blindness and mental retardation (Chen et al. 2024a, b).	Avoid raw or undercooked meat and unwashed vegetables. Handle cat litter with gloves (Verma and Singh 2020).	Charlier et al. (2020); Chen et al. (2024a, b)
Norovirus	A highly contagious virus spreads through contaminated food or water.	It can lead to dehydration and preterm labor. Severe vomiting or diarrhea in the mother can put the fetus at risk (Rani and Dhok 2023).	Practice good hand hygiene, avoid contaminated food and water, and avoid contact with infected individuals (Macias 2024).	Rani and Dhok (2023)
Escherichia coli (*E. coli*)	Caused by foodborne *E. coli* infection, especially from undercooked meat, leading to gastrointestinal issues.	It can result in severe gastrointestinal illness, preterm labor, and hemolytic uremic syndrome affecting both mother and fetus (Blakstad and Smith 2020).	Cook meats thoroughly and avoid unpasteurized dairy and improperly washed vegetables (Priya et al. 2021).	Blakstad and Smith (2020)

(continued)

Table 6.2 (continued)

Foodborne pathogen	Pathophysiology and risks	Effects on maternal and fetal health	Prevention strategies	References
Campylobacteriosis	Caused by *Campylobacter jejuni*, typically from contaminated poultry, unpasteurized milk, or contaminated water.	It can cause fever, vomiting, and abdominal pain in mothers, with possible fetal loss or preterm labor (Kraus Jr et al. 2024).	Avoid undercooked meat, especially poultry, and unpasteurized dairy products. Ensure food is cooked to the appropriate temperature (Macias 2024).	Neha (2021)
Listeriosis case study	Pregnant woman in New Zealand with *Listeria* infection, resulting in stillbirth.	The woman contracted *listeria* from contaminated deli meats and had a miscarriage.	Early recognition and prevention of *Listeria* infection through safe food practices (Jeffs et al. 2020).	Jeffs et al. (2020)
Salmonella case study	A pregnant woman in the U.S. experiences severe *Salmonella* infection after eating undercooked eggs.	Resulted in dehydration, preterm birth, and a low birth weight baby.	Safe food handling practices include cooking eggs thoroughly and avoiding raw or undercooked foods (Kumar et al. 2022).	Kumar et al. (2022)
Toxoplasmosis case study	Pregnant women in Europe contract *Toxoplasma* after handling cat litter.	The infant was born with congenital defects, including mental retardation and vision impairment.	Avoid handling cat litter without gloves, and cook meat thoroughly (Verma and Singh 2020).	Charlier et al. (2020)
Norovirus case study	A pregnant woman in a hospital outbreak contracts *norovirus*.	The mother experienced severe dehydration, leading to preterm labor and low birth weight.	Hand hygiene, isolation from infected individuals, and avoiding contaminated food or water (Rani and Dhok 2023).	Rani and Dhok (2023)
E. coli case study	A pregnant woman contracts *E. coli* after eating contaminated beef at a restaurant.	Resulted in severe abdominal pain, nausea, and preterm birth.	Ensure meat is cooked thoroughly and avoid raw or unwashed vegetables (Blakstad and Smith 2020).	Blakstad and Smith (2020)

(continued)

Table 6.2 (continued)

Foodborne pathogen	Pathophysiology and risks	Effects on maternal and fetal health	Prevention strategies	References
Listeria miscarriage	Case of the pregnant woman in New Zealand contracting *Listeria* from unpasteurized cheese.	Miscarriage occurred at 16 weeks due to maternal *listeria* infection.	Avoid unpasteurized dairy products and safely handle food (Neha 2021).	Jeffs et al. (2020)
Salmonella infected pregnant woman	A pregnant woman from the UK infected with *Salmonella* after consuming contaminated poultry.	Resulted in fever, diarrhea, and premature labor.	Ensure proper poultry cooking and proper hygiene in food preparation (Neha 2021).	Kumar et al. (2022)
Toxoplasmosis congenital infection	A woman in the U.S. contracted *Toxoplasma* from undercooked lamb.	The infant was born with vision and developmental issues due to congenital toxoplasmosis.	Cook meat thoroughly and avoid handling cat litter, especially when pregnant (Verma and Singh 2020).	Chen et al. (2024a, b)
Norovirus outbreak in pregnancy	A case where a pregnant woman became ill with *Norovirus* during the hospital stay.	Experienced vomiting, dehydration, and preterm birth due to *Norovirus* infection.	Practice good hygiene, especially in hospitals, and avoid food or water that could be contaminated (Macias 2024).	Rani and Dhok (2023)
E. coli in pregnancy	Pregnant woman contracts *E. coli* from undercooked hamburgers.	Results in gastrointestinal distress, dehydration, and premature delivery.	Thoroughly cook ground beef, wash vegetables well, and avoid unpasteurized dairy products (Blakstad and Smith 2020).	Blakstad and Smith (2020)
Listeria neonatal infection	Case of neonatal infection due to maternal *Listeria* transmission via placenta.	Infant diagnosed with neonatal listeriosis after birth. Symptoms included respiratory distress and fever.	Prevent infection by avoiding unpasteurized foods and practising food safety (Charlier et al. 2020).	Kraus Jr et al. (2024)

(continued)

Table 6.2 (continued)

Foodborne pathogen	Pathophysiology and risks	Effects on maternal and fetal health	Prevention strategies	References
Salmonella Pregnancy case	A pregnant woman develops *Salmonella* infection after handling raw eggs.	Resulted in fever, severe gastrointestinal distress, and miscarriage.	Handle eggs and poultry carefully; cook eggs until yolks and whites are firm (Priya et al. 2021).	Kumar et al. (2022)
Toxoplasmosis Fetal defects	A case where a pregnant woman unknowingly contracted *Toxoplasma* from eating raw beef.	Infants born with severe birth defects, including hearing loss and brain damage.	Safe meat handling, including cooking meats to safe temperatures (Verma and Singh 2020).	Charlier et al. (2020)
Norovirus in pregnant woman	Case of a pregnant woman contracting *Norovirus* at a family gathering.	Severe vomiting, dehydration, and a preterm birth.	Practice good hygiene, particularly when handling food and water in group settings (Macias 2024).	Rani and Dhok (2023)
E. Coli and preterm birth	A pregnant woman experiences *E. coli* infection after consuming contaminated spinach.	Resulting in preterm birth at 32 weeks.	Wash vegetables thoroughly before consumption and cook meat completely (Blakstad and Smith 2020).	Blakstad and Smith (2020)
Listeria sepsis in pregnancy	A pregnant woman contracts *Listeria* through food contamination and develops sepsis.	This resulted in a prolonged hospital stay and a stillbirth at 28 weeks.	Early detection, proper food safety measures, and antibiotics if necessary (Dunphy et al. 2022).	Dunphy et al. (2022)
Salmonella in Pregnancy case	A pregnant woman develops *Salmonella* infection from undercooked chicken.	Causes fever, vomiting, and the need for early labor induction.	Proper food safety measures to prevent contamination (Neha 2021).	Kumar et al. (2022)

6.6.2 Salmonellosis

Salmonella is a form of bacteria that may cause foodborne illness and often occurs with the consumption of raw or undercooked poultry, eggs, and unpasteurized milk. In the case of a *Salmonella* infection during pregnancy, the mother and the fetus can experience systemic disease and defects. Pregnancy worsens the infection because changes in the function of the gastrointestinal tract during this time can worsen the effects of *Salmonella.* In cases of severe maternal disease, like fever, dehydration, and septicemia, salmonellosis during pregnancy may lead to it, particularly when the infection extends beyond the gastrointestinal system. In some cases, *Salmonella* infection has been linked to spontaneous abortion or preterm labor (Kumar et al. 2022). Some of the fetal consequences, often indirect, include neonatal infection and potential long-term developmental issues following an early-life infection. The peril of complications due to dehydration, which might disturb pregnancy consequences, is a factor that exacerbates the threats to maternal health instigated by *Salmonella* infection (Chen et al. 2024a, b).

6.6.3 Toxoplasmosis

A protozoan parasite by the name *Toxoplasma gondii* is transmitted through contaminated food, especially raw meat or contaminated water (Fan et al. 2024). Another route of transmission might include contact with cat faeces. Since pregnancy suppresses the immune response, a pregnant woman is more prone to Toxoplasma. After the infection, *T. gondii* could cause severe damage to the fetus as it could pass through the placental barrier (Kraus Jr et al. 2024).

Toxoplasmosis causes severe congenital infections like hydrocephalus, chorioretinitis, and cerebral calcifications. The diseases are linked to developmental delays, intellectual incapacity, and possibly fetal death (Chen and Dong 2022). Preventive measures include avoiding undercooked food intake, thoroughly cleaning fruits and vegetables, and circumventing contact with cat litter, especially during pregnancy. Anti-parasitic drugs might be prescribed at times to pregnant women to decrease the risk of transmission to the fetus (Fan et al. 2024).

6.6.4 **Norovirus, Escherichia coli,** *and Their Risks During Pregnancy*

Escherichia coli (*E. coli*) and *Norovirus* are two of the most significant foodborne infections, which can also affect pregnant women. *Norovirus* is one of the viruses that can cause complicated pregnancies since it is frequently associated with outbreaks in food-handling environments. It can cause serious gastrointestinal symptoms, electrolyte abnormalities, and dehydration (Neha 2021). *E. coli* can also cause Hemolytic uremic syndrome (HUS), a potentially deadly disease that causes kidney failure and preterm delivery. *E. coli*, particularly some strains such as *E. coli* O157:H7, causes stark food poisoning (Kraus Jr et al. 2024).

6.7 Food Safety Practices to Prevent Infection

Proper handling, preparation, and food storage keep foodborne illnesses at bay during pregnancy (Gönenç et al. 2020). Expectant mothers should avoid unpasteurized dairy products, raw or undercooked meat, eggs, and shellfish. Infection risk decrease can be achieved by cooking food to appropriate temperatures, proper food storage at safe temperatures, and frequent hand washing with utensils and surfaces (D'Aleo et al. 2024). Public health enterprises and routine surveillance are also significant in warning the public about the dangers of foodborne illness during pregnancy. *Salmonella species, Toxoplasma gondii, Escherichia coli,* and *Listeria monocytogenes* infections pose a significant threat to both the mother's and the fetus's health. Precautionary practices such as appropriate food treatment and hygiene, education for pregnant women, and health promotion campaigns for the general population are essential for reducing the occurrence of these infections. Pregnancy outcomes for the health of the mother and the fetus can be improved by mitigating these perils using comprehensive anticipation programs.

6.7.1 Toxicants and Contaminants in Food: Effects on Pregnancy

Environmental and nutritional factors have greatly affected pregnancy outcomes; pregnancy is very important to the health of the fetus and mother, and there is a lot more danger to life from food toxicants and contaminants such as pesticide residues, heavy metals, and food additives together with endocrine disruptors. This chapter examines the conceivable consequences of these materials on pregnancy, concentrating on how they may affect fetal progress and the legal structures put in place to lessen the dangers.

6.7.1.1 Heavy Metals in Food

Environmental contaminants known as heavy metals, including mercury, lead, and arsenic, are commonly found in food, especially in grains, shellfish, and water supplies. Some metals can cause severe side effects during pregnancy (Chen et al. 2020a, b). Mercury is a neurotoxicant that can cause damage to the developing embryonic brain and is primarily found in fish. Exposure to mercury during pregnancy has been associated with neurodevelopmental delays in children, including cognitive deficits, glitches with motor function, and abnormalities in attention (Chen and Dong 2022). Lead acquaintance is also related to cognitive and developmental discrepancies in the fetus. It is particularly worrisome because it can damage the developing neurological system and pass across the placental barricade (Wang et al. 2021a, b, c). Other than low birth weight, developmental complications related to this element include issues involving cognitive impairments, often discovered within rice and also contaminated drinking water (Neha 2021). This exposure in pregnancy is minimized as the pregnant mother absorbs metal through the skin, causing direct exposure to the embryo (Duttaroy and Basak 2020). The World Health Organization and the U.S. Environmental Protection Agency have established maximum permissible limits for certain metals in food and water.

6.7.1.2 Pesticides and Food Additives

In agricultural methods, pesticides are continuously used for pest control and high productivity of crops. However, pesticide residues in food pose a highly serious threat to the pregnant woman and her developing fetus (Cavaliere et al. 2023). According to research, acquaintance with pesticide residues, particularly organophosphates, has resulted in lower birth weight, cognitive delays, and developmental difficulties (Craig et al. 2022). These chemicals also have neurotoxic effects on the developing fetus's brain. Even though food additives are generally considered safe when used within controlled limits, pregnant women can suffer from some of the adverse effects caused by these components. For example, specific food coloring and preservatives in certain animal studies have been associated with hormone disruption and developmental abnormalities (Suhani et al. 2021). The regulatory bodies responsible for monitoring and controlling pesticide residues in food include the Food and Drug Administration (FDA) in the United States and the European Food Safety Authority (EFSA), which set maximum permissible limits that ensure safety for pregnant women.

6.7.1.3 Endocrine Disruptors

These substances, known as endocrine-disrupting chemicals (EDCs), interfere with the body's hormone system and thereby negatively affect the development of the fetus in pregnancy. Probably the most studied substance in this group is the widespread EDC bisphenol A, also known as BPA, found in plastics, food cans, and paper receipts. It has been shown that BPA disrupts hormone signaling, particularly estrogen, which plays a critical role in the development of the embryo or fetus. Studies have associated BPA exposure during pregnancy with developmental defects, such as behavioral and earlier puberty, and a higher likelihood of obesity and metabolic diseases in later life, according to Khushboo et al. (2023). In addition, there is a relation between exposure to other EDCs, like pesticides and phthalates, and adverse pregnancy outcomes, which include preterm birth and low birth weight, according to Priya et al. (2021). Since the influences of EDCs both on the short-term health of the fetus and the offspring's long-term health, influences throughout pregnancy are particularly worrisome (Priya et al. 2021).

6.7.2 *Regulatory Frameworks*

Many regulatory regimes have been implemented worldwide to minimize the hazards associated with these toxicants (Vågsholm et al. 2020). For instance, the Codex Alimentarius of FAO and WHO provides international standards on food safety, including maximum residue levels for pesticides and heavy metals in food. FDA and EFSA are responsible for ensuring that the levels of pesticides, food additives, and other chemicals in foods do not exceed the threshold limit for consumption, especially among vulnerable groups like pregnant women (Rumph et al. 2022). Despite its regulatory base, enforcement and regulation present many challenges in third-world nations that often apply food safety less rigidly, especially in nations without strong economic or development capacities (Onyeaka et al. 2024). Second, lack of full awareness regarding the interactions among different exposure concentrations to some toxicants presents the complexity factor toward undertaking risk evaluation and subsequent policy formulation. Exposure to toxicants and pollutants in food during pregnancy poses serious risks to the fitness and growth of the fetus (Chopra et al. 2022). Heavy metals, pesticides, and endocrine disruptors can cause growth limits, long-term health issues, and interference with fetal neurodevelopment. Even though these frameworks are very important, they must be revised periodically to accommodate new threats and meet international food safety regulations. Policymakers should maintain their imperative to prioritize the development of more stringent food safety regulations. At the same time, pregnant women should be briefed on the risks posed by the toxins ((Duttaroy and Basak 2020).

6.8 Immune System Adaptation and Food Safety During Pregnancy

Being pregnant is the period when a woman's life experiences major physiological changes, affecting her immune system. These changes are needed to help the fetus develop inside the mother while keeping her safe from diseases and infections. However, these changes may impact the mother's susceptibility to foodborne infections, which explains why food safety becomes critical at such times. This part will cover maternal-fetal immunity and immunological conquest throughout pregnancy, focusing on how these characteristics relate to foodborne pathogens'dangers.

6.8.1 Immune Suppression During Pregnancy

Physiological changes cause an immune response to depression, which is an adaptation needed to prevent rejection of the genetically unique fetus by the mother's immune system. Changes in the T-helper cell proportion (Th1, Th2), cytokine production, and immune cell activity all contribute to the suppression (Suhani et al. 2021). Due to these changes, the immune system can tolerate the fetus and remain partially immunized against pathogens (Garsow et al. 2021). Due to this suppression, Pregnant females are generally more susceptible to infections that include foodborne pathogens. For instance, infection by *Salmonella*, *Escherichia coli*, or *Listeria monocytogenes* can harm the mother and the fetus, including foodborne illness, early delivery and fetal death (Arroyo et al. 2021). Food safety is more of a concern to expectant mothers because the immune system sometimes fails to surmount certain diseases. Moreover, environmental pollutants like cadmium could make it worse by causing the decline of the immunizer system (Suhani et al. 2021), and hence, why proper food safety measures, like proper cooking and handling, are particularly important when pregnant.

6.8.2 Maternal-Fetal Immunity

Pregnant maternal-fetal interactions are complicated and essential in the development of the fetus (Chopra et al. 2022). Besides acting as a barrier, the placenta enables antibodies that the mother receives to be transferred to the fetus, whereby the baby derives some passive immunity during the early life years that ward off diseases that may attack this baby (Fuhler 2020). The transfer of immunity through the placenta to the fetus is an important part of maternal-fetal immunity. It provides the child immunological protection against infections it would be exposed to after delivery (Gönenç et al. 2020). The immunological tolerance of the mother also significantly determines preventing problems from foodborne infections. The mother's immune system must balance its ability to protect the fetus from infections with protecting it against the same infections (Chopra et al. 2022). An underactive immune response might make the mother and fetus prone to infections, but an overactive immune response might lead to issues like preterm labor (PrabhuDas et al. 2021). It must adapt through pregnancy to limit the adverse effects of foodborne infections and maintain the immune responses necessary for protecting mother and child. Moreover, environmental pollutants and foodborne diseases can disturb this fine balance Beroukhim et al. 2022). It has been proven by researches those environmental pollutants, such as cadmium and certain xenobiotics in food, can alter the immunological rejoinders of the mother, which may influence the progress of the fetus and health of the postnatal period (Suhani et al. 2021; Arroyo et al. 2021). These pollutants increase the likelihood of issues from foodborne pathogens by weakening the mother's immune system from fighting off the infections (Huang et al. 2023).

6.8.3 The Role of Maternal Nutrition

An adequate level of nutrition is essential as it modulates maternal immunity and, consequently, the fetus's health (Duttaroy and Basak 2020). In addition to providing the fetus with the nutrients necessary for sound development, proper nutrition will ensure that the immune system functions optimally, thus enhancing the mother's ability to fight diseases (Mate et al. 2021). Key diet history deficiencies, especially zinc and vitamins A, C, and D, compromise the function of the immune system and expose the patient to diseases such as those caused by foodborne pathogens (Pahwa and Sharan 2022). Nevertheless, certain dietary factors can affect the danger of foodborne illnesses. According to Di Costanzo et al. (2022), a pregnant woman might be exposed to harmful organisms like *Listeria monocytogenes* while consuming raw or undercooked seafood, which would result in severe infections in both the mother and the fetus. Pregnancy calls for adopting suitable food safety trials, including avoiding harmful foods and serving food at the appropriate temperature (Di Costanzo et al. 2022).

6.8.4 Implications for Public Health

Due to elevated risks of foodborne illnesses when pregnant, pregnant women's food safety must be prioritized by initiatives that target the health of pregnant women (Vågsholm et al. 2020). For example, they must have more safe food education regarding intake and handling of their foods and personal hygiene (Gönenç et al. 2020). Furthermore, the risk of infection and dangerous acquaintances during pregnancy can be expressively abridged by ensuring pregnant women have admittance to safe food sources free of pollutants such as pesticides and heavy metals (Onyeaka et al. 2024). Being in a position where women know precisely how environmental contamination works and, thus, what the hazard is from foods with foodborne infections will make better dietary health choices (Al-Horani and Ayyad 2024). Dealing with the burden of maternal and fetal health issues cannot be understood, including the aspect concerning the changes seen in the system during pregnancy stages and how changes impact susceptibility to infections and foodborne infections (Cavaliere et al. 2023). Pregnancy has a wide range of alterations to the immune system, making it susceptible to several illnesses, including foodborne pathogens (Attini et al. 2022). Food safety is, therefore, part and parcel of prenatal care. The fetus is protected by maternal tolerance by the maternal immune system and antibodies that are shifted across the placenta. Still, the mother's impaired immune response also makes her prone to infections (Bai et al. 2021). Thus, proper food safety practices and awareness of the impact of environmental contaminants and maternal nutrition are important to ensure the health and well-being of both mother and child. Public health education programs can greatly recover pregnancy-related food safety outcomes by educating expectant mothers (Cavaliere et al. 2023).

6.9 Strategies for Ensuring Food Safety During Pregnancy

Maternal health entails food safety as changes due to pregnancy cause special physiological changes. Food safety is essential at this time because pregnancies cause heightened susceptibility and possibilities of infections and foodborne illnesses, and that is why appropriate nutrition must be maintained according to public health recommendations and safe food handling practices. Using relevant investigation from current studies, this chapter will address approaches for ensuring food safety throughout pregnancy, counting safe food handling procedures, policy and public health recommendations, and nutritional guidelines for expectant mothers.

6.9.1 Safe Food Handling Practices

With immune suppression essential in accepting the pregnancy, women during this period experience physiological changes, making them more prone to foodborne infections (Suhani et al. 2021). Strict food handling techniques, such as correct temperatures for cooking and storage and proper hygiene, must be followed to decrease the chance of illness for pregnant women.

6.9.1.1 Proper Cooking Temperatures

The killing of bacteria should be ensured through cooking food at the right temperature (Bai et al. 2021). For instance, to avoid getting ill from pathogens such as *Salmonella*, *Listeria*, and *Escherichia coli*, meat, poultry, and shellfish should be cooked to some temperatures inside them (Pahwa and Sharan 2022). The poultry should be roasted to an interior temperature of 165 °F or 74 °C, while the ground meat should be cooked to 160 °F or 71 °C. These temperatures help eliminate harmful bacteria, which may cause severe pregnancy difficulties, including preterm labor or miscarriage (Suhani et al. 2021).

6.9.1.2 Food Storage and Hygiene Practices

To reduce the possibility of contamination, pregnant women should know appropriate food storage techniques. Foods in the refrigerator must be kept at 40 °F (4 °C) or less, and leftovers must be eaten within two days to avoid the proliferation of bacteria. Cutting boards, cutlery, and countertops should be washed after contact with raw foods to avoid cross-contamination, the process through which raw foods make contact with food items ready to be consumed (Onyeaka et al. 2024). Reducing contact with pathogens is also achieved through good personal hygiene, such as thoroughly washing hands with soap and water before food preparation.

6.9.1.3 Avoiding Risky Foods

Foods should be avoided during pregnancy since there is a tendency to become contaminated (Garsow et al. 2021). Such foods comprise raw or undercooked meat and shellfish, which harbor hazardous germs and parasites, and unpasteurized dairy products, which may contain *Listeria* (Mate et al. 2021). Uncooked eggs also should be evaded by pregnant women since they harbor *Salmonella*. One must be very cautious while taking food decisions because all these harmful items can put both the mother's and the fetus's health at risk (Fuhler 2020).

6.9.2 Public Health and Policy Recommendations

Public health regulation is crucial in ensuring that the food eaten by pregnant women is safe. This is only possible with regulatory input from authoritative bodies such as WHO and the FDA regarding setting standards and guidelines for pregnant women.

6.9.2.1 Regulatory Guidelines for Food Safety During Pregnancy

The FDA provides pregnant women with special food safety guidelines that advise them on foods to avoid (Fair et al. 2020). For instance, due to the risk of developmental issues with the embryo, the FDA advises pregnant women to avoid high-mercury fish such as swordfish and sharks (Onyeaka et al. 2024). These efforts include guidance about maternal nutrition, coupled with the practice of food safety, focusing on implementing food safety protocols (Suhani et al. 2021; WHO 2020). These food safety practices must be standardized and up-scaled by different governments and international health sectors, incorporating the most contemporary scientific developments and findings (Brown and Wright 2020; WHO 2021).

6.9.2.2 Education and Awareness Campaigns

Educating expectant mothers on food safety through public health programs is crucial in encouraging safe food handling procedures and nutritional choices (Attini et al. 2022). Campaigns should provide easily understandable information to the expectant mother on what foods to avoid, how to store food, and the importance of vaccinations during pregnancy (Rizzo et al. 2021). Health care providers, including dietitians and obstetricians, should also regularly instruct patients on foodborne disease risks and offer personalized dietary counseling. Improved awareness can help reduce the incidence of foodborne diseases, which may make pregnancy outcomes more challenging (Dolin et al. 2021; WHO 2020).

6.9.3 Dietary Recommendations for Pregnant Women

Pregnancy nutrition is crucial in the expansion of the fetus and the mother's health. In addition to improving fetal growth and reducing the risk of complications related to foodborne infections, good nutrition can also improve immune response. However, pregnant women must also be informed about exposure to environmental contaminants, pesticides, and heavy metals through food.

6.9.3.1 Guidelines for Optimal Maternal Nutrition and Role of Supplements

A healthy, well-balanced diet full of vital nutrients will attend to both the emerging fetus and the hoping mother (Camara et al. 2022). Some important nutrients are calcium, which ensures bones are healthy; iron, to prevent anemia; folic acid, which stops the incidence of neural tube defects; and omega-3 fatty acids for fetal brain development (Di Costanzo et al. 2022). Prenatal supplements with folic acid, iron, and calcium are often recommended for women to ensure they meet their nutritional requirements, particularly when their diet intake might be insufficient (Jain et al. 2022). In any case, medical professionals must advise using supplements and not over-supplying, as this can prove harmful (Duttaroy and Basak 2020).

6.9.3.2 Best Practices for Minimizing Exposure to Harmful Substances in Food

Because these substances can potentially affect fetal development, pregnant women should minimize their acquaintance with dangerous substances such as pesticides, heavy metals, and endocrine-disrupting chemicals (Rolfo et al. 2020). One way to diminish exposure is by choosing organic vegetables wherever possible because they generally contain fewer pesticide residues. Exposure to hazardous chemicals may also be partial by limiting the use of plastic containers for food storage and cutting back on eating high-mercury seafood (Onyeaka et al. 2024). Pregnant women should also be invigorated to consume a variety of foods so that the risk of excessive acquaintance to any one toxin is dropped.

6.9.3.3 Dietary Considerations for Preventing Foodborne Illness

Unpasteurized dairy, raw or undercooked meats, and high-risk seafood should be circumvented in errand of wholesome, safe diets by pregnant women (Pahwa and Sharan 2022). Additionally, consuming antioxidant-rich foods like fruits and vegetables will enhance the immune system and prevent infections (Fuhler 2020). These dietary recommendations would thus decrease foodborne disease problems in pregnant women.

Food safety must be achieved during pregnancy to ensure the health of the mother and the fetus. Avoiding foodborne infections may be avoided by ensuring proper food handling procedures, such as avoiding dangerous foods, establishing hygiene, and ensuring adequate cooking temperatures. Public health policies and educational initiatives targeting expectant mothers provide guidelines and raise awareness to lower the risk of foodborne illness. Good maternal nutrition and best practices to limit exposure to dangerous substances support a healthy pregnancy and a lower chance of difficulties. Following these guidelines protects the health of the pregnant woman and her unborn child.

6.10 Conclusion and Future Directions

Pregnancy-related food safety is, therefore, crucial for both the mother's health and the fetus. Physiological changes during pregnancy put a woman at risk for exposure to various pollutants, toxicants, and foodborne pathogens, as highlighted in this chapter. These changes in immunological status, hormonal composition, and gastrointestinal and detoxification systems directly affect food safety and processing. Understanding physiological processes accompanying pregnancy is achieved with appropriate knowledge of how food safety becomes relevant throughout this period. The risk of foodborne illnesses, such as listeriosis, salmonellosis, and toxoplasmosis, becomes better managed, and the hazards associated with heavy metals, pesticides, and endocrine disruptors are controlled. Provide the mother with adequate maternal nutrition. Ensure appropriate safe food handling practices and adhere to public health guidelines. Pregnancies can, therefore, be much reduced in preterm births, developmental issues, and pregnancy-related illness if the women are healthy both during and at the time of pregnancy.

Although there is considerable scientific knowledge on pregnancy food safety today, many remain unanswered. It cannot be ascertained how serious the risk associated with maternal-fetal food safety is, mainly because of these new dietary toxins and changing environmental factors. Future research must then be directed at the long-term effects of foodborne infections, toxicants, and pollutants on maternal health and fetal development amid changing global food systems and perhaps new risks arising. Future research should focus on the additive effects of multiple exposures, the role of the microbiota in immune suppression during pregnancy, and the development of more effective preventive strategies. Improved public health campaigns are also necessary to instruct pregnant women and healthcare providers about safe food handling practices. We can better safeguard maternal and fetal health in an increasingly multifaceted and globalized food environment by substantial gaps.

References

Abo S, Smith D, Stadt M, Layton A (2022) Modelling female physiology from head to toe: impact of sex hormones, menstrual cycle, and pregnancy. J Theor Biol 540:111074

Ahmed S, Jiang X, Liu G, Yang H, Sadiq A, Yi D et al (2023) The protective role of maternal genetic immunization on maternal-fetal health and welfare. Int J Gynecol Obstet 163(3):763–777

Al Rubaye H, Adamson CC, Jadavji NM (2021) The role of maternal diet on offspring gut microbiota development: a review. J Neurosci Res 99(1):284–293

Al-Horani RA, Ayyad A (2024) Educating the public about toxic chemicals that we unknowingly consume: a potential important role for the practicing pharmacists. J Am Pharm Assoc 64(2):355–363

Apostolopoulou A, Tranidou A, Chroni V, Tsakiridis I, Tsekitsidi E, Kalaitzopoulou I et al (2024) Preconceptional micronutrient adequacy among women in Greece: a prospective epidemiological study. J Matern Fetal Neonatal Med 37(1):2343613

Areta JL, Taylor HL, Koehler K (2021) Low energy availability: history, definition and evidence of its endocrine, metabolic and physiological effects in prospective studies in females and males. Eur J Appl Physiol 121(1):1–21

Arroyo MC, Barrera TG, Leblic BC, Borrego AA, Collado MC (2021) A review of the impact of xenobiotics from dietary sources on infant health: early life exposures and the role of the microbiota. Environ Pollut 269:115994

Attini R, Leone F, Chatrenet A, Longhitano E, Casula V, Tomasi Cont A et al (2022) Plant-based diets improve maternal–fetal outcomes in CKD pregnancies. Nutrients 14(19):4203

Aworh OC (2021) Food safety issues in fresh produce supply chain with particular reference to sub-Saharan Africa. Food Control 123:107737

Bai X, Nakatsu CH, Bhunia AK (2021) Bacterial biofilms and their implications in pathogenesis and food safety. Food Secur 10(9):2117

Beroukhim G, Esencan E, Seifer DB (2022) Impact of sleep patterns upon female neuroendocrinology and reproductive outcomes: a comprehensive review. Reprod Biol Endocrinol 20(1):16

Blakstad MM, Smith ER (2020) Climate change worsens global inequity in maternal nutrition. Lancet Planet Health 4(12):e547–e548

Bohn MK, Adeli K (2022) Physiological and metabolic adaptations in pregnancy: importance of trimester-specific reference intervals to investigate maternal health and complications. Crit Rev Clin Lab Sci 59(2):76–92

Bordeleau M, Fernandez de Cossio L, Chakravarty MM, Tremblay MÈ (2021) From maternal diet to neurodevelopmental disorders: a story of neuroinflammation. Front Cell Neurosci 14:612705

Bowman CE, Arany Z, Wolfgang MJ (2021) Regulation of maternal–fetal metabolic communication. Cell Mol Life Sci 78:1455–1486

Brown B, Wright C (2020) Safety and efficacy of supplements in pregnancy. Nutr Rev 78(10):813–826

Camara B, Padoin C, Bolea B (2022) Relationship between sex hormones, reproductive stages and ADHD: a systematic review. Arch Womens Ment Health 25:1–8

Cavaliere AF, Perelli F, Mattei A, Dal Poggetto P, Marchi L, Vidiri A et al (2023) Case report: vertical transmission of Plesiomonas shigelloides. Is it time to strengthen information on safety concerns for raw seafood dietary exposure in pregnancy? J Matern Fetal Neonatal Med 36(1):2220061

Charlier C, Disson O, Lecuit M (2020) Maternal-neonatal listeriosis. Virulence 11(1):391–397

Chen B, Dong S (2022) Mercury contamination in fish and its effects on the health of pregnant women and their fetuses, and guidance for fish consumption—a narrative review. Int J Environ Res Public Health 19(23):15929

Chen J, Sun R, Pan C, Sun Y, Mai B, Li QX (2020a) Antibiotics and food safety in aquaculture. J Agric Food Chem 68(43):11908–11919

Chen M, Zeng J, Liu X, Sun G, Gao Y, Liao J et al (2020b) Changes in physiology and immune system during pregnancy and coronavirus infection: a review. Eur J Obstet Gynecol Reprod Biol 255:124–128
Chen L, Lu T, Yang D, Qin X, Huang L, Xu B et al (2024a) Clinical outcome and factors with dietary behaviors in pregnant women with *listeria monocytogenes*: a hospital-based case–control study in Shanghai. Foodborne Pathog Dis 21:431–439
Chen Q, Peng C, Xie R, Xu H, Su Z, Yilihan G et al (2024b) Placental and fetal enrichment of microplastics from disposable paper cups: implications for metabolic and reproductive health during pregnancy. J Hazard Mater 478:135527
Chopra C, Bhushan I, Mehta M, Koushal T, Gupta A, Sharma S et al (2022) Vaginal microbiome: considerations for reproductive health. Future Microbiol 17(18):1501–1513
Cipolla-Neto J, Amaral FG, Soares JM Jr, Gallo CC, Furtado A, Cavaco JE et al (2022) The crosstalk between melatonin and sex steroid hormones. Neuroendocrinology 112(2):115–129
Craig A, Federspiel J, Wein L, Thompson J, Dotters-Katz S (2022) Maternal and obstetric outcomes of *listeria* pregnancy: insights from a national cohort. J Matern Fetal Neonatal Med 35(25):10010–10016
Curulli A (2021) Electrochemical biosensors in food safety: challenges and perspectives. Molecules 26(10):2940
D'Aleo F, Tuscano A, Servello T, Tripodi M, Abramo C, Bonanno R et al (2024) Relevance of microbiological cultures of cord blood and placental swabs in the rapid diagnosis of preterm newborn infection due to *listeria monocytogenes*: a case report. Case Rep Women Health 43:e00638
Das PK, Mukherjee J, Banerjee D (2023) Female reproductive physiology. In: Textbook of veterinary physiology. Singapore, Springer Nature Singapore, pp 513–568
Declercq E, Zephyrin L (2020) Maternal mortality in the United States: a primer. Commonwealth Fund 10
Di Costanzo M, De Paulis N, Capra ME, Biasucci G (2022) Nutrition during pregnancy and lactation: epigenetic effects on infants' immune system in food allergy. Nutrients 14(9):1766
Dolin CD, Compher CC, Oh JK, Durnwald CP (2021) Pregnant and hungry: addressing food insecurity in pregnant women during the COVID-19 pandemic in the United States. Am J Obstetr Gynecol MFM 3(4):100378
Dunk CE, Serghides L (2022) Protease inhibitor-based antiretroviral therapy in pregnancy: effects on hormones, placenta, and decidua. Lancet HIV 9(2):e120–e129
Dunphy L, Polkampali M, Simmons W, Fowler G (2022) Maternal sepsis caused by *listeria monocytogenes* with a fatal fetal outcome. BMJ Case Rep CP 15(10):e249989
During WIHE (2022) Maternal nutrition. Creasy and Resnik's maternal-fetal medicine-E-Book: Principles and Practice, p 186
Duttaroy AK, Basak S (2020) Maternal dietary fatty acids and their roles in human placental development. Prostaglandins Leukot Essent Fat Acids 155:102080
Fair F, Raben L, Watson H, Vivilaki V, van Den Muijsenbergh M, Soltani H, ORAMMA Team (2020) Migrant women's experiences of pregnancy, childbirth and maternity care in European countries: a systematic review. PLoS One 15(2):e0228378
Fan M, Wu H, Sferruzzi-Perri AN, Wang YL, Shao X (2024) Endocytosis at the maternal-fetal interface: balancing nutrient transport and pathogen defense. Front Immunol 15:1415794
Fitzgerald E, Hor K, Drake AJ (2020) Maternal influences on fetal brain development: the role of nutrition, infection and stress, and the potential for intergenerational consequences. Early Hum Dev 150:105190
Fuhler GM (2020) The immune system and microbiome in pregnancy. Best Pract Res Clin Gastroenterol 44:101671
Gallo M, Ferrara L, Calogero A, Montesano D, Naviglio D (2020) Relationships between food and diseases: what to know to ensure food safety. Food Res Int 137:109414
Garsow AV, Campbell E, Closs G Jr, Kowalcyk BB (2021) Food safety challenges in refugee camps: what do we know? J Food Prot 84(5):876–884

Gete DG, Waller M, Mishra GD (2020) Effects of maternal diets on preterm birth and low birth weight: a systematic review. Br J Nutr 123(4):446–461

Gönenç İM, Yilmaz Sezer N, Yilmaz S (2020) Mycotoxin exposure and pregnancy. Crit Rev Toxicol 50(7):594–604

Gonya S, Kallmerten P, Dinapoli P (2024) Are infants and children at risk of adverse health effects from dietary Deoxynivalenol exposure? An integrative review. Int J Environ Res Public Health 21(6):808

Grant AD, Erickson EN (2022) Birth, love, and fear: physiological networks from pregnancy to parenthood. Comprehens Psychoneuroendocrinol 11:100138

Haggerty DK, Upson K, Pacyga DC, Franko JE, Braun JM, Strakovsky RS (2021) REPRODUCTIVE TOXICOLOGY: pregnancy exposure to endocrine disrupting chemicals: implications for women's health. Reproduction 162(5):F169–F180

Hassan S, Faiza N, Farooq MA, Egbuna C, Olatunde A (2020) Role of nutraceuticals in maternal nutrition. In: Functional foods and nutraceuticals: bioactive components, formulations and innovations. Springer, pp 527–541

Hlisníková H, Petrovičová I, Kolena B, Šidlovská M, Sirotkin A (2020) Effects and mechanisms of phthalates' action on reproductive processes and reproductive health: a literature review. Int J Environ Res Public Health 17(18):6811

Hu Z, Tylavsky FA, Kocak M, Fowke JH, Han JC, Davis RL et al (2020) Effects of maternal dietary patterns during pregnancy on early childhood growth trajectories and obesity risk: the CANDLE study. Nutrients 12(2):465

Huang P, Guo X, Duan M, Li H, Han C, Xue F (2023) Maternal infection with *listeria monocytogenes* in twin pregnancy. Infect Drug Resist 16:2511–2518

Ishaq MU, Kunwar D, Qadeer A, Komel A, Safi A, Malik A et al (2024) Effect of vitamin a on maternal, fetal, and neonatal outcomes: an overview of deficiency, excessive intake, and intake recommendations. Nutr Clin Pract 39(2):373–384

Jain S, Maheshwari A, Jain SK (2022) Maternal nutrition and fetal/infant development. Clin Perinatol 49(2):313–330

Jeffs E, Williman J, Brunton C, Gullam J, Walls T (2020) The epidemiology of listeriosis in pregnant women and children in New Zealand from 1997 to 2016: an observational study. BMC Public Health 20:1–8

Khushboo M, Sanjeev S, Murthy MK, Sunitadevi M, Dinata R, Bhanushree B et al (2023) Dietary phytoestrogen diosgenin interrupts metabolism, physiology, and reproduction of Swiss albino mice: possible mode of action as an emerging environmental contaminant, endocrine disruptor and reproductive toxicant. Food Chem Toxicol 176:113798

Kraus V Jr, Čižmárová B, Birková A (2024) *Listeria* in pregnancy—the forgotten culprit. Microorganisms 12(10):2102

Kumar M, Saadaoui M, Al Khodor S (2022) Infections and pregnancy: effects on maternal and child health. Front Cell Infect Microbiol 12:873253

Macias L (2024) Evolution of morning sickness and how it relates to maternal nutrition and fetal development

Mallott EK, Borries C, Koenig A, Amato KR, Lu A (2020) Reproductive hormones mediate changes in the gut microbiome during pregnancy and lactation in Phayre's leaf monkeys. Sci Rep 10(1):9961

Manta-Vogli PD, Schulpis KH, Dotsikas Y, Loukas YL (2020) The significant role of amino acids during pregnancy: nutritional support. J Matern Fetal Neonatal Med 33(2):334–340

Marshall NE, Abrams B, Barbour LA, Catalano P, Christian P, Friedman JE et al (2022) The importance of nutrition in pregnancy and lactation: lifelong consequences. Am J Obstet Gynecol 226(5):607–632

Mate A, Reyes-Goya C, Santana-Garrido Á, Sobrevia L, Vázquez CM (2021) Impact of maternal nutrition in viral infections during pregnancy. Biochim Biophys Acta Mol Basis Dis 1867(11):166231

Meaney S, Leitao S, Olander EK, Pope J, Matvienko-Sikar K (2022) The impact of COVID-19 on pregnant womens' experiences and perceptions of antenatal maternity care, social support, and stress-reduction strategies. Women Birth 35(3):307–316

Mills EG, Yang L, Nielsen MF, Kassem M, Dhillo WS, Comninos AN (2021) The relationship between bone and reproductive hormones beyond estrogens and androgens. Endocr Rev 42(6):691–719

Monfared Z (2024) The importance of nutrition for fetal health. Japanese J Med Sci 2(1):1–8

Morton A (2021) Physiological changes and cardiovascular investigations in pregnancy. Heart Lung Circulat 30(1):e6–e15

Neha SBH (2021) Overview of listeria monocytogens and its epidemiology: a potential threat for pregnant women (Doctoral dissertation, Brac University)

Nosek BA, Hardwicke TE, Moshontz H, Allard A, Corker KS, Dreber A et al (2022) Replicability, robustness, and reproducibility in psychological science. Annu Rev Psychol 73(1):719–748

Obeagu EI, Obeagu GU (2024a) Enhancing maternal and fetal well-being: the role of antioxidants in pregnancy. Elite J Med Sci 2(4):76–87

Obeagu EI, Obeagu GU (2024b) Hypoxia in pregnancy: implications for fetal development. Int J Curr Res Chem Pharm Sci 11(7):39–50

Obeagu EI, Obeagu GU, Obiezu J, Ezeonwumelu C, Ogunnaya FU, Ngwoke AO, Ugwu OP (2023) Immunomodulatory effects of transfusions on maternal immunity in pregnancy. Appl Sci 3(3):1–8

Oladunjoye AO, Awani-Aguma EU (2023) Foodborne illnesses: prevention and control. In Food safety and toxicology: present and future perspectives, p 149

Onyeaka H, Ghosh S, Obileke K, Miri T, Odeyemi OA, Nwaiwu O, Tamasiga P (2024) Preventing chemical contaminants in food: challenges and prospects for safe and sustainable food production. Food Control 155:110040

Pahwa H, Sharan K (2022) Food and nutrition as modifiers of the immune system: a mechanistic overview. Trends Food Sci Technol 123:393–403

Pannia E, Hammoud R, Simonian R, Kubant R, Anderson GH (2022) Folate dose and form during pregnancy may program maternal and fetal health and disease risk. Nutr Rev 80(11):2178–2197

Pattabhiramaiah M, Mallikarjunaiah S (2023) Foodborne pathogens and food safety regulations. In: Food microbial and molecular biology. Apple Academic Press, pp 179–212

Pipoyan D, Beglaryan M, Davtyan L, Stepanyan S, Mantovani A (2023) Risk assessment of dietary exposure to trace elements that are reproductive toxicants: lead, molybdenum and nickel. The case study of Armenia. Reprod Toxicol 118:108382

Plant TM, Zeleznik AJ (eds) (2014) Knobil and Neill's physiology of reproduction. Academic

PrabhuDas M, Piper JM, Jean-Philippe P, Lachowicz-Scroggins M (2021) Immune regulation, maternal infection, vaccination, and pregnancy outcome. J Women's Health 30(2):199–206

Priya K, Setty M, Babu UV, Pai KSR (2021) Implications of environmental toxicants on ovarian follicles: how it can adversely affect the female fertility? Environ Sci Pollut Res 28(48):67925–67939

Putri LA, Prabowo YD, Dewi DMM, Mumtazah Z, Adila FP, Fadillah G et al (2024) Review of noble metal nanoparticle-based colorimetric sensors for food safety monitoring. ACS Appl Nano Mater 7(17):19821–19853

Rajendran S, Giridhar S, Chaudhari S, Gupta PK (2021) Technological advancements in occupational health and safety. Measure Sens 15:100045

Randunu RS, Bertolo RF (2020) The effects of maternal and postnatal dietary methyl nutrients on epigenetic changes that lead to non-communicable diseases in adulthood. Int J Mol Sci 21(9):3290

Rani P, Dhok A (2023) Effects of pollution on pregnancy and infants. Cureus 15(1)

Rizzo DM, Lichtveld M, Mazet JA, Togami E, Miller SA (2021) Plant health and its effects on food safety and security in a one health framework: four case studies. One Health Outlook 3(1):6

Robinson, J. N., & Phelan, J. P. (2024). Pregnancy-induced physiologic alterations. Crit Care Obstetr, 49–75, Wiley

Rolfo A, Nuzzo AM, De Amicis R, Moretti L, Bertoli S, Leone A (2020) Fetal–maternal exposure to endocrine disruptors: correlation with diet intake and pregnancy outcomes. Nutrients 12(6):1744

Rumph JT, Stephens VR, Martin JL, Brown LK, Thomas PL, Cooley A et al (2022) Uncovering evidence: associations between environmental contaminants and disparities in women's health. Int J Environ Res Public Health 19(3):1257

Santander Ballestín S, Giménez Campos MI, Ballestin Ballestin J, Luesma Bartolomé MJ (2021) Is supplementation with micronutrients still necessary during pregnancy? A review. Nutrients 13(9):3134

Sarron E, Pérot M, Barbezier N, Delayre-Orthez C, Gay-Quéheillard J, Anton PM (2020) Early exposure to food contaminants reshapes maturation of the human brain-gut-microbiota axis. World J Gastroenterol 26(23):3145–3169

Schjenken JE, Green ES, Overduin TS, Mah CY, Russell DL, Robertson SA (2021) Endocrine disruptor compounds—a cause of impaired immune tolerance driving inflammatory disorders of pregnancy? Front Endocrinol 12:607539

Shukla V, Parvez S, Fatima G, Singh S, Magomedova A, Batiha GES et al (2024) Micronutrient interactions: magnesium and its synergies in maternal–fetal health. Food Sci Nutr 12(10):6913–6928

Speakman JR (2008) The physiological costs of reproduction in small mammals. Philos Trans R Soc Lond B Biol Sci 363(1490):375–398

Suhani I, Sahab S, Srivastava V, Singh RP (2021) Impact of cadmium pollution on food safety and human health. Curr Opin Toxicol 27:1–7

Takemoto ML, Menezes MO, Andreucci CB, Knobel R, Sousa LA, Katz L et al (2022) Maternal mortality and COVID-19. J Matern Fetal Neonatal Med 35(12):2355–2361

Thakur GK, Shankar H, Arora TK, Kulkarni B (2024) Role of mineral nutrients other than iron in pregnancy: under recognized opportunities to improve maternal/fetal outcomes: a literature review. Arch Gynecol Obstet 309(3):895–905

Tikkanen R, Gunja MZ, FitzGerald M, Zephyrin L (2020) Maternal mortality and maternity care in the United States compared to 10 other developed countries. Commonwealth Fund 10:22

Vågsholm I, Arzoomand NS, Boqvist S (2020) Food security, safety, and sustainability—getting the trade-offs right. Front Sustain Food Syst 4:16

Ventre S, Desai G, Roberson R, Kordas K (2022) Toxic metal exposures from infant diets: risk prevention strategies for caregivers and health care professionals. Curr Probl Pediatr Adolesc Health Care 52(10):101276

Verma S, Singh S (2020) An emerging food borne listeriosis (L. Monocytogense) in society, a mini review world wide and in India. Int J Res Appl Sci Biotechnol 7(5):337–342

Walter MH, Abele H, Plappert CF (2021) The role of oxytocin and the effect of stress during childbirth: neurobiological basics and implications for mother and child. Front Endocrinol 12:742236

Wang L, Tang J, Wang L, Tan F, Song H, Zhou J, Li F (2021a) Oxidative stress in oocyte aging and female reproduction. J Cell Physiol 236(12):7966–7983

Wang Z, Tao X, Liu S, Zhao Y, Yang X (2021b) An update review on *listeria* infection in pregnancy. Infect Drug Resist 14:1967–1978

Wang Y, Xie T, Wu Y, Liu Y, Zou Z, Bai J (2021c) Impacts of maternal diet and alcohol consumption during pregnancy on maternal and infant gut microbiota. Biomol Ther 11(3):369

Wastnedge EA, Reynolds RM, Van Boeckel SR, Stock SJ, Denison FC, Maybin JA, Critchley HO (2020) Pregnancy and COVID-19. Physiol Rev 101:303–318

Watterson A (2020) COVID-19 in the UK and occupational health and safety: predictable not inevitable failures by government, and trade union and nongovernmental organization responses. New Solut 30(2):86–94

World Health Organization (2020) Meeting report: the second global meeting of the FAO/WHO international food safety authorities network (INFOSAN), 9–11 December 2019, Abu Dhabi, United Arab Emirates

World Health Organization (2021) COVID-19 vaccines: safety surveillance manual: module on safety surveillance of COVID-19 vaccines in pregnant and breastfeeding women (No. WHO/MHP/RPQ/PVG/2021.1). World Health Organization

Yaw AM, McLane-Svoboda AK, Hoffmann HM (2020) Shiftwork and light at night negatively impact molecular and endocrine timekeeping in the female reproductive axis in humans and rodents. Int J Mol Sci 22(1):324

Young RE, Huh DD (2021) Organ-on-a-chip technology for the study of the female reproductive system. Adv Drug Deliv Rev 173:461–478

Yücel MA, Lühmann AV, Scholkmann F, Gervain J, Dan I, Ayaz H et al (2021) Best practices for fNIRS publications. Neurophotonics 8(1):012101

Zehravi M, Maqbool M, Ara I (2021) Correlation between obesity, gestational diabetes mellitus, and pregnancy outcomes: an overview. Int J Adolesc Med Health 33(6):339–345

Zhu L, Xie C, Chen L, Dai X, Zhou Y, Pan H, Tian K (2023) Transport of microplastics in the body and interaction with biological barriers, and controlling of microplastics pollution. Ecotoxicol Environ Saf 255:114818

Zohar Y (2021) Fish reproductive biology–reflecting on five decades of fundamental and translational research. Gen Comp Endocrinol 300:113544

Chapter 7
Endocrine Responses to Foodborne Challenges

Diksha Singh, Shalini Singh, and Swarnima Agnihotri

7.1 Introduction to Endocrine System

The endocrine system is a group of glands that release hormones into the blood. Salivary and sweat glands are among the other glands in the body that produce chemicals, but not into the circulation. They release them via ducts that transport them to adjacent body surfaces and are called exocrine glands. Endocrine hormones take time to reach the cells they impact because they must pass through the circulation. Since they are released into the circulation, endocrine hormones go throughout the body wherever blood circulates. Because of this, endocrine hormones can have an impact on a large number of cells and the entire body (Anonymous 2025).

The nervous system and the endocrine system are the two major communication systems in the body that control and coordinate the movement of the internal organs and muscles. Among the many body activities that the endocrine system controls are metabolic rate, immune system reactions, blood pressure, reproductive processes, and digestion. The thyroid, parathyroid, thymus, adrenal glands, pancreas, kidneys, gonads (the ovaries in females and the testes in men), and the hypothalamus and pituitary gland in the brain are the body's main endocrine glands. When an endocrine gland receives a stimulus, usually chemical rather than neurological, its primary job is to release a particular hormone (Regan and Fogarty 2005). The endocrine system functions through slower but more persistent hormonal messages, in contrast to the neurological system, which requires electrical signals for quick communication. Both work together to control several processes at once, making sure

D. Singh · S. Singh (✉)
Institute of Biosciences and Technology, Shri Ramswaroop Memorial University, Lucknow, Uttar Pradesh, India

S. Agnihotri
Swedish Centre for Resource Recovery, University of Borås, Borås, Sweden

T. Sarkar et al. (eds.), *Physiological Perspectives on Food Safety: Exploring the Intersection of Health and Nutrition*, https://doi.org/10.1007/978-3-031-94582-3_7

the body reacts to changes both within and outside the body. Comprehending the complicated relationships between these glands and hormones is crucial to understanding health, illness, and the delicate balance of the body.

7.2 Components of Endocrine System

A gland is a collection of cells that make chemicals and release them into the body using precursors found in the blood or cell. Certain glands produce chemicals that have a local effect; for instance, the salivary and sweat glands are exocrine glands that secrete substances into the mouth or onto the skin. The pituitary, pineal, adrenal, thyroid, parathyroid, pancreatic, and reproductive glands (ovaries and testes) are the main glands of the endocrine system (Table 7.1). The brain, heart, lungs, kidneys, liver, thymus, skin, and placenta are among the several non-endocrine organs that can produce hormones.

The pituitary gland is attached to the hypothalamus, by small capillaries. In reaction to the hypothalamus signal, it converts nervous system impulses into hormones that are produced and cause the pituitary to either create or block particular hormones. Both internal and external factors, such as fear, can cause signals that are processed by other parts of the brain and sent to the hypothalamus via neuronal signaling. The hypothalamus then translates these signals into hormones that either stimulate or inhibit the pituitary. Therefore, the hypothalamus is an essential component that connects the endocrine and neurological systems.

The pituitary (hypophysis) is sometimes referred to as the master gland, as it regulates the hormonal production of many other glands. It is located at the base of the brain just beneath the hypothalamus. It is situated just below the hypothalamus at the base of the brain. The anterior pituitary (adenohypophysis), and the posterior pituitary (neurohypophysis), are its two primary divisions. Hormones that impact the thyroid, adrenal, and reproductive glands are produced by the anterior pituitary. Antidiuretic hormone, which aids in regulating oxytocin and water balance, is produced by the posterior pituitary. Endorphins, which operate on the neurological system to lessen sensitivity to pain, are also produced by the pituitary.

In the centre of the brain lies a small gland called the pineal gland. Melatonin is produced by the pineal gland when the day/night cycle is at its darkest phase. The supra-chiasmatic nucleus, which keeps the body's biological clock in check, sends messages to the pineal in mammals. Even in a completely dark environment, the clock maintains an approximately 24-hour rhythm; nevertheless, it is synchronized with the external day/night cycle by signals from the eye. Around three months after birth, the pineal begins to produce melatonin. During childhood, the pineal grows bigger, and after puberty, both its size and melatonin production decrease.

The thyroid gland resembles a butterfly or bowtie and is situated in the front portion of the lower neck. Thyroid hormone synthesis serves as the thyroid's main means of controlling metabolism. The parathyroid glands are a group of four glands that work in tandem with the thyroid. It releases calcitonin, which works with the parathyroid hormone produced by the parathyroid glands to control blood calcium levels.

Table 7.1 Overview of Endocrine glands and their major functions

Endocrine gland	Hormone	Primary Hormone function
Hypothalamus	Corticotropin-releasing hormone (CRH)	Encourages the release of adrenocorticotropic hormone (ACTH) from the pituitary
	Gonadotropin-releasing hormone (GnRH)	Encourages the pituitary gland to release follicle-stimulating hormone (FSH) and luteinizing hormone (LH)
	Thyrotropin-releasing hormone (TRH)	Encourages the release of thyroid-stimulating hormone (TSH) by the pituitary
Anterior pituitary gland	ACTH	Encourages the adrenal cortex to release hormones
	LH	Encourages the ovaries and ovulation of women to produce sex hormones (i.e., oestrogens); in males, it encourages the testes to produce testosterone
Posterior pituitary gland	Vasopressin	Aids in regulating the body's electrolyte and water balance.
	Oxytocin	Stimulates milk ejection in nursing mothers and encourages uterine contraction during labour
Adrenal cortex	Cortisol	Aids in regulating the metabolism of fats, proteins, and carbohydrates; guards against stress
	Aldosterone	Aids in regulating the body's electrolyte and water balance
Testes	Testosterone	Promotes the growth of the male reproductive system, the creation of sperm, and the anabolism of proteins
Ovaries	Estrogen (produced by the follicle)	Promotes the development of women's reproductive organs
	Progesterone (produced by the corpus luteum)	Prepares the mammary glands for breastfeeding and the uterus for pregnancy
Thyroid gland	Thyroid hormone (i.e., thyroxine [T_4] and triiodothyronine [T_3])	Regulates all cells' metabolic functions
	Calcitonin	Reduces blood calcium levels and aids in the regulation of calcium metabolism
Parathyroid gland	Parathyroid hormone (PTH)	Raises blood calcium levels and aids in regulating calcium metabolism
Pancreas	Insulin	Aids in regulating the metabolism of carbohydrates, which decreases blood sugar levels
	Glucagon	Helps regulate the metabolism of carbohydrates, which raises blood sugar levels

Adapted from Sturmhöfel and Bartke (1998)

Each kidney has a triangular-shaped adrenal gland on top of it. The adrenal cortex, which is the outside section of the adrenal glands, and the adrenal medulla, which is the interior part, is physically and functionally separate. Both mineral corticoid and glucocorticoid are produced by the adrenal cortex. These hormones play a role in metabolism, immune system regulation, respond to stress, regulate salt and water balance, and are involved in sexual development and function. The catecholamines that are involved in the first line reaction to stress, epinephrine (adrenalin) and norepinephrine (noradrenalin), are produced by the adrenal medulla.

The pancreas produces the hormones insulin and glucagon which act together to maintain homeostasis in blood glucose levels. Glucagon synthesis rises to raise blood glucose if blood glucose levels fall below the set-point, whereas insulin production rises to transport glucose into cells, lowering blood glucose if blood glucose levels rise beyond the set-point.

The testes in men and the ovaries in women are the reproductive glands, or gonads. The ovaries, which are found in the pelvis, release the hormones progesterone and estrogen in addition to producing eggs. Pregnancy, menstruation, and healthy female sexual development are all influenced by these hormones. The scrotum contains the testes, which generate androgens essential for sperm production, sexual behavior, and healthy male sexual development (Watamura 2008).

7.3 Function of the Endocrine System

The main purpose of the endocrine system is to control the body's physiological functions via secreting hormones (Fig. 7.1). In order to maintain appropriate operation and homeostasis, these hormones act as chemical messengers, sending information from endocrine glands to specific organs and tissues. Important functions like metabolism, growth and development, reproduction, immunological response, and stress response are all regulated by hormones. For example, insulin and glucagon from the pancreas govern blood glucose levels, while the thyroid hormones (T3 and T4) produced by the thyroid gland control metabolic rate and energy expenditure.

The endocrine system is essential for controlling metabolism as well as the body's reaction to stress and environmental changes. Cortisol and adrenaline, which are released by the adrenal glands, mediate the 'fight or flight' response in stressful conditions by raising blood pressure, heart rate, and energy levels. Similar to how sex hormones (estrogen, progesterone, and testosterone) control sexual development and reproductive activities, growth hormone (GH) from the pituitary gland affects development and tissue growth. The endocrine system's complex feedback mechanisms, including positive and negative feedback loops, maintain hormone levels in balance and stop either too much or too little production. The significance of the endocrine system in preserving physiological equilibrium and reacting to internal and external stimuli is illustrated by these regulating mechanisms.

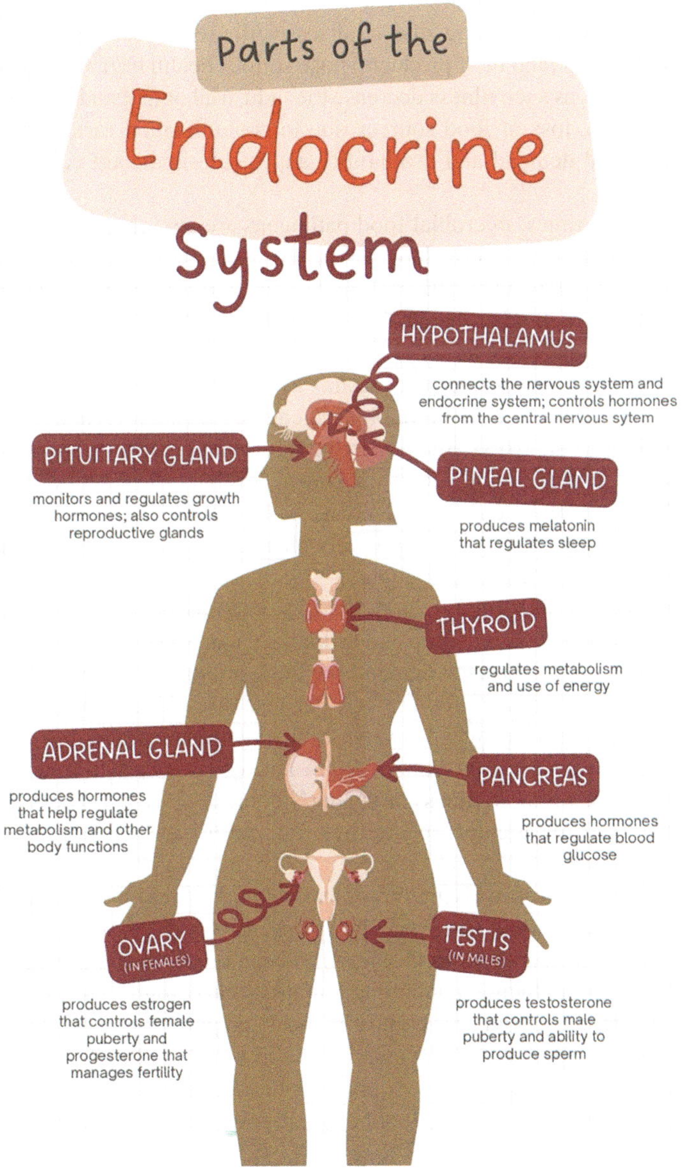

Fig. 7.1 Parts of the Endocrine system

7.4 Food Borne Challenges

Food borne challenges arising due to diseases/illness caused by contaminating microbial pathogens (and their products) pose serious health hazards and negatively impact economies, as such illness decrease the industrial output and human resource productivity. Also, loss of food sources due to contamination further increase the gap between food demand and availability and incur significant economic losses as well.

With changing times, microbial food pathogens are also changing. Food preferences and consumption patterns have also been witnessed and also influence food quality and safety (Nicolay et al. 2011). As the food borne infections are better understood today, methods to control known pathogens are becoming more effective. This has also added to another problem where the changing food environment is also giving rise to newer microbial strains as agents of food borne illness. Cases of rising strains resistant to antimicrobial agents are also appearing rapidly (Aziz et al. 2023; Tack et al. 2019). Such evolving microbial pathogens are also difficult to detect which further complicate control measures that can be applied (Růžičková et al. 2008). The world is changing fast with better connectivity and trade crossing borders, food supply chains have drastically changed. These changes are also to be crucially evaluated to ensure food quality.

Food production and supply chain changes, such as increased imports, might raise the risk of food contamination and sickness. Changes in consumer preferences and habits, such as an increased demand for fresh produce and ready-to-eat meals, can lead to new challenges in food safety. Because of underreporting and the difficulty in proving causal linkages between food contamination and disease, the burden of foodborne diseases on public health and economies has frequently been underestimated, resulting in illness or death (Nicolay et al. 2011). These challenges highlight the need for attempts to improve food safety and prevent the spread of foodborne pathogens.

The intricate and ever-evolving biological network within the human food supply suggests that we should anticipate the emergence of new pathogens. While significant progress in enhancing food safety has been made in recent years, much of it has focused on reducing contamination after harvesting. This has involved improving sanitation and process control in meat and poultry processing, as well as improving the management of processed foods to minimize foodborne pathogenic contamination. This study aims to examine prior studies and emphasize the introduction of new and re-emerging foodborne diseases, as well as the obstacles to ensuring food safety.

With changing environment and food habits, Emerging pathogens have added another dimension to food borne challenges as such pathogens pose additional significant burden on food and health management. Though improved understanding of pathogens is leading to better diseases control the emergence of new pathogens, and the rise of antimicrobial resistance, is bringing new challenges. Reports indicate a range of emerging foodborne pathogens, including Salmonella, Campylobacter,

Escherichia coli, Listeria monocytogenes, and Norovirus are few important examples and understanding of contributing factors will help in better control of health risk associated (Yilmaz et al. 2020; Onyeaka et al. 2022; Khanna et al. 2018, 2019, 2020, 2023; Sachan et al. 2020; Singh and Das 2017).

Apart from food specific illness, metabolic disorders and other health conditions, arising in the human body may alter endocrine functions. **Hunger** and **thirst** may be associated with food borne challenges as a part of exposure or consumption of unsafe food and water sources. The resulting effects, malnutrition and dehydration along with possible food borne illness, may trigger specific endocrine responses towards signals of resultant body stress. The individual responses may vary as people with weak immunity, children and pregnant women are prone to serious impacts. Ensuring food safety, water quality and nutrition education are some of the target areas to minimize these food borne challenges. **Health conditions** like, **diabetes** may also be recognized as food associated challenges and these may affect the working of the endocrine system. Endocrine system, on the other hand, may influence response to such health conditions. Increased or decreased levels of **calcium**, **phosphorus**, **Vitamin** deficiencies, metabolic disorders like, **diabetes**, and lifestyle disorders like obesity, associated with food, may also trigger stress conditions while manipulating endocrine functions and overall body health. **Diabetes** is a chronic metabolic disease. It is typified by elevated blood glucose levels brought on by either insulin resistance, inadequate insulin synthesis, or both (McKinley and Johnson 2004).

Endocrine-disruptors (EDCs) are substances (natural or man-made) that may mimic, block, or interfere with the body's hormones, which are part of the endocrine system and result in different health issues. EDCs may be acquired through air, food, skin, and water. Food and food packaging materials are important sources of such chemicals. The exposure to these disruptors can take place through air, skin, contaminated water and contaminated food and may be presented in pesticides applied in agricultural fields, food and beverages, cosmetics, etc. The chances of exposure and potential health risks can be mitigates by making informed choices. As the endocrine system responds to even small quantities of changes, through hormones, leading to significant, even very low doses of endocrine disruptors may cause serious outcomes and negatively impact biological developments.

Atrazine is a common herbicide used to control weeds in corn, sorghum, and sugarcane crops, and is considered to be a potential endocrine disruptor. Some other important xamples include the following: Bisphenol A (BPA) is used in manufacturing food packaging and it may be found in the lining of some canned foods and beverages. Perchlorate is an industrial chemical used to make rockets, explosives, and fireworks, which can be found in some groundwater. Per- and polyfluoroalkyl substances (PFAS), a large group of industrial chemicals is used widely in nonstick pans, paper manufacturing. Phthalates are used as liquid plasticizers and can be found food packaging. Phytoestrogens are natural, hormone-like substances found in some plants; they may have a similar effect to estrogen produced by the body. Soy foods, for example, contain phytoestrogens. Polybrominated diphenyl ethers (PBDE) are used to make flame retardants for products such as furniture foam and carpet. Polychlorinated biphenyls (PCBs) were used to make electrical equipment, such as transformers, and are in hydraulic fluids, heat transfer fluids, lubricants, and plasticizers. PCBs were mass-produced globally until they were banned in 1979. Triclosan is one of the ingredients in commercial products like, liquid body wash and soaps.

7.5 Endocrine Responses to Hunger and Thirst and Related Disorders

The feelings of appetite and satiety may pose another food associated challenge and involves interactions between hormones from the gastrointestinal tract to the hypothalamus and subsequent feedback looping. Specific hypothalamic regions release hormones that trigger sensations of appetite and satiety leading to food consumption and subsequent feeling of fullness and leading to energy homeostasis (McKinley and Johnson 2004). Ghrelin (hunger hormone) (Cummings et al. 2001), a growth-hormone-releasing peptide and leptin, a hormone regulating body weight (adipocyte-derived hormone), are the two most important hormones involved in responses to appetite an satiety. The two hormones send signals to various nuclei within the hypothalamus for food intake and imbalances of these hormones may significantly influence the body's responses to hunger and in turn, energy homeostasis. Ghrelin is also involved in regulating sleep-wake rhythms, taste sensation, and glucose metabolism (Poher et al. 2018; Inui 2001). With a response, just opposite to that of ghrelin, leptin acts as the body's satiety signal. As the levels of ghrelin increase before meals, it stimulates the lateral region of hypothalamus. The ventromedial region of the hypothalamus is stimulated by leptin is and influences satiety. Also, leptin inhibits stimulation of the lateral hypothalamus by ghrelin. Leptin is also involved in reproduction, blood pressure, and the immune system (Zhang et al. 2005).

An imbalance between ghrelin and leptin action may cause different pathophysiology and incorrect energy homeostasis. Obesity, eating disorders (like, Anorexia nervosa, bulimia nervosa), Prader-Willi syndrome (characterized by hyperphagia), Rheumatid arthritis, mood disorders are some important examples of disrupted energy imbalances arising due to abnormal function of these two hormones associated with digestive system (Yeung and Tadi 2025).

Thirst is an urge in humans and animals to drink fluids which is guided by body mechanisms that maintain fluid homeostasis of the body. It is of subjective perception in nature, includes habitual, cultural, and psychogenic drives and is guided by body mechanisms that maintain fluid homeostasis of the body. Hormonal changes (involvement of circulating concentration of some dipsogenic hormones) and hypertonicity of the extracellular fluid, also influence this body urge. The loss of water from body is associated with solute (NaCl) loss and thus, extracellular fluid loss is more significant when NaCl is simultaneously lost with. Thirst as food borne challenge, can trigger a range of compensatory responses are when depletion of either the intra- or extracellular water (and solute). Vasopressin secretion, activation of sympathetic system, stimulation of the renin-angiotensin-aldosterone system, and diminished renal solute and water excretion are some important examples of associated body responses which help in minimizing body fluid changes. However, fluid losses must be replenished to restore body fluids to original state. The immediate response to hypovolemia (conditions where the body has low levels of blood or fluid and including, conditions of loss of body fluids) is activation of components of the endocrine and autonomic nervous systems in a manner that reduces the consequences of reduced cardiac output and falling blood pressure.

Activation of the sympathetic nervous system contributes to increased vascular tone, venous return, heart rate and contractility, and renal sodium and water reabsorption. Elevated plasma vasopressin, renin-angiotensin, aldosterone, epinephrine/norepinephrine, adrenocorticotrophic hormone, and glucocorticoids act directly or indirectly to retain sodium and water or to redistribute blood and interstitial fluids in an attempt to maintain critical regional blood flows. However, eventually it is necessary to correct the absolute deficits in both water and extracellular solutes. This involves the generation of behaviors associated with the acquisition and ingestion of sodium (i.e., salt appetite) in addition to thirst. The generation of behaviors that correct extracellular deficits is similar to the sympathetic and endocrine responses to hypovolemia, in that all of these effector systems require that the central nervous system receive information reflecting the condition of blood and/or interstitial fluid volumes. Body-to-brain signaling of the status of extracellular fluids entails both hormonal and visceral afferent pathways.

7.6 Endocrine Responses to Calcium and Phosphorus and Vitamin Levels

The main endocrine organ involved in control of calcium and phosphate metabolism is parathyroid gland. The parathyroid hormone (PTH) increases calcium and decreases phosphate concentrations in extracellulasr fluids (ECFs) and exerts straight effects on bone and kidney metabolism of calcium and indirect effects on GI metabolism of calcium. The hormone also enhances calcium absorption in kidneys while decreasing renal phosphate reabsorption. The effect of the hormone in the kidney is also seen in activation of vitamin D. It mediates the absorption of calcium from the gut indirectly through its effect on vitamin D. Epinephrine, on the other hand, stimulates PTH release through activation of adrenergic receptors. Though less in effect, magnesium like, calcium also exerts its influence of PTH secretions.

Calcitonin, synthesized as a 32-amino acid long preprohormone, also affects calcium metabolism, acting as counterbalance to PTH and causing hypocalcemia and hypophosphatemia. Its effects are mainly seen in bones. Increased calcium concentrations cause increased secretion of calcitonin. In hypercalcemia, calcitonin secretion increases while that of PTH, decreases. Hypocalcemic conditions, on the other hand, inhibit calcitonin production while PTH restores normal calcium concentrations in the ECFs. The hormones secretin, gastrin, cholecystokinin, and glucagon, stimulate the release of calcitonin.

Vitamin D, a steroid like molecule, is crucial in calcium absorption from the gut and it's biologically inactive till transformed by liver and kidney to its active form. It also affects the bone facilitating calcium movement into ECFs and the resorption of bone, as well as enhancing the effects of PTH on bone metabolism of calcium (Greco 2012).

7.7 How Does Endocrine Respond Towards Food Borne Challenges?

Chemicals known as endocrine disruptors have the ability to affect endocrine systems when taken in excess. Cancerous tumors, birth abnormalities, and other developmental diseases can result from these disturbances. Hormone disruptors have the ability to throw off any hormone-regulated system in the body. Endocrine disruptors, in particular, may be linked to the emergence of learning difficulties, severe attention deficit disorder, and issues with cognition and brain development. Issues with sexual development, such as the feminization of men or the masculinizing effects on women; bodily deformities, such as breast, prostate, thyroid, and other cancers (Colborn 2003).

The endocrine system aids in preserving homeostasis during disturbances brought on by dangerous microbes, poisons, or nutritional imbalances, it is essential in reacting to foodborne illnesses. A variety of endocrine pathways can be activated by an immune response that is triggered when foodborne pathogens, such as bacteria, viruses, or parasites, enter the digestive tract. The gut-associated lymphoid tissue (GALT), which can signal the release of inflammatory cytokines and other immune mediators, is the body's first line of defense. The hypothalamus may be triggered by this immunological response, which causes the pituitary gland to release stress hormones from the adrenal glands, such as cortisol. By encouraging the creation of glucose, reducing inflammation, and adjusting the immune system to combat the infection, the steroid hormone cortisol aids in controlling the body's reaction to stress.

Foodborne illnesses can affect the endocrine system's control over digestion and metabolism in addition to immunological reactions. Hormones that affect appetite and fullness, including ghrelin and leptin, are produced in the gut. These hormones can become dysregulated during an infection or foodborne sickness. For example, the body may respond to infection by increasing the secretion of ghrelin, which increases appetite, in an effort to make up for energy loss from diarrhea or fever. On the other hand, metabolic alterations brought on by the illness may result in a drop in leptin levels, which indicate fullness. Though they can also add to the agony and exhaustion frequently felt with foodborne diseases, these hormone changes are intended to support the body's healing process and encourage energy balance.

7.8 Endocrine Responses to Diabetes and Obesity

Diabetes has a major effect on the endocrine system, which is essential for controlling metabolism (Matsuura 2023). as blood glucose levels are mostly controlled by the endocrine system. Cellular uptake of glucose from the bloodstream is influenced by insulin. Type 1 diabetes prevents the body from producing enough insulin for glucose metabolism. Consequently, blood glucose levels sharply increase, resulting in hyperglycemia. The pancreas responds by releasing glucagon, a hormone that encourages the liver to release stored glucose, in response to low blood sugar levels but the given reaction makes the blood sugar levels worse. Inadequate amount of insulin and counter-regulatory actions of hormones like glucagon and cortisol, further trigger a series of metabolic abnormalities. The situation may complicate into diabetic ketoacidosis (DKA) and long-term harm to important organs as long as blood glucose levels stay high. Diabetic nephropathy, retinopathy, and neuropathy can result from long-term hyperglycemia's effects. The endocrine system plays a crucial role in managing Type 1 diabetes since the dysregulation of glucose metabolism in this condition requires lifetime insulin replacement therapy to control blood glucose levels and avoid these serious consequences.

Obesity may cause significant imbalances in hormones that can lead to a number of metabolic diseases. Adipose tissue, understood to be an active endocrine organ secretes adipokines, a class of signaling chemicals like, leptin, resistin, and adiponectin, involved in controlling hunger, insulin sensitivity, and inflammation. These chemicals are produced in significantly high quantities by the extra fat tissue in obese people. For example, obese people may encounter leptin resistance which subsequently causes overeating. The obesity triggered overproduction of adipokines additionally raises the risk for type 2 diabetes and aggravates insulin. The resulting hormonal changes may also impact the thyroid, gonads, and pancreas. The reproductive health may also be compromised due to resultant imbalances in sex hormones. Poor testosterone levels and poor spermatogenesis in males, and irregular menstrual periods, polycystic ovarian syndrome (PCOS), and fertility issues in women may be linked to arising hormonal issues (Alvarez-Castro et al. 2012).

7.9 Effect of EDCs on Endocrine System

Erythrosine, a known synthetic red dye, used in various industrial applications, is known to significantly impact human body as it disrupts the normal functioning of endocrine glands, such as the thyroid gland or the adrenal glands, which leads to imbalances in physiological processes (Dees et al. 1997). Erythrosine may modulate the influences of estrogen in the body, which can disrupt hormonal balance and impact estrogen-dependent functions. It has also been found to have anti-estrogenic effects in certain cases (Dees et al. 1997; Hiasa et al. 1988). Suggestions on possible link between erythrosine exposure and DNA damage, cell proliferation in certain cell types and tumor development especially in the thyroid and adrenal glands, have also been made. It may affect the binding of thyroid hormones to their receptors, and disrupt associated gene expression and other physiological processes (Dees et al. 1997; Hiasa et al. 1988). Suggestions on possible link between erythrosine exposure and DNA damage, cell proliferation in certain cell types and tumor development especially in the thyroid and adrenal glands, have also been made (Hiasa et al. 1988). Erythrosine may induce hypersensitivity reactions. Individuals who are allergic to other food dyes or related compounds may also experience allergic reactions to erythrosine due to cross-reactivity (Hiasa et al. 1988; Sarah and Michael 2012).

Tartrazine is a yellow coloring agent widely used in food industries (Cohon 1975) that can potentially bind to hormone receptors and influence cellular functions and hormone-related processes. It can potentially modulate the activity of enzymes involved in hormone biosynthesis, leading to alterations in hormone synthesis or clearance (Axon et al. 2012; Mindang et al. 2022) and may interfere with hormone signaling pathways (Poul et al. 2009; Atli et al. 2017).

Bisphenol A (BPA) is widely employed to prevent corrosion and maintain the quality of the contents, which are clear, hard, and durable (Park et al. 2020) and may enter food from the coating. It can mimic or modulate estrogen activities, disrupt normal hormonal signaling and change physiological processes regulated by estrogen (Okugbe and Ohore 2019). It can also inhibit the production and transportation of thyroid hormones. BPA can also interact with androgen receptors and can disrupt androgen signaling and potentially affect processes regulated by androgens. It is also associated with changes in epigenetic modifications like, DNA methylation, etc. (Kawa et al. 2021). It is linked to alterations in the development and function of the reproductive system, can influence brain development, altered metabolic development in prenatal and early life. A greater risk of obesity, type-2 diabetes, and metabolic syndrome later in life has been linked to BPA exposure as well. Additionally, BPA exposure can affect susceptibility to breast cancer (La Merrill et al. 2020).

Artificial Sweeteners like aspartame and acesulfame potassium, have been suggested to influence influence hormones, such as insulin, leptin, and ghrelin (Karstadt 2006), affects functioning of the gut and gut microbiota, and influencing glucose metabolism (Birnbaum and Suzanne 2003).

Phthalates could also interfere with hormone signaling in the body are not intentionally used as additives in processed foods but can enter food from packaging materials or contamination during processing, storage, or transportation (Di Paola et al. 2022) especially under high temperatures or prolonged storage conditions. These substances can interact with hormone receptors, like, estrogen receptors (ERs), androgen receptors (ARs), and peroxisome proliferator-activated receptors (PPARs) and can mimic or completely stop the activities of natural hormones. Butylbenzylphthalate and diisononyl phthalate, have been shown exhibit anti-androgenic effects, leading to adverse effects on reproductive function and development (Chen et al. 2014; Olujimi et al. 2010). They can interfere with the activity of enzymes like, aromatase, or 5-alpha reductase causing hormone level disruptions. Phthalates can also affect the metabolism and transport hormones and binding proteins in the bloodstream (Olujimi et al. 2010). They may also influence the activation and signaling pathways mediated by hormone receptors. The hormone signaling pathways may also be affected leading to changes in gene expression (Casals et al. 2008; Boberg et al. 2010).

Parabens are produced by esterification of para-hydroxybenzoic acid with an alcohol and are used as preservatives in processed foods like, sauces and dressings, baked goods, and beverages (Boberg et al. 2010). Parabens inhibit the activity of enzymes involved in conversion of androgens to estrogens and those involved in hormone metabolism. They elicit weak estrogenic activity and mimic the effects of estrogens and can interfere with hormonal balance and estrogen-signaling pathways. Some parabens may inhibit aromatase activity, and interfere thyroid peroxidase activity (Nicolopoulou et al. 2015). They may affect the transport, distribution, and bioavailability of hormones in the body by interacting with certain hormone-transporting proteins like, sex hormone binding globulins (Jagne et al. 2016; Damáris et al. 2023; Macedo et al. 2023).

7.10 Conclusion

Food borne challenges, whether arising due to food borne illnesses (microbial or otherwise), can pose serious threat to the wellbeing of individuals. The presence of numerous substances directly in our foods or through packaging that are linked with endocrine disruption, add another dimension to food safety needs and efficient food management systems. Awareness about possible threats, recognition of metabolic disorders and lifestyle disorders as potent food borne challenges, is also needed overcome the challenges posed by these disorders. Identification of food additives as significant EDCs, and their possible health hazards is much needed so as to ensure minimal exposure to EDCs, thereby reducing the risk of hormonal disorders and improving the quality of life and environment. Making informed choices by consumers and leading a healthy lifestyle to better manage metabolic disorders/ lifestyle disorders are important steps in that direction.

References

Alvarez-Castro P, Sangiao-Alvarellos S, Cordido F (2012) In: El Solh AA (ed) Metabolic and endocrine physiology in obesity. https://doi.org/10.1002/9781119962083.ch4

Anonymous (2025). https://bio.libretexts.org/Courses/Cosumnes_River_College/Introductory_Anatomy_and_Physiology_(Aptekar)/08%3A_Endocrine_System/8.02%3A_Introduction_to_the_Endocrine_System). Accessed 2 Apr 2025

Atli SZ, Gunes B, Yedier SK, Sekeroglu V, Aydan B (2017) Effects of tartrazine on proliferation and genetic damage in human lymphocytes. Toxicol Mech Meth 27:370–375

Axon A, May FEB, Gaughan LE, Williams FM, Blain PG, Wright MC (2012) Tartrazine and sunset yellow are xenoestrogens in a new screening assay to identify modulators of human oestrogen receptor transcriptional activity. J Toxicol 298:40–51

Aziz S, Ameer T, Younus M, Nisa Qu NI, Ali U, Tayyaba A, Hanif AM, Munir B (2023) Salmonellosis: food borne plague. In: One health triad, vol 2. Unique Scientific Publishers, Faisalabad, pp 18–24

Birnbaum LS, Suzanne EF (2003) Cancer and developmental exposure to endocrine disruptors. Environ Health Perspect 111:389–394

Boberg J, Taxvig C, Christiansen S, Hass U (2010) Possible endocrine disrupting effects of parabens and their metabolites. Reprod Toxicol 30:301–312

Casals C, Cristina JN, Feige BD (2008) Interference of pollutants with PPARs: endocrine disruption meets metabolism. Int J Obes 32:53–61

Chen X, Xu S, Tan T, Lee ST, Cheng SH, Lee FWF, Xu SJL, Ho KC (2014) Toxicity and estrogenic endocrine disrupting activity of phthalates and their mixtures. Int J Environ Res Public Health 11:3156–3168

Cohon MS (1975) Tartrazine revisited. Drug Intell. Clin Pharm 9:198–203

Colborn T (2003) Neurodevelopment and endocrine disruption. Environ Health Perspect 112(9):944–949. https://doi.org/10.1289/ehp.6601

Cummings DE, Purnell JQ, Frayo RS, Schmidova K, Wisse BE, Weigle DS (2001) A preprandial rise in plasma ghrelin levels suggests a role in meal initiation in humans. Diabetes 50(8):1714–1719

Damáris BCA, Denilson de Sousa A, Paula S, Jones BG, D'Angelo Carlo M, Leandro M (2023) Environmental endocrinology: parabens hazardous effects on hypothalamic–pituitary–thyroid axis. Int J Mol Sci 24:15246

Dees C, Askari M, Garrett S, Gehrs K, Heneley D, Ardies CM (1997) Estrogenic and DNA damaging activity of Red No. 3 in human breast cancer cells. Environ Health Perspect 105:625–632

Di Paola D, Natale S, Iaria C, Crupi R, Cuzzocrea S, Spano N, Gugliandolo S, Peritore AF (2022) Environmental co-exposure to potassium perchlorate and cd caused toxicity and thyroid endocrine disruption in zebrafish embryos and larvae (Daniorerio). Toxics 10:198

Greco DS (2012) Endocrine causes of calcium disorders. Top Companion Anim Med 27:150–155

Hiasa Y, Ohshima M, Kitahori YL, Konishi N, Shimoyama Y, Sakaguchi YL, Hashimoto H, Minami S, Kato Y (1988) The promoting effects of food dyes, erythrosine (Red 3) and rose bengal B (Red 105), on thyroid tumors in partially thyroidectomized N-bis (2-hydroxypropyl)-nitrosamine-treated rats. Gan To Kagaku Ryoho 79:314–319

Inui A (2001) Ghrelin: an orexigenic and somatotrophic signal from the stomach. Nat Rev Neurosci 2(8):551–560

Jagne J, Dominique W, Felicia J (2016) Endocrine-disrupting chemicals: adverse effects of bisphenol A and parabens to women's health. Water Air Soil Pollut 227:1–10

Karstadt ML (2006) Testing needed for acesulfame potassium, an artificial sweetener. Environ Health Perspect 114:516–517

Kawa IA, Masood AL, Fatima Q, Mir SA, Jeelani H, Manzoor S, Rashid F (2021) Endocrine disrupting chemical Bisphenol A and its potential effects on female health. Diab Metab Syndr 15:803–811

Khanna C, Singh S, Vyas M (2018) Antimicrobial potential of *Tylophora indica* and its future considerations in health and food industry. Pharmacogn Rev 12(1):72–77

Khanna C, Singh S, Vyas M, Das S (2019) Biological potential of semi-purified enterocin of enterococcus sp. YT3 against selected food pathogens. Orient J Chem 35(5):1584–1596

Khanna C, Singh S, Vyas M, Das S (2020) Evaluation of antimicrobial behaviour of *Tylophora indica* leaf extracts against selected food microbial pathogens. Res J Pharm Technol 13(11):5224–5228

Khanna C, Singh S, Vyas M (2023) Nisin and plant extracts as promising antimicrobials for active food packaging formulations. In: International conference on materials for emerging technologies-2021. Lovely Professional University, AIP Publishing. Article ID: 0202021

La Merrill MA, Vandenberg LN, Smith MT, Goodson W, Browne P, Patisaul HB, Guyton KZ, Kortenkanp A, Cogliano VJ, Woodruff TJ, Rieswijk L, Sone H, Korach KS, Gore AC, Zeise L, Zoeller RT (2020) Consensus on the key characteristics of endocrine-disrupting chemicals as a basis for hazard identification. Nat Rev Endocrinol 16:45–57

Macedo S, Teixeira E, Gaspar TB, Boaventura P, Soares MA, Miranda-Alves L, Soares P (2023) Endocrine-disrupting chemicals and endocrine neoplasia: A forty-year systematic review. Environ Res 218:114869

Matsuura R (2023) Diabetes and its impact on the endocrine system: diabetes endocrinology. Diabetes Manag 13(10)
McKinley M, Johnson AK (2004) The physiological regulation of thirst and fluid intake. News Physiol Sci 19:1–6. https://doi.org/10.1152/nips.01470.2003
Mindang ELN, Awounfack CF, Ndinteh DT, Krause RWM, Njamen D (2022) Effects of tartrazine on some sexual maturation parameters in immature female Wistar rats. Int J Environ Res Public Health 19:10410
Nicolay N, Thornton L, Cotter S, Garvey P, Bannon O, McKeown P, Cormican M, Fisher I, Little C, Boxall N, Pinna ED, Peters TM, Cowden J, Salmon R, Mason B, Invine N, Rooney P, Flanagan DO (2011) Salmonella enterica serovar Agona European outbreak associated with a food company. Epidemiol Infect 2011(139):1272–1280
Nicolopoulou S, Polyxeni LH, Annie JS (2015) Cosmetics as endocrine disruptors: are they a health risk. Rev Endocr Metab Disord 16:373–383
Okugbe E, Ohore SZ (2019) Endocrine disrupting effects of bisphenol A exposure and recent advances on its removal by water treatment systems. Sci Afr 5:e00135
Olujimi O, Fatoki OS, Odendaal JP, Okonkwo JO (2010) Endocrine disrupting chemicals (phenol and phthalates) in the South African environment: a need for more monitoring. Water SA 36:671–682
Onyeaka H, Anumudu C, Okolo C (2022) A review of the top 100 most cited papers on food safety. Qual Assur Saf Crops Foods 14(4):91–104. https://doi.org/10.15586/qas.v14i4.1124
Park C, Song H, Choi J, Sim S, Kojima H, Park J, Iida M, Lee Y (2020) The mixture effects of bisphenol derivatives on estrogen receptor and androgen receptor. Environ Pollut 260:114036
Poher AL, Tschöp MH, Müller TD (2018) Ghrelin regulation of glucose metabolism. Peptides 100:236–242
Poul M, Jarry G, Elkhim MO, Poul J-M (2009) Lack of genotoxic effect of food dyes amaranth, sunset yellow and tartrazine and their metabolites in the gut micronucleus assay in mice. Food Chem Toxicol 47:443–448
Regan F, Fogarty B (2005) Endrocrine disrupting chemicals. Encyclopedia Anal Sci:495–502. https://doi.org/10.1016/B0-12-369397-7/00138-2
Růžičková V, Karpíšková R, Pantůček R, Pospíšilová M, Černíková P, Doškarˇ J (2008) Genotype analysis of enterotoxin H-positive Staphylococcus aureus strains isolated from food samples in The Czech Republic. Int J Food Microbiol 121:60–65
Sachan R, Das S, Singh S, Khanna C (2020) Food packaging formulation using chitosan and bacteriocin as an antimicrobial agents. Eur J Mol Clin Med 7(7):3053–3067
Sarah KF, Michael J (2012) Review: toxicology of food dyes. Int J Occup Environ Health 18:220–246
Singh S, Das S (2017) Evaluation of market curd for sanitary quality and bacteriocin-producing lactic acid bacteria for potential application as a natural, healthy food preservative. Res J Pharm Technol 10(4):1029–1033
Sturmhöfel SH, Bartke A (1998) The endocrine system: an overview. Alcohol Health Res World 22(3):153–164
Tack DM, Ray L, Griffin PM, Cieslak PR, Dunn J, Rissman T, Jervis R, Lathrop S, Muse A, Duwell M, Smith K, Tobin-D'Angelo M, Vugia DJ, Kufel JZ, Wolper BJ, Tauxe R, Payne DC (2019) Preliminary incidence and trends of infections with pathogens transmitted commonly through food – foodborne diseases active surveillance network, 10 U.S. sites, 2016–2019. Am J Transplant 19:1859–1863
Watamura SE (2008) Endocrine system. Encyclopedia Infant Early Child Dev:450–459. https://doi.org/10.1016/B978-012370877-9.00057-8
Yeung AY, Tadi P (2025) Physiology, obesity neurohormonal appetite and satiety control. StatPearls Publishing, Treasure Island
Yilmaz B, Terekeci H, Sandal S, Kelestimur F (2020) Endocrine disrupting chemicals: exposure, effects on human health, mechanism of action, models for testing and strategies for prevention. Rev Endocr Metab Disord 21(1):127–147. https://doi.org/10.1007/s11154-019-09521-z
Zhang F, Chen Y, Heiman M, Dimarchi R (2005) Leptin: structure, function and biology. Vitam Horm 71:345–372

Chapter 8
Developmental Physiology and Food Safety

Dipjyoti Sharma, Abhilash Bharadwaj, Pollobi Porasar, Siddharth Sankar Das, Tarun Kumar Dua, Md. Rocky Shaikh, Darshana Hazarika, Suman Lamichhane, Jitendra Pandey, and Koushik Nandan Dutta

8.1 Introduction to Developmental Physiology

Developmental physiology is a branch of biology that explores the ways in which organisms develop, grow and mature from birth through adulthood (Hanson and Gluckman 2014; Slack and Dale 2021). It seeks to understand the processes that control cellular differentiation, organogenesis, and general physiological function at various stages of life. Developmental physiology draws upon concepts from genetics, molecular biology and environmental sciences to elucidate how internal and external forces drive developmental processes (West-Eberhard 2003). It is imperative to have a thorough understanding of developmental physiology in health sciences due to the fact that several physiological systems in organisms experience

D. Sharma · A. Bharadwaj · P. Porasar · K. N. Dutta (✉)
Department of Pharmacognosy, NETES Institute of Pharmaceutical Science, NEMCARE Group of Institution, Guwahati, Assam, India

S. S. Das
Department of Pharmaceutical Chemistry, NETES Institute of Pharmaceutical Science, NEMCARE Group of Institution, Guwahati, Assam, India

T. K. Dua · M. R. Shaikh
Department of Pharmaceutical Technology, University of North Bengal, Darjeeling, West Bengal, India

D. Hazarika
Department of Community Health Nursing, Nemcare institute of Nursing Sciences, Mirza, Assam, India

S. Lamichhane
Department of Pharmacy, Chitwan Medical College, Chitwan, Nepal

J. Pandey
Department of Chemistry, University of Hawaii at Manoa, Honolulu, HI, USA

T. Sarkar et al. (eds.), *Physiological Perspectives on Food Safety: Exploring the Intersection of Health and Nutrition*, https://doi.org/10.1007/978-3-031-94582-3_8

decisive modifications during the initial stages of life (Miller and Davidson 2013). Not only do these alterations decide the immediate survival of the organism, but they also define long-term health consequences. For example, immune system, metabolic and neurological function development during early life heavily influences susceptibility to diseases later in life (Bilbo and Schwarz 2009). Developmental physiology contributes importantly to medical and biological sciences by offering an understanding of normal patterns of growth, congenital disease and adaptive reactions to environmental stimuli (Hanson and Gluckman 2014; Hochberg et al. 2011).

Organogenesis is determined by genetic, maternal and environmental influences (Breuninger and Lenhard 2010). Cardiovascular, nervous and skeleton development are subject to drastic modification during formation. Malformations or dysfunction within organ systems are caused by insufficient or imbalanced nutrients (Schwank and Basler 2010). Metabolic and developmental diseases like diabetes and thyroid disorders frequently have developmental causes (Hoffman et al. 2021). Growth hormone, insulin and thyroid hormones regulate metabolism and tissue growth (Fowden 1995). Endocrine-disrupting chemicals such as those in the womb can result in long-term reproductive and metabolic dysfunction. The hypothalamic-pituitary-adrenal axis forms during early life and is involved in stress response and metabolic processes (Panagiotidou 2024). Early-life stimulation has a great impact on neuronal connections and synaptic plasticity (Bishop and Rutter 2008). Deficits in necessary nutrients such as folic acid, iron and omega-3 fatty acids can affect brain development. Neurodevelopmental disorders such as ADHD, autism spectrum disorder and learning disability may have roots in early developmental stages (Heland et al. 2022). Epigenetic changes during development may have lifelong health effects (Hemberger and Dean 2023). Genetic susceptibilities when interacting with environmental exposures can decide the risk of chronic diseases in a person. Stress and environmental exposures to toxins such as air pollutants, pesticides, and industrial chemicals can adversely affect fetal and childhood development. The gut microbiome is also important in developmental physiology and is involved in immune function and metabolic wellness. Nutrition is one of the most powerful determinants of developmental physiology. The intake of nutrients during periods of crucial development such as prenatal and early childhood years can leave enduring impacts on well-being. Micronutrients (vitamins and minerals) and essential macronutrients (carbohydrates, fats, and proteins) need to be sufficiently supplied to facilitate correct growth and physiological processes (Norris et al. 2022).

There is a great effect of early nutrition on physiological development. Maternal malnutrition may result in conditions such as low birth weight which is linked with increased cardiovascular diseases and metabolic disorders and it may predispose to obesity, insulin resistance and hypertension in later life (Hoffman et al. 2021). Exposure to toxins like alcohol, nicotine, and endocrine disruptors may result in fetal developmental abnormalities such as neurological and metabolic disorders(Ross and Desai 2021). Poor diet during infancy leads to stunted growth, impaired immunity and cognitive dysfunction. Colostrum, the milk secreted first after birth contains antibodies that aid in immune function. Breast milk composition varies according to the diet of the mother, which influences the quality of nutrition that reaches the infants (Innis 2014). Dietary deficiencies or surfeits in early childhood affect growth patterns, risks for obesity and metabolic disorders like type 2 diabetes. Excessive use of processed food and sugar-sweetened drinks leads to childhood obesity and insulin resistance. Lack of iron, vitamin D and calcium results in anaemia, weak bone growth and poor cognition. Taste preference and eating behaviours in adulthood are shaped by early-childhood dietary patterns (Kadosh et al. 2021). Proper nutrition from whole foods, vegetables and lean proteins ensures optimal physical and mental development. Poor nutrition during childhood has been associated with higher chances of developing cardiovascular diseases, osteoporosis and immune dysfunction in later life.

Food safety plays a major role in guaranteeing that people especially infants and toddlers get clean and nutritious food. Exposure to toxic compounds during developmental periods can result in serious health complications such as developmental disorders and chronic diseases (Barouki et al. 2012). Early childhood lead exposure is associated with cognitive dysfunction and behavioural problems. Mercury, which is commonly present in contaminated fish may impact brain development. Cadmium exposure through contaminated water and food items may impair kidney function and bone density. Long-term exposure to heavy metals has been associated with neurodevelopmental disorders and lower IQ in children. Prevention of exposure includes eating foods with lower contamination rates and more stringent regulatory measures (Heng et al. 2022). Certain pesticides interfere with hormonal processes and reproductive health can cause neurotoxicity. Some food colourings and flavour enhancers can cause allergic reactions and changes in behaviour (Ukwo et al. 2022). Bacterial infection in food may causes diarrhoea, dehydration and malnutrition among young children. Phthalates in plastic packaging can migrate into food and affect endocrine function. Repeated exposure to endocrine disruptors can modify metabolic processes, which can lead to obesity and diabetes. Alternative materials for packaging and tighter regulations on food contact materials can minimize exposure (Duncan 2011).

8.2 Role of Nutrition in Developmental Physiology

Nutrients are essential for human growth and development, influencing various physiological and metabolic processes. Macronutrients, including proteins, fats and carbohydrates serve as the body's primary energy sources and structural components. Meanwhile, micronutrients such as vitamins and minerals though required in smaller quantities are crucial for cellular function, immune response and neural development. Proper nutrition is vital throughout all stages of life from prenatal development to adulthood as it affects physical growth, cognitive development and overall health outcomes(Savarino et al. 2021).

8.3 Prenatal, Infancy Nutrition, Adolescence and Nutritional Demands

During pregnancy adequate intake of micronutrients is essential for fetal growth and development. Folic acid supports DNA synthesis and cell division reducing the risk of neural tube defects while Iodine plays a key role in thyroid function and cognitive development (Koletzko et al. 1998). Iron is crucial for oxygen transport and brain development with deficiency leading to anaemia and developmental delays. While calcium and vitamin D are vital for fetal bone formation and skeletal growth. Breast milk provides an optimal balance of nutrients for rapid infant growth, supplying vitamins A, C and E along with calcium and phosphorus, to support immune function, tissue growth, and brain development. For infants who cannot be breastfed, fortified formulas help ensure proper nutrient intake (Savarino et al. 2021; Gaidhani et al. 2021). Adolescence is marked by rapid growth, hormonal changes, and increased metabolic demands. Specific nutrients play a crucial role (Savarino et al. 2021). A balanced diet during adolescence is essential for preventing nutrient deficiencies and ensuring proper physical and cognitive development (Tasginn 2017). During puberty, proper nutrient intake is essential for growth and development. Protein supports muscle and tissue growth, while zinc plays a key role in cell division and immune function, aiding growth spurts. Iron is particularly important for menstruating females to prevent anaemia, and calcium with vitamin D ensures peak bone mass acquisition and long-term skeletal health (Savarino et al. 2021; Koletzko et al. 1998; Gaidhani et al. 2021). Maternal nutrition includes both macronutrients and micronutrients, which each are needed in different proportions to maintain the mother's body and the growing foetus. The graphical representation of the Maternal diet on fetal development is shown in Fig. 8.1.

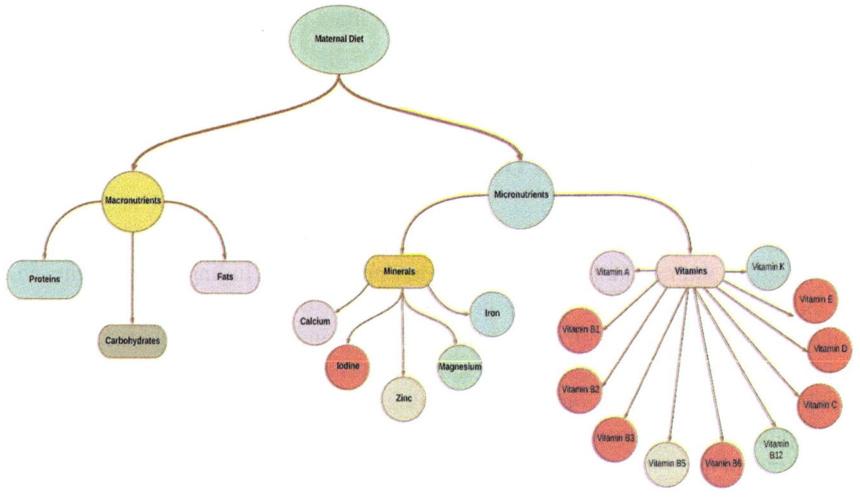

Fig. 8.1 Maternal diet on fetal development

1. **Macronutrients:** To sustain metabolic processes, construct tissues, and provide energy, pregnant women require larger amounts of the macronutrient's protein, carbohydrates, and fat.

   (a) Proteins: Tissue regeneration and cell division both rely on protein. A pregnant woman's protein requirements increase to accommodate the developing baby and placenta and the increased volume of her own blood. Dairy, lentils and lean meats are some examples of high-quality protein sources that may help meet these needs (Danley 2024).

   (b) Fats: Prenatal brain and nerve system development relies on good fats, particularly omega-3 fatty acids. Vegetable oils, nuts, seeds, and seafood all contain these fats. A pregnant woman's daily calorie intake should consist mostly of unsaturated fats and include no more than 10% saturated fats (Danley 2024).

   (c) Carbohydrates: In particular, carbs play an essential role in foetal brain development and serve as the body's primary source of energy. Fruits, vegetables, whole grains and other carbs should make up the bulk of a pregnant woman's diet. In order to keep blood glucose levels steady, 45–65% of your daily calories should come from complex carbs (Danley 2024).

2. **Micronutrients:** The body needs to get micro-minerals that are important for nutrition from food sources since it can't produce them on its own. The skeleton system relies on these inorganic nutrients, which also keep the plasma membrane polarity constant and act as second messengers in cell signalling. The regulation of intracellular and extracellular osmolality by minerals is critical for the preservation of cell viability, shape, and blood volume. Minerals have an important role in many biological processes, including electron transport, redox reactions, and enzyme and metallo protein cofactors (Wu et al. 2012).

   (a) Calcium: A person's teeth, cytoskeleton and bones all contain calcium (Villar et al. 2006). Several enzymes and proteins, including as a-amylase, phospholipases A2 and C, protein kinase C, phosphorylase kinase, calmodulin, troponin C, NO synthase, calcium-ATPase and blood clotting proteins are activated by calcium. Hence calcium has a crucial role in digestion, blood coagulation, intracellular proteolysis, cell-to-cell adhesion, and NO generation in several cell types, including endothelial cells and the uterus (Trumbo and Ellwood 2007; López-Jaramillo 1996).

   (b) Iodine: Iodine is crucial for thyroid hormones (triiodothyronine and thyroxine). Thyroid hormones control gene expression, energy consumption, growth, development, and reproduction. Thyroid hormones are necessary for foetal development, enabling radial neurone migration and mylenation (Vanderpas 2006). The iodine recommended daily intake (RDA) rises from 150 mg/day to 220 mg/day, an increase of almost 1/3 to accommodate the growing fetus's requirement for thyroid hormone(Labeling 2004).

   (c) Iron: The human body has more iron than any other trace element. It is an essential part of haemoglobin, myoglobin and cytochromes among other haemoglobin proteins (Winzerling and Law 1997). Iron deficiency anaemia, hypoxia, preterm delivery, intrauterine growth restriction (IUGR), and an increased risk of maternal mortality (Zimmermann 2006). The daily iron need almost quadruples during pregnancy, rising from 15 mg to 27 mg (Labeling 2004).

(d) Magnesium: Magnesium is involved in several metabolic processes as a cofactor for numerous enzymes. These routes include glycolysis, the pentose cycle, the urea cycle, gluconeogenesis, purine and pyrimidine synthesis, fatty acid oxidation, and a-ketoacid decarboxylation, among many others (Romani 2011). Clinical medicine has long made use of magnesium sulphate to treat pre-eclampsia, premature labour, and preterm delivery due to its ability to relax vascular smooth muscle cells and reduce the contraction of the uterine myometrium (Mercer and Merlino 2009).

(e) Zinc: Zinc is an essential element for many biochemical processes, including DNA and protein synthesis and food metabolism. Carbonic anhydrase, carboxypeptidases A and B, insulin, protein kinase C, phospholipase C, superoxide dismutase (cytosol), thymulin, and poly (ADP-ribose) polymerase are among the enzymes and proteins that rely on zinc. Consequently, zinc is essential for proper nutrition metabolism, neuronal function, immunity, growth, development, antioxidative responses, and DNA and protein synthesis (Yin et al. 2009; Merialdi et al. 2004).

- Vitamins: Due to their different levels of solubility, vitamins are categorised as either water-soluble or lipid-soluble. Both forms of vitamins are needed in trace quantities as cofactors in biological processes. In contrast to water-soluble vitamins, which are absorbed into the portal circulation by the small intestine, lipid-soluble vitamins can only be properly absorbed into the lymph circulation by the gut when normal fat absorption occurs. A brief description of various vitamins and their role in pregnancy is enlisted in Table 8.1 (Wu et al. 2012), (Meyyazhagan et al. 2023).

1. Vitamin A: Vitamin A and its metabolites are essential for proper immune system development and function because they stimulate various processes including lymphocyte proliferation and differentiation, T-cell activation for cytokine synthesis, tissue-specific homing of lymphocytes, production of antibodies specific to antigens, and the production of NO by activated macrophages, which is a mediator of the immune response (Mora et al. 2008).
2. Folic acid: Folate is crucial for the growth and development of embryos, foetus and infants. A shortage in this vitamin during early pregnancy might affect neural tube formation. Deficiencies in Folate also cause megaloblastiaemia by interfering with red blood cell formation (Koury and Ponka 2004; De-Regil et al. 2015).
3. Vitamin B_6: Vitamin B_6 is essential for the elimination of ammonia from the body, the creation of neurotransmitters, polyamines and histamine. Vitamin B_6 has been used to relieve nausea and vomiting linked with pregnancy, which is of particular importance to women (Merrill and Henderson 1987; Swallow 2010).

Table 8.1 The importance of vitamins with their sources and its significance during pregnancy

Vitamins	Source	Role in pregnancy
Vitamin A (retinol and carotenoids)	Whole milk, liver, eggs, dark-colored fruits and vegetables	Follicle formation and growth; embryonic and fetal survival; possible teratogenicity with high vitamin A consumption; regulation of hematopoiesis
Vitamin B_1 (thiamin)	Meat, unrefined cereal	Maternal health, fetal growth and development, and the brain's ATP production from glucose
Vitamin B_2 (riboflavin)	Meat, grains, legumes	Metabolism of nutrients; development of the fetus; health of the mother
Vitamin B_3 (niacin)	Meat, fresh vegetables	Maternal Well-being, nutrient metabolism, embryonic and fetal survival and fetal growth and development
Vitamin B_5 (pantothenic acid)	Meat, whole grain cereals, legumes, fresh vegetables	Metabolism of nutrients; development and growth of the fetus
Vitamin B_6 (pyridoxal phosphate)	Meat, dairy products, nuts, fresh vegetables	Muscle function, ammonia clearance and amino acid metabolism
Vitamin B_{12} (cobalamin)	Meat	Brain development, erythropoiesis and neural tube formation
Vitamin C	Fresh vegetables, fruits	Development of connective tissues; defense against oxidative stress in the mother and the developing baby
Vitamin D	Can be produced naturally by people when they are exposed to sunshine; found in fatty fish and fortified milk.	Ensures proper blood pressure regulation, bone development and calcium-phosphorus balance
Vitamin E	Vegetable oils; animal fats; meat; fresh vegetables	The lifespan of RBCs throughout embryonic and fetal development
Vitamin K	Vegan diets; meat; bacterial fermentation in the intestines	Quickly stop the bleeding if a blood artery is ruptured.

4. Vitamin B_{12}: Vitamin B_{12} is essential for converting N5-methyl-tetrahydrofolate into tetrahydrofolate, the active form of folate that plays a crucial role in DNA synthesis and homocysteine metabolism (De-Regil et al. 2015). The absorption of bile-derived vitamin B12 by the small intestine is generally sufficient, and acute deficiency is rare in women who begin pregnancy with adequate rich diet (Allen 1994).
5. Vitamin D: Vitamin D is converted into 1,25 dihydroxyvitamin D3 (calcitriol, a hormone) that is crucial for calcium and phosphate metabolism, thus influencing bone growth and mineralization (DeLuca 2004).

8.4 Food Safety and Developmental Risks

Food safety is a critical global concern as contamination with pathogens, toxins, or chemicals can disrupt developmental processes, particularly in vulnerable populations such as infants, children, pregnant women, and immunocompromised individuals (Rather et al. 2017), it is essential for healthy development, particularly during critical life stages. Contamination with pathogens, toxins, heavy metals, and pesticides poses serious risks to human health. Implementing regulatory measures, improving agricultural practices, and raising consumer awareness can help mitigate these risks, ensuring a safer food supply for future generations (Rather et al. 2017).

8.5 Food Safety Issues Affecting Development: Pathogen, Toxins and Chemical Contaminants

Food contamination can occur at various stages, from production to consumption. Contaminants include biological agents, heavy metals, pesticides, and endocrine-disrupting chemicals, all of which can interfere with normal growth and development (Thompson and Darwish 2019). These infections can lead to malabsorption, diarrhoea, and prolonged nutrient deficiencies, which are particularly harmful during early childhood development (Flynn et al. 2019).

Pathogen Contamination Foodborne hazards include bacteria, viruses, parasites, worms, and chemicals. Bacteria like *Salmonella, Campylobacter, E. coli, Listeria*, and *Vibrio cholerae* cause gastrointestinal illness, with sources including raw or undercooked foods, dairy, and water. Viruses like *Norovirus* and Hepatitis A spread through contaminated seafood and produce. Parasites such as *Echinococcus* and *Cryptosporidium* enter food via soil, water, or animals. Worms (*Cestodes, Nematodes, Trematodes*) are linked to poor sanitation (Fung et al. 2018).

Toxins and Chemical Contaminants Chemicals like mycotoxins, marine toxins, POPs, dioxins, and heavy metals contaminate food through pollution, leading to cancer, immune suppression, and organ damage (Fung et al. 2018). **Mycotoxins**: Produced by fungi, mycotoxins (e.g., aflatoxins, ochratoxins, and fumonisins) contaminate crops such as grains and nuts. Chronic exposure can cause immunosuppression, growth retardation, and neurodevelopmental impairments (Milićević et al. 2010). **Heavy Metals**: Lead, mercury, arsenic, and cadmium in food disrupt neurodevelopment and organ function. For instance, prenatal exposure to lead is linked to cognitive deficits and behavioural disorders (Thompson and Darwish 2019). **Pesticides**: Organophosphate and carbamate pesticides disrupt neurodevelopment, causing cognitive delays and hormonal imbalances (Fowden 1995). **Endocrine Disruptors**: Bisphenol A (BPA), dioxins, and phthalates interfere with hormonal regulation, impacting reproductive health and metabolic processes (Borchers et al. 2010). Systemic presentation of pathogens in food & their risk is enlisted in

Table 8.2 and parasites in different foods, incorporating all necessary details is enlisted in Table 8.3 while common chemical contaminants in drinking water are represented in Table 8.4.

Table 8.2 Pathogens in food & their risk

Pathogen	Occurrence	Main transmission routes	Symptoms	Possible complications
Campylobacter jejuni	Intestines of animals and birds, raw milk, untreated water, sewage sludge	Contaminated water, raw milk, raw/undercooked meat or poultry, shellfish	Fever, headache, muscle pain, severe diarrhea (sometimes bloody)	Reactive arthritis, neurological disorders (2–10% of cases)
Listeria monocytogenes	Intestines of humans and animals, soil, silage	Raw milk, soft cheeses, ice cream, raw meats, smoked fish	Flu-like symptoms, nausea, vomiting, diarrhea, meningitis, encephalitis	Spontaneous abortion, stillbirth, septicemia
Salmonella	Intestines and feces of animals; *S. enteritidis* in eggs	Raw/undercooked meat, poultry, eggs, milk	Fever, headache, nausea, vomiting, abdominal pain, diarrhea (1–7 days)	Reactive arthritis, Reiter's syndrome
Shigella	Human intestinal tract	Fecal-oral route, contaminated food by infected handlers	Dysentery, diarrhea, vomiting, cramps, fever	Hemolytic uremic syndrome (some strains have 10–15% fatality rate)
Escherichia coli O157:H7	Intestines of mammals, raw milk, unchlorinated water	Contaminated water, raw milk, undercooked meat, fresh produce	Bloody diarrhea, abdominal cramps, nausea, vomiting, fever	Hemolytic uremic syndrome, kidney failure (3–5% fatality rate)
Vibrio vulnificus	Shellfish and fish, coastal waters	Raw oysters, clams, crabs	Gastroenteritis within 16 h	Septicemia in chronic disease patients (50% mortality)
Yersinia enterocolitica	Soil, water, pigs, cats, dogs, birds, beavers	Meat, raw milk, shellfish	Fever, abdominal pain, gastroenteritis, diarrhea	Appendicitis-like symptoms, septicemia in chronic disease patients
Staphylococcus aureus	Skin, hair, nasal passages of humans and animals	Contaminated food (meat, poultry, eggs, salads, dairy products)	Nausea, vomiting, diarrhea, severe abdominal cramps within 1–6 h	Usually self-limiting, dehydration in severe cases
Clostridium botulinum	Soil, water, plants, intestines of animals/fish	Improperly canned foods, garlic in oil, tightly wrapped food	Blurred vision, difficulty swallowing/ speaking, respiratory distress	Paralysis, respiratory failure, fatal if untreated

Table 8.3 Parasites in different foods, incorporating all necessary details

Parasite	Type	Food sources	Transmission route	Health effects
Cryptosporidium spp.	Protozoa	Contaminated water, raw vegetables	Fecal-oral, waterborne	Diarrhea, dehydration, abdominal pain
Entamoeba histolytica	Protozoa	Contaminated food and water	Fecal-oral	Amoebic dysentery, liver abscess
Giardia lamblia	Protozoa	Contaminated water, raw produce	Fecal-oral, waterborne	Diarrhea, bloating, weight loss
Toxoplasma gondii	Protozoa	Undercooked meat, unpasteurized milk, contaminated soil	Ingestion of cysts in meat or water	Flu-like symptoms, fetal abnormalities (if congenital)
Echinococcus granulosus	Cestode (tapeworm)	Raw vegetables, contaminated water, undercooked meat	Fecal-oral (dog/sheep cycle)	Cystic hydatid disease (liver, lungs)
Echinococcus multilocularis	Cestode	Contaminated wild berries, water	Fecal-oral (fox/dog cycle)	Alveolar echinococcosis (tumor-like liver damage)
Taenia solium	Cestode	Undercooked pork	Ingestion of larvae in pork	Cysticercosis (brain, muscles), epilepsy
Taenia saginata	Cestode	Undercooked beef	Ingestion of larvae in beef	Mild gastrointestinal symptoms
Diphyllobothrium latum	Cestode	Raw or undercooked freshwater fish	Ingestion of larvae in fish	Vitamin B12 deficiency, anemia
Trichinella spiralis	Nematode	Undercooked pork, wild game meat	Consumption of infected meat	Muscle pain, fever, swelling
Ascaris lumbricoides	Nematode	Contaminated soil, raw vegetables	Fecal-oral	Intestinal obstruction, malnutrition
Ancylostoma duodenale / Necator americanus	Nematode (Hookworms)	Contaminated soil (barefoot contact)	Skin penetration	Iron-deficiency anemia, malnutrition
Strongyloides stercoralis	Nematode	Contaminated soil, water	Skin penetration	Persistent diarrhea, weight loss, hyperinfection in immunocompromised
Fasciola hepatica	Trematode	Water plants (e.g., watercress)	Ingestion of metacercariae	Liver fluke disease (hepatomegaly, jaundice)

(continued)

Table 8.3 (continued)

Parasite	Type	Food sources	Transmission route	Health effects
Opisthorchis viverrini / *Clonorchis sinensis*	Trematode	Raw or undercooked freshwater fish	Ingestion of metacercariae	Bile duct obstruction, cholangiocarcinoma (liver cancer)
Paragonimus westermani	Trematode	Raw or undercooked crustaceans	Ingestion of metacercariae	Lung fluke disease (chronic cough, hemoptysis)

Table 8.4 Common chemical contaminants in drinking water

Chemical contaminants	Diseases/health effects caused
Aluminum, arsenic	Skin, bladder, and prostate cancers, Alzheimer's, peripheral neuropathy, cardiovascular, reproductive, immunological, and neurological diseases
Disinfection by-products (Trihalomethanes, Dichloroacetic acid)	Leukemia, reproductive diseases, bladder and colon cancers
Fluoride	Osteosarcoma, skeletal fluorosis
Lead	Occupational cancers, hemoprotein degradation, intellectual disability, antisocial behavior, high blood pressure, heart disease, kidney disease, reduced fertility
Nitrate	Stomach, esophagus, bladder, brain, colon, rectum, pancreas, ovarian, and kidney cancers, adverse pregnancy outcomes, diabetes, thyroid disorders
Pesticide residues (2,4-D, Malathion, Diazinon, Fenpropimorph)	Leukemia, reproductive, immunological, and neurological cancers
Radon	Lung cancer
Sulfate (Gypsum, Anhydrite, Barite, Celestine)	Diarrhea, laxative effect

8.6 Prenatal Risks, Early Childhood Vulnerabilities, and Long-Term Effects

Exposure to environmental contaminants such as mercury, lead, dioxins, and PCBs during prenatal development, infancy, and adolescence can have severe health consequences. Mercury in contaminated seafood affects fetal brain development and increases the risk of neurodevelopmental disorders. Persistent organic pollutants like dioxins and PCBs accumulate in fatty tissues and breast milk, causing endocrine disruption in newborns (Thompson and Darwish 2019). Infants consuming contaminated formula or weaning foods face a high risk of infections and toxic exposure (Fung et al. 2018). Lead, present in contaminated water and food, lowers IQ levels and delays cognitive development (Thompson and Darwish 2019). Exposure to pesticide residues in food can interfere with puberty and metabolic functions (Borchers et al. 2010). Long-term exposure to heavy metals through contaminated food results in neurological disorders and immune system suppression (Thompson and Darwish 2019). Examples of food contamination with different chemicals and health effects of food contaminants are depicted in Tables 8.5 and 8.6 respectively.

8.7 Measures to Promote Food Safety

To mitigate food contamination risks, strategies must be implemented at multiple levels. Regulatory measures include setting maximum residue limits (MRLs) for pesticides, heavy metals, and food additives(Thompson and Darwish 2019), conducting routine surveillance and testing for microbial and chemical contaminants(Rather et al. 2017), and enforcing clear labelling on food products regarding pesticide use, organic certification, and allergens (Borchers et al. 2010). Agricultural and industrial practices such as organic farming help reduce pesticide contamination (Milićević et al. 2010), while using safe irrigation water prevents heavy metal accumulation in crops (Fung et al. 2018). Additionally, proper drying and storage techniques can prevent mycotoxin contamination in food (Borchers et al. 2010). At the consumer level, food safety can be enhanced through proper handling and hygiene, including washing produce, cooking meat thoroughly, and avoiding cross-contamination (Fung et al. 2018). Reducing heavy metal exposure by choosing seafood with lower mercury content, such as salmon over tuna, minimizes health risks (Thompson and Darwish 2019). Finally, limiting processed food consumption helps reduce exposure to food additives, preservatives, and endocrine disruptors (Borchers et al. 2010).

8.7.1 Strategies to Identify Unintended Effects

Certain undesirable changes do occur in the makeup of genetically modified (GM) foods. The development of methods for evaluating the safety of genetically modified (GM) foods has benefited greatly from these alterations. The intended effects may be obscured by unintended alterations to metabolic pathways brought about by gene knock-out and knock-in techniques introduced into the host genome. Nutrient concentrations, secondary metabolite profiles, and the generation of novel toxins are all susceptible to change when metabolic pathways undergo these kinds of modifications(Cassells et al. 2003). Although unforeseen consequences do emerge with genetically modified food crops, it is important to note that this occurs often in traditional plant breeding methods such as chromosomal recombination and point mutations (Mohan 2001). There are established and widely-accepted methods for assessing the food safety of genetically modified crops. To better understand the similarities and differences between genetically modified (GM) and non-GM food

Table 8.5 Examples of food contamination with different chemicals

Food contaminants	Foods affected	Country
Metals/metalloids		
Pb, Hg	Grains and vegetables	China
Pb, Cd	Livestock organs	Morocco
Pb	Sheep livers	Spain
Pb, Cd, Cu, Zn	Agricultural crops	Romania
Cd	Locally produced foods	Belgium
Tl	Lettuce and chard	Germany
Pb, Cd	Soybeans	Argentina
Cu, Zn, Mn, Fe	Fish	Turkey
Mycotoxins		
Deoxynivalenol, Zearalenone, T2 toxin, HT-2 toxin	Wheat, barley, Japanese retail foods	Japan
Aflatoxin, Ochratoxin	Wheat flour	China
Fumonisins	Maize	South Africa
Nivalenol	Cereals and cereal products	Tunisia
Aflatoxins	Groundnut oil	Sudan
Antimicrobials		
Various antimicrobials	Pork meat	Madagascar
Various antimicrobials	Table eggs	Sudan
Various antimicrobials	Milk	Peru
Various antimicrobials	Beef	Nigeria
Various antimicrobials	Meat	Brazil
Polycyclic aromatic hydrocarbons (PAHs)		
Benzo[a]pyrene	Barbecued foods	Sweden
Chrysene, Anthracene, Fluoranthene	Yogurt	Italy
19 PAHs	Grains, flour, and bran	Poland
Total PAHs	Oyster	Japan
Pesticides and polychlorinated biphenyls (PCBs)		
Chlorpyrifos	Catfish	Australia
Chlorpyrifos	Vegetables	China
Chlorpyrifos	Food plant	Algeria
DDTs and other OCPs	Edible offal	Egypt
DDTs and other OCPs	Milk, chicken products	South Africa
DDTs and other OCPs	Milk	Ethiopia
DDTs and other OCPs	Fish	Mozambique
PCBs and OCPs	Baby foods	Korea
OCPs and pyrethroids	Honey	Egypt
PCBs and OCPs	Cereals	Poland
PCBs and OCPs	Milk, yak muscle, and liver	Tibet plateau
Radioactive substances		
Radioactive substances	Water and food	Japan

(continued)

Table 8.5 (continued)

Food contaminants	Foods affected	Country
Radionuclides	Seafood	India
Uranium isotopes	Food	Balkans
Radioactive substances	Water and food	Switzerland

Table 8.6 Examples of health effects of food contaminants

Food contaminant	Health effects
Heavy metals	
Lead (Pb)	Neurotoxicity, anemia, hypertension, kidney damage
Mercury (Hg)	Neurotoxicity, immune dysfunction
Cadmium (Cd)	Kidney toxicity, bone demineralization
Arsenic (As)	Cancer, skin lesions, cardiovascular disease
Mycotoxins	
Aflatoxins	Liver cancer, immunosuppression
Ochratoxin A	Kidney toxicity, carcinogenicity
Fumonisins	Neural tube defects, esophageal cancer
Pesticides & PCBs	
Organochlorine Pesticides (DDT, DDE)	Endocrine disruption, reproductive toxicity
PCBs	Neurodevelopmental toxicity, immune suppression
Polycyclic aromatic hydrocarbons (PAHs)	
Benzo[a]pyrene	Carcinogenicity, DNA damage

crops, which should be evaluated for their possible effects on human and animal health, the notion of significant equivalence was developed (Directorate DE 1993; Joint FAO 2000; Kuiper et al. 2001). Examining the areas around the trans gene to see whether the insertion happened within or near an endogenous gene is the simplest approach to foretell any side effects. Sequencing the site of insertion(s) will be increasingly significant as a technique to anticipate phenotypic changes in the changed organism, even though genome information and gene expression regulation are still constrained (Rogan et al. 2000). The targeted technique involves the examination of a specific known chemical, including micro-nutrients, macronutrients, and anti-nutrients or toxins. These anti-nutrients may impede the digestion of macronutrients and inhibit the gastrointestinal function of important elements. This evaluation of chemical constituents highlights a critical safety examination of genetically modified foods. This evaluation serves as a superior instrument for obtaining representations of the primary components (carbohydrates, proteins, and lipids) as well as the secondary constituents such as vitamins, minerals and toxins such like glycol alkaloids in addition to the structure of proteins (Kim et al. 2008; Jelenic 2005; Yabor et al. 2010). The targeted analysis method; nevertheless, is limited to examining just a narrow spectrum of substances. Furthermore, this method cannot distinguish between anti-nutrients and unidentified poisons. These profiling strategies are proposed as ways to characterise changes in the contents of genetically modified (GM) plants to enhance the likelihood of identifying undesirable results (Lehrer and Bannon 2005). As a tool for characterising changes in the composition of GM plants, profiling approaches have been suggested to boost the possibilities of discovering unexpected consequences (Directorate DE 1993). The insertion of several genes into a crop improves its nutritional value or adds health benefits; this may be especially relevant for genetically modified (GM) food crops. This non-targeted method enables "unbiased" assessment of potential alterations in the host organism's physiology and metabolism at various levels of cellular integration via the use of DNA/RNA microarray technology, proteomics, and hybrid analytical methods. Below, we have covered the possibilities of various tools for investigating plant physiology and molecular genetics (Fiehn et al. 2001). The graphical representation of the various strategies to identify unintended effects is depicted in Fig. 8.2.

8.7.2 *Profiling Methods*

1. **Proteomics:** This technological field investigates genes' proteins in cells, organisms, and tissues under controlled conditions. The current method uses (2DGE) two-dimensional gel electrophoresis to break down proteins by eliminating areas from the gel and utilising particular proteases and examine proteins using MS mass spectrometry and identify fragments using databases (Andersen and Mann

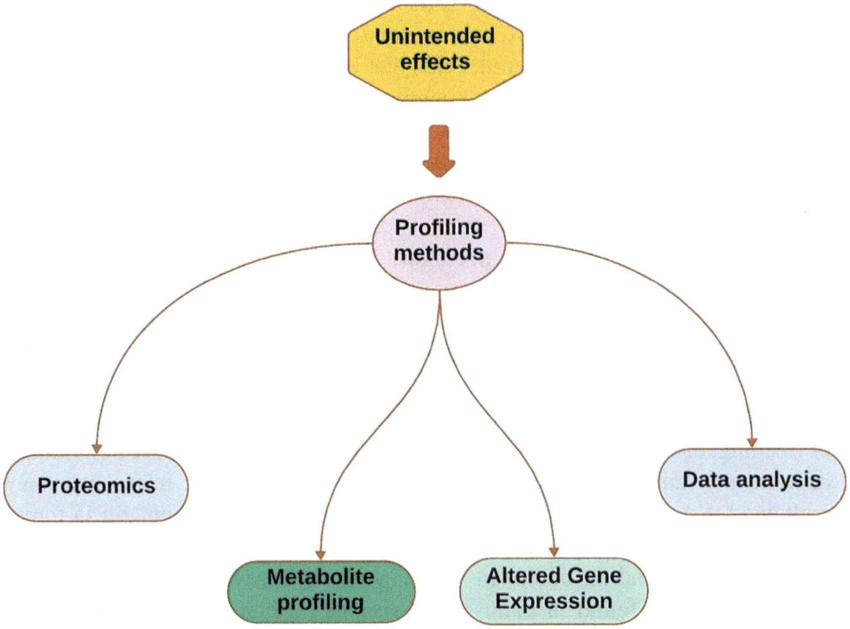

Fig. 8.2 Strategies to identify unintended effects

2000). Conducting relative analysis of protein configurations using diverse proteomic approaches would serve as a crucial instrument for food safety assessment. This approach examines alterations in post-translational modifications and quantifies protein concentrations resulting from genetic mutations or environmental influences (Cánovas et al. 2004; Xie et al. 2007). The measurement of proteins is one of the most difficult tasks. 2DGE methods can only identify proteins with a high level of expression and they have a restricted dynamic range for measurement. Quantification using isotope-coded affinity tags or multidimensional liquid chromatography linked to tandem mass spectrometry are two alternatives that might be considered (Gygi et al. 2000; Jensen et al. 2000; Link et al. 1999; Zhang and McElvain 2000).

2. **Metabolite profiling:** It is ideal to analyse as many individual chemicals as possible in a genetically modified food crop in order to detect changes in the concentration of cellular components including sugars, lipids, acids, and other necessary metabolites. In order to achieve this goal, methods using gas chromatography (GC), high-performance liquid chromatography (HPLC), MS, nuclear magnetic resonance (NMR), or Fourier-transform near-infrared spectroscopy are being developed (Raamsdonk et al. 2001; Johnson et al. 2000; Taylor et al. 1998). In a single sample, these approaches can detect, identify, resolve and quantify a large variety of chemicals. Use is either entirely non-biased or directed towards metabolites in critical metabolic pathways (Fraser et al. 2000). Researchers have developed a metabolite profiling system that uses rice as a model crop. The process entails separating out different parts of the rice extract, which allows for the examination of many different components. Through the use of gas chromatography (GC) in conjunction with flame ionisation detection, we are able to get profiles of molecules that have been silylated or methylated. GC-MS allows us to identify these compounds (Frenzel et al. 2002). Genomic tomato cultivars (modified with antisense RNA exogalactanase to delay ripening) and their non-modified counterparts were analysed using off-line coupling of liquid chromatography to nuclear magnetic resonance (NMR)(Noteborn et al. 2000).
3. **Detection of altered Gene Expression:** Microarray technology is the most recent advance in the evolution of gene expression analysis. This method allows one to track the expression of thousands of distinct genes concurrently and straight to the underlying gene(s) (Watson et al. 1998). Many research initiatives on potato and tomato as model crops have been started in order to provide useful arrays on food crops. These have distinct cDNA levels for red and green stages of ripening. cDNA libraries utilised for the building of arrays in case of tomatoes have been obtained. Whereas the library with green-specific microarray may include cDNAs related with the synthesis of natural poisons, tomatin in tomatoes, the red-specific microarray library may have cDNAs involved with metabolism of vitamins and flavonoids in tomatoes (Arabi et al. 2015). It is possible to compare changes found in plants from various generations, such as those in genetically modified (GM) plants and those from naturally occurring parent organisms with commonly anticipated background configurations. Research is

underway to determine the presence of inherent differences in distinct plant tissues; the findings will demonstrate the use of microarray technology in distinguishing GM food types according to their effects. Efficiently observing predicted results in genetically modified (GM) tomato varieties and identifying undesirable consequences of plant breeding processes are both possible using this method (Lawrie et al. 2007).

4. **Data analysis:** Using profiling methods on even a small sample size generates an enormous amount of data. To make an informed decision on the safety of a genetically modified (GM) food compared to its non-GM equivalent, it is necessary to consider all possible changes. This is why multivariate methods like hierarchical cluster analysis (HCA) and principal component analysis (PCA) are widely recognized (Fiehn et al. 2001), (Roessner et al. 2001). It may not always be able to distinguish between intended and unexpected impacts when using multivariate approaches, despite their usefulness. The massive amount of data produced by profiling techniques necessitates a thorough evaluation for its biological significance. For this reason, it is crucial to establish a network of databases that represent various environmental conditions and phases of development via gene expression, protein, and metabolite profiles (Baxevanis 2001).

8.8 Long-Term Health Implications

Complex chronic respiratory disorders in particular asthma and chronic obstructive pulmonary disease (COPD) are important noncommunicable illnesses that should be investigated throughout the human lifetime in order to develop preventative techniques that may be applied before these diseases have a foothold in the population. In point of fact, chronic obstructive respiratory disorders tend to have their origins in the early stages of life despite the fact that they may not become clinically evident until much later in life (Carraro et al. 2014). The significant advancements in scientific understanding regarding the early origins of adult chronic diseases especially cardiovascular, metabolic and respiratory diseases which revitalised the epidemiological life course perspective in the final quarter of the twentieth century. By the early twenty-first century, life cycle epidemiology has expanded to include several dimensions of health and ageing becoming increasingly integrated within a comprehensive multidisciplinary life course framework. Health during development and ageing (post maturity) is progressively evaluated by functional testing at the individual, systemic and molecular levels enabling the examination of early life drivers of subsequent function and functional alterations, frequently serving as precursors to chronic illness (Kuh and Ben-Shlomo 2016). The notion that adult health may be affected by early life developmental processes is an old one recognised at least since the era of Hippocrates. In experimental science the research conducted by Dörner et al., corroborated this while Forsdahl's study (Forsdahl 1977), demonstrated how adverse events in early childhood persist in people, influencing their later risks. The uses and consequences of modern treatment were well shown by the research of *Barker* et al., (Ktavtzoff et al.

1992; Hall and McKeigue 1999). This resulted in the notion of foetal roots of adult illness. Numerous foundational results in the discipline pertained to the correlation between low birth weight and subsequent illness risk findings that have been well reproduced but remain subjects of ongoing debate. The notion was mechanistically captured in the 'thrifty phenotype' theory (Hanson and Gluckman 2008). Some early clues such as the connections between early developmental occurrences and subsequent adult vulnerabilities have been previously observed sometimes in surprising ways. Kawahata and Sakamoto et al., observed that troops sent in tropical regions during World War II, particularly those originating from warmer climes have a greater number of sweat glands and exhibited a lower susceptibility to heat stroke compared to their counterparts from colder climates. Additionally, Roland et al., presented anecdotal data indicating that World War II prisoners of war who were smaller in stature presumably due to lesser birth size had a higher likelihood of surviving the harsh circumstances of captivity including malnutrition as compared to their greater counterparts (White et al. 2019). Epidemiological studies globally have linked certain environmental influences in early life to modifications in the expression of individual's genetic predispositions resulting in distinct patterns of health and illness. In several animal species environmental factors affect the foetus via the mother/placenta or the young through nursing hence resulting in physiological modifications that enhance the individual's survival prospects in that environment. These predictive adaptive responses in contrast to the thrifty phenotype, lack immediate adaptive value but forecast long-term adaptation aimed at ensuring survival at a minimum until reproductive age. Maternal stress indicates a detrimental external environment for the foetus resulting in persistent hyperactivity of the hypothalamus-pituitary-adrenal axis (HPA) in the person and hence enhancing alertness and increasing survival probabilities (Silveira et al. 2007). One initiative spearheaded by the United States was the increasing focus on evaluating cardiovascular risk in cohort studies of children (e.g., the Bogalusa study established in 1972), (Berenson 2002). Moreover, both the selection of nutritionally sound foods and their balanced inclusion in regular meals are part of the behaviour set that constitutes healthy eating. Consequently, thus further categorisation involves healthy behaviours elements into two groups: weekly food choice and daily meal adequacy. A number of studies have shown that in contrast to obese people, normal and underweight people spread out their caloric intake more equally throughout the day (Bellisle et al. 1988). Despite this, research has shown that consumers often fail to recognise their part in preventing food poisoning and tend to understate both the prevalence of food poisoning and the severity of its effects Habitual activity differs significantly from non-habitual behaviour according to the research. When compared to non-habitual activity which is motivated by conscious control seems to be more influenced by environmental factors. This suggests that conventional methods of health education may not be enough to bring about lasting behavioural changes (Losasso et al. 2012). Sustaining life, encouraging good health and economic progress are all directly tied to the availability of enough quantities of safe and nutritious food. As in case of children, they are lack in the developmental maturity to make informed decisions about what they eat and their immune systems are still maturing, so they aren't always prepared to fight infections. Periodically new rules and regulations are necessary to ensure a safe and healthy food supply for humans. To sum up there is a tight relationship between nutritional value and food safety. Malnutrition and diseases affecting newborns, young children, the old and the ill are perpetuated by contaminated food. Governments, manufacturers, suppliers, distributors and consumers must all work together to guarantee food safety in the twenty-first century due to the fact that food supply chains traverse several national and regional boundaries (Fung et al. 2018; Scharff 2012). Thus, Food safety and adequate nutrition are essential factors in mitigating the risk of non-communicable diseases (NCDs) and promoting optimum health throughout development. These issues are interrelated and impacting public health outcomes via dietary choices and food safety protocols. Adhering to appropriate dietary habits, including the intake of nutrient-dense foods and the regulation of calorie consumption is crucial in the prevention of non-communicable illnesses, including cardiovascular diseases, type 2 diabetes and certain malignancies. A diet abundant in fruits and vegetables and low in saturated fats is advised to reduce these risks (Bidlack 1996). Nutrition is essential throughout several life phases from gestation to senescence. Proper diet promotes healthy growth in youngsters and diminishes the likelihood of chronic illnesses in adults. Particular nutrients are vital for development, cognitive performance and illness prevention. Optimal nutrition enhances the likelihood of favourable pregnancy outcomes, appropriate childhood development and healthy ageing while simultaneously reducing the risk of prevalent diet-related chronic illnesses such as obesity, cardiovascular disease, cancer and type 2 diabetes (Neuhouser et al. 2023). Concerns over food safety have escalated due to reasons such as globalisation, urbanisation, rising disposable income and evolving buying habits that are altering eating patterns globally. In low- and middle-income countries (LMICs), the majority of foodborne illnesses continue to arise from the improper handling and preparation of fresh foods, including animal-derived products, fruits and vegetables. In these areas, food safety issues often affect dietary habits resulting in less intake of fresh produce and heightened dependence on processed foods. Enhancing food safety infrastructure and education may improve access to safe and healthy food (Liguori et al. 2022). Integrated policies are necessary to simultaneously address food safety and nutrition in order to enhance public health outcomes. This encompasses the augmentation of food safety protocols across supply chains and the advocacy of nutrition education to foster better dietary choices. Thus when it comes to lowering the number of cases of foodborne illnesses and improving general health it is essential to implement strategies that aim to enhance consumer knowledge and behaviour (Losasso et al. 2012).

8.9 Challenges and Innovations in Food Safety

Contamination of food is a major challenge for food safety, quality, economic impact and public health. Food contamination along with foodborne diseases has put a lot of pressure on various countries, especially in infants and other susceptible population and moreover has burdened locals and global markets. Awareness for the importance of food safety is continuously growing especially in the developing countries. However, key challenges to globally grow food safety is research on evidence-based decisions and leverage on existing capacity. In today's globalized food supply chain, major challenges lies with bacterial pathogens, antibiotic resistance, mycotoxins and pesticides, that are caused due to lack of food safety knowledge and suitable laws enforced by the government (Abebe 2020; Uyttendaele et al. 2016; Unnevehr 2022). Some of the challenges in food safety are shortly described below.

8.10 Microbial Contamination

Contamination caused by contagious microbes like bacteria, virus, fungi, protozoa and yeast are called microbial contamination. Intake of food containing pathogens or intake of toxins produced by pathogens in food products can lead to foodborne diseases(Thakali and MacRae 2021; Chatterjee and Abraham 2018). To get a food product from the farm to the dining table includes various steps and these steps are collectively called food production chain (Alum et al. 2016). Contamination of fresh produced food is caused due to contaminated irrigation water and soil, use of immature manure, manure stockpiles on the farm, pesticides, livestock and human movements. Plant based-food gets easily contaminated during pre-harvest phase due to unhygienic practices of the workers. Faecal contamination is one of the major cause of contamination in milk at farms (Thakali and MacRae 2021; Chatterjee and Abraham 2018). Dairy sector is one of the most essential parts of a country's economy as thousands of people, including peasants and smallholders depends on this sector. Therefore, contamination in milk is a serious food safety concern (Uyttendaele et al. 2016). Microbial contamination in food also occurs during processing and preparation of food, when it comes in contact with equipment that may harbour microbes and introduce it to the food (Alum et al. 2016).

E. Coli and *Salmonella* are the most common pathogens that causes foodborne diseases globally. When food is contaminated by pathogens it can lead to an outbreak causing high mortalities. More than 800 outbreaks were documented in US in 2017 that led 21 deaths and 800+ hospitalised. *E. coli, Salmonella, Norovirus, Campylobacter* and *Bacillus* were the common pathogens responsible for these outbreaks. In the same year, *Norovirus* caused more outbreaks than any other pathogens whereas *Salmonella* caused more deaths. *Salmonella* is commonly found in livestock and sometimes in fish and derived animal products, therefore during the post-slaughter processing, it may spread to other food material while food handling by cross-contamination. Even though both animal and non-animal origin foods are the vehicle for *Salmonella* infection in human, it is difficult to determine if all the infection cases are directly or indirectly from a particular type of food (Uyttendaele et al. 2016; Thakali and MacRae 2021).

8.11 Anti-biotic Resistance

In agriculture, antibiotics are used for treatment and prevention of diseases and for enhancing feed proficiency and growth of the plant. Genes that grant antibiotic resistance are called antibiotic resistance gene. These genes can be encoded on mobile genetic elements or in the genome as they are acquired through transformation (acquired from the environment), mutation, transduction (transferred by viral infection) or conjugation (directly transferred from different organism). Food act as a transferring vector connecting the use of antibiotics in agriculture and human pathogens resistance (Thakali and MacRae 2021). Antibiotic resistance in the microorganisms like bacteria other foodborne pathogens has emerged and spread in recent decades posing a great threat to the human health. This surge in antibiotic resistance in foodborne pathogens is due to uncontrolled and inappropriate use of antibiotics in agricultural veterinary practices (Abebe 2020; Van et al. 2007). This has effect on the agroeconomic system, ecosystem of the soil, vegetable production system, and quality of ground water. Unhygienic practices by the vendors and workers also contribute to the contamination of fresh fruits and vegetables with antibiotic resistance bacteria. This challenge can be overcome by maintaining personal hygiene among the locals and workers and preventing contamination from air and soil (Rahman et al. 2022).

8.12 Heavy metals Contamination

Heavy metal contamination disrupts the ecosystem and creates serious risk regarding human health (Munir et al. 2021). Some of the heavy metals such as, Zinc, copper, Cobalt, Manganese, chromium and iron in low concentration are edible and essential to the body where as in high concentration they are toxic to humans, plants and animals. However, barium, lithium, tin, aluminium, mercury, cadmium and lead are toxic and non-essential heavy metals that causes harm to the body. Heavy metal contamination is primarily caused due to pollution, usage of metal-based pesticides, transportation, harvesting process and industrial emission. They have a density greater than 4.5 g/sm^3. Prolonged exposure to these harmful heavy metals causes toxic effect as they are not metabolized and accumulates in various tissues such as kidney, central nervous system, liver and peripheral neuronal system. This can lead to gastro intestinal disorders, pneumonia, stomach aches, diarrhoea, ataxia, paralysis, depression, vomiting, stomatitis, haemoglobinuria and convulsion. Moreover, it can damage brain, liver, bone marrow, lungs, kidney and other organs in the body (Thakali and MacRae 2021; Hezbullah et al. 2016; Onakpa et al. 2018). Heavy metal contamination in vegetables is mostly caused by air and soil pollution, when the plant comes in contact with polluted air or contaminated soil. Therefore, a well-planned approach is required to reduce the risk of heavy metal contamination and to improve food safety (Munir et al. 2021).

8.13 Adulteration

Food adulteration is the process of intentionally adding inferior or foreign substance to the authentic food product for various reasons. Adulteration is done in the form of mixing, hiding poor quality substance in packaging material, substitution, selling decomposed food, addition of toxicants and giving false labels or misbranding (Anagaw et al. 2024). Foods and drinks are mostly targets for adulteration and they are harmful to consumers that could lead to various diseases if not managed appropriately. Adulterants are added in the food and drinks to increase their volume and reduce production cost. However, it is also done for deceptive and malicious purposes. It is done because of seasonality in supply or due to high demand of the food productor drinks. Adulteration can be done intentionally or unknowingly. It is mostly done intentionally but sometimes it occurs unknowingly due to lack of hygiene in production site. Food products such as fruits, vegetables, flour, oil, nuts, pulses, milk, spices, milk, chocolates, coffee, tea, honey, juice etc. are easy to adulterate as they are sold in unbranded loose forms. As a result of this customers feels cheated and falls victim to serious illness like cancer, asthma, ulcer etc. Adulteration hence poses a great threat to public health and wellbeing. Therefore, stringent regulations and strict enforcement must rules should be regulated on adulteration of food products (Ayza and Belete 2015; Akhtar 2015).

These factors challenge food quality and safety that could easily affect public health. There are various scientists that have researched and developed new methods and techniques to enhance food safety measures and improve public health. Moreover, these tools can help in reducing the time required for risk assessment and enhancing management of food safety (Allende et al. 2022). Some of the new innovations are mentioned below:

8.13.1 3D Food Printing

3D food printing is building the final product layer by minute layer that works by offering infinite possibilities for the texture, shape, taste and composition of the food products. 3D food printing lets customer customise the end food product to their specific taste and demands. Its similar to cooking food for oneself but without work. This technique helps in producing high-tech healthy food and greatly reducing waste production (Xu et al. 2022).

8.13.2 Surface-Enhanced Raman Scattering (SERS)

SERS technique uses powerful molecular spectroscopy to detect chemical, harmful metal and biological substances. It also helps in detection and identification of food poisons and allergies by utilizing label-free or label-based techniques based on semiconductors and metals as substates. This technique can also be used for food planting, packing, breeding, storage, processing and transportation. It can be used to improve food safety and quality in a rapid, easy, reliable and non-destructive manner (Xu et al. 2022).

8.13.3 Biofertilizers

Biofertilizers uses microorganisms, that are obtained from roots of the plant, to help in promoting the plant growth which further enhances the supply of nutrients to the host plant when administered to seed, soil or plant. Biofertilizers help in increasing microbial activities that enhance the nutrient availability for the plant to absorb and also increases the soil fertility through fixing the atmospheric N_2 in the atmosphere. Moreover, they are helpful in solubilizing the insoluble phosphates in the soil that produces growth promoting plant chemicals, which in turn increases crop yield. It also helps in protecting the plant from diseases and pest, and enhance the survival of seedling and eliminate any harmful chemicals. Biofertilizers are a safe alternative to synthetic or chemical fertilizers that negatively effects the plant and contaminates the soil. A plant cultivated in a safe environment and ecosystem will produce safer food products, therefore, biofertilizers contributes to not only sustainable ecosystem but also overall human well-being (Daniel et al. 2022).

8.13.4 Bio/Chemosensory Devices

Biosensoristic and chemosensoristic devices, collectively known as bio/chemosensoristic devices, helps in analysing environmental and food samples for identification of new safety and quality marker indexes in real matrices and for determination of various effects of the bioactive compounds. These devices are cheap, portable, simple and require less time-consuming protocols. Biosensor devices have become the primary tool for assessing biomolecular interaction in the field of life science and food safety (Dragone et al. 2017).

8.13.5 Nanotechnology in Food Processing

Nanotechnology helps in modifying colour, taste and nutritional properties of the food based on a person's need. It is also used in food production, processing, monitoring as well as packaging. It lowers the cost of food ingredients and increases the shelf-life of the food products. Nanotechnology is used to make smart packaging that could repair small tears in the packages, respond to changes in the temperature and moisture and warn customers about contaminated foods. Nanotechnology also used in food analytical methods that helps in detection of microbes/pathogens and chemical contaminants in food (Adeyeye and Fayemi 2019).

These technologies have revolutionized food safety and quality management that has enabled rapid detection of contamination, reduction in waste, enhance nutrient availability and improving food customization while reducing health risk. They promote sustainable agriculture and safer food production by replacing harmful chemicals and production optimization. Current safety challenges can be addressed by integrating these advancements into global food supply chains, that would ensure healthy consumption and long-term environmental benefits. However, widespread adoption of these technologies requires proper regulatory rules and public awareness. Science driven solutions help in protecting public health and meets evolving consumer demands.

8.14 Conclusion

Food safety is a global concern that directly impacts public health, especially those within vulnerable developmental stages which includes prenatal life, infants and children. The relationship between nutrition, food safety and developmental physiology requires the need for strict measures to ensure good quality and safe foods are available to the consumers. During these periods, exposure to pathogens, chemical contaminants and heavy metals can lead to severe illnesses such as metabolic dysfunction, neural disorders and other chronic diseases. This chapter highlights the importance of maternal nutrition, dietary habits in early childhood, role of macronutrients and micronutrient in optimizing growth and long-term health outcomes. There are various challenges in food safety which includes microbial contamination, antibiotic resistance, heavy metal contamination and food adulteration. Microbes such as *Salmonella, E. coli* and *Norovirus* are threat to humans that causes various foodborne diseases and outbreaks. Antibiotic resistance due to misuse of antibiotics in agriculture reduces the efficacy of the treatments for bacterial infections. Severe toxicological risks are caused due to the accumulation of heavy metals like lead, mercury, arsenic and cadmium, in the food chain. Furthermore, food quality and safety are compromised due to intentional and unintentional adulteration. To overcome these challenges robust regulatory rules and framework is necessary to protect the consumers. Addressing these challenges, new innovative technologies are emerging to enhance food safety and quality. Tools such as 3D food printing, Surface-enhanced Raman Scattering (SERS), biofertilizers, biosensors and chemo sensors and nanotechnology has revolutionized the food industry. Food safety is the collective responsibility that requires global cooperation. A safer, healthier and good quality food supply chain can be achieved only through a holistic approach.

References

Abebe GM (2020) The role of bacterial biofilm in antibiotic resistance and food contamination. Int J Microbiol 2020:1–10

Adeyeye SAO, Fayemi OE (2019) Nanotechnology and food processing: between innovations and consumer safety. J Culin Sci Technol 17(5):435–452

Akhtar S (2015) Food safety challenges-a Pakistan's perspective. Crit Rev Food Sci Nutr 55(2):219–226

Allen LH (1994) Vitamin B12 metabolism and status during pregnancy, lactation and infancy. In: Nutrient regulation during pregnancy, lactation, and infant growth [Internet]. Springer, pp 173–186. Available from: http://link.springer.com/10.1007/978-1-4899-2575-6_14

Allende A, Bover-Cid S, Fernández PS (2022) Challenges and opportunities related to the use of innovative modelling approaches and tools for microbiological food safety management. Curr Opin Food Sci [Internet] 45:100839. Available from: https://linkinghub.elsevier.com/retrieve/pii/S2214799322000418

Alum AE, Urom CSMO, Ben ACMA (2016) Microbiological Contamination of Food: the mechanisms, impacts and prevention. Int J Sci Technol Res 5(3):65–78

Anagaw YK, Ayenew W, Limenh LW, Geremew DT, Worku MC, Tessema TA et al (2024) Food adulteration: causes, risks, and detection techniques—review. SAGE Open Med 12:20503121241250184

Andersen JS, Mann M (2000) Functional genomics by mass spectrometry. FEBS Lett 480(1):25–31

Arabi MIE, Al-Daoude A, Shoaib A, Jawhar M (2015) Accumulation of transcripts abundance after barley inoculation with cochliobolus sativus. Plant Pathol J 31(1):72–77

Ayza A, Belete E (2015) Food adulteration: its challenges and impacts. Food Sci Qual Manag [Internet] 41:50–56. Available from: www.iiste.org

Barouki R, Gluckman PD, Grandjean P, Hanson M, Heindel JJ (2012) Developmental origins of non-communicable disease: implications for research and public health. Environ Health A Glob Access Sci Source 11(1):1–9

Baxevanis AD (2001) The molecular biology database colletion: an updated compilation of biological database resources. Nucleic Acids Res 29(1):1–10

Bellisle F, Rolland-Cachera MF, Deheeger M, Guilloud-Bataille M (1988) Obesity and food intake in children: evidence for a role of metabolic and/or behavioral daily rhythms. Appetite 11(2):111–118

Berenson GS (2002) Childhood risk factors Oredict adult risk Bogalusa. Am J Cardiol [Internet] 90(02):3–7. Available from: https://pdrives.manchester.ac.uk/horde/gollem/view.php?actionID=view_file&type=pdf&file=childhood+risk+factors+oredict+adult+risk+bogalusa.pdf&dir=%2Froot_vdm_9%2Ffs_02_04%2FUSERS%2Fmogcjsk4%2Fsophia&driver=ftp%5Cnpapers2://publication/uuid/F23BC90F-00A8-

Bidlack WR (1996) Interrelationships of food, nutrition, diet and health: the national association of state universities and land grant colleges white paper. J Am Coll Nutr 15(5):422–433

Bilbo SD, Schwarz JM (2009) Early-life programming of later-life brain and behavior: a critical role for the immune system. Front Behav Neurosci 3(AUG):1–14

Bishop D, Rutter M (2008) Neurodevelopmental disorders: conceptual issues. In: Rutter's child Adolesc psychiatry. Wiley, pp 32–41

Borchers A, Teuber SS, Keen CL, Gershwin ME (2010) Food safety. Clin Rev Allergy Immunol 39(2):95–141

Breuninger H, Lenhard M (2010) Control of tissue and organ growth in plants. Curr Top Dev Biol 91:185–220

Cánovas FM, Dumas-Gaudot E, Recorbet G, Jorrin J, Mock HP, Rossignol M (2004) Plant proteome analysis. Proteomics 4(2):285–298

Carraro S, Scheltema N, Bont L, Baraldi E (2014) Early-life origins of chronic respiratory diseases: understanding and promoting healthy ageing. Eur Respir J [Internet] 44(6):1682–1696. Available from: https://publications.ersnet.org/lookup/doi/10.1183/09031936.00084114

Cassells AC, Joyce SM, O'Herlihy EA, Perez-Sanz MJ, Walsh C (2003) Stress and quality in in vitro culture. Acta Hortic [Internet] 625(625):153–164. Available from: https://www.actahort.org/books/625/625_16.htm

Chatterjee A, Abraham J (2018) Microbial contamination, prevention, and early detection in food industry. Microbial contamination and food degradation. Elsevier Inc., pp 23–62. Available from: https://doi.org/10.1016/B978-0-12-811515-2/00002-0

Daniel AI, Fadaka AO, Gokul A, Bakare OO, Aina O, Fisher S et al (2022) Biofertilizer: the future of food security and food safety. Microorganisms 10(6):1220

Danley S (2024) The influence of maternal nutrition on fetal development 2025;200:62–102

DeLuca HF (2004) Overview of general physiologic features and functions of vitamin D. Am J Clin Nutr 80(6 Suppl):1689–1696

De-Regil LM, Peña-Rosas JP, Fernández-Gaxiola AC, Rayco-Solon P (2015) Effects and safety of periconceptional oral folate supplementation for preventing birth defects. Cochrane Database Syst Rev 2015(12):CD007950

Directorate DE (1993) Science DD for, safety DG of NE on. Safety evaluation of foods derived by modern biotechnology: concepts and principles.

Dragone R, Grasso G, Muccini M, Toffanin S (2017) Portable bio/chemosensoristic devices: innovative systems for environmental health and food safety diagnostics. Front Public Heal 5(MAY):1–6

Duncan TV (2011) Applications of nanotechnology in food packaging and food safety: barrier materials, antimicrobials and sensors. J Colloid Interface Sci 363(1):1–24

Fiehn O, Kloska S, Altmann T (2001) Integrated studies on plant biology using multiparallel techniques. Curr Opin Biotechnol 12(1):82–86

Flynn K, Villarreal BP, Barranco A, Belc N, Björnsdóttir B, Fusco V, et al. (2019) An introduction to current food safety needs. Trends Food Sci Technol [Internet] 84:1–3. Available from: https://doi.org/10.1016/j.tifs.2018.09.012

Forsdahl A (1977) Are poor living conditions in childhood and adolescence an important risk factor for arteriosclerotic heart disease? Br J Prev Soc Med 31(2):91–95

Fowden AL (1995) Endocrine regulation of fetal growth. Reprod Fertil Dev 7(3):351–363

Fraser PD, Pinto ES, M, Holloway DE, Bramley PM. (2000) Application of high-performance liquid chromatography with photodiode array detection to the metabolic profiling of plant isoprenoids. Plant J 24(4):551–558

Frenzel T, Miller A, Engel KH (2002) Metabolite profiling – a fractionation method for analysis of major and minor compounds in rice grains. Cereal Chem 79(2):215–221

Fung F, Wang HS, Menon S (2018) Food safety in the 21st century. Biomed J [Internet] 41(2):88–95. Available from: https://doi.org/10.1016/j.bj.2018.03.003

Gaidhani KA, Harwalkar M, Bhambere D, Nirgude PS (2021) World journal of pharmaceutical research formulation. SJIF J [Internet] 2(5):1685–1703. Available from: https://hdl.handle.net/10669/87006

Gygi SP, Corthals GL, Zhang Y, Rochon Y, Aebersold R (2000) Evaluation of two-dimensional gel electrophoresis-based proteome analysis technology. Proc Natl Acad Sci USA 97(17):9390–9395

Hall EF, McKeigue PM (1999) Fetal origins of coronary heart disease. Evid Based Cardiovasc Med [Internet] 3(4):87–88. Available from: https://linkinghub.elsevier.com/retrieve/pii/S1361261199902683

Hanson MA, Gluckman PD (2008) Developmental origins of health and disease: new insights. Basic Clin Pharmacol Toxicol 102(2):90–93

Hanson MA, Gluckman PD (2014) Early developmental conditioning of later health and disease: physiology or pathophysiology? Physiol Rev [Internet] 94(4):1027–1076. Available from: https://www.physiology.org/doi/10.1152/physrev.00029.2013

Heland S, Fields N, Ellery SJ, Fahey M, Palmer KR (2022) The role of nutrients in human neurodevelopment and their potential to prevent neurodevelopmental adversity. Front Nutr 9:992120

Hemberger M, Dean W (2023) The placenta: epigenetic insights into trophoblast developmental models of a generation-bridging organ with long-lasting impact on lifelong health. Physiol Rev 103(4):2523–2560

Heng YY, Asad I, Coleman B, Menard L, Benki-Nugent S, Were FH, et al. (2022) Heavy metals and neurodevelopment of children in low and middle-income countries: a systematic review. PLoS One [Internet] 17(3 March):1–25. Available from: https://doi.org/10.1371/journal.pone.0265536

Hezbullah M, Sultana S, Chakraborty SR, Patwary MI (2016) Heavy metal contamination of food in a developing country like Bangladesh: an emerging threat to food safety. J Toxicol Environ Heal Sci 8(1):1–5

Hochberg Z, Feil R, Constancia M, Fraga M, Junien C, Carel JC et al (2011) Child health, developmental plasticity, and epigenetic programming. Endocr Rev 32(2):159–224

Hoffman DJ, Powell TL, Barrett ES, Hardy DB (2021) Developmental origins of metabolic diseases. Physiol Rev 101(3):739–795

Innis SM (2014) Impact of maternal diet on human milk composition and neurological development of infants. Am J Clin Nutr 99(3):734S–741S

Institute of Medicine (US) Committee on Use of Dietary Reference Intakes in Nutrition Labeling (2004) Dietary reference intakes: guiding principles for nutrition labeling and fortification [Internet]. National Academies Press. Available from: https://books.google.co.in/books?id=XzjYhuCGrL0C&lpg=PP18&ots=Cbm4YR1Hd_&dq=Dietaryreferenceintakes%3A guiding principles for nutrition labeling and fortification&lr&pg=PP1#v=onepage&q=Dietary reference intakes: guiding principles for nutrition labeling

Jelenic S (2005) Food safety evaluation of crops produced through genetic engineering- how to reduce unintended effects? Arh za Hig Rada I Toksikol Ind Hyg Toxicol 56(2):185–193

Jensen PK, Paša-Tolić L, Peden KK, Martinović S, Lipton MS, Anderson GA et al (2000) Mass spectrometic detection for capillary isoelectric focusing separations of complex protein mixtures. Electrophoresis 21(7):1372–1380

Johnson HE, Gilbert RJ, Winson MK, Goodacre R, Smith AR, Rowland JJ et al (2000) Explanatory analysis of the metabolome using genetic programming of simple, interpretable rule. Genet Program Evolvable Mach 1(3):143–258

Joint FAO (2000) Safety aspects of genetically modified foods of plant origin: report of a Joint FAO/WHO expert consultation on foods derived from biotechnology. WHO headquarters, Geneva. Available from: https://coilink.org/20.500.12592/t788xw

Kadosh KC, Muhardi L, Parikh P, Basso M, Mohamed HJJ, Prawitasari T et al (2021) Nutritional support of neurodevelopment and cognitive function in infants and young children—an update and novel insights. Nutrients 13(1):1–26

Kim MY, Son CW, Shim HJ, Lee JH, Lee KJ, Sok DE et al (2008) Effect of cultivars and cooking methods on the trypsin inhibitor activities of potatoes. Food Sci Biotechnol 17(1):161–165

Koletzko B, Aggett PJ, Bindels JG, Bung P, Ferré P, Gil A et al (1998) Growth, development and differentiation: a functional food science approach. Br J Nutr 80(S1):S5–S45

Koury MJ, Ponka P (2004) New insights into erythropoiesis: the roles of folate, vitamin B 12, and iron. Annu Rev Nutr [Internet] 24(1):105–131. Available from: https://www.annualreviews.org/doi/10.1146/annurev.nutr.24.012003.132306

Ktavtzoff R, Desbois I, Doinel C, Colombat P, Lamagnere JP, Chassaigne M et al (1992) Immunological response to L-Asparaginase loaded into red blood cells. In: The use of resealed erythrocytes as carriers and bioreactors [Internet]. Springer US, Boston, pp 175–182. Available from: http://link.springer.com/10.1007/978-1-4615-3030-5_21

Kuh D, Ben-Shlomo Y (2016) Early life origins of adult health and aging. In: Handbook of aging and the social sciences [Internet]. Elsevier, pp 101–122. Available from: https://doi.org/10.1016/B978-0-12-417235-7.00005-6

Kuiper HA, Kleter GA, Noteborn HPJM, Kok EJ (2001) Assessment of the food safety issues related to genetically modified foods. Plant J 27(6):503–528

Lawrie CH, Soneji S, Marafioti T, Cooper CDO, Palazzo S, Paterson JC et al (2007) MicroRNA expression distinguishes between germinal center B cell-like and activated B cell-like subtypes of diffuse large B cell lymphoma. Int J Cancer 121(5):1156–1161

Lehrer SB, Bannon GA (2005) Risks of allergic reactions to biotech proteins in foods: perception and reality. Allergy Eur J Allergy Clin Immunol 60(5):559–564

Liguori J, Trübswasser U, Pradeilles R, Le Port A, Landais E, Talsma EF et al (2022) How do food safety concerns affect consumer behaviors and diets in low- and middle-income countries? A systematic review. Glob Food Sec 32:32

Link AJ, Eng J, Schieltz DM, Carmack E, Mize GJ, Morris DR et al (1999) Direct analysis of protein complexes using mass spectrometry. Nat Biotechnol 17(7):676–682

López-Jaramillo P (1996) Prevention of preeclampsia with calcium supplementation and its relation with the L-arginine:nitric oxide pathway. Braz J Med Biol Res = Rev Bras Pesqui medicas e Biol [Internet] 29(6):731–741. Available from: http://europepmc.org/abstract/MED/9070385

Losasso C, Cibin V, Cappa V, Roccato A, Vanzo A, Andrighetto I, et al. (2012) Food safety and nutrition: improving consumer behaviour. Food Control [Internet] 26(2):252–258. Available from: https://doi.org/10.1016/j.foodcont.2012.01.038

Mercer BM, Merlino AA (2009) Magnesium sulfate for preterm labor and preterm birth. Obstet Gynecol [Internet] 114(3):650–668. Available from: www.greenjournal.org

Merialdi M, Caulfield LE, Zavaleta N, Figueroa A, Costigan KA, Dominici F et al (2004) Randomized controlled trial of prenatal zinc supplementation and fetal bone growth. Am J Clin Nutr 79(5):826–830

Merrill AH, Henderson JM (1987) Diseases associated with defects in vitamin B6 metabolism or utilization. Annu Rev Nutr [Internet]. 7(1):137–156. Available from: https://www.annualreviews.org/doi/10.1146/annurev.nu.07.070187.001033

Meyyazhagan A, Kuchi Bhotla H, Tsibizova V, Pappuswamy M, Chaudhary A, Arumugam VA, et al. (2023) Nutrition paves the way to environmental toxicants and influences fetal development during pregnancy. Best Pract Res Clin Obstet Gynaecol [Internet] 89:102351. Available from: https://doi.org/10.1016/j.bpobgyn.2023.102351

Milićević DR, Škrinjar M, Baltić T (2010) Real and perceived risks for mycotoxin contamination in foods and feeds: challenges for food safety control. Toxins (Basel) 2(4):572–592

Miller CJ, Davidson LA (2013) The interplay between cell signalling and mechanics in developmental processes. Nat Rev Genet 14(10):733–744

Mohan JS (2001) Tissue culture-derived variation in crop improvement. Euphytica 118(2):153–166

Mora JR, Iwata M, Von Andrian UH (2008) Vitamin effects on the immune system: vitamins a and D take Centre stage. Nat Rev Immunol 8(9):685–698

Munir N, Jahangeer M, Bouyahya A, El Omari N, Ghchime R, Balahbib A et al (2021) Heavy metal contamination of natural foods is a serious health issue: a review. Sustainability [Internet] 14(1):161. Available from: https://www.mdpi.com/2071-1050/14/1/161

Neuhouser ML, Prentice RL, Tinker LF, Lampe JW (2023) Enhancing capacity for Food and nutrient intake assessment in population sciences research. Annu Rev Public Health 44:37–54

Norris SA, Frongillo EA, Black MM, Dong Y, Fall C, Lampl M et al (2022) Nutrition in adolescent growth and development. Lancet 399(10320):172–184

Noteborn HPJM, Lommen A, Van Der Jagt RC, Weseman JM (2000) Chemical fingerprinting for the evaluation of unintended secondary metabolic changes in transgenic food crops. J Biotechnol 77(1):103–114

Onakpa MM, Njan AA, Kalu OC (2018) A review of heavy metal contamination of food crops in Nigeria. Ann Glob Heal 84(3):488–494

Panagiotidou EE (2024) Study of the effects of prenatal exposure to mixtures of endocrine disruptors, on the reproductive and nervous system of mice. Εθνικό και Καποδιστριακό Πανεπιστήμιο Αθηνών (ΕΚΠΑ). Σχολή Επιστημών Υγείας ...

Raamsdonk LM, Teusink B, Broadhurst D, Zhang N, Hayes A, Walsh MC et al (2001) A functional genomics strategy that uses metabolome data to reveal the phenotype of silent mutations. Nat Biotechnol 19(1):45–50

Rahman M, Alam MU, Luies SK, Kamal A, Ferdous S, Lin A et al (2022) Contamination of fresh produce with antibiotic-resistant bacteria and associated risks to human health: a scoping review. Int J Environ Res Public Health 19(1):1–15

Rather IA, Koh WY, Paek WK, Lim J (2017) The sources of chemical contaminants in food and their health implications. Front Pharmacol 8(NOV):830

Roessner U, Luedemann A, Brust D, Fiehn O, Linke T, Willmitzer L et al (2001) Metabolic profiling allows comprehensive phenotyping of genetically or environmentally modified plant systems. Plant Cell 13(1):11–29

Rogan GJ, Bookout JT, Duncan DR, Fuchs RL, Lavrik PB, Love SL et al (2000) Compositional analysis of tubers from insect and virus resistant potato plants. J Agric Food Chem [Internet] 48(12):5936–5945. Available from: https://pubs.acs.org/doi/10.1021/jf000742b

Romani AMP (2011) Cellular magnesium homeostasis. Arch Biochem Biophys [Internet] 512(1):1–23. Available from: https://doi.org/10.1016/j.abb.2011.05.010

Ross MG, Desai M (2021) Fetal origins of adult health and genet disord fetus diagnosis, Prev Treat 214

Savarino G, Corsello A, Corsello G (2021) Macronutrient balance and micronutrient amounts through growth and development. Ital J Pediatr 47(1):1–14

Scharff RL (2012) Economic burden from health losses due to foodborne illness in the United States. J Food Prot [Internet] 75(1):123–131. Available from: https://doi.org/10.4315/0362-028x_jfp-11-058

Schwank G, Basler K (2010) Regulation of organ growth by morphogen gradients. Cold Spring Harb Perspect Biol 2(1):1–17

Silveira PP, Portella AK, Goldani MZ, Barbieri MA (2007) Origens desenvolvimentistas da saúde e da doença (DOHaD). J Pediatr 83(6):494–504

Slack JMW, Dale L (2021) Essential developmental biology [Internet]. Wiley. Available from: https://books.google.co.in/books?id=RPBLEAAAQBAJ

Swallow BL (2010) Nausea and vomiting in pregnancy. Psychologist 23(3):206–209

Tasginn E (2017) Macronutrients and micronutrients in nutrition. Int J Innov Res Rev [Internet] 1(1):10–15. Available from: http://www.injirr.com/article/view/8

Taylor J, Goodacre R, Wade WG, Rowland JJ, Kell DB (1998) The deconvolution of pyrolysis mass spectra using genetic programming: application to the identification of some Eubacterium species. FEMS Microbiol Lett 160(2):237–246

Thakali A, MacRae JD (2021) A review of chemical and microbial contamination in food: what are the threats to a circular food system? Environ Res [Internet] 194:110635. Available from: https://doi.org/10.1016/j.envres.2020.110635

Thompson LA, Darwish WS (2019) Environmental chemical contaminants in food: review of a global problem. J Toxicol 2019:1–14

Trumbo PR, Ellwood KC (2007) Supplemental calcium and risk reduction of hypertension, pregnancy-induced hypertension, and preeclampsia: an evidence-based review by the US Food and Drug Administration. Nutr Rev 65(2):78–87

Ukwo SP, Udo II, Ndaeyo N (2022) Food additives: overview of related safety concerns. Food Sci Nutr Res 5(1):1–10

Unnevehr LJ (2022) Addressing food safety challenges in rapidly developing food systems. Agric Econ (United Kingdom) 53(4):529–539

Uyttendaele M, De Boeck E, Jacxsens L (2016) Challenges in food safety as part of food security: lessons learnt on food safety in a globalized world. Procedia Food Sci [Internet] 6(Icsusl 2015):16–22. Available from: https://doi.org/10.1016/j.profoo.2016.02.003

Van TTH, Moutafis G, Tran LT, Coloe PJ (2007) Antibiotic resistance in food-borne bacterial contaminants in Vietnam. Appl Environ Microbiol 73(24):7906–7911

Vanderpas J (2006) Nutritional epidemiology and thyroid hormone metabolism. Annu Rev Nutr [Internet] 26(1):293–322. Available from: https://www.annualreviews.org/doi/10.1146/annurev.nutr.26.010506.103810

Villar J, Abdel-Aleem H, Merialdi M, Mathai M, Ali MM, Zavaleta N et al (2006) World Health Organization randomized trial of calcium supplementation among low calcium intake pregnant women. Am J Obstet Gynecol 194(3):639–649

Watson A, Mazumder A, Stewart M, Balasubramanian S (1998) Technology for microarray analysis of gene expression. Curr Opin Biotechnol 9:609–614

West-Eberhard MJ (2003) Developmental plasticity and evolution [Internet]. Oxford University Press. Available from: https://books.google.co.in/books?id=7DQNTPYaHlYC

White SW, Hillman SL, Newnham JP (2019) The developmental origins of health and disease. Fetal Med Basic Sci Clin Pract:572–579.e2

Winzerling JJ, Law JH (1997) Comparative nutritionof iron and copper. Annu Rev Nutr [Internet] 17(1):501–526. Available from: https://www.annualreviews.org/doi/10.1146/annurev.nutr.17.1.501

Wu G, Imhoff-Kunsch B, Girard AW (2012) Biological mechanisms for nutritional regulation of maternal health and fetal development. Paediatr Perinat Epidemiol 26(SUPPL. 1):4–26

Xie J, Ouyang XZ, Xia KF, Huang YF, Pan WB, Cai YP et al (2007) Chloroplast-like organelles were found in enucleate sieve elements of transgenic plants overexpressing a proteinase inhibitor. Biosci Biotechnol Biochem 71(11):2759–2765

Xu ML, Gao Y, Han XX, Zhao B (2022) Innovative application of SERS in Food quality and safety: a brief review of recent trends. Food Secur 11(14):1–12

Yabor L, Valle B, Carvajal C, Aragón C, Hernández M, González J et al (2010) Characterization of a field-grown transgenic pineapple clone containing the genes chitinase, AP24, and bar. Vitr Cell Dev Biol – Plant 46(1):1–7

Yin J, Li X, Li D, Yue T, Fang Q, Ni J, et al. (2009) Dietary supplementation with zinc oxide stimulates ghrelin secretion from the stomach of young pigs. J Nutr Biochem [Internet] 20(10):783–790. Available from: https://doi.org/10.1016/j.jnutbio.2008.07.007

Zhang Z, McElvain JS (2000) De novo peptide sequencing by two-dimensional fragment correlation mass spectrometry. Anal Chem 72(11):2337–2350

Zimmermann MB (2006) The influence of iron status on iodine utilization and thyroid function. Annu Rev Nutr [Internet] 26(1):367–389. Available from: https://www.annualreviews.org/doi/10.1146/annurev.nutr.26.061505.111236

Chapter 9
Aging Physiology and Susceptibility to Foodborne Illness

Rukshana Irani and Suchandra Datta

9.1 Introduction

The aging of the population is a global concern, and thus, understanding the physiological changes associated with aging and its health implications is extremely important. Aging involves biological changes that affect nearly every organ system of the body, causing alterations in the functioning of the immune and digestive systems as well. These changes significantly affect the ability of older adults to fight infections, particularly foodborne illnesses. As a result, they are at a much higher risk of developing disabilities and diseases (Aalami et al. 2003).

According to the Centers for Disease Control and Prevention (CDC), individuals who are aged 65 years and older experience increased rates of hospitalisation, morbidity and mortality due to foodborne illnesses in comparison to younger adults. It is largely due to their diminished ability to combat pathogens as a result of immunosenescence and gut changes. The proficiency of the immune system to recognise and neutralise pathogens decreases, and thus, the susceptibility to various infections increases manifold (Assistant Secretary for Public Affairs (ASPA) 2023).

In addition, the alterations in the physiology of the alimentary system, like decreased gastric acid secretion and slower intestinal transit time, hamper the body's natural defence against foodborne pathogens. A comprehension of the association between aging and the heightened risk of foodborne illnesses can help in the formulation of effective prevention strategies. The relationship between the physiological changes of aging and their impact on susceptibility to foodborne infections must be given utmost importance. Interventions should be taken to enhance food safety

R. Irani (✉)
Department of Food & Nutrition, Raidighi College, Raidighi, West Bengal, India

S. Datta
Department of Home Science, University of Calcutta, Kolkata, West Bengal, India

T. Sarkar et al. (eds.), *Physiological Perspectives on Food Safety: Exploring the Intersection of Health and Nutrition*, https://doi.org/10.1007/978-3-031-94582-3_9

amongst this at-risk, vulnerable population (Aalami et al. 2003; Boss and Seegmiller 1981).

Every country is witnessing a rise in its elderly population. A gradual increase in both the number and percentage of older individuals is a critical issue worldwide. This inevitable demographic shift increases the demand for advancements in health and medical care systems. Increased longevity and reduced birth rates have increased the number of individuals aged 60 years and older. This trend of population ageing has a significant impact on the socioeconomic and health status of the country (NSO 2021).

Improvements in the fields of nutrition, medical science, socioeconomic status and public health have contributed to reduced mortality and morbidity rates and thus significantly increased life expectancy or longevity (United Nations Department of Economic and Social Affairs, Population Division 2023).

In the year 2023, a study stated that about 1.1 billion individuals worldwide were aged 60 years or above. It constituted approximately 14.16% of the world's population. A significant increase in the population size will occur in the coming decades as well. The number of people aged 60 years and above is expected to double in size in terms of population by the year 2050. The population has been estimated to increase in size by approximately 2.1 billion, i.e., 22% of the global population, by 2050. A major proportion of the world's population will be composed of older adults in the coming years. This demographic shift will profoundly impact various aspects of society and the economy (Kumar and Kujur 2024).

Although India has the highest number of young individuals, the rate of ageing is progressing quite rapidly. 2023 data stated that the elderly population of 153 million (aged 60 and above) is expected to reach 347 million by 2050. A report by the United Nations Population Fund's India Ageing Report-2023 estimated that the percentage of the elderly population in India is anticipated to grow from 10.5% in 2022 to 20.8% by 2050 (Wojnar 2023).

Foodborne diseases (FBD) are a significant problem that is closely associated with agriculture and food systems. It is characterized by illnesses caused by the intake of unsafe and unhygienic food. It may be a result of contamination from external sources, or they may be inherently hazardous (Grace 2023).

Diseases and infections pose a significant and concerning public health challenge worldwide. Amongst many health problems, foodborne diseases remain a major cause of morbidity and mortality. It is responsible for causing profound socio-economic losses in the world. In developing and underdeveloped countries, the chances of food becoming contaminated are very high due to several factors. It includes limited access to clean water for the preparation of food, inadequate transportation facilities, improper storage conditions, poor personal hygiene, and wrong handling practices. These issues altogether increase the susceptibility of people to foodborne illnesses (Bisht et al. 2021).

Foodborne diseases have emerged as a critical public health problem that can cause morbidity and mortality in the general population. It affects the economy of the countries adversely. Any disease that may be caused by infection or intake of toxins is referred to as a Foodborne disease. The medium of entry into the human body can be food or water (Sudershan et al. 2014).

Elderly people are the most vulnerable group to the development of Foodborne illness amongst all other socio-demographic groups (Jackson and Mesh 2017). The incidence of foodborne infections has been increasing over time. It has emerged as a significant public health problem worldwide, with over 600 million people suffering every year due to these ailments. Many foodborne pathogens contribute to thousands of infections and pose risks to human health and the economy. Pathogenic microbes can contaminate food products at various stages, such as manufacturing, processing, storage, transportation and consumption. The gastrointestinal disorders caused by foodborne pathogens increase the financial and health burdens of the country. Foodborne microorganisms can result in the occurrence of a broad spectrum of symptoms that can range from mild gastroenteritis to severe disorders such as meningitis. Along with children, older adults are also very vulnerable due to their compromised immunity (Elbehiry et al. 2023).

Age has been a prominent host factor responsible for heightened susceptibility. This vulnerability to foodborne illnesses has been linked with immunologically suppressed physiology. When these people get affected by foodborne diseases, treatment and recovery become more challenging. Pathogens cause severe illness to these elderly people because of their weakened immune systems (Lund and O'Brien 2011).

Research studies show that the greater susceptibility of elderly people to morbidity and mortality from foodborne diseases is due to many factors, which include declining immunity, the presence of chronic diseases and unsafe food handling practices (Berger et al. 2022).

9.2 Physiological Changes in Aging

With aging, various physiological changes occur in the body. These changes enhance the susceptibility to infection due to the consumption of contaminated food. Altered gastrointestinal functionality and decreased immunocompetence are chief factors behind diminishing the body's ability to combat pathogens. The deranged function of the liver and kidneys also decreases the body's ability to get rid of pathogenic microorganisms. The presence of comorbidities like diabetes and cancer also complicates the situation when they get affected by foodborne illnesses. It is extremely important to understand the relationship between aging and the heightened risk of foodborne illness. It can help to formulate and develop appropriate health guidelines and interventions for this vulnerable population (Assistant Secretary for Public Affairs (ASPA) 2023).

Aging is an inevitable physiological change that can cause a lowering in the number of cellular tissues, a reduction in the metabolic rate, an increase in disease incidence, and a loss of adaptability of the body system. These changes also vary from person to person, depending on the condition of their organs. Aging involves degenerative change associated with biological characteristics. The environmental factors associated with lifestyles, such as stress, exercise, smoking, and exposure to sunlight, influence the progression of aging (Park and Yeo 2013).

Here, the critical physiological changes associated with aging that contribute to increased susceptibility to foodborne illnesses have been discussed.

9.2.1 Gastrointestinal System

Deranged functioning of the gastrointestinal tract leads to heightened susceptibility to foodborne infections in the elderly. Aging affects motility, secretion of digestive enzymes, and the integrity of the intestinal barrier. All of these have been linked to an enhanced risk of infections from contaminated food (Bhutto and Morley 2008).

One of the primary changes in the gastrointestinal system due to aging is the decline in gastrointestinal motility. The contractions of the oesophagus and intestines slow down. It leads to delayed transit times for food and waste, i.e., the time it takes for food and waste to stay in the intestine increases. As a result, the foods remain in the gut for prolonged periods, which increases the chances of bacterial overgrowth (Dumic et al. 2019).

Older adults also suffer from constipation and other bowel dysfunctions, which makes pathogens in the food proliferate more before being eliminated from the body. The greater retention time for food also complicates the digestive processes, which affects the destruction of potentially harmful micro-organisms (Conaway 2012).

The propulsive activity of the colon is reduced as a consequence of aging. This increases the chance of diverticula formation. The diverticula are structures that can develop in the wall of the intestine (Bhutto and Morley 2008). It can serve as a reservoir for bacteria. It causes diverticulitis, which is a condition characterized by inflammation and infection. It occurs when diverticula pouches get blocked or filled with ingested food particles (Conaway 2012). The secretion of digestive enzymes and gastric acid reduces due to aging. The decreased production of gastric acid causes a condition called achlorhydria (Nigam and John Knight 2017).

Inadequate secretion of gastric acid not only affects the digestion process but acts as a defence mechanism against many pathogens as well. The lowered pH improves the survival of bacteria like *Salmonella* or *Escherichia coli* in the upper gastrointestinal tract. Alterations in enzyme secretion hinder the absorption of nutrients, which causes malnutrition and deficiency diseases. Malnutrition adversely affects the immune system. Thus, the body's ability to fight pathogens that enter through food also diminishes, which increases their susceptibility to foodborne diseases (Assistant Secretary for Public Affairs (ASPA) 2023).

The integrity of the intestinal barrier prevents pathogens from entering the bloodstream. Aging has been associated with a gradual decline in the strength of the mucosal barrier, which makes it "leaky". Increased permeability allows bacteria and toxins to move from the gut into systemic circulation, which elevates the risk of sepsis and other serious health complications aggravated by foodborne pathogens.

Reduction in Peyer's patches also occurs as a result of aging. Peyer's patches are integral components of gut-associated lymphoid tissue that are responsible for the gut's immune defences. Lesser Peyer's patches seriously impact the immune function of the gut. Alteration in the composition of gut microbiota composition also occurs with age. It can exacerbate the body's inflammatory response towards dietary antigens. The disruption of the homeostasis of the microbiome weakens the body's ability to initiate an effective immune response against invading pathogens, making older adults very susceptible to infections due to consuming contaminated foods (Nigam and John Knight 2017).

Changes in the characteristics and functioning of the oral cavity of the alimentary tract influence the food consumption patterns of the elderly. Xerostomia or dry mouth caused due to impaired secretion and composition of saliva due to aging significantly reduces food consumption. This is one of the most common oral sensorial complaints that impact their diet and food habits.

Dysgeusia or hypogeusia, i.e. complete loss of sense of taste or decreased loss of taste, respectively, are very common in the elderly. It is caused due to aging as well as side effects of medications related to this age group. It influences food intake in the elderly, and because they cannot perceive proper taste, the urge to eat and prepare food also decreases.

Oropharyngeal dysphagia is characterised by difficulty in chewing or swallowing food. It is very common in older adults suffering from multiple sclerosis, Parkinson's disease, dementia or after a stroke. These conditions decrease food consumption in the elderly, which predisposes them to malnutrition.

Presbyesophagus is a term used to describe the aging oesophagus and the changes that occur during this process. Conditions like oesophagal dysphagia, odynophagia, gastroesophageal reflux disease (GERD), Barrett's columnar-lined oesophagus (CLE), and pill esophagitis associated with oesophagus have been associated with discomfort in food intake, leading to low consumption, decreased appetite and malnutrition (Dumic et al. 2019).

Age-related changes also occur in the stomach, which has serious consequences on the food intake of the elderly and their overall health. Significant changes occur in the gastric microbiota with a reduced number of beneficial microbes and an increased number of pro-inflammatory bacteria. This alteration in the microbiota affects the immune system, leading to an inflammatory state in the body, which increases susceptibility to many diseases in the elderly (Wu et al. 2021).

The protective mechanism of the gastric mucosa also gets deranged due to aging. With age, the ability of mucosa to protect itself from injury or damage decreases. This decrease has been associated with reduced prostaglandin secretion and reduced mucus and bicarbonate secretion (Kang et al. 2010).

Decrease in the gastric blood flow, reduced gastric motility, delayed gastric emptying, hypochlorhydria or achlorhydria, compromised healing mechanism and decrease in postprandial contractile force of the stomach increase the risks of gastric ulcer, atrophic gastritis, and peptic ulcer disease in the elderly.

Diseases and disorders associated with the small bowel, i.e., Celiac disease, Mesenteric ischemia, and Small intestinal bacterial overgrowth, are very common in older adults, which also affects their appetite and health.

Constipation is very common among the elderly and affects their food intake and quality of life. A decrease in physical activity, mobility, impaired cognition, and comorbidities lead to severe discomfort and health consequences in older people. Irritable bowel syndrome, Clostridium difficile colitis, and inflammatory bowel disease affect the large intestine as well as the overall gut health (Dumic et al. 2019).

All these factors cause lowered food intake, leading to malnutrition and reduced immunity to fight against pathogens, including causing foodborne illnesses.

9.2.2 Changes in the Immune System

As a major portion of the population is aging, it is important to understand the immunological changes associated with it, especially related to the susceptibility towards infection. Immunosenescence is an important change that occurs during the process of aging where alterations occur in both qualitative and quantitative immune responses. Declining immunocompetence as a result of aging increases the vulnerability of older subjects towards the development of various infections. Notable changes occur in acquired immunity with aging, which includes T cell dysfunction. The dysfunctional T cell can occur at any stage, from the naive stage to the end stage of differentiation during an immune response. The thymus gland, which is responsible for generating T cells, undergoes involution and atrophy, which leads to reduced production of new T cells against novel pathogens (Minato et al. 2020).

A decrease in CD4+ T cell function as a result of immunosenescence reduces the stimulation of T cell receptors (Akha and Miller 2005).

Although the T memory cells are greater in the elderly, they adopt a senescence phenotype, which leads to reduced proliferation of immune cells, decreased cytokine production, and thus diminished immune response (Bulut et al. 2020).

An increase in the number of regulatory T cells has been observed in the elderly, but they have a suppressive activity, which hampers the overall immune response. It also triggers chronic inflammation and reduces responsiveness towards infections. A reduction in the length of telomere in memory T cells, terminally differentiated T cell subsets, and the total T cell was observed. The expression of cytotoxic molecules like granzyme B and perforin was found to be lower in older adults.

Aging also affects B cells significantly. Decreased activity of bone marrow's hematopoietic stem cells causes a decline in the production of both B1 and B2 cells, which decreases the antibody production (Martinez et al. 2021).

IL-10 + and IL-35 + bregs, regulatory B cells, play a critical role in protecting against my infections, but their functions get impaired due to aging. The Age-associated B-cells accumulate within the body as a result of aging, which increases the chances of autoimmune diseases. Aging causes noteworthy negative changes in the memory cells or marginal zone B cells, and it also increases the risk of autoimmune diseases (Yu et al. 2024).

Innate immunity is also a very important and critical part of the entire immune system. Alterations in the innate immune response due to aging increase the susceptibility of the elderly to pathogens. Impaired chemotaxis by neutrophils, a decline in phagocytic activity, and decreased oxidative burst capability reduce the immune response in the elderly. This increases the chances of colonization and biofilm formation by the bacteria (Solana et al. 2006).

Aging hampers the innate immune system, and as a result, irregular inflammatory reactions occur in elderly people. Toll-like receptors (TLR receptors) are an important class of protein which is an important part of innate immunity.

Aging reduces the activity of TLR receptors in macrophages and monocytes, and the population of dendritic cells (DC) also decreases. Lowered innate immunity aggravates many inflammatory processes in the body (Shaw et al. 2013).

An increase in the number of monocytes occurs as a result of aging, but their phagocytic and inflammatory properties decrease. A lowered number of natural killer cells and their decreased cytotoxicity have been observed due to aging. The number of mast cells in various tissues increases during aging, which causes chronic inflammation in older adults (Gao et al. 2024).

9.2.3 Cellular and Organ System Changes

Aging causes significant alteration in the structure and function of cells, which influences organ systems and thereby increases the susceptibility to foodborne infections in older subjects. Aging leads to cellular senescence, where the cells cease to divide and enter a pro-inflammatory state. The two chief causes of cellular senescence are DNA damage and oxidative stress. Tissue homeostasis is affected due to aging, which promotes chronic inflammation and also impairs tissue functions (McHugh and Gil 2018).

The gastrointestinal system also undergoes many changes. Alterations in gastrointestinal motility, diminished secretion of digestive juices, and changes in the composition of gut microbiota increase the risk of foodborne infections in elderly people (Rémond et al. 2015).

The immune system also experiences immunosenescence, which is characterized by a decline in immune cell numbers and functions. Immunosenescence reduces the ability of the body to trigger effective responses against infections. Weakened immune response in older adults makes them more vulnerable to foodborne pathogens (Aiello et al. 2019).

The efficiency and functioning of the liver and kidneys are also affected by aging. These organs are involved in regulating detoxification processes that are crucial for protecting the body from foodborne illnesses. Deranged renal function reduces the excretion of toxins from the body. A decline in the liver enzyme activity hampers the metabolism of harmful toxic chemicals. These changes further increase the risk of infections in the elderly due to the improper clearance of pathogens and toxins from the body (Kim et al. 2015).

Reduction in blood flow rate into the kidney, glomerular filtration rate and kidney mass have been found to decrease by approximately 10% every 10 years after the age of 30 years. This not only impairs the body's ability to excrete wastes and toxins but also substantially affects the ability to eliminate harmful bacteria or toxins in the contaminated food. Consequently, elderly people are at a greater risk of developing complications due to foodborne pathogenic agents (Amarya et al. 2015).

Sarcopenia is a very common occurrence in the elderly. It can be defined as the age-related decline in muscle mass and strength. Reduced activity and mobility as a result of sarcopenia hampers the ability of the elderly to procure, prepare, cook and consume safe and hygienic food. Immobility reduces access to fresh, nutritious and healthy food, and as a result, older people rely on packaged foods that are potent causes of food-borne illnesses (Rémond et al. 2015).

Nutritional status is a vital factor that affects the susceptibility of elderly people to foodborne infections. Common problems of old age, like reduced appetite, altered taste, and difficulty in swallowing, lead to malnutrition, which impairs immune function and increases vulnerability. The majority of elderly people consume an inadequate diet lacking essential nutrients, which further weakens the immune system. Cellular senescence, changes in the organ system, and nutrition-related challenges due to aging increase susceptibility to foodborne infections (Fig. 9.1) (Norman et al. 2021).

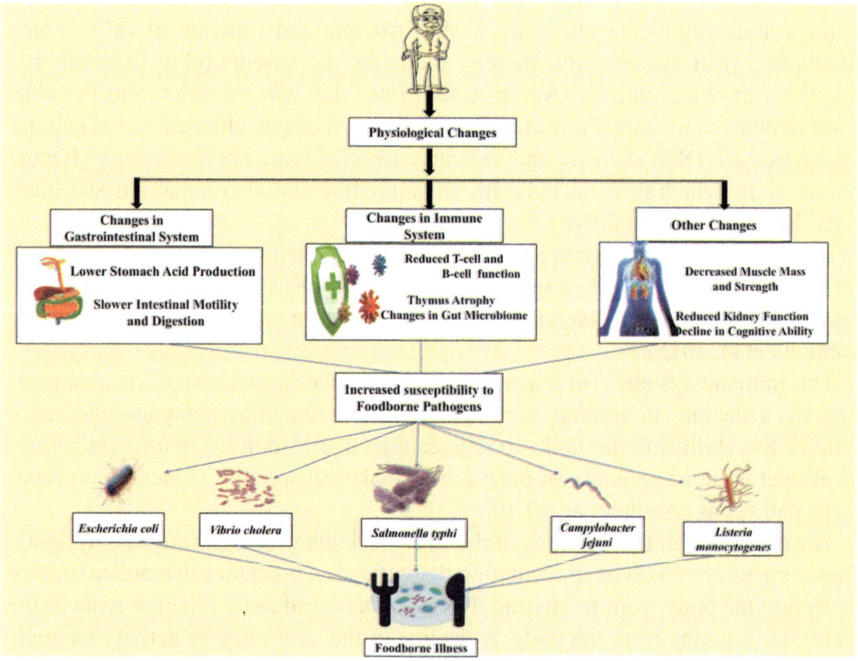

Fig. 9.1 Interrelationship between aging and higher susceptibility to foodborne illnesses

9.3 Common Foodborne Illness: Types and Causes

Foodborne illnesses are caused when pathogens are ingested along with the food, and it gets established by multiplication in the host or when a toxin produced by the pathogen in the food is consumed by the host. Therefore, foodborne illnesses can generally be classified into two types. When the pathogen enters the body, multiplies and produces illness, it is called a foodborne infection. When the ingested toxin that has been generated by the pathogen causes signs and symptoms, it is called foodborne intoxication. In foodborne infection, microbes take a longer time to produce symptoms than food intoxication, as in infection, microbes enter the body, multiply and then cause symptoms, which is also known as the incubation period. The incubation period greatly varies depending upon the kind and virulence of the pathogen, infectious dose, and host factors like age, immunity, nutritional status and overall health, etc. The foodborne diseases can be of many types, their number may reach 200 or more. The severity of these diseases depends on the age, immunity of the host, and the higher dose of microbes that enter into the host. Older people, infants, children, and immunocompromised people are more at risk, and the impact of illnesses in these groups is much higher in comparison to the others. These factors also influence the recovery process from illnesses (Bintsis 2017).

The most common foodborne illness and their characteristics that occurs commonly in older adults have been described briefly in Table 9.1.

9.4 Impact of Aging on Food Safety Practices

Aging causes physiological and cognitive changes in elderly people, affecting their food safety practices. Older adults have compromised immune systems, diminished sensory abilities, and cognitive decline, which makes them more vulnerable to foodborne illnesses. The changes in the microbiome of the intestine affect digestion and the body's ability to fight infection.

The sensory perception of older adults also gets affected adversely. A decline in the detection of smell and taste makes them vulnerable to many conditions. Due to their inability to perceive smell, they fail to detect spoiled or unsafe food. They consume the foods that should be discarded, which increases the risk of foodborne illnesses (Anderson et al. 2011).

Due to the diminished perception of taste in older adults, their food intake decreases gradually. Inadequate intake of nutrients affects their health and immunity and makes them more susceptible to foodborne illnesses. They fail to enjoy their meals and detect hazardous food items. Cognitive development is also a major factor that affects their food intake and food practices. Forgetfulness or their inability to recall makes them eat expired products and unfollow hygienic food-handling practices (Blackburn et al. 2014).

Table 9.1 Name of pathogens, incubation period, symptoms and associated food for the foodborne illnesses prevalent in the elderly

Name of pathogen	Incubation period	Predominant symptoms	Associated foods	References
Bacillus cereus toxin	1–6 hours	Vomiting, diarrhoea, abdominal cramps, nausea	Meats, stews, gravy, improperly refrigerated cooked rice.	Elbehiry et al. (2023)
Staphylococcus aureus toxin	1–8 hours	Nausea, vomiting, retching, diarrhoea, abdominal pain, prostration	Cream desserts and pastries, potato salad	Almaary (2023)
Norovirus (Viral gastroenteritis)	12–48 hours	Nausea, vomiting, watery non-bloody diarrhoea, dehydration	Poorly cooked shellfish, ready-to-eat foods touched by an infected worker	Almaary (2023)
Clostridium perfringens toxin	6–24 hours	Abdominal cramps, watery diarrhoea and nausea	Meats, poultry, gravy, dried or precooked foods	Elbehiry et al. (2023)
Escherichia coli (Urinary tract infection)	3–4 days	Diarrhoea (often bloody), abdominal cramps	Improperly cooked beef, unpasteurised milk and juice, sprouts and contaminated water	Sabbithi et al. (2024)
Hepatitis A	2–7 weeks	Jaundice, fatigue, anorexia, nausea	Raw or poorly cooked seafood harvested from contaminated waters, ready-to-eat foods handled by an infected food handler	Sabbithi et al. (2024)
Salmonella typhi (Typhoid fever)	7–28 days	Headache, fever, fever, cough, nausea, vomiting, constipation, abdominal pain, chills, rose spots, bloody stools	Undercooked poultry, raw egg desserts and mayonnaise, sprouts	Priyanka et al. (2016)
Clostridium botulinum and its neurotoxins (Botulism)	2 hours to 6 days, usually 12–36 hours	Vertigo, double or blurred vision, loss of light reflex, difficulty in swallowing, dry mouth, weakness, respiratory paralysis	Improperly home-canned low-acid foods, honey (especially for infants), garlic in oil, baked potatoes wrapped in foil, and fermented or smoked fish	Priyanka et al. (2016)

(continued)

Table 9.1 (continued)

Name of pathogen	Incubation period	Predominant symptoms	Associated foods	References
Vibrio cholerae (Cholera)	6 hours to 5 days	Abdominal cramps, diarrhoea, vomiting, fever, malaise, nausea, headache, dehydration	Raw or undercooked seafood, particularly shellfish and crustaceans, as well as contaminated water and food	Priyanka et al. (2016)
Campylobacter jejuni (Campylobacteriosis)	2–5 days, but can range from 1 to 10 days.	Diarrhoea (frequently bloody), abdominal pain, fever, and occasionally nausea and vomiting	Raw or undercooked poultry or other foods, including seafood and meat	Buzby and Roberts (2009)
Listeria monocytogenes (Listeriosis)	Few days to over 90 days, with a median of around 3 weeks	Fever, muscle aches, headache, stiff neck, and in severe cases, confusion, loss of balance, and convulsions	Unpasteurized milk, soft cheeses, deli meats, hot dogs, smoked seafood, and certain ready-to-eat foods, especially those refrigerated.	Buzby and Roberts (2009)

9.4.1 Dietary Habits and Risk Factors

Changes in the perception of taste and smell affect their dietary habits. Research indicates that a reduction in appetite has been observed with aging, with a preference towards specific food textures and flavours and reduced appetite results in lower food intake, which causes nutrient deficiencies and malnutrition. Lower food intake has been associated with lower intakes of calcium, zinc, iron, B vitamins, vitamin E, etc. It affects their overall health and increases their susceptibility to illness. They consumed meals with lesser nutrient density due to slow gastric emptying, lowered basal metabolic rate, altered hormonal response, hyposmia, hypogeusia, socio-economic status, etc. (Drewnowski and Shultz 2001).

Older adults suffer from chronic illnesses like diabetes, hypertension, cardiovascular diseases, etc., which further limit their food choices. Strict dietary restriction makes meal planning more complicated. They also take different kinds of medicines that affect the ability of the body to properly metabolise food items. The higher the number of co-morbidities, the greater the risk of malnutrition. Disability and reduced performance in older adults adversely affect their ability to access and prepare safe food. Depression is a common factor in older adults, which influences their dietary habits and pushes them towards malnutrition (Donini et al. 2020).

Cognitive decline and impaired memory affect their food preparation and consumption safely and hygienically. They struggle to remember the expiry dates of the food items, proper cooking temperatures, consumption of perishable food items at the right time, proper storage of leftovers, etc. These practices, when not followed properly, lead to enhanced vulnerability towards food pathogens. Deteriorated oral health, wrong attitudes and practices, bereavement, cut off from social networking, loneliness, financial crunch, etc., impact their dietary habits and thus increase their chances of food-borne illnesses both directly and indirectly (Govindaraju et al. 2022).

9.4.2 Challenges in Food Handling and Preparation

As people age, physiological changes affect their food-handling and preparation abilities. Physical and cognitive challenges make it difficult for the elderly to prepare food that is safe for consumption, which makes them vulnerable to foodborne infection. Food preparation is a skillful work that demands focus, strength, motor skills and flexibility. Physical restrictions and immobility challenge older people to cook or procure safe food. Lack of access to transportation also influences their ability to procure proper food from the market. Financial instability is very common among older adults. They have fixed incomes, and they have to cope with the expenditures in this time of rising costs. Social isolation with reduced access to support and resources limits their ability to have good food (Jason 2024).

Arthritis is very common among older people. It hinders their ability to perform normal day-to-day tasks, and as a result, they become more dependent upon convenience foods that may not be prepared following proper food safety practices. Convenient or packaged foods are deficient in vitamins, minerals, and fibre and higher in sodium, sugar, etc. Consumption of such a diet makes them prone to malnutrition, which in turn affects their immunity. It not only delays their recovery from various illnesses but also aggravates the comorbidities that they suffer from (Nicole 2023).

Older adults are more prone to the development of nutrient deficiency due to physiological, psychosocial, environmental, economic factors, etc. Chronic disorders, obesity, frailty, and sarcopenia predispose them to deficient or unhealthy food intake. Cognitive impairment, swallowing difficulties, and eating/chewing issues are also among the major contributory factor that causes malnutrition in adults (Porter Starr et al. 2015).

9.5 Preventive Measures and Interventions

The risk of contracting food-borne illnesses in the elderly results from many physiological changes that include the gastrointestinal and immune systems. The decreased strength and overall health not only decrease their strength but also affect the healing process adversely. Some strategies or interventions can be followed to reduce the incidence of foodborne illnesses in the elderly.

9.5.1 Awareness Campaigns and Educational Programs

Food safety is an extremely vital and critical aspect of public health and human welfare. The integrity of the food supply chain must be maintained to prevent foodborne illnesses. It not only protects the health of the consumers but also improves the trust of people in the food industries. A better understanding of food safety can be helpful to prevent infections and intoxications. Many activities were initiated to educate people regarding food safety. The activities included organizing awareness campaigns, enhancing knowledge, and inculcating behavioral changes in consumers. Public education for elderly people to inform, educate and empower them regarding food safety can be provided by the government, non-profitable organizations, healthcare centers, and other stakeholders as well. They should be taught about proper food handling techniques, the importance of storage at the right conditions, correct preparation methods, and safe consumption. Heterogenous audiences can be addressed by organizing campaigns, programs, workshops, etc. A variety of communication strategies can be employed to educate the elderly. Audio-visual aids, traditional media, community outreach services, educational programs, and social media can be highly effective in generating awareness on food safety (Eruaga 2024).

9.5.2 Addressing Misconception

The psychometric approach for the perception of risk reveals that elderly people underestimate the dangers of the issues associated with food safety in their home environment. They think that foodborne illnesses occur due to improper food handling at manufacturing industries and restaurants. They are unaware of the fact that their kitchen can be a potent source of foodborne infections and intoxications. So, educational programs must be organized to correct these misconceptions and should be encouraged about the significance of safe food handling within the home itself.

9.5.3 Promoting Behaviour Change

As elderly people believe that the food handling behaviors that they follow are not responsible for causing food-borne illnesses, they should be made aware of the critical steps where the risk of infection or illness are more. Food safety education for the elderly should be focused on detecting those steps and showing them the correct practices. They sometimes do not understand the consequences of foodborne illnesses and are demotivated to make necessary changes in their lives and behaviors. Motivating and encouraging them for behavioral change is also a key step of this entire process.

So, these factors need to be taken into consideration while formulating a Food Safety Education Program for aged people. The common food sources linked with specific pathogenic agents and the mistakes that normally happen during food handling and preparation should be given extra emphasis. The materials and curriculum of the educational programs for the aged should focus on five critical factors for controlling pathogens.

- Awareness regarding personal hygiene
- Thoroughly cooking food
- Adopting measures to prevent cross-contamination
- Storing foods at safe temperatures, avoiding the danger zone while cooking and storing
- Avoiding consumption of food from unknown and unsafe sources that include ready-to-eat foods, processed foods, convenience foods, etc.

It is not enough to simply provide information; programs should also incorporate strategies to motivate elderly persons to adopt and maintain safe food-handling practices. This may involve using effective communication techniques, providing practical tips and tools, and creating opportunities for social support and reinforcement.

Information circulation is sufficient, but integrated strategies must also be included in the educational program to encourage the elderly to opt for and follow safe food handling practices. Effective communication is an integral part of this program. Additionally, offering practical tips and resources and providing social support and encouragement can be extremely helpful to the elderly in this regard.

9.5.4 *Effective Delivery Methods*

Health campaigns for reducing the risk of foodborne illnesses must take into consideration the special needs and limitations of this age group. Studies indicate that older adults engage in various incorrect food safety practices in comparison to the people belonging to the general population. So, it is vital to formulate the food safety messages in a way that applies to their situation. Age, gender, cultural background, health, socioeconomic status, mental state, etc., of the elderly should all be taken into account while giving them any informative message. Printed materials like brochures, flyers, and pamphlets can be beneficial. In addition to these, multimedia resources like videos and announcements on the radio can also be used. Interactive workshops and method demonstrations on healthy cooking practices can be highly advantageous to the elderly. Studies suggest that the role of physicians and trusted sources in educating older adults about safe food safety techniques and practices is critical and profound (Gettings 2017).

An integration of targeted interventions, innovative technologies and adaptive policies is required to ensure food safety in older adults. The food practices of vulnerable populations need to be monitored carefully using data-driven policy adaptation. The government should make stringent rules on the regulation of food safety, improving food labelling, encouraging targeted approach for educational programs, developing technology-based problem solving, and establishing a specialised task force for food safety to prioritize prevention of food hazards in the elderly (Albert et al. 2024).

9.6 Conclusion

Aging causes a gradual but significant loss in the physiological function that impairs the body's ability to resist pathogens, and thus the elderly population are more susceptible to the development of several ailments, including foodborne illnesses.

A broad spectrum of alteration in the function of the entire system occurs that makes them highly vulnerable. An altered microbiome, reduced acid secretion in the stomach, decreased intestinal motility, and lowered smell and taste perception, along with immunosenescence, create a favourable environment which facilitates the entry and proliferation of pathogens. Innate and acquired immunity are profoundly impacted due to aging with deranged production and functioning of both cellular and humoral immunity. In addition, the age-associated comorbidities like diabetes, cardiovascular diseases, cancer, etc. and the different medications like antacids, immunosuppressant, etc., further make them an at-risk group. The decreased appetite, reduced digestibility and lowered absorption causes undernutrition and nutrient deficiency in the elderly population, which further aggravates the situation.

Cognitive changes due to aging, like dementia, Alzheimer's Disease, Parkinson's disease, etc., make them incapable of recognising spoiled food, properly handling foods, and adhering to correct food safety practices. Immobility and physical disability impair their ability to access safe and fresh food, and therefore, they depend on prepackaged and processed convenience foods, which carry a higher risk of contamination.

Socioeconomic factors like limited income, loneliness, social isolation, and inadequate access to health care facilitate the risk of making incorrect dietary choices. These factors also prevent them from cooking, preparing, and storing foods in appropriate conditions.

Stringent regulations on food safety need to be implemented to mitigate this risk. The government should intervene by enforcing food safety strategies like enhanced food labelling, targeted education programs, and proper standard maintenance rules, etc. Special food services for the elderly need to be encouraged along with fortification policies to correct the micronutrient deficiencies in the elderly. A collaboration of various stakeholders, i.e., regulatory bodies, food industries, healthcare providers, and community organisations, is essential to take effective steps. This multifaceted approach can lower the incidence of foodborne illnesses among elderly persons and improve their quality of life. As the global population ages, the increased occurrence of foodborne illnesses is placing a financial burden on healthcare systems. Policymakers must prioritize the health and safety of the elderly and should encourage evidence-based interventions to protect and empower this vulnerable group.

References

Aalami OO, Fang TD, Song HM, Nacamuli RP (2003) Physiological features of aging persons. Arch Surg 138(10):1068–1076. https://doi.org/10.1001/archsurg.138.10.1068

Aiello A, Farzaneh F, Candore G, Caruso C, Davinelli S, Gambino CM, Ligotti ME, Zareian N, Accardi G (2019) Immunosenescence and its hallmarks: how to oppose aging strategically? A review of potential options for therapeutic intervention. Front Immunol 10:2247. https://doi.org/10.3389/fimmu.2019.02247

Akha AAS, Miller RA (2005) Signal transduction in the aging immune system. Curr Opin Immunol 17(5):486–491. https://doi.org/10.1016/j.coi.2005.07.004

Albert V, Ramamurthy T, Das S, Dolma KG, Majumdar T, Baruah PJ, Hazarika SC, Apum B, Das M (2024) Comprehending the risk of foodborne and waterborne disease outbreaks: current situation and control measures with Special reference to the Indian Scenario. Heliyon 10(16):e36344. https://doi.org/10.1016/j.heliyon.2024.e36344

Almaary KS (2023) Food-borne diseases and their impact on health. Biosci Biotechnol Res Asia 20(3):745–755. https://doi.org/10.13005/bbra/3129

Amarya S, Singh K, Sabharwal M (2015) Changes during aging and their association with malnutrition. J Clin Gerontol Geriatr 6(3):78–84. https://doi.org/10.1016/j.jcgg.2015.05.00

Anderson AL, Verrill LA, Sahyoun NR (2011) Food safety perceptions and practices of older adults. Public Health Rep 126(2):220–227. https://doi.org/10.1177/003335491112600213

Assistant Secretary for Public Affairs (ASPA) (2023) People at risk: older adults. FoodSafety.gov. https://www.foodsafety.gov/people-at-risk/older-adults

Berger N, Koch S, Jungnickel K, Böl GF (2022) Food safety in the aging population: qualitative findings on what to communicate and how. Risk Anal 43(9):1843–1854. https://doi.org/10.1111/risa.14069

Bhutto A, Morley JE (2008) The clinical significance of gastrointestinal changes with aging. Curr Opin Clin Nutr Metab Care 11(5):651–660. https://doi.org/10.1097/MCO.0b013e32830b5d37

Bintsis T (2017) Foodborne pathogens. AIMS Microbiol 3(3):529–563. https://doi.org/10.3934/microbiol.2017.3.529

Bisht A, Kamble MP, Choudhary P, Chaturvedi K, Kohli G, Juneja VK, Sehgal S, Taneja NK (2021) A surveillance of foodborne disease outbreaks in India: 2009–2018. Food Control 121:107630. https://doi.org/10.1016/j.foodcont.2020.107630

Blackburn ML, Bruhn CM, Miller LS, Ganthavorn C, Ober B (2014) Seniors, and their food handlers and caregivers, need food safety and nutrition education. Calif Agric 68(1–2):30–37. https://doi.org/10.3733/ca.v068n01p30

Boss GR, Seegmiller JE (1981) Age-related physiological changes and their clinical significance. West J Med 135(6):434. https://pmc.ncbi.nlm.nih.gov/articles/PMC1273316/

Bulut O, Kilic G, Domínguez-Andrés J, Netea MG (2020) Overcoming immune dysfunction in the elderly: trained immunity as a novel approach. Int Immunol 32(12):741–753. https://doi.org/10.1093/intimm/dxaa052

Buzby JC, Roberts T (2009) The economics of enteric infections: human foodborne disease costs. Gastroenterology 136(6):1851–1862. https://doi.org/10.1053/j.gastro.2009.01.074

Conaway B (2012) Aging and digestive health. WebMD. https://www.webmd.com/digestive-disorders/features/digestive-health-aging

Donini LM, Stephan BCM, Rosano A, Molfino A, Poggiogalle E, Lenzi A, Siervo M, Muscaritoli M (2020) What are the risk factors for malnutrition in older-aged institutionalized adults? Nutrients 12(9):2857. https://doi.org/10.3390/nu12092857

Drewnowski A, Shultz JM (2001) Impact of aging on eating behaviors, food choices, nutrition, and health status. J Nutr Health Aging 5(2):75–79

Dumic I, Nordin T, Jecmenica M, Lalosevic MS, Milosavljevic T, Milovanovic T (2019) Gastrointestinal tract disorders in older age. Can J Gastroenterol Hepatol 2019:6757524. https://doi.org/10.1155/2019/6757524

Elbehiry A, Abalkhail A, Marzouk E, Elmanssury AE, Almuzaini AM, Alfheeaid H, Alshahrani MT, Huraysh N, Ibrahem M, Alzaben F, Alanazi F, Alzaben M, Anagreyyah SA, Bayameen AM, Draz A, Abu-Okail A (2023) An overview of the public health challenges in diagnosing and controlling human foodborne pathogens. Vaccine 11(4):725. https://doi.org/10.3390/vaccines11040725

Eruaga MA (2024) Assessing the role of public education in enhancing food safety practices among consumers. Int J Sch Res Sci Technol 4(1):22–31. https://doi.org/10.56781/ijsrst.2024.4.1.0023

Gao H, Nepovimova E, Adam V, Heger Z, Valko M, Wu Q, Kuca K (2024) Age-associated changes in innate and adaptive immunity: role of the gut microbiota. Front Immunol 15:1421062. https://doi.org/10.3389/fimmu.2024.1421062

Gettings MA (2017) Food safety and older people. In: Food for the aging population, Woodhead Publishing series in food science, technology and nutrition. Elsevier. https://doi.org/10.1016/B978-0-08-100348-0.00015-9

Govindaraju T, Owen AJ, McCaffrey TA (2022) Past, present and future influences of diet among older adults – a scoping review. Ageing Res Rev 77:101600. https://doi.org/10.1016/j.arr.2022.101600

Grace D (2023) Burden of foodborne disease in low-income and middle-income countries and opportunities for scaling food safety interventions. Food Secur 15:1475–1488. https://doi.org/10.1007/s12571-023-01391-3

Jackson P, Mesh A (2017) Re-assessing vulnerability to foodborne illness: pathways and practices. Crit Public Health 28(1):81–93. https://doi.org/10.1080/09581596.2017.1285008

Jason (2024) Understanding senior hunger: a closer look at the challenges and solutions. https://feedingnefl.org/closer-look-at-senior-hunger/

Kang JM, Kim N, Kim JH, Oh E, Lee BY, Lee BH, Shin CM, Park JH, Lee MK, Nam RH, Lee HE, Lee HS, Kim JS, Jung HC, Song IS (2010) Effect of aging on gastric mucosal defense mechanisms: ROS, apoptosis, angiogenesis, and sensory neurons. Am J Physiol Gastrointest Liver Physiol 299(5):G1147–G1153. https://doi.org/10.1152/ajpgi.00218.2010

Kim IH, Kisseleva T, Brenner DA (2015) Aging and liver disease. Curr Opin Gastroenterol 31(3):184–191. https://doi.org/10.1097/MOG.0000000000000176

Kumar N, Kujur A (2024) India's population is also greying. The Hindu, 5 Oct 2024. https://www.thehindu.com/opinion/op-ed/indias-population-is-also-greying/article68719178.ece

Lund BM, O'Brien SJ (2011) The occurrence and prevention of foodborne disease in vulnerable people. Foodborne Pathog Dis 8(9):961–973. https://doi.org/10.1089/fpd.2011.0860

Martinez F, Novarino J, Mejía JE, Fazilleau N, Aloulou M (2021) Ageing of T-dependent B cell responses. Immunol Lett 233:97–103. https://doi.org/10.1016/j.imlet.2021.03.012

McHugh D, Gil J (2018) Senescence and aging: causes, consequences, and therapeutic avenues. J Cell Biol 217(1):65–77. https://doi.org/10.1083/jcb.201708092

Minato N, Hattori M, Yoko Hamazaki Y (2020) Physiology and pathology of T-cell aging. Int Immunol 32(4):223–231. https://doi.org/10.1093/intimm/dxaa006

Nicole (2023) Food preparation for elderly patients at home, nutritious & delicious meals seniors will want to eat. Chefs for Seniors. https://chefsforseniors.com/blog/food-for-elderly-patients-at-home/

Nigam Y, John Knight J (2017) Anatomy and physiology of ageing 3: the digestive system. Nursing Times, 27 Mar 2017. https://www.nursingtimes.net/older-peoples-nursing/anatomy-and-physiology-of-ageing-3-the-digestive-system-27-03-2017/

Norman K, Haß U, Pirlich M (2021) Malnutrition in older adults-recent advances and remaining challenges. Nutrients 13(8):2764. https://doi.org/10.3390/nu13082764

NSO (2021) Elderly in India 2021. National Statistical Office, Ministry of Statistics & Programme Implementation, Government of India, New Delhi. https://mospi.gov.in/sites/default/files/publication_reports/Elderly%20in%20India%202021.pdf

Park DC, Yeo SG (2013) Aging. Korean J Audiol 17(2):39–44. https://doi.org/10.7874/kja.2013.17.2.39

Porter Starr KN, McDonald SR, Bales CW (2015) Nutritional vulnerability in older adults: a continuum of concerns. Curr Nutr Rep 4(2):176–184. https://doi.org/10.1007/s13668-015-0118-6

Priyanka B, Patil RK, Dwarakanath S (2016) A review on detection methods used for foodborne pathogens. Indian J Med Res 144(3):327–338. https://doi.org/10.4103/0971-5916.198677

Rémond D, Shahar DR, Gille D, Pinto P, Kachal J, Peyron MA, Dos Santos CN, Walther B, Bordoni A, Dupont D, Tomás-Cobos L, Vergères G (2015) Understanding the gastrointestinal tract of the elderly to develop dietary solutions that prevent malnutrition. Oncotarget 6(16):13858–13898. https://doi.org/10.18632/oncotarget.4030

Sabbithi A, Naveen Kumar R, Gavaravarapu SR, Nagalla B, Vemula SR (2024) Prevalence and characterization of selected foodborne pathogens in frequently consumed street foods of Hyderabad, India: an exploratory study. Indian J Microbiol Res 11(4):260–268. https://doi.org/10.18231/j.ijmr.2024.046

Shaw AC, Goldstein DR, Montgomery RR (2013) Age dependent dysregulation of innate immunity. Nat Rev Immunol 13(12):875–887. https://doi.org/10.1038/nri3547

Solana R, Pawelec G, Tarazona R (2006) Aging and innate immunity. Immunity 24(5):491–494. https://doi.org/10.1016/j.immuni.2006.05.003

Sudershan RV, Kumar RN, Kashinath L, Bhaskar V, Polasa K (2014) Foodborne infections and intoxications in Hyderabad India. Epidemiol Res Int 163:1–5. https://doi.org/10.1155/2014/942961

United Nations Department of Economic and Social Affairs, Population Division (2023) World population ageing: challenges and opportunities of population ageing in the least developed countries. UN DESA/POP/2023/TR/NO.5. https://www.un.org/development/desa/pd/sites/www.un.org.development.desa.pd/files/undesa_pd_2024_wpa2023-report.pdf#:~:text=Population%20ageing%20is%20an%20inevitable%20outcome%20of,much%20larger%20as%20share%20of%20the%20total

Wojnar A (2023) India's ageing population: why it matters more than ever. UNFPA India, 23 Dec 2023. https://india.unfpa.org/en/news/indias-ageing-population-why-it-matters-more-ever#:~:text=The%20current%20elderly%20population%20of,a%20complex%20and%20intricate%20issue

Wu YL, Xu J, Rong XY, Wang F, Wang HJ, Zhao C (2021) Gut microbiota alterations and health status in aging adults: from correlation to causation. Aging Med 4(3):206–213. https://doi.org/10.1002/agm2.12167

Yu Y, Lu C, Yu W, Lei Y, Sun S, Liu P, Bai F, Chen Y, Chen J (2024) B cells dynamic in aging and the implications of nutritional regulation. Nutrients 16(4):487. https://doi.org/10.3390/nu16040487

Chapter 10
Olfactory and Gustatory Responses to Food Quality and Safety

Neetu Gupta, Mahesh Jaiwar, Akash Deep Shukla, Yashi Srivastava, and Sandeep Singh

10.1 Introduction

The human chemosensory system involves olfactory and gustatory senses to judge the food quality and determine its safeness (Fig. 10.1). Where, Olfaction is recognized as one of the five fundamental human senses, regarded as one of the most primitive. Prehistoric societies relied on it for survival, and it still has an impact on how humans connect with one another, form bonds, and pass their species on. The sense of smell is significantly associated with overall quality of life and health (Han et al. 2023). While, gustation is a chemical sensation that is made possible by taste receptor cells in the larynx, throat, and epiglottis, in addition to taste buds on the tongue's surface (which are actually taste papillae) identifies five distinct types of taste perception that are sweet, salty, bitter, umami, salty (Rudnitskaya and Lvova 2025).

The prior assumption regarding humans' inferior olfactory capabilities stemmed from the relatively small size of olfactory bulbs in relation to the overall brain volume. Studies indicate that humans possess a sophisticated olfactory system characterized by intricate networks and pathways, enabling them to perform comparably or even superiorly to certain other mammals (McGann 2017). Humans are capable to differentiate a minimum of 1 trillion olfactory stimuli (Bushdid et al. 2014). Olfactory perception significantly impacts behavior, with decisions potentially guided by smell. Olfaction initiates in utero, with the developing fetus responding to olfactory stimuli approximately at 30 weeks, thereby becoming acquainted with

N. Gupta · M. Jaiwar · A. D. Shukla · Y. Srivastava (✉)
Department of Applied Agriculture, Central University of Punjab, Village Ghudda, Bathinda, Punjab, India

S. Singh
The University of Texas, MD Anderson Cancer Center, Houstan, TX, USA

T. Sarkar et al. (eds.), *Physiological Perspectives on Food Safety: Exploring the Intersection of Health and Nutrition*, https://doi.org/10.1007/978-3-031-94582-3_10

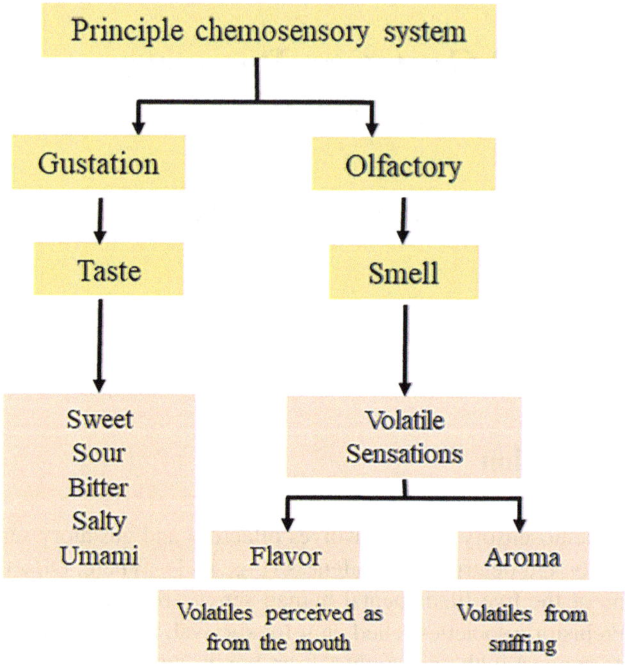

Fig. 10.1 Principle chemosensory system. (Adapted from (Lawless and Heymann 2010))

elements of the maternal diet. This development is essential for the bonding process between infants and mothers (Sarnat et al. 2017). Research has also proposed the existence of additional gustatory taste modalities, encompassing receptors for fat that respond to free fatty acids and carbonic anhydrase 4, an enzyme found in sour-sensing cells. The research posits that humans do not possess taste receptors for metallic substances, calcium ions (Ca21), and starch. Sour and salty taste receptors are selective ion channels for Na + and H+, respectively, whereas G protein-coupled receptors are for sweet, bitter, and umami taste profile. Umami receptors respond to L-amino acids and monosodium L-glutamate, whereas sweet receptors may detect d-amino acids, natural sugars, and artificial sweeteners. There are around 30 bitter taste receptors in mammals. Some of these receptors are sensitive to a broad variety of bitter chemicals, while others are more compound specific (Rudnitskaya and Lvova 2025).

The senses are integral to daily functioning, and the loss of any sense can result in psychological difficulties and practical complications. Vision and hearing impairments are notably impacted, while the Covid-19 pandemic underscored the significance of olfactory function in daily activities. Common causes of olfactory deficits, which include hyposmia and annosmia, include aging, sinonasal illness, upper respiratory tract infections, head trauma, and neurological diseases; they impact 3–20% of the population overall (Lipp et al. 2025). The sense of smell helps identify dangers, provides nutritional information, and regulates food intake. It also plays subtle roles in social behavior, such as sweat containing chemical signals conveying personal identity, health status, and emotional state (Boesveldt and Parma 2021). A person's quality of life is greatly affected by olfactory impairment, which increases the risk of accidents and decreases appetite, interest in food-related activities, and social contacts. Also associated with it are feelings of sadness, isolation, and changes in sexual orientation (Kohli et al. 2016). Efforts to develop solutions, such as new prosthesis technologies, are emerging to address this issue, catering to at least a third of anosmic patients (Besser et al. 2019).

Nature serves as a source of inspiration for individuals seeking to enhance their lives. Global food safety relies on reducing food fraud, safeguarding public health, and increasing consumer trust; all of which can be achieved only by making sure food is legitimate and can be traced. Due to the intricate nature of food supply chains, it is essential to incorporate AI, IoT, and blockchain technology to increase transparency, improve detection accuracy, and address major obstacles in food traceability (Liu et al. 2025). Advanced materials, structures, tools, processes, and algorithms have been developed through the integration of scientific methods with natural mechanisms and capacities. Biomimetics represents a developing field that has the potential to drive significant technological progress. Electronic noses and tongues, or bioelectronic counterparts, are instruments that replicate human olfactory and gustatory systems, demonstrating significant potential for application in food research and technology (Gebeshuber and Drack 2008).

The first practical odor detection system, electronic noses, was developed in the early 1960s. The inception of electronic noses as intelligent multisensor systems occurred in 1982, with the electronic tongue or taste sensor being the initial development. Both systems replicate the organizational structure of human chemosenses, comprising nonspecific or cross-sensitive sensors alongside pattern recognition techniques. The initial electronic tongues resembled artificial olfaction systems, employing potentiometry as a transducer. Subsequently, electronic tongues advanced towards their natural equivalents by integrating selective and cross-sensitive sensors, various types of transducers, and biosensors within a single instrument (Rudnitskaya and Lvova 2025). Cell-based biosensors present a viable approach for flavor evaluation, characterized by their specificity, rapid response capabilities, and sensitivity. By mimicking the functioning of actual cells-like those in the nose and taste buds, these biosensors outperform more conventional approaches. Microelectrode array systems that use taste receptor cells to detect bitter, sweet, and umami substances in real-time and improved cell immobilization techniques to detect complex odorant profiles are examples of recent breakthroughs. Signal stability, scalability, cell cultivation, and selective detection are still major obstacles. Artificial intelligence (AI) integration with mobile technology has the potential to expand their usefulness (Rashed et al. 2025).

Consumer tests and trained analytical panels are the two types of data that are most commonly used in scientific evaluations of sensory quality. The relationship between the two allows for the identification of sensory profiles that align with the product quality concept in the target market, facilitating big firms in implementing control operations, enhancing quality, and innovating new products. Concerning the viability of this approach for smaller manufacturers of goods with common sensory traits (Elortondo et al. 2007). Also, sensory evaluation technology employs human sensory organs to facilitate analysis, comparison, and quality control within the food industry. Fatigue and the subjectivity of sensory evaluators compromise its objectivity. Electronic tongues and noses are advanced analytical instruments designed to replicate human taste and smell. Their objective is to standardize and systematize sensory evaluation, progressively supplanting traditional methods (Lu et al. 2022).

Biomimetics is the study of incorporating natural solutions into new technologies, with the goal of creating cleaner and greener solutions. The term originates from the Greek words bios, signifying life, and mimesis, denoting imitation. Bionics, bionik, and biomimetics refer to the study, imitation, and learning derived from natural systems. The term originates from the Greek words "bios," signifying life, "in," denoting like, and "mimeistai," meaning to imitate. Bionics is defined as the integration of biological systems and electronic technology. Electronic tongues and noses, considered biomimetic systems, possess the capability to emulate the mechanisms of the human sensory system, suggesting their increasing application in food control. (Ghasemi-Varnamkhasti et al. 2010). Food processing faces challenges in quality and safety control, with trained human sensory panels being a common method. However, these panels have drawbacks such as human fatigue, time consumption, and cost. To address these issues, food industry engineers are looking for more consistent and cost-effective methods for objective food product measurement. Biometics systems like electronic tongue and electronic nose could be suitable for this purpose (Ghasemi-Varnamkhasti et al. 2010). These advancements find applications in prosthetics, robotics, human-machine interfaces, health monitoring, disease diagnosis, fitness, and therapy. The trend in electronic skin development encompasses enhancements in properties, the capacity to capture diverse bio-signals, and multifunctional features including display, augmented reality/virtual reality, therapeutic applications, self-powering, and integration (Wang 2025).

This chapter explores the interaction between olfactory and gustatory systems in food perception and their role in assessing food quality and safety. It aims to highlight the mechanisms underlying sensory responses, their influence on consumer preferences and behavior, and recent technological advancements in bioelectronic sensors.

10.2 Fundamentals of Olfactory and Gustatory Systems

The deep interaction between organisms and their environments is exemplified by the complexity of the human sensory system. The capacity for humans to sense, engage with, and comprehend their immediate physical and social environments is a product of a long evolutionary history. Perception, from the most primitive detecting mechanisms used by prehistoric creatures to the highly developed sensory equipment of contemporary humans, is crucial for the maintenance of life and the development of new skills (Pellegrino et al. 2017; Swain et al. 2021). The development of senses has been influenced by the necessity to address nature challenges, including food acquisition, predator detection, and terrain navigation. Early humans exhibited advanced auditory and olfactory capabilities, essential for hunting and gathering (Frost 2022), while sophisticated visual systems enhanced their ability to identify threats and opportunities across varied environments (Yilmaz Balban et al. 2021). Touch establishes a direct link to the physical environment, crucial for the utilization of tools and the formation of social bonds, thereby facilitating the evolution of sophisticated communication and cultural advancement. The complexity of human sensory abilities surpasses basic survival needs, enabling us to perceive the world in intricate detail, thereby influencing our emotions, perceptions and memories. The senses offer a complex range of sensory information that the brain integrates and processes, facilitating navigation of complex decisions, extraction of insights from experiences, and engagement with others in complex and significant interactions (Della Longa et al. 2022).

Human sensing capabilities are crucial in culture and society, influencing various aspects such as dietary preferences and artistic expression. They also enable empathy and communication, and are essential for knowledge and tradition transmission (Della Longa et al. 2022). The study of fundamental natural laws pertaining to the human senses deepens our understanding of biology and has far-reaching consequences for the fields of technology, medicine, and psychology. The workings of the human senses provide the basis for artificial intelligence, sensory enhancement devices, and prosthetics. Medical advances in our understanding of how the senses work have led to better therapies for those with sensory impairments, which has improved the quality of life for many (Albustanji et al. 2023; Kulkarni et al. 2024; Qian et al. 2024). As technological advancements potentially exceed our natural senses, it is essential to reflect on our evolutionary origins and the intricacies of our sensory systems. This understanding deepens our comprehension of the natural world and our capacities to perceive, adapt, and thrive within it. (Haick and Arnab 2025).

Exposure to particular odor-taste combinations during regular food intake affects the perception of odors when encountered in isolation, especially when paired with sweet flavors such as strawberry or vanilla. This interaction influences flavor perceptions, which are assessed and evaluated hedonically based on their similarity to a known food item. (Fondberg et al. 2021). This is grounded in three phenomena: First, the extent of familiarity with an odor-taste mixture influences its hedonic rating, as familiar combinations are more palatable than unfamiliar ones (Fondberg et al. 2018). Second, the intensity of an odor can be amplified when paired with a taste, provided that this specific combination has been frequently experienced previously. As an example, adding sucrose to a vanillin solution makes the retronasal vanilla aroma seem stronger, but salt has no such effect (Linscott and Lim 2016). Thirdly, while smelling vanilla custard, one could experience the olfactory location illusion, also known as odor referral, which happens when one believes that the aroma component of a flavor is coming from one's lips. The occurrence of this perceptual illusion is more prevalent with familiar odor-taste combinations compared to unfamiliar ones (Fondberg et al. 2018; Stevenson et al. 2011).

10.2.1 Olfactory

The olfactory system, for example, is able to translate a wide variety of environmental cues into nerve impulses with varied frequencies because it uses a variety of receptors activated by different inputs. A separate nerve conduction pathway carries the sensory signal from the nerve impulse to the main pallium sensory location. The taste bud's biological membrane acts as an epidermal cell that takes in chemical substances as part of the taste bud's reception mechanism. The membrane potential undergoes alterations during chemical reception. Olfactory cells are responsible for odorant reception and relay information to the brain, generating spike trains upon activation of olfactory cilia (Ghasemi-Varnamkhasti et al. 2010).

The olfactory system comprises three stages (Fig. 10.2): the initial stage occurs in the nasal cavity, where odorous molecules bind to receptors; the next phase is in the olfactory bulb. Finally, brain areas like the piriform cortex, amygdala, orbitofrontal cortex, insula, and hippocampus are involved in the final stage. Odorous molecules interact with these receptors, resulting in neuron depolarization and the generation of an electrical signal. Olfactory sensation can be artificially induced in all three stages (Gunder et al. 2023; Mai et al. 2023).

The oral cavity experiences simultaneous sensations linked to a single flavor perception. Researchers consider taste and odor, mouthfeel, appearance, and sound during mastication, but most agree they are flavor components (Gotow and Kobayakawa 2023). Repeated exposure to a flavor can alter its perceptual and hedonic features, with associative learning being crucial in this process, and odor

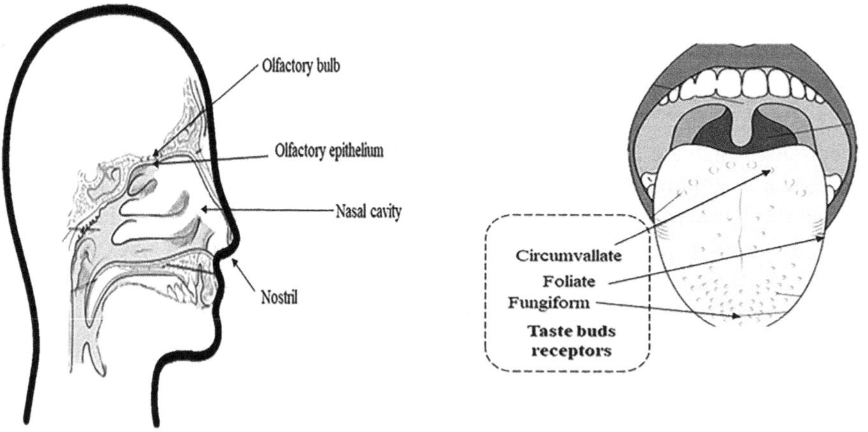

Fig. 10.2 Schematic representation of olfactory receptors and gustatory (taste buds) receptors

itself can also influence changes (Fondberg et al. 2021). Odorant molecules are chemical senses that are detected by olfactory receptor neurons, functioning as remote chemoreceptors. Neurons detect odorant molecules via two pathways: the orthonasal pathway, in which odorants traverse the nasal vestibule, and the retrograde pathway, where they are conveyed through the nasopharynx (Burdach and Doty 1987). Upon breakdown, these molecules bind to olfactory receptor neurons, leading to depolarization and the propagation (Ronnett and Moon 2002). Further, transmitted to the olfactory bulb and olfactory cortex via the olfactory tract (Cleland and Linster 2019).

The olfactory system functions as a direct and swiftly responsive sensory modality, triggering emotions and memories via synaptic connections between olfactory receptor neurons and neurons in the main olfactory bulb. Mitral and tufted cells transmit their axons to the primary olfactory cortex via the olfactory peduncle, subsequently relaying signals to the primary olfactory cortex (Wilson et al. 2014). Multiple layers make up the cortical networks that are responsible for receiving signals from the olfactory bulb. An area of the brain known as the orbitofrontal cortex, which is part of the secondary olfactory cortex, combines visual information with information about flavor and odor. (Han et al. 2023; Rolls et al. 1996). The orbitofrontal cortex, which is responsible for perceiving flavors, combines visual and tactile information. It includes the gustatory and secondary olfactory cortex. This method creates flavor by combining the senses of smell, taste, sight, and touch. Associative learning facilitates training through the integration of olfactory stimuli and taste. When we smell anything, our noses attach to certain neurons called receptors, and then our brains remap that information with other relevant data. The olfactory-to-taste link is most clearly demonstrated when one tastes wine. Flavor is the result of the integration of fragrance and taste (De Araujo et al. 2003).

10.2.2 Gustatory

The fundamental taste attributes comprise salty, sweet, bitter, and sour. Umami, a fifth taste sensation, is elicited by glutamate, specifically monosodium glutamate, which is employed to augment food flavor. In humans, all populations of taste buds are capable of perceiving all taste qualities. Taste is a sensory experience initiated by chemical compounds, primarily derived from taste buds that house gustatory receptor cells, located in the tongue, pharynx, larynx, and soft palate. In contrast to the extra-lingual taste buds found on the epithelium's surface, the papillae of the tongue contain the taste buds. A group of spindle-shaped cells composed of basal cells, taste receptor cells, and edge cells make up the taste bud. Three types of receptor cells exist: dark, light, and intermediate. Light cells lack numerous granules in the apical area, whereas dark cells do. Basal cells exhibit proliferative activity and continuously replace taste receptor cells during the turnover process (Hadley et al. 2004).

A central pore, structured around concentric cells, connects the taste buds to the oral cavity through the epithelium. To allow gustatory cues to engage with the receptors, apical microvilli that carry membrane receptors extend toward the central pore. Synapses are formed between receptor cells and afferent nerve fibers upon entering the taste bud through its basal surface. In vertebrates, taste receptor cells are specialized epithelial cells that display a plethora of neural traits. The tongue's taste buds are linked to three types of papillae: fungiform, foliate, and circumvallate (Fig. 10.2). In humans, there are around 200–300 fungiform papillae that contain around 1600 taste buds. Additionally, there are around 250 taste buds in each of the foliate and circumvallate papillae. The fourth category, filiform papillae, is anatomically unrelated to gustation and devoid of taste receptors.

Saliva plays a crucial role in taste transduction by removing stimuli following stimulation and aiding the passage of taste molecules to taste receptor cells' membranes. Saliva is produced by main salivary glands and immerses taste buds in fungiform papillae, although foliate and circumvallate papillae have deep, narrow clefts that isolate them from saliva (Brescia and Seiden 2009).

10.2.3 Trigeminal

The perception of chemical agents, including perfumes and fruit scents, is associated with memory and can elicit positive emotions. They are essential in nutrition and food choice. The detection of chemicals can fulfill protective or defensive roles, including serving as deterrents to animals in search of food (Viana 2011). The trigeminal somatosensory system is crucial in chemosensation and food flavor. Its sensory endings, located in the face, nasal and oral cavities, and eye, can be activated by physical stimuli and chemical agents (Bryant and Silver 2000), causing sensations of touch, temperature, and pain. This cutaneous chemosensation, which includes trigeminal endings, is known as chemesthesis. Chemosensitive glossopharyngeal and vagal nerve fibers also play a role in chemosensation. Oral chemesthesis accounts for the pungent or sharp sensations experienced with certain foods and spices, including chili peppers, horseradish, wasabi roots, and Szechuan pepper. Sneezing, salivation, tears, coughing, and respiratory depression are some of the defensive responses triggered by trigeminal activation. Medications applied topically often cause unpleasant side effects such as stinging or burning pain (Viana 2011).

10.3 Role of Olfaction and Gustation

Millions of people in developed countries suffer from food allergies, which affect 8% of American children (Gupta et al. 2011). The incidence of food allergies has risen consistently over the last 20 years, suggesting that contemporary environmental factors are influencing illness susceptibility (Iweala and Nagler 2019). Dietary changes, reduced exposure to micro- and macro parasites, changes in the gut flora and the immune system of the intestines, and chemical additions in processed foods, detergents, clothes, and other home items are all factors. Animals have developed advanced mechanisms to assess food quality, utilizing various chemosensory pathways that involve gustatory, olfactory, and gut chemosensory systems. To improve the quality of food, these pathways trigger physiological and behavioral reactions. These control systems optimize the processes of consumption, digestion, and absorption of beneficial components while minimizing detrimental components. The ability to sense food quality comprises both innate and learned elements, enabling animals to enhance the quality of their dietary choices (Florsheim et al. 2021).

Food enters the gastrointestinal system subsequent to traversing olfactory and gustatory checkpoints. Specialized intestinal epithelial cells, including Enteroendocrine cells (EECs), Enterochromaffin cells (ECCs), and tuft cells, are mainly involved in the gut chemosensory systems that carry out the evaluation (Gribble and Reimann 2016). These cells possess a range of nutrient and xenobiotic sensors, including taste receptors. Upon detection of these stimuli, signals are produced that influence sensory neurons and immune system cells, such as myeloid and lymphoid cells. This process facilitates the effective digestion and absorption of nutrients from food (Gribble and Reimann 2019). EECs are essential for the digestion, absorption, and metabolism of food. Located all across the digestive system, they detect nutrients, microbial metabolites, bile acids, and other substances in the luminal contents. Electronic endothelial cells (EECs) communicate with resident gut mucosal cells, afferent neurons, and distal gastrointestinal organs via secreting gut hormones in response to luminal stimuli. These cells are essential for the coordination of these processes. The presence of EECs in the gastrointestinal tract is crucial for the maintenance of effective digestion and nutrient absorption (Gribble and Reimann 2016, 2019).

10.4 Olfactory and Gustatory Indicators of Food Quality

The quality of food is assessed based on safety, sensory characteristics, chemical composition, and nutritional value. The sensory quality of food is determined by its appearance, color, flavor, and texture. Assessing food quality, authenticity, and adulteration is a critical issue for both consumers and regulatory bodies. This activates the senses of sight and smell, which are markers of food quality. For example, skatol, methane thiol, putrescine, and cadaverine are signs of polluted or rotting food, while ethanol and ethyl esters are signs of fruit sugar content for primates. People either eat more or less of certain foods depending on how they perceive their aroma (Kavaliers et al. 2004; Nevo and Valenta 2018). During ingestion, the gustatory system detects amounts of carbohydrates, amino acids, salts, simple acids, and possibly dangerous chemicals, thereby evaluating food quality. The five main tastes are sweet, umami, salty, sour, and bitter; engage particular receptors to produce pleasing behavior and food acceptance. On the other hand, the activation of bitter taste receptors causes natural adverse reactions and food rejection, even if tiny amounts of some bitter chemicals could be tolerated and valued (Breslin 2013). Evolution may select for particular innate taste preferences, as bitter tastes are frequently associated with toxins, whereas a heightened preference for sweetness may direct individuals toward safe energy sources. Odor perception is intricate, constraining the potential for evolution to favor particular patterns. A flexible olfactory system may provide evolutionary advantages that surpass the drawbacks of failing to categorize specific odor classes as safe or hazardous (Yeomans 2006).

Olfactory perception related to food significantly influences appetite regulation; however, limited research has investigated the learning processes associated with food-related odors. By frequently combining new food scents with sweet and sour tastes on the subsequent perception of the odor when presented in isolation, Stevenson and colleagues established a model to examine olfactory-based learning in humans. The surprising outcome showed that smells conditioned in this way acquired the sensory traits of the corresponding tastant. Although there is strong evidence of sensory changes, hedonic evaluations of the related smells stayed mostly unaltered in all these trials (Stevenson et al. 1995, 1998, 2000). This result is especially unexpected given the compelling data suggesting an intrinsic inclination for sweet tastes in humans (Steiner et al. 2001). The consistent association of a novel odor stimulus with a taste of significant hedonic value is expected to lead to a modification in the hedonic assessment of the conditioned stimulus (Looy et al. 1992; Looy and Weingarten 1991).

Taste is frequently associated with food quality, as undesirable flavors may signify deterioration (Fig. 10.3). The interplay among taste, flavor, and consumer preference is essential for achieving customer satisfaction (Khan and Shafiur Rahman

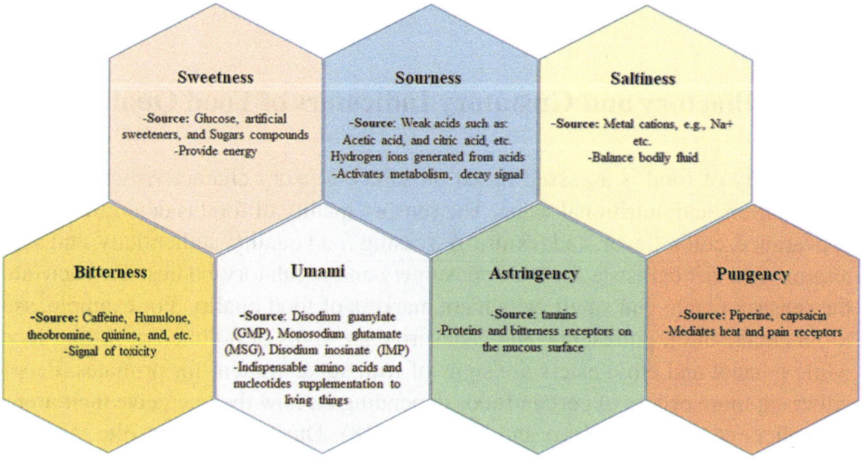

Fig. 10.3 Gustatory responses and their feature. (*Adapted from Toko et al. 2013)

2021). Olfactory and gustatory indicators are crucial in determining food quality, signaling freshness, ripeness, and desirability. Sensory perception detects volatile compounds influencing consumer preferences and food safety. The sense of smell distinguishes fresh from spoiled food by detecting volatile compounds associated with freshness or decay. Sweetness arises from sugar, glucose, and artificial sweeteners, contributing to energy production. Cations such sodium ions cause saltiness; anions also contribute to its sense. Sourness comes from hydrogen ions drawn from acids such citric acid, hydrochloric acid, and acetic acid. Sourness, a signal of decay, is activated by metabolism and can indicate toxicity or medicine efficacy. Bitter substances like caffeine, theobromine, quinine, and isohumulone can also signal toxicity. Its production is mediated by heat and pain receptors (bitterness receptors) on the mucosal surface induce proteins to produce it. Umami (savories) is a taste that supplies essential amino acids and nucleotides to living things. Compounds that contribute to characteristic umami flavor are disodium inosine and disodium guanylate. An electronic measurement-based taste sensor utilizing lipid/polymer membranes is capable of detecting, classifying, and quantifying taste (Toko et al. 2013).

10.5 Food Quality Assessment

Food quality assessment based on sensory aspects involves evaluating products through the five senses—sight, touch, smell, taste, and hearing, since sensory characteristics like color, shape, flavor, and texture heavily influence consumer perception and acceptance. Sensory evaluation measures responses to stimuli generated during food consumption and widely used to determine consumer preferences and satisfaction (Lawless and Heymann 2010). Furthermore, it plays a crucial role in designing quality systems, providing technical support for quality assurance during food production. By offering valuable feedback, sensory evaluation aids decision-making and facilitates modifications to enhance product quality (Meilgaard et al. 2007.

Flavor perception is primarily driven by aroma and taste, which interact cognitively and are shaped by learned associations (Tournier et al. 2007). Studies suggest that aroma can enhance or modify taste perception, influencing overall flavor acceptability. Texture also plays a significant role in sensory perception, affecting how aroma and taste are experienced over time. Firmer textures, for instance, can decrease the perceived intensity of both aroma and taste (Saint-Eve et al. 2011). The sequence of flavor experiences may be influenced by texture, as demonstrated in confectionery products where differing textures generate unique flavor profiles. While aroma, taste and texture form the triumvirate of sensory experience, their integration determines the consumer's satisfaction. The researchers discovered that specific pairings of texture and aroma in preserved sea fennel substantially affected perceived quality, with some eliciting higher ratings than others (Radman et al. 2023). Furthermore, sensory cues impact both energy intake and a feeling of fullness, underlining that the sensory elements are not merely about pleasure but carry physiological consequences as well. Sensory signals speak to both mind and body (Forde 2016).

10.6 Food Safety Evaluation

Unpleasant tastes and smells are crucial indicators of food contamination or spoiling, warning consumers of possible health risks. Signs like sour or rancid smells often point to bacterial contamination or chemical spoilage, making them essential indicators for ensuring food safety. As the two main senses of flavor, aroma and taste interact in ways that are primarily influenced by long-term cognitive associations (Tournier et al. 2007). According to research, aroma has a substantial impact on total flavor acceptance by enhancing or changing taste perception. Texture also significantly influences sensory perception, with firmer consistencies often reducing the perceived strength of both flavor and aroma (Saint-Eve et al. 2011). For example, different textures in confections produce unique flavor profiles that influence how flavors are released and interpreted over time (Saint-Eve et al. 2011).

In addition to providing sensory enjoyment, aspects like flavor, texture, and scent affect food preferences and customer satisfaction. According to a study on preserved sea fennel, texture and aroma had a big influence on how good it was rated, with some combinations scoring better (Radman et al. 2023). Sensory cues also control energy intake and satiety, indicating that their impact goes beyond pleasure to physiological reactions (Forde 2016). However, unbalanced sensory components may result in unfavorable customer experiences. For example, an excessively strong aroma may overshadow other flavors, reducing the overall appeal of a dish (Radman et al. 2023). Thus, achieving a balance among sensory components is essential for ensuring an optimal food experience.

10.7 Innovations in Sensory Analysis for Food Quality and Safety

10.7.1 Electronic Noses (E-noses)

The electronic nose has an emerged pivotal tool used for the food sensory evaluation, providing an alternative to traditional human olfactory systems. It is the apparatus designed to analyze, recognize and identifies complex gases (Zhong 2019) and mimic the human olfactory cells. By mimicking human olfactory cells assess the flavor and aroma profiles of the food products as well some other sensitive volatile compounds, provides rapid evaluations which are important for the quality control and for the new product development in the food industry (Nayak et al. 2020).

The core components of the e- nose system are basically involve three principle parts: Sample delivery system, detection system, computing system (Dhar et al. 2018). The first system is usually more crucial for extracting the volatile compounds, aroma, odor from the sample. The proper extraction of the volatile compounds from the headspace of the sample is significant challenge. There are some common methods which are used for the extraction of the volatile compounds including static headspace, dynamic headspace and the solid- phase micro extraction (SMPE) (Majchrzak et al. 2018). Among all three methods, Dynamic headspace provides quicker results and recovery whereas static headspace is an early method which is used with sensitive detectors and SMPE used rarely due to its high economy and complexity. After that second system is detection system also called sensory assay which have a collection of the different gases, odor molecules that detects smells from the sample. These sensors response like human olfactory senses and transform the different odor molecules on their surface into a detectable signal and convert into the digital value. A variety of sensors have been developed according to their materials such as metal oxides, intrinsically conducting polymers, conducting polymer composites. However, conductive, optical, surface acoustic wave, gas-sensitive field effect transistors, and quartz can be used as the detection principle for the sensory array. There are various emerging technologies including micro electromechanical systems (MEMS) and nanotechnologies. And the last system is computing system known as signal processing system. This system is responsible for processing the information obtained from the sensory array and performing feature extraction. It mimics human olfactory nerve systems and judges the received signals by using multivariate pattern analysis techniques. After data collection from the individual sensors, software is used to analyze and for the identification of the volatile compounds to operators (Fig. 10.4). A classification scheme using multivariate statistics is applied to interpret complex data sets (Zhong 2019).

Electronic nose are rapidly used for the quality assessment of the food products in the various processing sectors including meat and fresh produce, where they monitor freshness and detect spoilage (Zhou and Hai 2009). (Modesti et al. 2021)

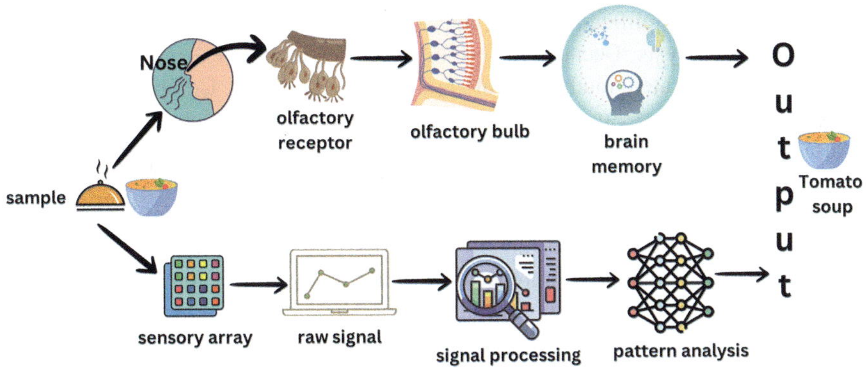

Fig. 10.4 Schematic diagram of an electronic nose for the sensory evaluation

demonstrated assessment of products such as virgin olive oil for checking their effectiveness in terms of stability and shelf life. With regard to fish and meat, electronic noses are mostly used for quality assessment, spoilage identification, taint and off-flavor detection, and bacterial strain categorization. Furthermore, the methods of preparing and the distinctions between them have been studied (Loutfi et al. 2015). Also evaluate odor quality of food packaging material and detect food adulteration in various food (Loutfi et al. 2015). Despite their advantages electronic nose have several disadvantages like limited sensitivity and required more database for the accurate result (Nayak et al. 2020).

10.7.2 Electronic Tongues (E-tongues)

The electronic tongue serves as a sophisticated instrument designed to broadly mimic the human taste sensory organs. The e-tongue emerged in 1995, drawing an analogy to the human tongue which uses receptor molecules to trigger nerve signals when exposed to different tastes (Titova and Nachev 2020). E-tongue technology has been developed to mimic the taste sense of man through a selection of sensors which interact with chemical substances in a sample to provide distinct response profiles. Contrary to very selective sensors which will detect particular chemicals, e-tongues have cross-selective sensors that recognize many substances all at once, giving a general impression of the taste attributes (Lvova 2016). The principle behind an e-tongue is fundamentally one of signal transduction, in that the interaction between taste compounds and the sensors leads to changes in sensor properties, which in turn can be converted into suitable electrical signals by methods such as potentiometry and voltammetry. This process makes it easier to process and analyze the data obtained. In the process of measurement, a sensor array is submerged in a sample and in a reference solution, which provides a comparative analysis. Additionally, several systems use changes in membrane potential to identify adsorption effects, offering more thorough information on a substance's flavor profile.

After the signals are created, they are subjected to data acquisition and preprocessing, where amplification, filtering, and noise reduction methods make them clear before analysis. Because the e-tongue creates a complex, multi-dimensional dataset, sophisticated mathematical models like Principal Component Analysis (PCA), Partial Least Squares (PLS), and Artificial Neural Networks (ANN) are employed to analyze the responses (Sliwinska et al. 2014). These models are useful for classifying, predicting, and confirming taste properties with good accuracy. Several kinds of sensors are used in e-tongues, such as potentiometric sensors like Ion-Selective Electrodes (ISEs) and Ion-Sensitive Field-Effect Transistors (ISFETs), which record potential differences for the analysis of ion concentrations. Voltammetric sensors work by varying the potential applied and recording the currents generated, providing high sensitivity and durability. Potentiometric sensors, particularly Ion-Selective Electrodes (ISEs), are widely employed, detecting the difference in potential between a sensing and a reference electrode (Fig. 10.5). Surface

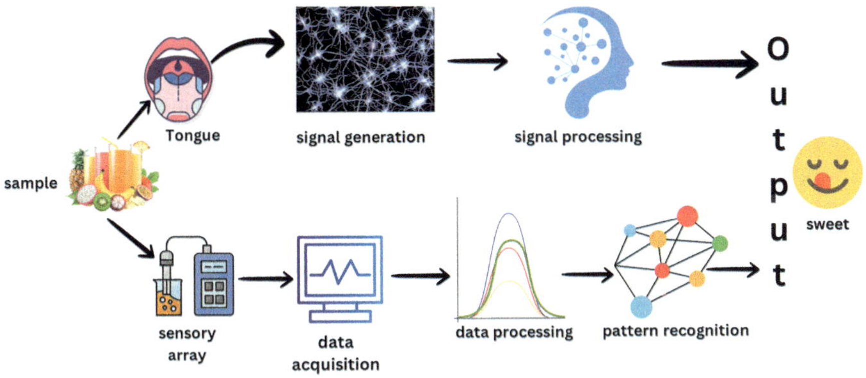

Fig. 10.5 Schematic diagram of an electronic tongue for the sensory evaluation

Acoustic Wave (SAW) devices and optical chemical sensors have also been investigated for use in e-tongue applications (Lvova 2016).

Electronic tongues are used to check the quality of many food categories, such as fruits, dairy goods, meat items, and oils (Yadav et al. 2025). E- tongue can identify delicate flavor differences that may be missed by human panels (Cho and Moazzem 2022) also useful for the determination of possibly harmful components (Imam and Nagpal 2023). Though their merits, e-tongue has various demerits like they have less sensitivity and stability compared to other analytical methods like gas chromatography-mass spectrometry and interpreting data is tricky with techniques like voltammetry with special mathematical models being required (Titova and Nachev 2020).

10.7.3 E-eyes

Electronic eyes devices that mimic human vision that analyze color attributes of food matrices, contributing to quality evaluation by providing visual data (Calvini and Pigani 2022; Galvan et al. 2022). The principle of an electronic eye (E-eye) is based on converting optical images using an image sensor to objectively assess visual characteristics, eliminating the subjectivity of human vision. This advanced computer vision technology integrates multiple components, including mechanics, optics, electromagnetic detection, colorimetry, spectrophotometry, digital video, and image processing, to ensure precise analysis. Image acquisition is the initial step, in which an optical image of a food product is taken by an image sensor. Through advanced algorithms and camera-based inspection, the E-eye analyzes major visual characteristics such as shape, size, color, and texture for a quick, precise, and non-destructive measurement (Xu 2019). By eliminating human subjectivity, this technology ensures reproducible and objective measurements, making it highly reliable in quality control applications. Additionally, the E-eye processes and analyzes captured images in-depth, enabling correlation with sensory panel assessments, thus providing a comprehensive evaluation of a product's visual properties (Rodrigues et al. 2024).

There are many application of the electronic eyes including dairy, vegetables and fresh food to ensure quality as well to play role in shelf life evaluation by detecting the spoilage (Grassi et al. 2023; Ordoñez-Araque et al. 2020). Extensive research has demonstrated the effectiveness of the E-eye in color assessment, marbling grading, quality prediction, defect detection, and sorting. Since color is a key indicator of freshness, discrepancies can lead to rejection (Rodrigues et al. 2024). Studies have shown strong correlations between E-eye measurements and colorimeters, yielding accurate results for lightness and reasonable regression coefficients for redness and yellowness in chicken meat (Barbin et al. 2016). Despite their application, electronic eye several limitations such as high initial investment in equipment, requirement of the calibration and maintenance. Additionally data interpretation is difficult and complex (Rodrigues et al. 2024).

10.7.4 Neurogastronomy

Neurogastronomy is a multidisciplinary field that explores how the brain constructs our perception of flavor, integrating biological, emotional, and cultural factors. It emphasizes that flavor is not solely determined by taste but is a complex experience created in the brain, combining sensory inputs from taste, smell, sight, sound, and touch (Robinson 2015). The retronasal olfactory signal, where aroma compounds travel from the mouth to the nose, plays a key role by merging with taste signals in the orbitofrontal cortex, explaining why food tastes different when one's nose is blocked. Beyond the five basic tastes—sweet, salty, sour, bitter, and umami—research explores additional taste elements like fat, calcium, and kokumi, (Yılmaz and Altuntaş 2022) a proposed flavor enhancer. External factors such as music, texture, temperature, and food presentation also influence perception, making dining a multi-sensory experience. Technological advancements, including virtual reality, further demonstrate how altering sensory stimuli can change how the brain perceives food, such as modifying the visual appearance of coffee to affect its perceived creaminess. Unlike traditional approaches that focus solely on modifying food itself, neurogastronomy explores ways to "reinvent" brain perception, potentially making healthier foods more appealing. This emerging field bridges neuroscience, nutrition, psychology, food science, and gastronomy, uniting scientists and culinary experts to better understand how the brain processes flavor and the broader experience of eating (Tokat and Yilmaz 2023).

10.8 Applications in Food Science

Examining sensory responses is important in the field of food science as it plays a crucial role in food product development, safety monitoring, and quality assurance. Sensory view is key to improvements and rules in the food industry since it affects what consumers like, how they eat, and what they buy. (Chambers IV 2019).

While objective analytical techniques like assessing chemical composition and microbial analysis yield quantifiable results regarding the properties of food, sensory evaluation adds another layer of human perception and gives a fuller picture. Sensory testing combined with instrumental analysis ensures that the food products have scientific justification for safety as well as market appeal.

The use of sensory analysis especially in relation to smell and taste has become more important in food science. Sensory evaluation is very beneficial in understanding consumer behavior towards preferences, food quality, and safety. This enables food scientists to develop products to address the needs of consumers. (Gámbaro and McSweeney 2020).

Sensory analysis has found great use in the food industry placing much emphasis on food quality, nutrition, and health. The combination of sensory evaluation with microbiological and chemical methods has enhanced the credibility and precision of food quality evaluation (Sirangelo 2019).

Evaluative smell and taste go further than just product innovation; they are part of measuring the safety and quality control processes. Smell and taste can offer clues of spoilage, off-flavors, or contaminants of some sort, which can help in flagging food safety issues beforehand. Certain trained panels can sense the changes in the food that is offered and thus mark if further explorations, changes, or corroborative actions need to be done, like further analyses tests, which entail foremost examination of analytical testing.

The use of smell and taste in food science has numerous implications as well as benefits. Food scientists rely on equipment with calibrated sensors that mark odour and flavor indicators. In these cases, scientists are well equipped in their attempts to measure and make sure they capture the changes that are being made in food safety, quality, and acceptability alongside innovation, because in the development of any food products, safeguarding quality becomes essential in achieving novel aims (Sirangelo 2019; Gámbaro and McSweeney 2020).

10.8.1 Role of Sensory Responses in New Product Development (NPD)

Analyzing the aroma and taste of a product is central to its acceptance in the market, especially during new product development (NPD) by ensuring consumer acceptance and optimizing food formulations. Sensory attributes such as aroma and flavor are among the most critical factors influencing consumer preference, guiding the reformulation of existing products and the innovation of new food offerings. Sensory evaluation methods, including hedonic testing and descriptive analysis, provide valuable insights into the intensity and acceptability of flavors, textures, and aromas, helping manufacturers refine their products before launch (Fiorentini et al. 2020). Without sensory analysis, even the most nutritionally optimized or technologically advanced products may fail due to poor acceptance by consumers (Świąder and Marczewska 2021).

Development of a new product involves a range of techniques such as descriptive analysis, affective testing, and instrumental sensory methods. In the case of descriptive analysis, a trained sensory panel can evaluate specific attributes like sweetness which leads to further detailing in product feedback in relation to its sensory profile. Affective testing focuses on the outcome of consumer studies on overall liking, preference, and intent to purchase towards a particular product. This set of techniques aids in developing products that are not only nutritious, but also critical to the well-being of the consumers (Guiné et al. 2020).

With regard to the above, the latest updates in the area of sensory evaluation include bioelectronic tongues and noses, which greatly resemble human sensory receptors and allow for the more accurate automated assessment of taste and smell (Trigo et al. 2019). Such technology aids in the development of plant-based and alternative protein products by maintaining meat's texture, mouthfeel, and umami flavor (Fiorentini et al. 2020). Additionally, machine learning is now added to sensory analysis which aids in modeling through predictive analysis optimizing the formulation of flavor preferences of the targeted consumers.

The trending diets that favor health and sustainability have increased the need for more thorough sensory evaluation. This field of study aids the testing of new products like reduced-fat dairy, sugar free beverages, and high protein snacks to ensure that their desirable sensory properties remain intact. This form of testing helps to meet dietary as well as legal benchmarks. The application of sensory science into designing new products not only fulfills the consumer needs but also increases market competition (Świąder and Marczewska 2021).

10.8.2 Sensory Responses in Food Quality Assurance

In the food industry, ensuring uniform quality is one of the most important goals because any change in taste, texture, or smell can negatively affect consumers and the brand. Quality monitoring and management measures are kept in check through food Sensory evaluation, which helps in monitoring quality by detecting gaps in various sensory attributes, checking ingredient uniformity or consistency, and confirming processes (Sirangelo 2019).

Methods of measuring food quality include modern instrumental approaches and conventional sensory techniques. Some of the accepted traditional methods are hedonic testing, which captures consumer preferences, and quantitative descriptive analysis involving measuring specific sensory attributes by trained panels (Chambers

IV 2019). These evaluations enable manufacturers to create standardized recipes and correct inconsistencies in flavor, texture, and other components of the product.

The development of electronic tongues and noses has greatly facilitated food quality measurement. Electronic noses research the volatile compounds causing aroma, while electronic tongues analyze sweetness, bitterness, and acidity to assess the taste components (Trigo et al. 2019). These bioelectronic sensors provide real-time quality control data, ultimately reducing the time and reliance on human panels and thus ensuring consistency in large-scale production.

Furthermore, sensory evaluation plays a crucial role in detecting product degradation over time. Changes in texture, loss of aroma, or development of off-flavors are early indicators of reduced shelf life or improper storage conditions. By integrating sensory analysis with chemical and microbiological testing, food scientists can identify and address quality concerns before products reach consumers (Świąder and Marczewska 2021).

In traditional and artisanal food types, sensory evaluation ensures that the products maintain their authentic nature and cultural value. A good example would be cheese production, where sensory panels have to assess the texture, ripeness, and flavor complexity of cheese so that it retains its unique features competitively. Other industries like wine and coffee have to rely on sensory profiling to guarantee consistency in aroma and mouthfeel. Tasting also becomes an essential description of these products (Sirangelo 2019).

Maintaining product appeal and palatability in line with consumer expectations requires an unrestrained degree of control which makes sensory evaluation crucial in hygiene management. The use of advanced technology alongside traditional methods provides food producers with reduced variability in their controlled processes while achieving their intended qualitative goals.

10.8.3 Sensory Evaluation in Food Safety Assurance

Food safety remains the utmost priority within the food sector, since sensory evaluation is useful in the detection and diagnosis of potential hazards such as spoilage, contamination, and adulteration of food products. The human senses are the first and finest detectors of issues related to food safety, for instance, off-odors, sour tastes, and changes in texture often points towards microbial growth or chemical deterioration (Chambers IV 2019).

The initial stage to identify food spoilage and contamination is sensory testing. Specialized trained panels alongside consumer testers can discern very delicate changes in aroma and taste that signal bacterial growth, oxidation, or chemical contamination. Furthermore, the primary scope targets the development of an electronic sensing device that offers safety evaluations of an objectively assured level of precision. Spoilage detection is accomplished by artificial nasal devices that monitor the presence of volatile organic compounds that mark as spoilage, while electronic tongues monitor notable changes in pH and other chemicals associated with microbial activity (Trigo et al. 2019).

In compliance with policies, these processes of monitoring ensure food safety as they check whether pasteurization, fermentation, and sterilization overstep sensory boundaries. This is critical in dairy, meat, and beverage industries where any changes in sensory elements could indicate inadequately processed or contaminated products (Sirangelo 2019).

Additionally, the sensory evaluation is of great importance for monitoring the food packaging and storage condition. Sometimes the packaging can leach certain odor or interact with the product's flavor affecting safety and acceptability of the product. By incorporating various instruments of sensory evaluation, including gas chromatography coupled with mass spectrometry (GC-MS) and microbial analysis, manufacturers can identify numerous safety concerns at the early stages of the production cycle (Świąder and Marczewska 2021).

Shifts and advancements in food safety are likely to come with the application of sensory evaluation alongside digital monitoring systems as well as artificial intelligence (AI) which will facilitate better real-time monitoring. Automated sensors can assess the possibility of contamination quickly and therefore, contribute to the enhanced safety of the food supply chain globally.

10.8.4 Role of Sensory Analysis with Objective Testing Methods in Consumer-Centered Food Evaluation

Sensory analysis and objective testing methods play complementary roles in food science, ensuring that food products are not only scientifically validated but also acceptable to consumers. Objective testing relies on quantitative, instrument-based assessments for the analysis of food composition, texture, and safety, while sensory analysis captures the human experience by integration of subjective factors such as taste, aroma, texture, and appearance. Although objective methods provide precise, repeatable measurements, they do not account for the complexities of consumer perception, which significantly influence food acceptance and market success (Wang et al. 2024a, b).

Techniques employed in sensory analysis include descriptive sensory analysis, hedonic testing, and time-intensity analysis, which focus on the examination of foods' defining characteristics. These strategies allow food science and industry professionals to adjust the ingredient blend, processing techniques, or product to better fit the customer's needs. An illustrative example is provided by descriptive sensory analysis, which enables trained panels to quantify specific attribute intensities, like sweetness or bitterness, further providing an overall sensory profile. Consumer preference and emotional reaction is assessed through hedonic testing registries which evaluate consumer emotional response, capturing more overarching preference patterns along with potential market value (Chen et al. 2024). Sensory analysis is of utmost importance in the design process of plant-based and functional foods because the products must meet the consumer's expectation regarding texture, flavor combination, and aftertaste. Moreover, the application of AI and machine learning technologies has greatly impacted sensory research by enabling the prediction of consumer preferences from sensory panel data, thereby improving formulation efficiency and reducing development time (Fuentes et al. 2021; Zatsu et al. 2024).

Objective testing, in contrast, is essential for ensuring food quality, safety, and regulatory compliance. Techniques such as gas chromatography-mass spectrometry (GC-MS), texture profile analysis (TPA), electronic noses (E-noses), and near-infrared spectroscopy (NIRS) provide scientific validation of food properties. For example, GC-MS is used to identify and quantify volatile compounds responsible for food aroma, while TPA assesses food firmness, elasticity, and chewiness. Similarly, E-noses and E-tongues can detect flavor compounds, spoilage markers, and adulterants, allowing for rapid quality control and safety monitoring (Wang et al. 2024a, b). These methods are highly reproducible and precise, making them indispensable for ensuring consistency in large-scale food production. However, objective methods alone cannot analyse human taste perception or predict how consumers will react to a product, making sensory analysis a crucial complement to instrumental testing.

The combining of sensory and objective testing yields a better method for assessing food. While objective testing guarantees product safety and regulation compliance, sensory analysis bridges the gap between scientific evaluation and market success. For instance, when determining yogurt's sensory attributes, researchers combine GC-MS analysis of volatile compounds with hedonic testing to measure consumer preference. Such methodologies make it possible for manufacturers to modify formulations while ensuring that a product meets quality and consumer expectations (Wang et al. 2024a, b).

In the context of the integration of objective and sensory testing, the use of AI in modeling sensory assessment, VR sensor labs, and blockchain for sensory traceability will be beneficial for food science. AI models can predict consumer acceptances based on chemical and sensory data, while VR sensory environments can simulate consumption locations. Additionally, blockchain technologies provide verifiable information in sensory-based quality control, enabling consistency in products for various regions around the world (Chen et al. 2024).

Ultimately, sensory and objective testing are not competing methods, but complementary tools that together provide a holistic understanding of food quality, safety, and consumer satisfaction. Sensory evaluation captures the human experience, while objective methods ensure scientific accuracy. By integrating these approaches, food scientists can validate product characteristics, minimize errors, and enhance consumer trust. With continuous advancements in AI, bioelectronics, and sensory-driven data analytics, the future of food quality assessment will be more precise, efficient, and consumer-focused than ever before (Zhu and Wang 2023).

10.9 Trends in Food Sensory Science

Recent advancements in food sensory science reflect a profound shift toward technologically integrated, consumer-centered, and data-enhanced practices. Traditional sensory analysis—which historically relied on human panels evaluating basic parameters like taste and aroma—is evolving into a far more complex and dynamic discipline. Today, sensory science integrates cutting-edge technology, behavioral neuroscience, personalized nutrition, and digital systems to create immersive, reliable, and adaptable food experiences. This chapter explores the four most transformative trends in this space: multi-sensory integration, personalized nutrition, the application of blockchain and IoT technologies, and the development of hybrid sensory panels that combine human expertise with artificial intelligence and biosensors (Sempionatto et al. 2021).

10.9.1 Incorporating Multi-sensory Experiences in Product Development

Food science has increasingly recognized that eating involves multiple senses, not just taste. As research in neuroscience and consumer psychology has advanced, scientists and product developers have started incorporating a wider range of sensory inputs—such as visual appearance, sound, texture, and environmental factors—to influence how consumers perceive food. (Spence 2023). For example, the color and shape of a plate can alter taste perception; foods served on white plates are often rated sweeter or more intense than the same food on black or blue plates (White et al. 2019). Similarly, the ambient music in a restaurant or the packaging sound of a crisp bag can subtly alter how we perceive the food inside. Recent studies have used Virtual Reality (VR) and Augmented Reality (AR) environments to simulate different dining settings, allowing researchers to test food perception in controlled but realistic environments (Torrico et al. 2022; Spence 2023). This "immersive sensory testing" enables a more holistic understanding of consumer preferences and opens the door to innovation in product formulation. For example, using AR to visually change the appearance of a beverage can influence how it tastes to the consumer, even if the chemical composition remains unchanged (Crofton et al. 2019; Spence 2023). Understanding how these sensory elements interact allows for the creation of more appealing and satisfying food products (Velasco et al. 2018).

This integration of sensory cues across modalities is grounded in crossmodal perception—the way in which stimuli from one sense can influence how stimuli from another sense are interpreted. The application of this principle has practical implications in everything from food marketing to therapeutic nutrition, where appetite and food enjoyment can be enhanced through strategic sensory stimulation. In short, multi-sensory product design is revolutionizing how we approach food development by emphasizing experience over mere consumption (Świąder and Marczewska 2021).

10.9.2 Impact of Personalized Nutrition on Sensory Science

Another key trend in food sensory science is the move toward personalized nutrition, where food design and delivery are tailored to the individual's genetic makeup, health status, sensory preferences, and lifestyle habits (Berciano et al. 2022). This shift has been driven by the convergence of nutrigenomics, biosensor technologies, and machine learning, enabling researchers to assess not only what nutrients a person needs but also how they prefer to consume them.

The integration of sensory profiling into personalized nutrition is a critical development. It recognizes that a healthy food option is only effective if it is acceptable and enjoyable to the consumer. For instance, a person may have a genetic predisposition toward bitter taste perception and therefore avoid nutrient-rich bitter foods like kale or Brussels sprouts. Personalized strategies can counter this by recommending alternative foods, incorporating masking agents, or developing products that align with the individual's taste preferences.

The development of personalized flavor profiles involves detailed sensory mapping for individual consumers (Nolden and Feeney 2020). This involves creating detailed models that predict liking and preference based on sensory attributes, physiological responses, and psychological factors. Advanced technologies like wearable biosensors, combined with AI algorithms, enable real-time assessment of an individual's response to food, paving the way for dynamic and adaptive nutrition plans.

Advancements in wearable biosensors and smart utensils now allow real-time monitoring of eating behaviors and physiological responses, enabling dynamic adjustment of dietary plans (Atanasova et al. 2022). These technologies can track everything from chewing frequency to glucose spikes after meals, feeding data into machine learning models that then generate personalized diet recommendations. Such systems are often paired with mobile apps, making them accessible and user-friendly for end-users.

Moreover, personalized nutrition intersects with behavioral psychology to support long-term adherence. By aligning dietary interventions with sensory satisfaction, food companies and healthcare providers can significantly enhance user compliance and health outcomes. This data-driven personalization not only improves the effectiveness of nutrition plans but also elevates consumer satisfaction and loyalty (Adams et al. 2019).

10.9.3 Use of Blockchain and IoT to Enhance Sensory Data Accuracy in Food Supply Chains

The adoption of blockchain and Internet of things (IoT) technologies has positively impacted the food sensory science and quality control technologies by disrupting the capturing, validation, and application of sensory data across the food value chain. These digital technologies improve supply chain transparency, traceability, and real-time decision-making. They bolster the integrity and reliability of sensory evaluations by ensuring data is secure, easy to obtain, and verifiable. This fosters consumer trust in the quality and safety of food products by providing a transparent, traceable record of the sensory assessment process. Blockchain technology offers a decentralized and unchangeable record keeping system for sustaining the sensory data from farm to table (Awan 2019).

10.9.3.1 Blockchain for Sensory Data Integrity

Blockchain technology can be understood as a decentralized ledger system in which information is saved in blocks that have been timestamped and secured using cryptographic methods. Data cannot be modified after it has been added which guarantees reliable documentation of recordings and environmental conditions. This is critical to preventing the fraudulent manipulation of sensory data so that the data used for product development and quality control is trustworthy (Rejeb et al. 2020). Blockchain makes it possible to document every sensory evaluation, ingredient batch, and processing step, which can be accessed and verified without modification. Its fundamental properties such as visibility and lack of modification make it a strong infrastructure for monitoring food products at all levels of the supply chain to ensure full traceability from the source to the final consumer (Ellahi et al. 2023).

For example, blockchain systems have been implemented in the coffee industry, where flavor notes, bean processing methods, and origin-specific attributes are certified and stored in a blockchain ledger. Consumers can scan QR codes on packaging and access the entire sensory profile history, offering transparency that traditional labeling cannot match.

The application of blockchain technology significantly fortifies food safety protocols by enabling rapid and precise tracking of products across the supply chain, expediting the identification and containment of contamination sources, and bolstering accountability among stakeholders (Kshetri and DeFranco 2020; Balcıoğlu et al. 2024). Companies are increasingly investing in automated food software to facilitate real-time tracking and tracing of products within the supply chain, thereby enabling the rapid detection of production anomalies, precise determination of product origin, efficient management of code and expiry information, and comprehensive tracking of order numbers (Kumar et al. 2022).

10.9.3.2 IoT-Based Sensory Monitoring

A network of physical devices with sensors and software that can automatically collect and transfer data is known as Internet of Things (IoT). Considering food sensory science, IoT devices are utilized in monitoring the environmental parameters like temperature, humidity, pH, gas composition, texture, aroma profiling, and others that affect the sensory quality of food. There devices can be applied throughout the entire supply chain, which includes production, processing, storage, and even during transportation, helping to ensure that food products retain their sensory properties until the consumer receives them.

Smart packaging systems that are equipped with temperature and spoilage indicators are one of the most developed IoT applications. These systems cover the environment's exposure and track in real time, transmitting the data directly to mobile applications or central platforms. For example, IoT-enabled utensils and wearable devices capable of tracking the eating behavior, flavor perception, and emotional response during consumption have been developed in one study. Their system utilizes biosensors that measure chewing frequency and salivation, delivering dynamic sensory feedback and personalized food experiences (Atanasova et al. 2022).

The integration of IoT devices with cloud-based platforms and machine learning algorithms allows for predictive modeling of food quality and shelf life. By analyzing real-time data from sensors, these models can forecast potential spoilage or degradation of sensory attributes, enabling proactive interventions such as adjusting storage conditions or rerouting shipments to minimize waste.

Furthermore, It was highlighted that virtualization of food supply chains via IoT allows not only automation of quality control processes but also integration with ERP and logistics systems, ensuring that deviations in sensory-critical factors (like freezing, overcooking, or excessive humidity) are caught in real-time and corrected before reaching the consumer. This enhances both quality assurance and consumer trust, as the final product is more likely to meet expected sensory standards. (Verdouw et al. 2016).

10.9.3.3 Synergizing Blockchain and IoT

Exploration of synergy between blockchain and IoT reveals where the true advancement lies. Providing sensors of IoT and linking them to IoT devices gives automated real-time access to information, while automation of information stores true and unlockable. The two components create a self-sufficient framework that can be designed and operated singularly for exquisite food products such as seafood, dairy, or gourmet food. In these products, sensory qualities such as taste, aroma, and texture describe the products themselves and the worth of the market. Blockchain technology linked with IoT guarantees comprehensive monitoring and provides real-time reports devoid of unauthorized modifications to sensory information, ensuring the data integrity for the entire supply chain.

As an example, the use of IoT and blockchain will advance us from having only smart farms to developing the international system of smart farms and increase the control mechanisms in supply-chain networks (Torky and Hassanein 2020). This is a new concept that gives feedback at once on every environmental sensor linked and other sensory devices utilized.

While the initial capital expenditure for integrating blockchain and IoT solutions may be considerable, the long-term benefits, including improved traceability, increased transparency, enhanced data security, streamlined supply chain operations, and strengthened consumer confidence, make it a worthwhile investment.

This integration also plays a critical role in food recalls and audits. If a batch of food is found to have sensory defects or quality issues, blockchain records allow companies to trace the issue back to its source—whether it was a deviation in cooking temperature, prolonged storage, or inaccurate sensory profiling. This improves accountability and reduces the time and cost of investigating quality issues (George et al. 2019).

So the integration of blockchain and IoT into food sensory science represents a foundational shift in how data integrity, sensory accuracy, and transparency are ensured throughout the food supply chain. These technologies enhance the depth, reliability, and predictive capacity of sensory evaluations, enabling not only improved quality control but also consumer trust and personalized food experiences.

10.9.4 Development of Hybrid Sensory Panels Combining Human and Artificial Sensing

The integration of human and artificial intelligence in sensory evaluation—termed hybrid sensory panels—represents a transformative trend in food science, biotechnology, and quality assurance. These hybrid panels blend the nuanced perceptual abilities of human panels with the precision, objectivity, and reproducibility of machine-based sensing technologies. This approach is gaining traction across multiple disciplines, enhancing accuracy, reducing sensory fatigue, and enabling real-time data acquisition (Feng et al. 2020).

At the heart of hybrid sensory panels is the synergistic collaboration between traditional human sensory analysis and emerging artificial sensing systems, including electronic noses (e-noses), electronic tongues (e-tongues), and machine learning-based vision systems. These artificial sensing technologies can collect quantitative data and process it faster than human sensory panels. On the other hand, human panels excel at capturing the nuanced interplay of flavors, aromas, and textures that define the overall sensory experience. However, human assessments can be affected by psychological, physiological, and environmental factors, potentially leading to inconsistencies in the data.

Incorporating machine learning into artificial sensing significantly enhances the discriminative power of these technologies. In a research study, it was demonstrated a hybrid system was demonstrated where deep learning algorithms were used to process data from multi-sensory inputs to mimic human decisions in flavor profiling of tea. This not only improved consistency but also enabled predictive analytics in flavor development. (Wang et al. 2023).

From a quality assurance perspective, hybrid panels have become indispensable in food manufacturing. A human-machine systems were used to detect adulteration and spoilage, particularly in dairy and meat. Artificial sensors offer early spoilage detection through volatile compound analysis, which complements human sensitivity to subtle textural changes. (Loutfi et al. 2015).

The development of hybrid sensory panels is transforming sensory science. By combining human perception with AI-based technologies, industries can improve accuracy, efficiency, and consumer focus. As these systems advance, their influence will likely expand beyond food science into other areas like pharmaceuticals, perfumes, and environmental monitoring, leading to a future where machine-assisted human judgment is the standard.

10.10 Challenges and Future Directions

Sensory evaluation, the scientific discipline that measures, analyzes, and interprets human responses to food products through the senses of sight, smell, taste, touch, and hearing, has long been a cornerstone of food science and product development (Stone and Sidel 2009). Historically, its primary role was to ensure product consistency and quality control, such as detecting off-flavors in canned goods or optimizing the texture of baked goods (Mihafu et al. 2020). However, as the global food industry undergoes rapid transformation—driven by trends like plant-based diets, functional foods, personalized nutrition, and sustainability—the scope and complexity of sensory evaluation have expanded exponentially. Today, sensory scientists are tasked not only with validating product stability but also with decoding multisensory consumer experiences, navigating cultural preferences, and addressing the psychological and physiological mechanisms that underpin food choices (Byrne and Heymann 2024).

This evolution has exposed critical gaps in traditional methodologies. For instance, conventional descriptive analysis panels struggle to characterize hybrid products like lab-grown meats or 3D-printed foods, which combine novel textures and flavors (Świąder and Marczewska 2021). Similarly, the rise of "clean label" movements and ethical consumption has shifted consumer priorities, making extrinsic factors like packaging claims and brand narratives as influential as intrinsic product attributes (Ruiz-Capillas et al. 2021). Compounding these challenges are advancements in neurobiology and biometrics, which reveal that sensory perception is not merely a passive response to stimuli but a dynamic interplay of cognitive, emotional, and contextual factors.

10.10.1 Enhancing Sensory Methods for Complex Food Matrices

Contemporary food innovation prioritizes sustainability and health, leading to products such as algae-based proteins, insect flour, and probiotic-infused snacks. These matrices often defy conventional sensory frameworks due to their heterogeneous structures and unfamiliar sensory profiles. For example, plant-based meats must replicate the fibrous texture of animal muscle while masking off-notes from legume proteins (Mihafu et al. 2020). Similarly, functional foods fortified with vitamins or polyphenols risk compromising palatability through bitter or metallic aftertastes.

A key limitation lies in the disconnect between instrumental measurements (e.g., rheology, gas chromatography) and human perception. While instruments quantify properties like viscosity or volatile compounds, they cannot predict how these elements interact dynamically during consumption. For instance, a texture analyzer might confirm that a gluten-free bread has comparable firmness to wheat-based bread, yet consumers may still reject it due to a "crumbly" mouthfeel unaccounted for in instrumental data (Ruiz-Capillas et al. 2021).

Innovative Approaches

1. Multivariate Analysis and Machine Learning: Combining chemometric data (e.g., electronic tongue outputs) with sensory panel results through techniques like partial least squares regression (PLS-R) can identify "sensory markers" in complex matrices. A 2023 study on pea protein isolates used PLS-R to correlate specific amino acid profiles with bitter aftertastes, enabling targeted enzymatic treatments to improve flavor (Byrne and Heymann 2024).
2. Rapid Profiling Techniques: Methods like Flash Profile and Polarized Sensory Positioning (PSP) allow faster screening of novel ingredients. In a case study, PSP successfully differentiated six commercial cricket protein powders based on their "umami intensity" and "mouth-coating" properties, streamlining supplier selection for snack manufacturers (Świąder and Marczewska 2021).
3. Standardization for SMEs: Small-scale producers often lack resources for rigorous sensory testing. Initiatives like the European Institute of Innovation and Technology's (EIT) Regional Innovation Scheme provide SMEs with subsidized access to portable e-nose devices and virtual training modules, democratizing quality control.

10.11 Conclusion

Despite these encouraging findings, the complex interaction between the gustatory and olfactory systems is central to human food perception, directly affecting food selection, liking, and safety assessment. As important biological sensors, they allow humans to identify and food quality based on integrated sensory perceptual information including odor, taste, texture, and flavor. By combining biological receptors with nano-electronic signal transducers, recent advancements in bioelectronic noses, tongues, eyes, and neurogastronomy offer novel technologies that mimic human sensory systems. These have demonstrated remarkable promise in accurately identifying and evaluating food quality indicators, enhancing the precision and effectiveness of food safety inspections. Although these developments open up new avenues for consumer protection, food quality monitoring, and product innovation, there are still obstacles to overcome in terms of scalability, cost-effectiveness, and regulatory authority acceptability. However, continued multidisciplinary research and technical advancements are probably going to drive the usage of bioelectronic sensory tools in the future, setting new benchmarks for sensory evaluation and food quality assurance.

References

Adams SH, Anthony JC, Carvajal R, Chae L, Khoo CS, Latulippe ME, Matusheski NV, McClung HL, Rozga M, Schmid CH, Wopereis S, Yan W (2019) Perspective: guiding principles for the implementation of personalized nutrition approaches that benefit health and function (Review of perspective: guiding principles for the implementation of personalized nutrition approaches that benefit health and function). Adv Nutr 11(1):25

Albustanji R, Elmanaseer S, Alkhatib A (2023) Robotics: five senses plus one—an overview. Robotics 12(3):68. https://doi.org/10.3390/robotics12030068

Atanasova GL, Atanasov BN, Atanasov NT (2022) Fully textile dual-band logo antenna for IoT wearable devices. Sensors 22(12):4516

Awan SH (2019) Role of internet of things (IoT) with blockchain technology for the development of smart farming. J Mech Continua Math Sci 14(5):170–188

Balcıoğlu YS, Çelik AA, Altındağ E (2024) Integrating blockchain technology in supply chain management: a bibliometric analysis of theme extraction via text mining. Sustain For 16(22):10032

Barbin DF, Mastelini SM, Barbon S Jr, Campos GF, Barbon APA, Shimokomaki M (2016) Digital image analyses as an alternative tool for chicken quality assessment. Biosyst Eng 144:85–93

Berciano S, Figueiredo JP, Brisbois TD, Alford S, Koecher K, Eckhouse S, Ciati R, Kussmann M, Ordovás JM, Stebbins K, Blumberg JB (2022) Precision nutrition: maintaining scientific integrity while realizing market potential (Review of precision nutrition: maintaining scientific integrity while realizing market potential). Front Nutr 9:979665

Besser G, Liu DT, Renner B, Hummel T, Mueller CA (2019) Olfactory implant: demand for a future treatment option in patients with olfactory dysfunction. Laryngoscope 129(2):312–316. https://doi.org/10.1002/lary.27476

Boesveldt S, Parma V (2021) The importance of the olfactory system in human well-being, through nutrition and social behavior. Cell Tissue Res 383(1):559–567. https://doi.org/10.1007/s00441-020-03367-7

Brescia AI, Seiden AM (2009) The anatomy and physiology of olfaction and gustation. In: Stucker FJ, De Souza C, Kenyon GS, Lian TS, Draf W, Schick B (eds) Rhinology and facial plastic surgery. Springer, Berlin Heidelberg, pp 91–96. https://doi.org/10.1007/978-3-540-74380-4_7

Bryant, & Silver (2000) Chemesthesis: the common chemical sense. In: Finger TE (ed) The neurobiology of taste and smell, 2nd edn https://catalogimages.wiley.com/images/db/pdf/y0471257214.fm.pdf

Breslin PAS (2013) An Evolutionary Perspective on Food and Human Taste. Current Biology, 23(9):R409–R418. https://doi.org/10.1016/j.cub.2013.04.010
Burdach KJ, Doty RL (1987) The effects of mouth movements, swallowing, and spitting on retronasal odor perception. Physiol Behav 41(4):353–356. https://doi.org/10.1016/0031-9384(87)90400-8
Bushdid C, Magnasco MO, Vosshall LB, Keller A (2014) Humans can discriminate more than 1 trillion olfactory stimuli. Science 343(6177):1370–1372. https://doi.org/10.1126/science.1249168
Byrne DV, Heymann H (2024) Sensory evaluation: Future perspectives. Foods 13(18):2940
Calvini R, Pigani L (2022) Toward the development of combined artificial sensing systems for food quality evaluation: a review on the application of data fusion of electronic noses, electronic tongues and electronic eyes. Sensors 22(2):577
Chambers E IV (2019) Analysis of sensory properties in foods: a special issue. Food Secur 8(8):291
Chen Z, Wang M, Wu Y, Tan J, Li Z (2024) High value utilization of waste in food processing. Front Sustain Food Syst 8:1423046
Cho S, Moazzem MS (2022) Recent applications of potentiometric electronic tongue and electronic nose in sensory evaluation. Prev Nutr Food Sci 27(4):354–364
Cleland TA, Linster C (2019) Central olfactory structures. In: Handbook of clinical neurology, vol 164. Elsevier, pp 79–96. https://doi.org/10.1016/B978-0-444-63855-7.00006-X
Crofton E, Botineştean C, Fenelon MA, Gallagher E (2019) Potential applications for virtual and augmented reality technologies in sensory science. Innovative Food Sci Emerg Technol 56:102178
De Araujo IET, Rolls ET, Kringelbach ML, McGlone F, Phillips N (2003) Taste-olfactory convergence, and the representation of the pleasantness of flavour, in the human brain. Eur J Neurosci 18(7):2059–2068. https://doi.org/10.1046/j.1460-9568.2003.02915.x
Della Longa L, Valori I, Farroni T (2022) Interpersonal affective touch in a virtual world: feeling the social presence of others to overcome loneliness. Front Psychol 12:795283. https://doi.org/10.3389/fpsyg.2021.795283
Dhar P, Kashyap P, Jindal N, Rani R (2018) Role of electronic nose technology in food industry. In: Emerging sustainable technologies in food processing, 1–2
Ellahi RM, Wood LC, Bekhit AEA (2023) Blockchain-based frameworks for food traceability: a systematic review (Review of Blockchain-based frameworks for food traceability: a systematic review). Foods 12(16):3026
Elortondo FJP, Ojeda M, Albisu M, Salmerón J, Etayo I, Molina M (2007) Food quality certification: an approach for the development of accredited sensory evaluation methods. Food Qual Prefer 18(2):425–439. https://doi.org/10.1016/j.foodqual.2006.05.002
Feng D, Schütz C, Rosenbaum L, Hertlein H, Gläser C, Timm F, Wiesbeck W, Dietmayer K (2020) Deep multi-modal object detection and semantic segmentation for autonomous driving: datasets, methods, and challenges. IEEE Trans Intell Transp Syst 22(3):1341–1360
Florsheim EB, Sullivan ZA, Khoury-Hanold W, Medzhitov R (2021) Food allergy as a biological food quality control system. Cell 184(6):1440–1454. https://doi.org/10.1016/j.cell.2020.12.007
Fiorentini M, Kinchla AJ, Nolden AA (2020) Role of sensory evaluation in consumer acceptance of plant-based meat analogs and meat extenders: a scoping review. Food Secur 9(9):1334
Fondberg R, Lundström JN, Blöchl M, Olsson MJ, Seubert J (2018) Multisensory flavor perception: the relationship between congruency, pleasantness, and odor referral to the mouth. Appetite 125:244–252. https://doi.org/10.1016/j.appet.2018.02.012
Fondberg R, Lundström JN, Seubert J (2021) Odor–taste interactions in food perception: exposure protocol shows no effects of associative learning. Chem Senses 46:bjab003. https://doi.org/10.1093/chemse/bjab003
Forde CG (2016) Flavor perception and satiation. In: Flavor. Elsevier, pp 251–276
Frost P (2022) Humans and the olfactory environment: a case of gene-culture coevolution? Psych 4(2):301–317. https://doi.org/10.3390/psych4020027
Fuentes S, Tongson E, Viejo CG (2021) Novel digital technologies implemented in sensory science and consumer perception. Curr Opin Food Sci 41:99–106

Galvan D, Aquino A, Effting L, Mantovani ACG, Bona E, Conte-Junior CA (2022) E-sensing and nanoscale-sensing devices associated with data processing algorithms applied to food quality control: a systematic review. Crit Rev Food Sci Nutr 62(24):6605–6645

Gámbaro A, McSweeney MB (2020) Sensory methods applied to the development of probiotic and prebiotic foods. In: Advances in food and nutrition research, vol 94. Academic, pp 295–337

Gebeshuber IC, Drack M (2008) An attempt to reveal synergies between biology and mechanical engineering. Proc Inst Mech Eng C J Mech Eng Sci 222(7):1281–1287. https://doi.org/10.1243/09544062JMES890

Gupta RS, Springston EE, Warrier MR, Smith B, Kumar R, Pongracic J, Holl JL (2011) The Prevalence, Severity, and Distribution of Childhood Food Allergy in the United States. Pediatrics 128(1):e9–e17. https://doi.org/10.1542/peds.2011-0204

Gribble FM, Reimann F (2016) Enteroendocrine Cells: Chemosensors in the Intestinal Epithelium. Annual Review of Physiology 78(1):277–299. https://doi.org/10.1146/annurev-physiol-021115-105439

George RV, Harsh HO, Ray P, Babu AK (2019) Food quality traceability prototype for restaurants using blockchain and food quality data index. J Clean Prod 240:118021

Ghasemi-Varnamkhasti M, Mohtasebi SS, Siadat M (2010) Biomimetic-based odor and taste sensing systems to food quality and safety characterization: an overview on basic principles and recent achievements. J Food Eng 100(3):377–387. https://doi.org/10.1016/j.jfoodeng.2010.04.032

Gotow N, Kobayakawa T (2023) Olfactory–gustatory simultaneity judgments: a preliminary study on the congruency-dependent temporal window of multisensory binding. Brain Behav 13(1):e2821. https://doi.org/10.1002/brb3.2821

Grassi S, Benedetti S, Casiraghi E, Buratti S (2023) E-sensing systems for shelf life evaluation: a review on applications to fresh food of animal origin. Food Packag Shelf Life 40:101221

Guiné RPF, Florença SG, Barroca MJ, Anjos O (2020) The link between the consumer and the innovations in food product development. Food Secur 9(9):1317

Gunder N, Dörig P, Witt M, Welge-Lüssen A, Menzel S, Hummel T (2023) Future therapeutic strategies for olfactory disorders: electrical stimulation, stem cell therapy, and transplantation of olfactory epithelium—an overview. HNO 71(S1):35–43. https://doi.org/10.1007/s00106-022-01249-8

Hadley K, Orlandi RR, Fong KJ (2004) Basic anatomy and physiology of olfaction and taste. Otolaryngol Clin N Am 37(6):1115–1126. https://doi.org/10.1016/j.otc.2004.06.009

Haick H, Arnab M (2025) Auditory based electronic sensory system. In: Nature-inspired sensors. Elsevier, pp 429–439. https://doi.org/10.1016/B978-0-443-15684-7.00034-8

Han SA, Kim JK, Cho D-Y, Patel ZM, Rhee C-S (2023) The olfactory system: basic anatomy and physiology for general Otorhinolaryngologists. Clin Exp Otorhinolaryngol 16(4):308–316. https://doi.org/10.21053/ceo.2023.00185

Iweala OI, Nagler CR (2019) The Microbiome and Food Allergy. Annual Review of Immunology, 37(1):377–403. https://doi.org/10.1146/annurev-immunol-042718-041621

Imam M, Nagpal K (2023) The electronic tongue: an advanced taste-sensing multichannel sensory tool with global selectivity for application in the pharmaceutical and food industry. Pharm Dev Technol 28(3–4):318–332

Kavaliers M, Choleris E, Ågmo A, Pfaff DW (2004) Olfactory-mediated parasite recognition and avoidance: Linking genes to behavior. Hormones and Behavior 46(3):272–283. https://doi.org/10.1016/j.yhbeh.2004.03.005

Khan MS, Shafiur Rahman M (eds) (2021) Techniques to measure food safety and quality: microbial, chemical, and sensory. Springer International Publishing. https://doi.org/10.1007/978-3-030-68636-9

Kohli P, Soler ZM, Nguyen SA, Muus JS, Schlosser RJ (2016) The association between olfaction and depression: a systematic review. Chem Senses 41(6):479–486. https://doi.org/10.1093/chemse/bjw061

Kshetri N, DeFranco JF (2020) The economics behind food supply blockchains. Computer 53(12):106–110

Kulkarni PG, Paudel N, Magar S, Santilli MF, Kashyap S, Baranwal AK, Zamboni P, Vasavada P, Katiyar A, Singh AV (2024) Overcoming challenges and innovations in orthopedic prosthesis design: an interdisciplinary perspective. Biomed Mater Devices 2(1):58–69. https://doi.org/10.1007/s44174-023-00087-8

Kumar N, Upreti K, Mohan D (2022) Blockchain adoption for provenance and traceability in the retail food supply chain. Int J E-Bus Res 18(2):1–17

Lawless HT, Heymann H (2010) Sensory evaluation of food: principles and practices. Springer Science & Business Media

Linscott TD, Lim J (2016) Retronasal odor enhancement by salty and umami tastes. Food Qual Prefer 48:1–10. https://doi.org/10.1016/j.foodqual.2015.08.004

Lipp C, Laamari L, Bertsch A, Podlesek D, Bensafi M, Hummel T, Brugger J (2025) Devices for the electrical stimulation of the olfactory system: a review. Biosens Bioelectron 271:117063. https://doi.org/10.1016/j.bios.2024.117063

Liu Z, Yu X, Liu N, Liu C, Jiang A, Chen L (2025) Integrating AI with detection methods, IoT, and blockchain to achieve food authenticity and traceability from farm-to-table. Trends Food Sci Technol 158:104925. https://doi.org/10.1016/j.tifs.2025.104925

Looy H, Weingarten HP (1991) Effects of metabolic state on sweet taste reactivity in humans depend on underlying hedonic response profile. Chem Senses 16(2):123–130. https://doi.org/10.1093/chemse/16.2.123

Looy H, Callaghan S, Weingarten HP (1992) Hedonic response of sucrose likers and dislikers to other gustatory stimuli. Physiol Behav 52(2):219–225. https://doi.org/10.1016/0031-9384(92)90261-Y

Loutfi A, Coradeschi S, Mani GK, Shankar P, Rayappan JBB (2015) Electronic noses for food quality: a review. J Food Eng 144:103–111

Lu L, Hu Z, Hu X, Li D, Tian S (2022) Electronic tongue and electronic nose for food quality and safety. Food Res Int 162:112214. https://doi.org/10.1016/j.foodres.2022.112214

Lvova L (2016) Electronic tongue principles and applications in the food industry. In: Electronic noses and tongues in food science. Elsevier, pp 151–160

Mai Y, Vogel C, Thiele J, Hölscher T, Hummel T (2023) Abnormal visual and olfactory sensations during radiation therapy: a prospective study. Strahlenther Onkol 199(10):936–949. https://doi.org/10.1007/s00066-023-02095-5

Majchrzak T, Wojnowski W, Dymerski T, Gębicki J, Namieśnik J (2018) Electronic noses in classification and quality control of edible oils: a review. Food Chem 246:192–201

McGann JP (2017) Poor human olfaction is a 19th-century myth. Science 356(6338):eaam7263. https://doi.org/10.1126/science.aam7263

Meilgaard M, Civille G, Carr B (2007) Sensory evaluation techniques, 4th edn. CRC Press LLC, New York

Mihafu FD, Issa JY, Kamiyango MW (2020) Implication of sensory evaluation in food product development. Curr Res Nutr Food Sci 8(3):1–14

Modesti M, Taglieri I, Bianchi A, Tonacci A, Sansone F, Bellincontro A et al (2021) E-nose and olfactory assessment: teamwork or a challenge to the last data? The case of virgin olive oil stability and shelf life. Appl Sci 11(18):8453

Nevo O, Valenta K (2018) The Ecology and Evolution of Fruit Odor: Implications for Primate Seed Dispersal. International Journal of Primatology 39(3):338–355. https://doi.org/10.1007/s10764-018-0021-2

Nayak SL, Arora B, Sethi S, Sharma R (2020) Electronic nose (E-nose) application in the food industry. In: Sensor-based quality assessment systems for fruits and vegetables. Apple Academic Press, pp 143–167

Nolden AA, Feeney EL (2020) Genetic differences in taste receptors: implications for the food industry (Review of genetic differences in taste receptors: implications for the food industry). Annu Rev Food Sci Technol 11(1):183

Ordoñez-Araque R, Rodríguez-Villacres J, Urresto-Villegas J (2020) Electronic nose, tongue and eye: their usefulness for the food industry. Vitae 27(3)
Pellegrino R, Sinding C, De Wijk RA, Hummel T (2017) Habituation and adaptation to odors in humans. Physiol Behav 177:13–19. https://doi.org/10.1016/j.physbeh.2017.04.006
Qian Y, Alhaskawi A, Dong Y, Ni J, Abdalbary S, Lu H (2024) Transforming medicine: artificial intelligence integration in the peripheral nervous system. Front Neurol 15:1332048. https://doi.org/10.3389/fneur.2024.1332048
Radman S, Brzović P, Radunić M, Rako A, Šarolić M, Ninčević Runjić T et al (2023) Vinegar-Preserved Sea fennel: chemistry, color, texture, aroma, and taste. Food Secur 12(20):3812
Rashed MS, Abdelkarim EA, Elsamahy T, Sobhy M, El-Mesery HS, Salem A (2025) Advances in cell-based biosensors: transforming food flavor evaluation with novel approaches. Food Chem X 26:102336. https://doi.org/10.1016/j.fochx.2025.102336
Rejeb A, Keogh JG, Zailani S, Treiblmaier H, Rejeb K (2020) Blockchain technology in the food industry: a review of potentials, challenges and future research directions (Review of blockchain technology in the food industry: a review of potentials, challenges and future research directions). Logistics 4(4):27
Robinson N (2015) Double pub sales with neurogastronomy. The Publican's Morning Advertiser 6
Rodrigues SS, Dias LG, Teixeira A (2024) Emerging methods for the evaluation of sensory quality of food: technology at service. Curr Food Sci Technol Rep 2(1):77–90
Rolls ET, Critchley HD, Mason R, Wakeman EA (1996) Orbitofrontal cortex neurons: role in olfactory and visual association learning. J Neurophysiol 75(5):1970–1981. https://doi.org/10.1152/jn.1996.75.5.1970
Ronnett GV, Moon C (2002) G proteins and olfactory signal transduction. Annu Rev Physiol 64(1):189–222. https://doi.org/10.1146/annurev.physiol.64.082701.102219
Rudnitskaya A, Lvova L (2025) Gustatory-based electronic tongues. In: Nature-inspired sensors. Elsevier, pp 467–486. https://doi.org/10.1016/B978-0-443-15684-7.00037-3
Ruiz-Capillas C, Herrero AM, Pintado T, Delgado-Pando G (2021) Sensory analysis and consumer research in new meat products development. Food Secur 10(2):429
Saint-Eve A, Déléris I, Panouillé M, Dakowski F, Cordelle S, Schlich P, Souchon I (2011) How texture influences aroma and taste perception over time in candies. Chemosens Percept 4(1):32–41
Sarnat HB, Flores-Sarnat L, Wei X-C (2017) Olfactory development, part 1: function, from fetal perception to adult wine-tasting. J Child Neurol 32(6):566–578. https://doi.org/10.1177/0883073817690867
Sempionatto JR, Montiel VR, Vargas E, Teymourian H, Wang J (2021) Wearable and Mobile sensors for personalized nutrition. ACS Sensors 6(5):1745–1760
Sirangelo TM (2019) Sensory descriptive evaluation of food products: a review. J Food Sci Nutr Res 2(4):354–363
Sliwinska M, Wisniewska P, Dymerski T, Namiesnik J, Wardencki W (2014) Food analysis using artificial senses. J Agric Food Chem 62(7):1423–1448
Spence C (2023) Digitally enhancing tasting experiences. Int J Gastron Food Sci 32:100695
Steiner JE, Glaser D, Hawilo ME, Berridge KC (2001) Comparative expression of hedonic impact: affective reactions to taste by human infants and other primates. Neurosci Biobehav Rev 25(1):53–74. https://doi.org/10.1016/S0149-7634(00)00051-8
Stevenson RJ, Prescott J, Boakes RA (1995) The acquisition of taste properties by odors. Learn Motiv 26(4):433–455. https://doi.org/10.1016/S0023-9690(05)80006-2
Stevenson RJ, Boakes RA, Prescott J (1998) Changes in odor sweetness resulting from implicit learning of a simultaneous odor-sweetness association: an example of learned synesthesia. Learn Motiv 29(2):113–132. https://doi.org/10.1006/lmot.1998.0996
Stevenson RJ, Boakes RA, Wilson JP (2000) Resistance to extinction of conditioned odor perceptions: evaluative conditioning is not unique. J Exp Psychol Learn Mem Cogn 26(2):423–440. https://doi.org/10.1037/0278-7393.26.2.423
Stevenson RJ, Mahmut MK, Oaten MJ (2011) The role of attention in the localization of odors to the mouth. Atten Percept Psychophys 73(1):247–258. https://doi.org/10.3758/s13414-010-0013-6

Stone H, Sidel J (2009) Sensory science and consumer behavior. In: Global issues in food science and technology. Academic, pp 67–77

Swain A, Hoffman T, Leyba K, Fagan WF (2021) Exploring the evolution of perception: an agent-based approach. Front Ecol Evol 9:698041. https://doi.org/10.3389/fevo.2021.698041

Świąder K, Marczewska M (2021) Trends of using sensory evaluation in new product development in the food industry. Food Secur 10(2):446

Titova T, Nachev V (2020) "Electronic tongue" in the food industry. Food Sci Appl Biotechnol 3(1):71–76

Tokat P, Yilmaz I (2023) Neurogastronomy: factors affecting the taste perception of food. Int J Gastron Res 2(1):1–10

Toko K, Onodera T, Tahara Y (2013) Nano-biosensors for mimicking gustatory and olfactory senses. In: Bagchi D, Bagchi M, Moriyama H, Shahidi F (eds) Bio-nanotechnology, 1st edn. Wiley, pp 270–291. https://doi.org/10.1002/9781118451915.ch15

Torky M, Hassanein AE (2020) Integrating blockchain and the internet of things in precision agriculture: analysis, opportunities, and challenges. Comput Electron Agric 178:105476

Torrico DD, Mehta A, Borssato AB (2022) New methods to assess sensory responses: a brief review of innovative techniques in sensory evaluation (Review of new methods to assess sensory responses: a brief review of innovative techniques in sensory evaluation). Curr Opin Food Sci 49:100978

Tournier C, Sulmont-Rossé C, Guichard E (2007) Flavour perception: aroma, taste and texture interactions. Food 1(2):246–257

Trigo JP, Alexandre EMC, Saraiva JA, Pintado ME (2019) High value-added compounds from fruit and vegetable by-products—characterization, bioactivities, and application in the development of novel food products. Crit Rev Food Sci Nutr 60:1388–1416

Velasco C, Obrist M, Petit O, Spence C (2018) Multisensory Technology for Flavor Augmentation: a mini review (Review of multisensory technology for flavor augmentation: a mini review). Front Psychol 9:26

Verdouw C, Wolfert S, Beulens A, Rialland A (2016) Virtualization of food supply chains with the internet of things. J Food Eng 176:128–136

Viana F (2011) Chemosensory properties of the trigeminal system. ACS Chem Neurosci 2(1):38–50. https://doi.org/10.1021/cn100102c

Wang Y (2025) Electronic skin. In: Nature-inspired sensors. Elsevier, pp 487–504. https://doi.org/10.1016/B978-0-443-15684-7.00038-5

Wang Y, Ren Z, Chen Y, Lu C, Deng WW, Zhang Z, Ning J (2023) Visualizing chemical indicators: spatial and temporal quality formation and distribution during black tea fermentation. Food Chem 401:134090

Wang X, Wang L, Wei X, Xu C, Cavender G, Lin W, Sun S (2024a) Invited review: advances in yogurt development: microbiological safety, quality, functionality, sensory evaluation, and consumer perceptions across different dairy and plant-based alternative sources. J Dairy Sci 108:33–58

Wang T, Patel S, Kim J (2024b) Integration of sensory and instrumental methods in meat quality evaluation. Meat Sci

White TL, Thomas-Danguin T, Olofsson J, Zucco GM, Prescott J (2019) Thought for food: cognitive influences on chemosensory perceptions and preferences. Food Qual Prefer 79:103776

Wilson DA, Xu W, Sadrian B, Courtiol E, Cohen Y, Barnes DC (2014) Cortical odor processing in health and disease. In: Progress in brain research, vol 208. Elsevier, pp 275–305. https://doi.org/10.1016/B978-0-444-63350-7.00011-5

Xu C (2019) Electronic eye for food sensory evaluation. In: Evaluation technologies for food quality. Elsevier, pp 37–59

Yadav U, Khurana S, Rawal K, Arora B, Yadav A (2025) Integrating AI with electronic tongue (E-tongue) for real-time detection in the food industry. In: Artificial intelligence in the food industry. CRC Press, pp 277–296

Yeomans MR (2006) Olfactory influences on appetite and satiety in humans. Physiol Behav 87(4):800–804. https://doi.org/10.1016/j.physbeh.2006.01.029
Yılmaz İ, Altuntaş N (2022) Kokumi tat algısı üzerine bir değerlendirme. Gıda ve Yem Bilimi Teknolojisi. *Dergisi* 27:13–19
Yilmaz Balban M, Cafaro E, Saue-Fletcher L, Washington MJ, Bijanzadeh M, Lee AM, Chang EF, Huberman AD (2021) Human responses to visually evoked threat. Curr Biol 31(3):601–612.e3. https://doi.org/10.1016/j.cub.2020.11.035
Zatsu V, Shine AE, Tharakan JM, Peter D, Ranganathan TV, Alotaibi SS, Mugabi R, Muhsinah AB, Waseem M, Nayik GA (2024) Revolutionizing the food industry: the transformative power of artificial intelligence-a review (Review of revolutionizing the food industry: the transformative power of artificial intelligence-a review). Food Chem X 24:101867
Zhong Y (2019) Electronic nose for food sensory evaluation. In: Evaluation technologies for food quality. Elsevier, pp 7–22
Zhou Y, Hai W (2009) Electronic nose and the application in meat sensory evaluation. Meat Res 8:55–58
Zhu J, Wang G (2023) Artificial intelligence technology for food nutrition. Nutrients 15(21):4562

Chapter 11
Renal Physiology and Detoxification of Food Contaminants

Raja Chakraverty, Tatini Debnath, and Iryna Vlasenko

11.1 Renal Physiology

Renal physiology refers to the examination of kidney function. This field covers all renal activities, which include the preservation of acid-base equilibrium; management of fluid homeostasis; regulation of sodium, potassium, and other electrolytic substances; detoxification; reabsorption of glucose, amino acids, and various small molecules; control of blood pressure; synthesis of multiple hormones such as erythropoietin; and the activation of vitamin D (Taheur et al. 2019).

A significant portion of renal physiology is analysed at the nephron level, recognized as the fundamental functional unit of the kidney. Each nephron initiates with a filtration segment that processes the blood entering the organ. This filtrate subsequently travels through the nephron's entirety, which is a tubular entity lined by a monolayer of specialized cells and encased by capillary networks. The primary roles of these epithelial cells involve the reabsorption of water and small solutes from the filtrate back into the bloodstream, as well as the excretion of metabolic waste from the blood into the urine.

The effective operation of the kidney necessitates its capability to receive and thoroughly filter blood. This task is accomplished at the microscopic scale by

R. Chakraverty (✉)
Department of Critical Care Medicine, Institute of Post Graduate Medical Education and Research, Kolkata, West Bengal, India

T. Debnath
Department of Pharmaceutical Technology, Maulana Abul Kalam Azad University of Technology, Haringhata, West Bengal, India

I. Vlasenko
International Diabetes Federation, Kyiv, Ukraine

T. Sarkar et al. (eds.), *Physiological Perspectives on Food Safety: Exploring the Intersection of Health and Nutrition*, https://doi.org/10.1007/978-3-031-94582-3_11

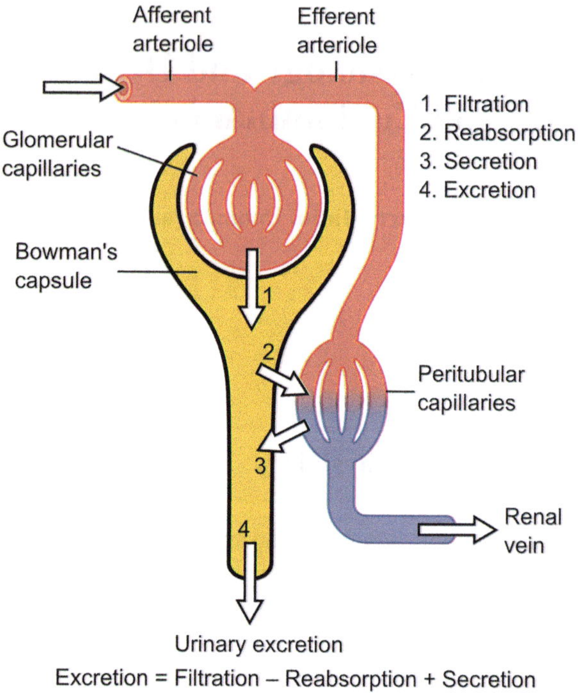

Fig. 11.1 Showing basic renal physiology

numerous filtration units known as renal corpuscles, each comprising a glomerulus and a Bowman's capsule. A comprehensive evaluation of renal performance is frequently determined by calculating the filtration rate, referred to as the glomerular filtration rate (GFR) as represented in Fig. 11.1.

11.1.1 Formation of Urine

The capacity of the kidney to execute various functions is contingent upon three essential processes: filtration, reabsorption, and secretion, collectively referred to as renal clearance or renal excretion (Savi et al. 2015; Temba et al. 2016). Specifically:

$$\text{Urinary excretion rate} = \text{filtration rate} - \text{reabsorption rate} + \text{secretion rate}$$

While the term excretion, in its most precise definition concerning the urinary system, pertains specifically to urination, renal clearance is also traditionally identified as excretion (for instance, in the established term fractional excretion of sodium).

11.1.2 Filtration

The renal system employs nephrons as its fundamental operational units for blood filtration. Each nephron initiates within a renal corpuscle, which comprises a glomerulus encased by Bowman's capsule. Through a mechanism known as ultrafiltration, cells, proteins, and other substantial molecules are extricated from the glomerulus, resulting in an ultrafiltrate that is akin to plasma (albeit with minimal plasma proteins) that then enters Bowman's space. The process of filtration is governed by Starling forces (Šegvić Klarić et al. 2009).

Subsequently, the ultrafiltrate traverses the proximal convoluted tubule, the loop of Henle, the distal convoluted tubule, and a succession of collecting ducts, ultimately culminating in the formation of urine.

11.1.3 Reabsorption

Tubular reabsorption refers to the mechanism through which solutes and water are extracted from the tubular fluid and subsequently conveyed into the bloodstream. The term reabsorption is employed (as opposed to absorption) due to the fact that these compounds have previously undergone absorption (notably within the intestines) and also because the organism is retrieving them from a post-glomerular fluid pathway that is nearing the transformation into urine (indicating that they will soon be excreted unless they are recuperated). Reabsorption shown in Fig. 11.2.

The process of reabsorption is characterized by two distinct phases, commencing with the active or passive translocation of substances from the tubular fluid into the renal interstitial matrix (the connective tissue encasing the nephrons), followed by the conveyance of these substances from the interstitium into the circulatory system. The mechanisms governing these transport processes are influenced by Starling forces, diffusion, and active transport (Schatzmayr et al. 2006; Bočarov-Stančić et al. 2011).

11.1.4 Indirect Absorption

In certain instances, the process of reabsorption occurs indirectly. For instance, bicarbonate ($HCO3-$) lacks a dedicated transporter; thus, its reabsorption necessitates a sequence of biochemical reactions within the tubular lumen and the tubular epithelium. The process is initiated by the active secretion of a hydrogen ion ($H+$) into the tubular fluid through a sodium/hydrogen exchanger:

In the lumen The $H+$ ion reacts with $HCO3-$ to generate carbonic acid ($H2CO3$) Luminal carbonic anhydrase catalyzes the conversion of $H2CO3$ into $H2O$ and $CO2$ diffuses freely into the epithelial cell In the epithelial cell Cytoplasmic carbonic

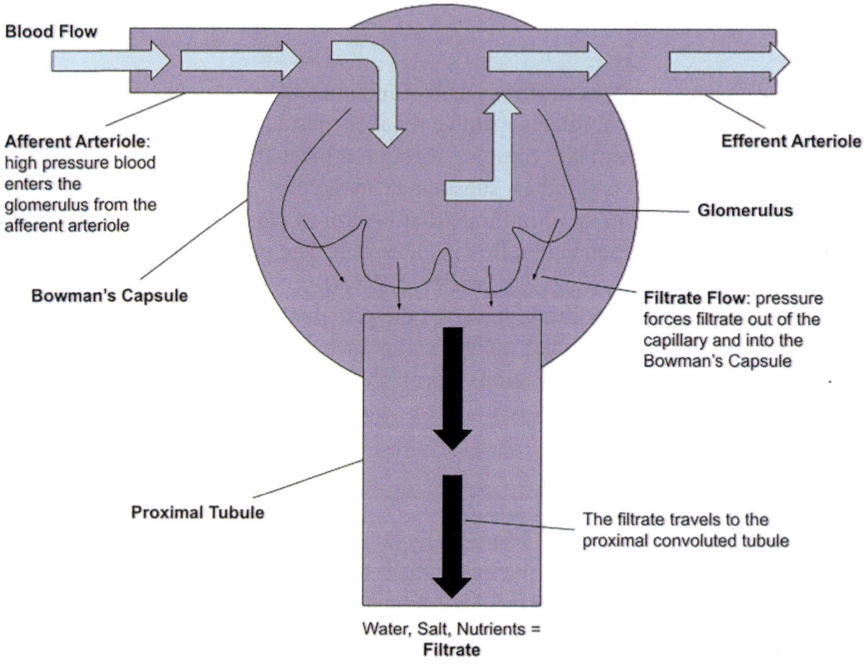

Fig. 11.2 Representation of reabsorption

anhydrase transforms the CO2 and H2O (which is plentiful within the cell) into H2CO3 H2CO3 readily dissociates into H+ and HCO3− HCO3− is then facilitated across the basolateral membrane of the cell (He et al. 2010).

11.1.5 Hormone Influence

Certain pivotal regulatory hormones pertinent to the process of reabsorption encompass: aldosterone, which promotes the active reabsorption of sodium (thereby facilitating water retention) antidiuretic hormone, which encourages passive reabsorption of water.

11.1.5.1 Both Hormones Primarily Influence the Collecting Ducts

Concurrently, tubular secretion transpires during the reabsorption of filtrate. This process involves the excretion of substances typically synthesized by the organism or metabolic by-products that may attain toxic levels in elevated concentrations, as well as certain pharmaceuticals (when administered). These compounds are secreted into the lumen of the renal tubule (Li et al. 2020; Huang et al. 2018). Tubular secretion may occur through active or passive mechanisms, including co-transport. The predominant substances secreted into the renal tubule comprise H+, K+, NH3, urea, creatinine, histamine, and pharmaceuticals such as penicillin. Tubular secretion is facilitated at both the Proximal Convoluted Tubule (PCT) and the Distal Convoluted Tubule (DCT); for instance, within the proximal convoluted tubule, potassium ions are secreted via the sodium-potassium pump, hydrogen ions are expelled through active transport and co-transport mechanisms, specifically antiporters, and ammonia diffuses into the renal tubule.

11.1.6 Other Functions

The kidneys are responsible for the secretion of several hormones, such as erythropoietin, calcitriol, and renin. Erythropoietin is secreted in response to hypoxic conditions (low oxygen availability at the tissue level) within the renal vasculature. This hormone promotes erythropoiesis (the formation of red blood cells) within the bone marrow. Calcitriol, which is the bioactive form of vitamin D, facilitates the intestinal uptake of calcium and the renal reabsorption of phosphate. Renin functions as an enzyme that modulates the levels of angiotensin and aldosterone (Afsah-Hejri et al. 2020).

11.2 Detoxification

Detoxification, or detoxication (commonly abbreviated as detox), refers to the physiological or therapeutic expulsion of harmful substances from a living organism, including humans, predominantly facilitated by the liver. Furthermore, it may denote the phase of substance withdrawal during which an organism re-establishes homeostasis following prolonged exposure to an addictive agent. In the realm of medicine, detoxification can be accomplished through the decontamination of ingested poisons and the administration of antidotes, in addition to techniques such as dialysis and, in select instances, chelation therapy. Numerous practitioners of alternative medicine advocate for various forms of detoxification, including detoxification diets. However, Sense about Science, a charitable organization based in the UK, has concluded that the majority of claims regarding these dietary "detox" methods are devoid of empirical support. The liver and kidneys possess inherent capabilities for detoxification, as do intracellular proteins, specifically those located within the inner membrane of mitochondria or in the endoplasmic reticulum of cells, such as cytochrome P450 enzymes. In instances of renal failure, dialysis serves as a substitute for kidney function; similarly, organ transplantation procedures are employed in cases of liver and kidney failure, respectively (Zhu et al. 2017; Lyagin and Efremenko 2019).

11.2.1 Pathways of Detoxification

Nutrients derived from food sources have been subjects of extensive investigation regarding their influence on the modulation of metabolic pathways pertinent to detoxification processes. Numerous scholarly articles have utilized cellular, animal, and clinical research methodologies to illustrate that components and nutrients extracted from food can influence the mechanisms of toxin transformation and subsequent elimination from the organism. Broadly, the implications of these findings suggest that particular foods may enhance or optimally regulate metabolic pathways to facilitate the biotransformation of toxins and their eventual expulsion (Sato et al. 2012).

Although these foods warrant attention, the scientific discourse in this dynamic field of research is continually advancing, uncovering new insights about the impact of food-derived nutrients on health. Consequently, the objective of this review article is to synthesize the current scientific landscape concerning the effects of whole foods, with a particular emphasis on phytonutrients and other food-derived constituents, on specific metabolic detoxification pathways. This includes an examination of phase I cytochrome enzymes, phase II conjugation enzymes, antioxidant defense mechanisms, and the upregulation of metallothionein for the metabolism of heavy metals. In light of this contemporary scientific evidence, the article will conclude with clinical recommendations that may be tailored to individual patients at the discretion of a licensed healthcare professional (Liuzzi et al. 2017).

The discourse surrounding physiological mechanisms for detoxification has predominantly focused on phase I and phase II enzymatic systems. This review will examine phase I cytochrome P450 enzymes alongside phase II enzymes, specifically UDP-glucuronosyl transferases, glutathione S-transferases, amino acid transferases, N-acetyl transferases, and methyltransferases. It is important to acknowledge the existence of other significant categories of phase I enzymes, particularly hydroxylation and reduction, which will not be addressed in this review. While these crucial enzymes warrant consideration, this analysis of the influence of dietary components on detoxification will also encompass additional pathways, including strategies to enhance the expression of genes related to antioxidant enzymes and metallothionein, an endogenous protein that functions as a carrier for heavy metals. Each of these four categories of detoxification-related pathways will be explored in relation to nutrients (Monachese et al. 2012).

11.2.1.1 Cytochrome P450 Enzyme

Initially, the "stages" of detoxification were delineated as functionalization (or phase I), which involves the incorporation of oxygen to create a reactive site on the toxic substance, and conjugation (phase II), which refers to the addition of a hydrophilic group to this newly reactive site. The cytochrome P450 superfamily of enzymes (CYP450) designated as "Phase I" represents the primary mechanism utilized by the organism to biotransform xenobiotics, steroid hormones, and pharmaceuticals. These microsomal membrane-associated, heme-thiolate proteins, predominantly found in the liver but also present in enterocytes, kidneys, lungs, and even the brain, facilitate the oxidation, peroxidation, and reduction of various endogenous and exogenous substrates. Specifically, the role of CYP450 enzymes is to introduce a reactive moiety such as a hydroxyl, carboxyl, or amino group via oxidation, reduction, and/or hydrolysis reactions. These preliminary reactions possess the potential to induce oxidative damage within cellular systems due to the resultant generation of reactive electrophilic species (Rajasekharan et al. 2018).

It is acknowledged that any fluctuations in the quantity of CYP450 enzymes could yield advantages and/or repercussions regarding an individual's reaction to the effects of specific toxins. The clinical application of insights regarding these phases I CYP450 enzymes has predominantly been explored within pharmacology to elucidate drug interactions, adverse effects, and interindividual variability in drug metabolism. The capacity of an individual to metabolize 90% of currently utilized pharmaceuticals will significantly depend on the genetic expression of these enzymes. It is well-established that many of these CYP450 genes are prone to genetic polymorphisms, leading to modified expression and functionality of individual enzymes. Presently, various laboratory assessments exist to detect these genetic variants. It is plausible that awareness of foods and their specific (phyto) nutrients, particularly in the context of dietary supplements and functional foods, could be advantageous for clinicians to contemplate for patients engaged in a polypharmacy regimen. Moreover, as nutritional strategies grow increasingly personalized, it appears that this information could be integrated with a patient's identified CYP450 polymorphisms to ascertain the optimal approach for enhancing health outcomes (Wang and Xie 2020).

11.2.1.1.1 Function of CYP1 Enzyme

The CYP1A enzyme family plays a crucial role in the metabolism of procarcinogenic compounds, hormones, and pharmaceuticals. It is particularly recognized for its involvement in the carcinogenic bioactivation of polycyclic aromatic hydrocarbons (PAHs), heterocyclic aromatic amines/amides, polychlorinated biphenyls (PCBs), and various environmental toxins. For instance, diminished activity of CYP1A2 has been associated with an increased risk of testicular cancer. Nevertheless, the heightened activity of CYP1A enzymes, in the absence of sufficient phase II metabolic support, may exacerbate the detrimental effects of environmental procarcinogens due to their swift transformation into highly reactive intermediates. Notably, genetic polymorphisms within this cytochrome family have been proposed as potential biomarkers for susceptibility to specific cancers. Furthermore, CYP1 enzymes are instrumental in the synthesis of clinically significant estrogen metabolites: CYP1A1/1A2 facilitate the 2-hydroxylation and 4-hydroxylation of estrogens, respectively. The prospective involvement of 4-hydroxyestradiol in estrogen-related carcinogenesis, through the generation of free radicals and associated cellular damage, has stimulated research into the factors influencing CYP1 enzyme activity (Liu et al. 2018).

Numerous food items and phytonutrients modulate CYP1 activity. Cruciferous vegetables have been documented, in human studies, to serve as inducers of CYP1A1 and 1A2, while animal research also indicates an upregulation of CYP1B1. The inductive effect of cruciferous vegetables on CYP1A2 appears particularly well substantiated. Clinical investigations also reveal that resveratrol and foods rich in resveratrol act as enhancers of CYP1A1. In contrast, berries and their bioactive polyphenol, ellagic acid, may mitigate CYP1A1 hyperactivity, while apiaceous vegetables and quercetin could reduce excessive CYP1A2 activity. Both cruciferous vegetables and berries have been proposed as potential modulators of estrogen metabolites: berries for their inhibitory impact on CYP1A1 (Bočarov-Stančić et al. 2011) and cruciferous vegetables for their comparatively stronger induction of CYP1A relative to 1B1 enzymes. Chrysoeriol, found in rooibos tea and celery, selectively inhibits CYP1B1 in vitro and may be particularly pertinent for individuals exhibiting CYP1B1 overactivity. However, additional research is necessary to validate this observation (Al-Nussairawi et al. 2020).

11.2.1.1.2 CYP2A-E Enzymes

The extensive CYP2 enzyme family plays a crucial role in the biotransformation of pharmaceuticals, xenobiotic substances, hormones, and other endogenous molecules including ketones, glycerol, and fatty acids. Significant genetic polymorphisms have been identified within the CYP2C and CYP2D subfamilies, resulting in the categorization of certain individuals as "poor metabolizers" of a range of medications: for instance, warfarin and CYP2C9, antiarrhythmic drugs such as metoprolol and propafenone linked to CYP2D6, as well as phenytoin, cyclobarbital, omeprazole, and CYP2C19, among others. Furthermore, polymorphisms in CYP2D may have correlations with Parkinson's disease and lung carcinoma. There is clinical evidence suggesting that quercetin and broccoli can induce CYP2A6. In animal studies, chicory has demonstrated the capability to induce CYP2A enzymes, while rosemary and garlic may enhance CYP2B activity. Additionally, clinical investigations involving resveratrol and garden cress have revealed CYP2D6 inhibition (Silva et al. 2020).

The CYP2E1 enzymes have garnered significant attention due to their involvement in a variety of diseases. CYP2E1 is responsible for the metabolism of neuroactive agents such as halothane, isoflurane, chlorzoxazone, and ethanol, as well as the bioactivation of procarcinogenic nitrosamines and aflatoxin B1. This enzyme generates free radicals independent of the substrate, and genetic polymorphisms in CYP2E1 have been linked to an increased susceptibility to coronary artery disease and gastric cancer. Furthermore, oxidative stress induced by CYP2E1 has been implicated in the impairment of insulin action through the downregulation of GLUT4 expression. Therefore, mitigating the overactivity of CYP2E1 may represent a crucial strategy for individuals at elevated risk.

Inhibitors of CYP2E1 in humans include watercress and garlic. Additionally, in vivo studies indicate that compounds such as N-acetyl cysteine, ellagic acid, green tea, black tea, dandelion, chrysin, and medium chain triglycerides (MCTs) possess the potential to downregulate CYP2E1. Notably, MCT oil may specifically reduce the ethanol-mediated upregulation of CYP2E1 and the resultant production of mitochondrial 4-hydroxynonenal, which serves as a biomarker for oxidative stress (Marin et al. 2013).

11.2.1.1.3 CYP3A Enzymes

The presence of various CYP3A isoforms exhibits tissue specificity. Rooibos tea, garlic, and fish oil seem to enhance the functionality of CYP3A, specifically 3A1 and 3A2.Potentially inhibitory dietary constituents include green tea, black tea, and quercetin. The enzyme of greatest clinical significance is CYP3A4, predominantly expressed in the liver and, to a lesser degree, in the kidney. Substrates for the CYP3A4 enzyme system include caffeine, testosterone, progesterone, and androstenedione, alongside various procarcinogens such as polycyclic aromatic hydrocarbons (PAHs) and aflatoxin B1. Nonetheless, the primary impetus for investigating CYP3A4 has stemmed from its involvement in the metabolism of more than 50% of all medications. The potential for drug interactions mediated by this singular enzyme, combined with significant interindividual variability in enzymatic activity, poses a degree of risk when administering elevated dosages and multiple medications, in addition to food-drug and herb-drug interactions. Grapefruit juice is likely the most recognized dietary inhibitor of this enzyme, though resveratrol and garden cress, which belongs to the cruciferous vegetable family, seem to exhibit comparable effects in humans, although at intake levels exceeding those anticipated without substantial supplementation. Curcumin may enhance 3A4 activity (Khan et al. 2020).

11.2.1.1.4 CYP4 Enzymes

The literature provides limited insights into this enzyme family, as its contribution to drug metabolism is perceived to be relatively minor. Nonetheless, it is recognized as a predominantly extrahepatic family of cytochromes that can be induced by clofibrate and ciprofibrate (which are hypolipidemic agents), NSAIDs, prostaglandins, and various toxicants including phthalate esters. The CYP4B1 isoform plays a significant role in the metabolism of medium chain triglycerides (MCTs) and is also implicated in the bioactivation of pneumotoxic and carcinogenic substances (Loi et al. 2017).

Variants and overexpression within this subgroup may correlate with the incidence of bladder cancer and colitis. A study conducted by Ye and team investigated the association between colitis and CYP4B1 activity, revealing that the enhancement of CYP4B1 activity by caffeic acid (present in caffeinated consumables) was linked to decreased inflammation and disease severity. Animal studies suggest that green tea possesses the potential to induce CYP4A1. Further investigation is required to elucidate the dietary impacts on this family of enzymes (Gil-Serna et al. 2020).

11.2.1.2 Phase II Conjugation Enzymes

Following the transformation of a xenobiotic into a hydrophilic entity through enzymatic reactions mediated by CYP450 enzymes, its reactive locus may undergo conjugation with an endogenous hydrophilic moiety. This biochemical transformation is frequently characterized as "phase II detoxification." Conjugation encompasses the transfer of various hydrophilic compounds (mediated by their respective enzymes), such as glucuronic acid (catalyzed by glucuronyl transferases), sulfate (by sulfotransferases), glutathione (utilized by glutathione transferases), amino acids (facilitated by amino acid transferases), an acetyl moiety (transferred by N-acetyl transferases), and a methyl group (transferred by N- and O-methyltransferases). The cumulative action of these enzymatic processes results in an augmented hydrophilicity of the metabolite, which theoretically promotes enhanced elimination via bile and/or urine. Analogous to the CYP450 enzymes, genetic polymorphisms can significantly affect the activity of these conjugating enzymes, with potential ramifications concerning the aetiology of various cancer types (Vanhoutte et al. 2016).

11.2.1.3 UDP Glucuronyltransferases

This category of enzymes, consisting of various proteins and subfamilies, is crucial for facilitating the excretion of biotransformed toxins through urine and faeces, as well as for the metabolism of steroid hormones and bilirubin. Their primary role involves catalyzing the covalent attachment of glucuronic acid derived from UDP-glucuronic acid to an accepting functional group on the substrate, a biochemical reaction known as glucuronidation. Although glucuronidation predominantly occurs in the liver, it can also take place in other tissues, including the small intestine. Specifically, bilirubin is primarily conjugated by UGT1A1 in hepatocytes before being excreted via bile into the gastrointestinal tract. It has been estimated that between 40% and 70% of all pharmaceuticals undergo glucuronidation processes in humans, highlighting the importance of this family of conjugation enzymes. Given that UDP-glucuronosyltransferases (UGTs) also play a role in the metabolism of phytochemicals, variations in their activity may arise from genetically downregulated enzyme function; flavonoids are conjugated with glucuronide and sulfate, thus polymorphisms in UGT or sulfotransferase (SULT) may lead to discrepancies in the clearance and effectiveness of phytochemicals (Winter and Pereg 2019).

11.2.2 Clinical Perspectives

With the persistent and growing body of empirical evidence that underscores the significant role that various toxins play in the etiology and progression of numerous chronic diseases, it has become increasingly imperative for healthcare practitioners and clinicians to cultivate a comprehensive understanding of the various therapeutic modalities available that can effectively mitigate the toxic burden experienced by patients. In this scholarly paper, a multitude of rigorous studies examining the profound impact of dietary choices and the incorporation of food-derived nutrients on the intricate physiological systems responsible for detoxification have been meticulously presented and analyzed. Based on the wealth of current information that has been thoroughly reviewed and synthesized, the following key concepts have been identified and are poised for effective translation into practical applications within the clinical environment (Adebo et al. 2017).

One of the prominent limitations that emerges prominently in this compilation of scholarly investigations is the fact that the information presented, in numerous instances, is predominantly restricted to research conducted exclusively within the context of cellular systems or animal models. This raises significant questions regarding the extent to which analogous effects might manifest in human subjects when exposed to moderate and reasonable dosages of the substances under investigation. Within the realm of cellular studies, it becomes exceedingly challenging to predict outcomes due to the inherent lack of pleiotropic activity that characterizes the intricate dynamics of a living organism, where multiple detoxification mechanisms operate concurrently and interactively. In a similar vein, studies involving animal models often present considerable challenges when it comes to extrapolating findings to human individuals, primarily due to the substantial variability that is observed in both genotype and environmental phenotype within the diverse spectrum of the human population. Consequently, at this juncture, it is prudent to exercise caution in firmly endorsing specific foods or food-derived nutrients that currently possess support solely from cellular or animal research data. It is advisable to depend instead on the clinical studies that have been published to date, which provide a more robust foundation for making definitive recommendations regarding dietary practices. This cautious approach underscores the necessity for further research that bridges the gap between laboratory findings and real-world human applications, thereby fostering a more comprehensive understanding of the potential health impacts of various nutritional elements. Ultimately, the translation of knowledge gained from cellular and animal studies into applicable human health recommendations requires a nuanced consideration of the complexities inherent in human biology and the diversity of individual responses to dietary interventions. Hence, it is imperative that future research endeavors prioritize clinical trials that encompass a broader spectrum of human subjects to validate the findings obtained from preliminary studies in more controlled experimental settings (Afshar et al. 2020; Nazhand et al. 2020).

11.2.3 Detoxification of Food Contaminants

The ingestion of food products that are contaminated with TCNs, which are toxic compounds, has the potential to significantly impact various vital organs and systems within the human body, including but not limited to the gastrointestinal tract, the kidneys, the liver, and the immune system, thereby leading to a wide array of health complications that manifest as symptoms such as a loss of appetite, episodes of vomiting, discomfort in the abdominal region, and disturbances in cardiovascular function, as indicated by the literature. TCNs, which are classified as tetracyclic sesquiterpenoid epoxides, represent a diverse group encompassing more than 200 distinct toxins that share a common core structural feature known as the atricyclic 12,13-epoxytrichothec-9ene (EPT). Based on their respective structural characteristics, the toxins that comprise this family are categorized into four distinct groups or types, namely types A, B, C, and D, and it is particularly noteworthy that types A and B have garnered considerable attention within the scientific community due to their pronounced levels of toxicity. Among the representative toxins that belong to type ATCNs, one can find the notorious T-2 toxin (referred to simply as T-2), the HT-2 toxin (designated as HT-2), and neosolaniol, commonly abbreviated as NEO, whereas the representative toxins from type BTCNs are exemplified by deoxynivalenol (often abbreviated as DON) and its various acetyl derivatives. The presence of the 12,13-epoxide ring, alongside the 9–10 double bond and the acylated side groups that are characteristic of trichothecenes, plays an essential role in facilitating the toxicological effects that are responsible for the inhibition of critical biological processes such as protein synthesis, as well as the synthesis of RNA and DNA. In general terms, the complete detoxification of TCNs presents a formidable challenge to researchers and practitioners in the field, primarily due to the inherently stable and complex nature of their molecular structures, which complicates the development of effective remediation strategies (Polak-Śliwińska and Paszczyk 2021).

11.2.4 Mycotoxin Detoxification

The comprehensive examination of the biological detoxification processes pertaining to mycotoxins by probiotic bacteria was conducted many years ago, as documented in various scholarly articles and publications. Probiotics, which encompass a diverse array of microorganisms, particularly focus on lactic acid bacteria (LAB) and certain yeasts belonging to the Saccharomyces genus, are known to effectively eliminate mycotoxins through two primary mechanisms: surface adsorption and biodegradation. The removal of toxins via surface adsorption by microorganisms is characterized as a rapid and reversible process, which intriguingly does not induce any chemical alterations in the substrate that is being treated. In stark contrast, the process of biodegradation, while leading to a permanent removal of the toxins, unfortunately results in the generation of undesirable metabolites that may pose additional challenges. This dichotomy between surface adsorption and biodegradation highlights the complex interplay of efficacy and potential side effects associated with the detoxification processes employed by these beneficial microorganisms. Consequently, understanding these mechanisms in depth is crucial for optimizing the application of probiotics in mitigating mycotoxin contamination in various substrates (Hodges and Minich 2015).

The process of biodegradation pertaining to mycotoxins, which are toxic compounds produced by fungi that can be converted into harmless metabolites, has recently surfaced as a promising alternative approach for the effective control of safety in food and animal feed. Nevertheless, it poses significant challenges, as it is often extremely difficult, if not outright impossible, to achieve complete detoxification of certain intricate pollutants, such as various aromatic compounds and specific mycotoxins, utilizing only single-species microbial strains. Consequently, the utilization of microbial consortia for the biodegradation of these toxins has garnered a growing interest in the scientific community, primarily due to their remarkable capability to undertake more complex biochemical processes than those managed by isolated strains or species. Furthermore, it has been observed that microbial consortia exhibit greater resilience and adaptability to fluctuations in environmental conditions when compared to individual microbial species. For example, the development of microbial consortia specifically engineered for the degradation of polycyclic aromatic hydrocarbons (PAHs) demonstrates a significantly enhanced potential for the bioremediation of contaminated sites, attributable to their superior tolerance levels and improved adaptive mechanisms. While microbial consortia have increasingly become a predominant method for the degradation of soil pollutants, the precise mechanisms underlying this degradation process remain largely elusive and inadequately understood in the current literature. Existing research suggests that the implementation of microbial consortia may offer a novel and innovative solution for the biodegradation of mycotoxins, a multifaceted process that necessitates the coordination of two or more tasks, which, in certain instances, could present insurmountable challenges for any single organism striving for effective detoxification. It is imperative to continue exploring the intricate dynamics of microbial consortia in order to fully elucidate the mechanisms of action involved in the biodegradation of these hazardous substances. Through further investigation, we may uncover valuable insights that could lead to the development of more efficient strategies for ensuring food safety and mitigating the risks posed by mycotoxins and other environmental contaminants (Liu et al. 2022).

11.3 Conclusion

Over the course of the last 10 years, there has been an extensive and rigorous investigation into the intricate relationships and interactions between nutrigenomic and epigenetic factors that are influenced by various food constituents and their profound effects on the development and progression of chronic diseases, as documented in numerous studies. In a similar vein, a multitude of research endeavors have elucidated that both exposure to environmental toxins and their subsequent accumulation within the human body play a remarkably significant role in the onset and advancement of several major health issues, including but not limited to cardiovascular disease, type 2 diabetes, and obesity. Consequently, it can be posited that an individual's dietary intake, in conjunction with various environmental influences, may exert a substantial impact on the overall incidence and prevalence of chronic diseases. Indeed, the ramifications of these influences extend beyond individual health, potentially affecting not only the individual but also multiple subsequent generations, owing to the transgenerational inheritance mechanisms associated with epigenetic modifications. Thus, it appears that the formulation of clinical recommendations aimed at optimizing the beneficial effects of specific food items while simultaneously mitigating the adverse impacts of environmental toxins is of paramount importance in contemporary healthcare practices. Nevertheless, it is crucial to approach the development of any detoxification protocol with a degree of caution and a rigorous application of critical thinking, ensuring that such protocols are carefully crafted and implemented by qualified and trained clinicians. There are still a multitude of unresolved questions and complex issues that remain regarding the precise mechanisms by which particular foods may modulate the processes of detoxification within the body, highlighting the need for further research in this vital area of study. As such, the intersection of diet, toxins, and health outcomes warrants a concerted effort to deepen our understanding of these intricate relationships. This knowledge is essential for advancing our clinical practices and fostering more effective strategies for disease prevention and management through dietary modifications. Ultimately, the goal of this ongoing research is to elucidate the complex interplay between nutrition, environmental factors, and chronic disease, thereby paving the way for more informed and effective public health recommendations.

References

Adebo OA, Njobeh PB, Sidu S, Adebiyi JA, Mavumengwana V (2017) Aflatoxin B1 degradation by culture and lysate of a Pontibacter specie. Food Control 80:99–103

Afsah-Hejri L, Hajeb P, Ehsani RJ (2020) Application of ozone for degradation of mycotoxins in food: a review. Compr Rev Food Sci Food Saf 19(4):1777–1808

Afshar P, Shokrzadeh M, Raeisi SN, Ghorbani-HasanSaraei A, Nasiraii LR (2020) Aflatoxins biodetoxification strategies based on probiotic bacteria. Toxicon 178:50–58

Al-Nussairawi M, Risa A, Garai E, Varga E, Szabó I, Csenki-Bakos Z et al (2020) Mycotoxin biodegradation ability of the Cupriavidus genus. Curr Microbiol 77:2430–2440

Bočarov-Stančić A, Adamović M, Salma N, Bodroža-Solarov M, Vučković J, Pantić V (2011) In vitro efficacy of mycotoxins adsorption by natural mineral adsorbents. Biotechnol Anim Husb 27(3):1241–1251

Gil-Serna J, Vázquez C, Patiño B (2020) Mycotoxins in functional beverages: a review. Beverages 6(3):52

He J, Zhou T, Young JC, Boland GJ, Scott PM (2010) Chemical and biological transformations for detoxification of trichothecene mycotoxins in human and animal food chains: a review. Trends Food Sci Technol 21(2):67–76
Hodges RE, Minich DM (2015) Modulation of metabolic detoxification pathways using foods and food-derived components: a scientific review with clinical application. J Nutr Metab 2015(1):760689
Huang W, Chang J, Wang P, Liu C, Yin Q, Zhu Q et al (2018) Effect of the combined compound probiotics with mycotoxin–degradation enzyme on detoxifying aflatoxin B1 and zearalenone. J Toxicol Sci 43(6):377–385
Khan GI, Akbar A, Anwar M, Shafee M, Sadia H, Razzaq A, Rehman FU (2020) Eco-friendly bacterial biodegradation of mycotoxins. Pure Appl Biol 9(3):1708–1717
Li P, Su R, Yin R, Lai D, Wang M, Liu Y, Zhou L (2020) Detoxification of mycotoxins through biotransformation. Toxins 12(2):121
Liu B, Peng X, Meng X (2018) Effective biodegradation of mycotoxin patulin by porcine pancreatic lipase. Front Microbiol 9:615
Liu L, Xie M, Wei D (2022) Biological detoxification of mycotoxins: current status and future advances. Int J Mol Sci 23(3):1064
Liuzzi VC, Fanelli F, Tristezza M, Haidukowski M, Picardi E, Manzari C et al (2017) Transcriptional analysis of Acinetobacter sp. neg1 capable of degrading ochratoxin A. Front Microbiol 7:2162
Loi M, Fanelli F, Liuzzi VC, Logrieco AF, Mulè G (2017) Mycotoxin biotransformation by native and commercial enzymes: present and future perspectives. Toxins 9(4):111
Lyagin I, Efremenko E (2019) Enzymes for detoxification of various mycotoxins: origins and mechanisms of catalytic action. Molecules 24(13):2362
Marin S, Ramos AJ, Cano-Sancho G, Sanchis V (2013) Mycotoxins: occurrence, toxicology, and exposure assessment. Food Chem Toxicol 60:218–237
Monachese M, Burton JP, Reid G (2012) Advances in ureteral stent technology. Appl Environ Microbiol 78(18):6397–6404
Nazhand A, Durazzo A, Lucarini M, Souto EB, Santini A (2020) Characteristics, occurrence, detection and detoxification of aflatoxins in foods and feeds. Foods 9(5):644
Polak-Śliwińska M, Paszczyk B (2021) Trichothecenes in food and feed, relevance to human and animal health and methods of detection: a systematic review. Molecules 26(2):454
Rajasekharan SK, Lee JH, Zhao Y, Lee J (2018) The mycotoxin zearalenone hinders Candida albicans biofilm formation and hyphal morphogenesis. Indian J Microbiol 58:19–27
Sato I, Ito M, Ishizaka M, Ikunaga Y, Sato Y, Yoshida S et al (2012) Thirteen novel deoxynivalenol-degrading bacteria are classified within two genera with distinct degradation mechanisms. FEMS Microbiol Lett 327(2):110–117
Savi GD, Piacentini KC, Scussel VM (2015) Ozone treatment efficiency in Aspergillus and Penicillium growth inhibition and mycotoxin degradation of stored wheat grains (Triticum aestivum L.). J Food Process Preserv 39(6):940–948
Schatzmayr G, Zehner F, Täubel M, Schatzmayr D, Klimitsch A, Loibner AP, Binder EM (2006) Microbiologicals for deactivating mycotoxins. Mol Nutr Food Res 50(6):543–551
Šegvić Klarić M, Cvetnić Z, Pepeljnjak S, Kosalec I (2009) Co-occurrence of aflatoxins, ochratoxin A, fumonisins, and zearalenone in cereals and feed, determined by competitive direct enzyme-linked immunosorbent assay and thin-layer chromatography. Arh Hig Rada Toksikol 60(4):427–433
Silva LJ, Pereira AM, Pena A, Lino CM (2020) Citrinin in foods and supplements: a review of occurrence and analytical methodologies. Foods 10(1):14
Taheur FB, Kouidhi B, Al Qurashi YMA, Salah-Abbès JB, Chaieb K (2019) Biotechnology of mycotoxins detoxification using microorganisms and enzymes. Toxicon 160:12–22
Temba BA, Sultanbawa Y, Kriticos DJ, Fox GP, Harvey JJ, Fletcher MT (2016) Tools for defusing a major global food and feed safety risk: nonbiological postharvest procedures to decontaminate mycotoxins in foods and feeds. J Agric Food Chem 64(47):8959–8972

Vanhoutte I, Audenaert K, De Gelder L (2016) Biodegradation of mycotoxins: tales from known and unexplored worlds. Front Microbiol 7:561

Wang J, Xie Y (2020) Review on microbial degradation of zearalenone and aflatoxins. Grain Oil Sci Technol 3(3):117–125

Winter G, Pereg L (2019) A review on the relation between soil and mycotoxins: effect of aflatoxin on field, food and finance. Eur J Soil Sci 70(4):882–897

Zhu Y, Hassan YI, Lepp D, Shao S, Zhou T (2017) Strategies and methodologies for developing microbial detoxification systems to mitigate mycotoxins. Toxins 9(4):130

Chapter 12
Respiratory Response to Food Safety Hazards

Shimaa N. Edris, Aya Tayel, and Ahmed Hamad

12.1 Introduction

Food safety hazards encompass a wide spectrum of biological, chemical, and physical agents, which can compromise human health through diverse pathophysiological mechanisms. Although gastrointestinal manifestations are commonly associated with foodborne hazards, an emerging body of evidence indicates significant respiratory system implications (Jeebhay et al. 2001).

Food safety hazards pose significant risks to public health, with bacterial toxins causing approximately 10% of the reported foodborne outbreaks in Europe (Rajkovic et al. 2020). The most relevant toxins from gram-positive pathogens include those produced by Bacillus cereus, Clostridium botulinum, Clostridium perfringens, and Staphylococcus aureus (Rajkovic et al. 2020). These toxins vary in size, structure, toxicity, and stability, which influence the characteristics of the required detection methods.

Interestingly, while microbiological hazards are encountered more frequently in dairy products than chemical and physical hazards, chemical hazards, such as aflatoxin M1, dioxins, and veterinary drug residues, are also significant concerns (Van Asselt et al. 2016). Additionally, consumer perceptions of food safety risks can be quite high, with some estimating that the annual fatality rate due to pesticide residues on conventionally grown food is similar in magnitude to the annual mortality risk from motor vehicle accidents in the United States (Williams and Hammitt 2001).

These respiratory effects may arise through various routes of exposure, including ingestion, inhalation, or occupational contact with food components, additives, or

S. N. Edris (✉) · A. Tayel · A. Hamad
Department of Food Hygiene and Control, Faculty of Veterinary Medicine, Benha University, Banha, Egypt
e-mail: Shimaa.edrees@fvtm.bu.edu.eg

T. Sarkar et al. (eds.), *Physiological Perspectives on Food Safety: Exploring the Intersection of Health and Nutrition*, https://doi.org/10.1007/978-3-031-94582-3_12

contaminants. The respiratory system, with its extensive epithelial surface area and rich vascular network, is vulnerable to food-related toxicants (Corradi et al. 2015).

This comprehensive chapter examines the epidemiology, pathophysiological mechanisms, clinical manifestations, and preventive strategies related to food safety hazards that affect respiratory health.

12.2 Food Allergens and Respiratory Pathophysiology

Food allergens represent a significant category of food safety hazards, with established respiratory implications (Fig. 12.1). The immunological mechanisms underlying food-induced respiratory symptoms primarily involve immunoglobulin E (IgE) antibody-mediated responses, although non-IgE-mediated mechanisms may also contribute to the symptomatology (Sicherer and Sampson 2018). When a susceptible individual ingests an allergenic protein, the immune system produces specific IgE antibodies that bind to the mast cells and basophils. Upon subsequent exposure, allergen cross-linking of cell-bound IgE triggers degranulation and the

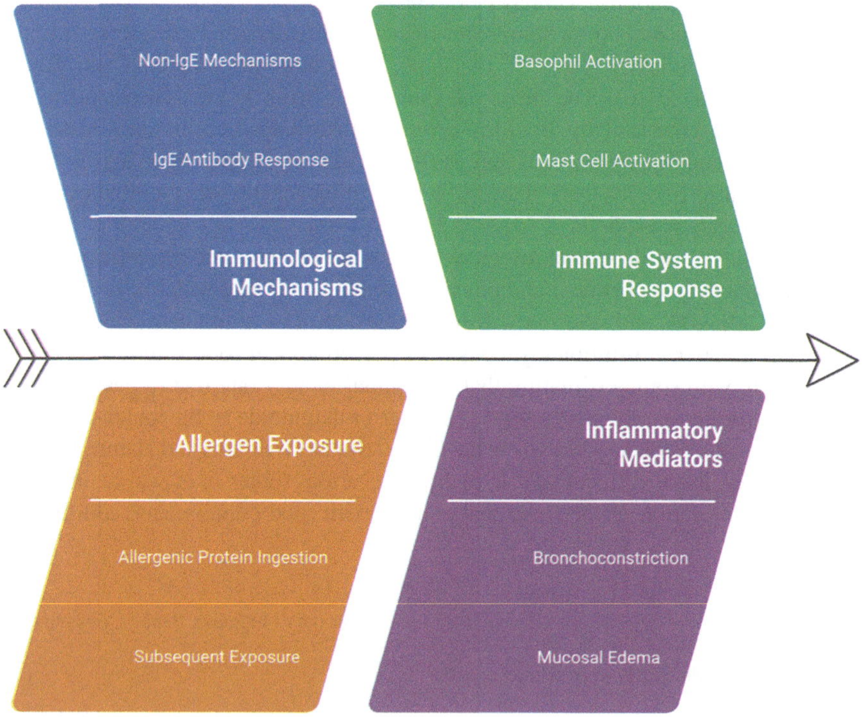

Fig. 12.1 Analyzing food allergens and respiratory symptoms

release of inflammatory mediators, resulting in bronchoconstriction, mucosal edema, and increased secretion (Bousquet et al. 2020).

The clinical manifestations of food-induced respiratory reactions range from mild rhinitis to life-threatening anaphylaxis. Upper respiratory symptoms include nasal congestion, rhinorrhea, and sneezing, whereas lower respiratory manifestations include wheezing, coughing, and dyspnea (Sicherer and Sampson 2018). Research indicates that approximately 25–30% of food allergic reactions involve respiratory symptoms, with a higher prevalence among individuals with pre-existing asthma (Allen et al. 2019). Particularly concerning is the risk of anaphylaxis, a severe systemic reaction characterized by rapid onset of airway compromise and potential circulatory collapse (Sampson et al. 2006). The prevalence of food-induced anaphylaxis has increased significantly in recent decades, with estimates suggesting a 377% increase in certain populations between 2007 and 2016 (Warren et al. 2021).

Children with persistent food sensitization for over a year in early childhood have a significantly higher risk of developing allergic rhinitis (3.4-fold) and asthma (5.5-fold) than those with transient sensitization (Kulig et al. 1998). This persistent food sensitization, especially in combination with a family history of atopy, is a strong predictor for the development of allergic airway diseases, with risks of up to 50% for allergic rhinitis and 67% for asthma by the age of 5 years (Kulig et al. 1998).

The relationship between food allergies and respiratory conditions appears bidirectional. Gastroesophageal reflux disease (GERD), often associated with food allergies, may induce changes in the mucosal immune system that favor the development of allergic sensitization to both food and aeroallergens (Hait and McDonald 2018). Additionally, the inflammatory response in allergic rhinitis involves IgE-mediated mast cell activation and a late-phase response characterized by the recruitment of eosinophils, basophils, and T cells expressing Th2 cytokines, which are also implicated in the pathogenesis of food allergy (Broide 2010).

12.3 Food Additives and Preservatives: Respiratory Implications

12.3.1 Sulfites and Respiratory Function

Sulfites comprise a group of compounds, including sulfur dioxide (SO_2) and various sulfite salts that are widely utilized as preservatives in foods and beverages. These additives have been conclusively linked to adverse respiratory effects, particularly in individuals with underlying asthma (Vally and Misso 2012). Epidemiological data indicate that sulfites can trigger bronchospasm in 5–10% of asthmatics, with reactions ranging from mild wheezing to severe life-threatening asthma exacerbations (Vreck et al. 2021).

The pathophysiological mechanisms underlying sulphite-induced respiratory symptoms are multifactorial. When foods containing sulfites are consumed, sulfur dioxide gas may be released and subsequently inhaled, directly irritating the airways and triggering bronchoconstriction (Vally et al. 2009). Additionally, some individuals exhibit reduced sulfite oxidase enzyme activity, limiting their capacity to effectively metabolize sulfites (Vally and Misso 2012). This enzymatic deficiency may contribute to the enhanced susceptibility to certain asthmatics. Moreover, evidence suggests potential IgE-mediated mechanisms in sulfite hypersensitivity, although this appears to be less common than the direct irritant effects (Fig. 12.2) (Peroni et al. 2020).

12.3.2 Artificial Food Colorings and Respiratory Health

Artificial food coloring, particularly azo dyes such as tartrazine (Yellow 5) and carmoisine (Red 3), have been associated with respiratory symptoms in susceptible individuals (Vojdani and Vojdani 2015). While the evidence remains less robust than that for sulfites, multiple case reports and small clinical studies suggest that these compounds may exacerbate asthma symptoms and contribute to rhinitis in sensitive populations (Feketea and Tsabouri 2017).

Research on the mechanisms underlying colorant-induced respiratory symptoms indicates potential non-IgE-mediated hypersensitivity reactions, including direct mast cell degranulation, inhibition of prostaglandin metabolism, and alterations in leukotriene pathways (Vojdani and Vojdani 2015). A meta-analysis by McCann et al. (2007) demonstrated significant associations between artificial food coloring and behavioral changes in children, raising questions about the broader systemic effects of these compounds, although respiratory outcomes were not specifically assessed.

In conclusion, although food additives and preservatives are essential for food safety and longevity, their respiratory implications vary significantly. The potential for both harmful and beneficial effects underscores the need for careful

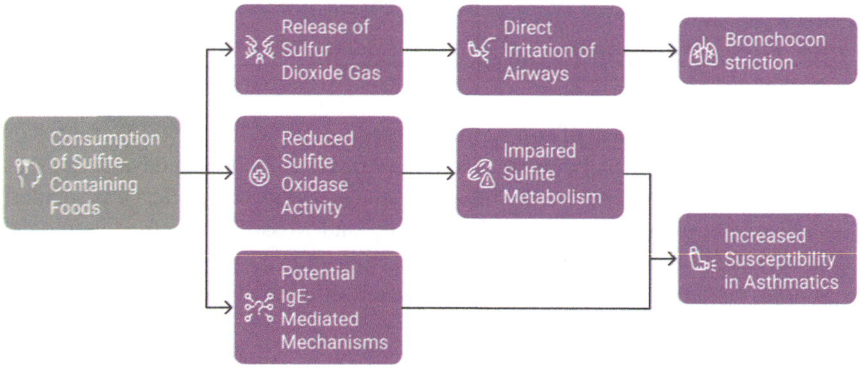

Fig. 12.2 Sulphite-induced respiratory symptoms flowchart

consideration of their use and consumption, particularly in individuals with preexisting respiratory conditions or sensitivities. Further research is needed to fully understand the mechanisms of food additive-induced respiratory symptoms and develop safer alternatives (Sulieman et al. 2023).

12.4 Occupational Respiratory Hazards in Food Processing

Occupational exposure to food in processing environments can lead to respiratory hazards and diseases. Up to 25% of occupational asthma and rhinitis cases are attributed to food-related exposures (Jeebhay et al. 2019). The main sources of inhalant allergens include aerosolized animal and vegetable high-molecular-weight proteins, additives, preservatives, antioxidants, and food contaminants. These agents typically cause IgE-mediated allergic reactions, resulting in a distinct form of food allergy known as Class 3 food allergy (Jeebhay et al. 2019).

Interestingly, the risk factors for developing respiratory issues during food processing extend beyond allergen exposure. Allergenicity of food proteins, exposure levels, and individual atopy play important roles (Jeebhay et al. 2019). Additionally, the processing methods themselves can influence the risk, with raw versus cooked food potentially having different impacts on worker health (Jeebhay et al. 2019).

12.4.1 Baker's Asthma and Flour Dust Exposure

Baker's asthma is one of the most prevalent occupational respiratory disorders, affecting approximately 1–10% of bakery workers globally (Fig. 12.3) (Jeebhay and Quirce 2007). This condition develops through IgE-mediated sensitization to airborne flour proteins, fungal α-amylase enzymes, or storage mites present in flour dust (Quirce and Diaz-Perales 2013). Longitudinal studies have demonstrated that symptom onset typically occurs after a latency period of exposure, with risk factors including atopic predisposition, genetic susceptibility, and dust concentrations in the workplace environment (Houba et al. 1998).

A survey conducted in a modern British bakery revealed that 35% of the participants reported chest symptoms, with 13% experiencing work-related issues, highlighting the significant health risks associated with flour dust inhalation (Crivellaro et al. 2020).

The pathophysiology of baker's asthma involves airway inflammation, characterized by eosinophilic infiltration, epithelial damage, and bronchial hyperresponsiveness (Quirce and Diaz-Perales 2013). Immunological research has identified multiple allergenic proteins in wheat, particularly water-soluble albumins, globulins, and α-amylase inhibitors (Salcedo et al. 2011). Clinical manifestations typically include work-related rhinoconjunctivitis, which often precedes asthma

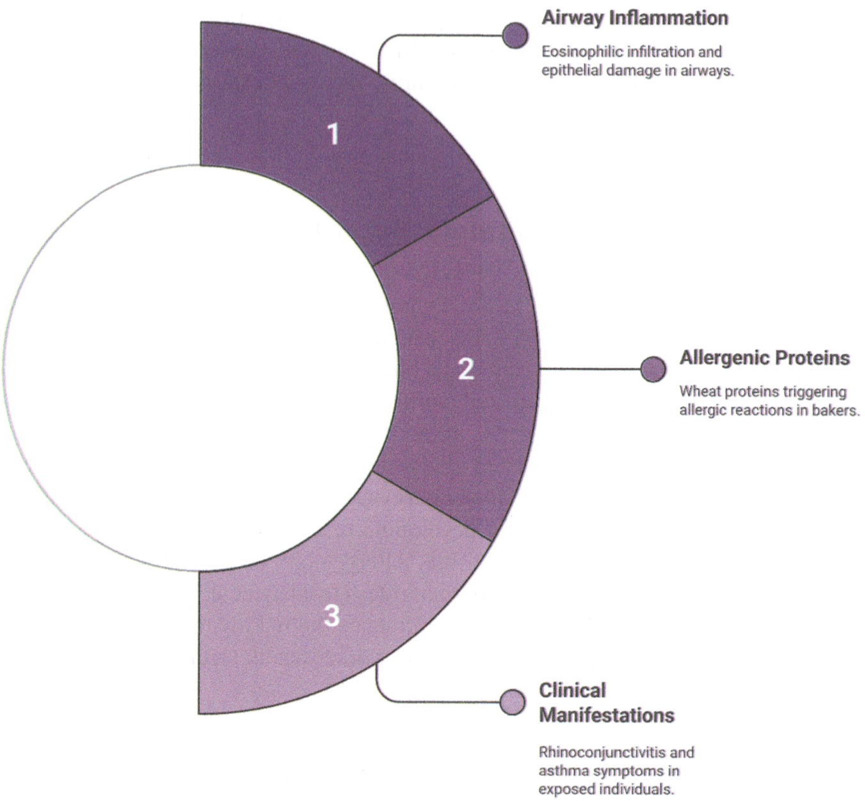

Fig. 12.3 Unraveling the complexities of Baker's Asthma

development, followed by progressive respiratory symptoms that may persist even after the cessation of exposure (Houba et al. 1998).

Various control measures have been implemented in bakery environments to mitigate the risk of asthma. Engineering controls such as local exhaust ventilation at flour release points and process modifications like refitting mixer lids have been effective in reducing flour dust exposure. Administrative controls, including training on safe work practices and the use of personal protective equipment, further contribute to minimizing exposure. A study in South African supermarket bakeries demonstrated that combining these interventions resulted in an 80% reduction in flour dust exposure (Jeebhay and Baatjies 2020; Baatjies et al. 2014a, b).

Innovative approaches, such as substituting regular flour with low-dust flour, have also proven beneficial. For instance, Bakkavor Bread introduced low-dust flour into its Twister Bread production line, leading to a significant decrease in airborne flour dust levels. This change not only reduced the need for respiratory protective equipment but also improved product quality and decreased cleaning time. These findings underscore the importance of implementing comprehensive strategies to protect bakery workers from the adverse health effects of flour dust exposure (Baatjies et al. 2014a, b; Health and Safety Executive 2018).

12.4.2 *Diacetyl Exposure and Bronchiolitis Obliterans*

Diacetyl (2,3-butanedione) and related α-diketones, which are widely used in artificial butter flavorings, have emerged as significant occupational respiratory hazards in the food industry (Kreiss 2007). These compounds are associated with bronchiolitis obliterans, a rare and potentially fatal form of obstructive lung disease, characterized by irreversible airflow limitation (Kreiss et al. 2002). This condition was initially identified among microwave popcorn manufacturing workers, leading to the colloquial term "popcorn lung," but has subsequently been documented across various food processing sectors (Hubbs et al. 2008).

Toxicological studies have demonstrated that diacetyl vapors induce epithelial damage, primarily affecting the bronchiolar region, where the compound is absorbed and metabolized (Hubbs et al. 2008). The pathophysiological process involves necrosis of bronchiolar epithelial cells followed by aberrant repair mechanisms, resulting in excessive fibroproliferation and airway obstruction (Palmer et al. 2011). Animal models have confirmed dose-dependent effects, with histopathological changes mirroring human disease when exposed to diacetyl concentrations comparable to those measured in food processing facilities (Morgan et al. 2012).

Epidemiological investigations have revealed a significantly elevated prevalence of respiratory symptoms and pulmonary function abnormalities among workers exposed to flavoring agents (Kreiss et al. 2002). A cross-sectional study by Kreiss et al. (2002) found an 11-fold increased risk of airway obstruction among workers in high-exposure areas of a microwave popcorn plant compared to national population standards, with dose-response relationships between exposure metrics and lung function decrements.

Exposure to diacetyl and its substitute 2,3-pentanedione has been linked to respiratory symptoms, spirometry abnormalities, and fixed airflow obstruction (Bailey et al. 2015; Harber et al. 2006; Holden and Hines 2016).

Interestingly, diacetyl exposure is not limited to the occupational setting. It has been found in flavored nicotine liquids used in e-cigarettes and is present in mainstream cigarette smoke at levels far exceeding the occupational exposure for most food/flavoring workers who smoke (Holden and Hines 2016; Pierce et al. 2014). This finding suggests that previous claims of a significant exposure-response relationship between diacetyl inhalation and respiratory disease in food/flavoring workers may have been confounded by smoking history (Pierce et al. 2014).

12.4.3 Organic Dust and Bioaerosols in Food Processing

Food processing environments frequently contain organic dust and bioaerosols that are capable of inducing respiratory inflammation through various immunological and non-immunological mechanisms. Organic dust toxicity syndrome (ODTS) is a non-allergic inflammatory response to inhaled organic dust containing bacterial endotoxins, fungi, and other bioactive components (Seifert et al. 2003). Unlike hypersensitivity pneumonitis, ODTS does not require prior sensitization and typically manifests as an acute self-limiting syndrome characterized by fever, malaise, and respiratory symptoms following high-level exposure (May et al. 2012).

Bioaerosols, including organic dust, are a significant concern in food-processing environments because of their potential health impacts on workers and food safety implications. These airborne particles contain various biological components such as bacteria, fungi, viruses, and their byproducts, which can pose risks of infection, allergies, and respiratory issues (Douwes et al. 2003; Su et al. 2020).

In food-processing facilities, bioaerosols can originate from raw materials, production processes, and workers themselves. Activities such as grinding, cutting, and packaging can generate organic dust and aerosolize microorganisms. The concentration and composition of bioaerosols can vary depending on the type of food being processed, environmental conditions, and ventilation system (Shailaja et al. 2023; Su et al. 2020).

Interestingly, while bioaerosols are generally considered harmful, some studies have suggested that certain microbial exposures may have protective effects against atopy and atopic conditions (Douwes et al. 2003). However, this potential benefit does not outweigh the overall risks associated with bioaerosol exposure in food processing environments.

Research on the pathophysiological mechanisms of ODTS indicates that bacterial endotoxins (lipopolysaccharides) play a central role in triggering inflammatory cascades through the activation of the toll-like receptor 4 (TLR4) and subsequent nuclear factor-κB (NF-κB) signaling pathways (Poole et al. 2011). This process leads to the production of pro-inflammatory cytokines, including tumor necrosis factor-alpha (TNF-α), interleukin-1 (IL-1), and IL-6, resulting in neutrophil recruitment and activation within the lungs (May et al. 2012).

Epidemiological studies have demonstrated substantial respiratory health risks associated with exposure to organic dust across various food-processing sectors. Zajda et al. (2020) reported a dose-dependent relationship between dust exposure levels and respiratory symptoms among food industry workers, with the highest prevalence in grain processing (27–42%), animal feed production (23–38%), and spice processing (18–32%) (Fig. 12.4).

To mitigate these risks, proper management and monitoring of bioaerosols in food-processing facilities are crucial. This includes implementing effective ventilation systems, regular cleaning and disinfection protocols, and providing appropriate personal protective equipment for workers. Additionally, emerging technologies, such as real-time PCR-based bioaerosol sensors and artificial intelligence-driven

Fig. 12.4 Understanding bioaerosols in food processing

monitoring systems, show promise for improving the detection and control of bioaerosols in food processing environments (Shammi et al. 2021).

In conclusion, organic dust and bioaerosols present significant challenges for food processing. Continued research is needed to better understand their health impact, develop more effective detection methods, and establish clear exposure limits. This will help to create safer working conditions and ensure food safety in the industry (Douwes et al. 2003; Shailaja et al. 2023).

12.5 Mycotoxins and Respiratory Health

Mycotoxins, secondary metabolites produced by certain fungi, primarily affect health through food contamination but may also affect respiratory function through multiple exposure routes (Fromme et al. 2016). Although ingestion represents the primary exposure pathway, inhalation of mycotoxin-containing dust during food handling and processing constitutes an important occupational hazard with potentially significant respiratory implications (Jarvis and Miller 2005).

Interestingly, while mycotoxins are primarily associated with ingestion through contaminated food, inhalation and dermal contact are also important routes of exposure, particularly in occupational settings (Mayer et al. 2008). This highlights the need for comprehensive workplace risk assessments and appropriate safety measures to protect workers from potential mycotoxin exposure (Mayer et al. 2008). Additionally, mycotoxin exposure can affect not only the respiratory system but also the ocular surface, and studies have shown that toxins, such as aflatoxin B1 and gliotoxin, can impact human corneal epithelial cells (Bossou et al. 2017).

Aflatoxins, ochratoxins, and trichothecenes have demonstrated respiratory toxicity in both experimental models and epidemiological studies (Fromme et al. 2016). Chen et al. (2015) identified dose-dependent pulmonary inflammation, oxidative stress, and epithelial damage following aflatoxin exposure in a murine model. Human studies corroborate these findings, with Jarvis and Miller (2005) documenting associations between occupational mycotoxin exposure and an increased prevalence of respiratory symptoms, including chronic cough, wheezing, and chest tightness.

The mechanisms underlying mycotoxin-induced respiratory toxicity involve multiple pathways, including direct cytotoxicity to pulmonary epithelial cells, disruption of protein synthesis, generation of reactive oxygen species, and immunomodulatory effects (Bennett and Klich 2003). These processes can result in impaired mucociliary clearance, compromised epithelial barrier function, and altered immune response to respiratory pathogens (Fig. 12.5) (Fromme et al. 2016).

In conclusion, mycotoxin exposure poses a significant risk to respiratory health, particularly in indoor environments with fungal contaminants. The effects can range from acute respiratory illnesses to chronic conditions, and may even extend to other systems, such as the eyes. Given the potential health hazards, further research is needed to develop better clinical tools for evaluating mycotoxin-related illnesses and to establish more effective prevention and control strategies in occupational and residential settings (Mayer et al. 2008; Trout et al. 2001).

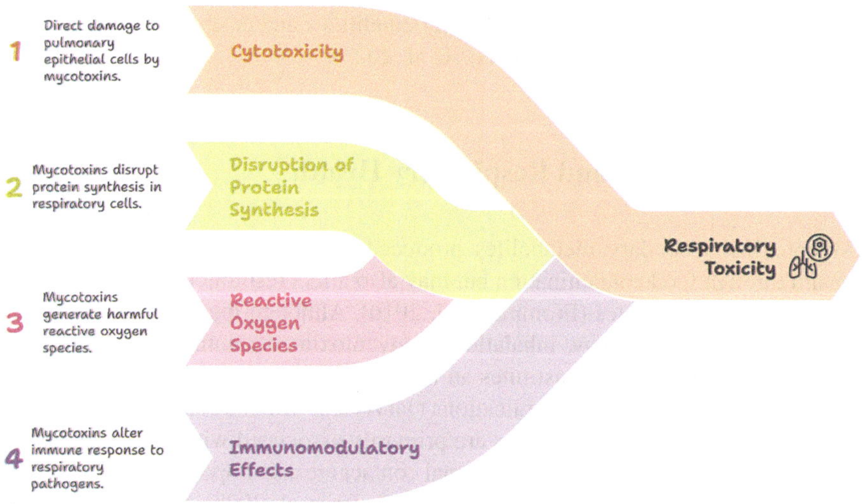

Fig. 12.5 Pathways to respiratory impact

12.6 Heavy Metals and Respiratory Toxicity

Heavy metals, including Pb, Cd, As, and Hg, can contaminate food through environmental pollution, agricultural practices, and processing methods (Jaishankar et al. 2014). Although typically associated with neurological, renal, and hepatic toxicity, emerging evidence indicates significant respiratory implications following both acute and chronic exposure to these contaminants (Kim et al. 2015).

Cd demonstrates particularly pronounced respiratory toxicity, with both occupational and environmental exposures linked to decreased lung function, chronic bronchitis, and increased risk of chronic obstructive pulmonary disease (COPD) (Nordberg et al. 2015). A longitudinal study by Lampe et al. (2008) found that blood cadmium levels were independently associated with an accelerated decline in forced expiratory volume in 1 s (FEV1) among individuals with normal baseline spirometry, suggesting potential contributions to COPD pathogenesis.

The pathophysiological mechanisms underlying heavy-metal-induced respiratory toxicity include oxidative stress, inflammation, and disruption of cellular homeostasis (Fig. 12.6) (Jaishankar et al. 2014). Cd exposure induces mitochondrial dysfunction and reactive oxygen species generation in pulmonary epithelial cells, leading to apoptosis and tissue remodeling (Kim et al. 2015). Additionally, heavy metals may interfere with essential element metabolism, disrupt zinc-dependent enzymes that are critical for antioxidant defense, and alter gene expression patterns in respiratory tissues (Nordberg et al. 2015).

The toxicity of heavy metals depends on factors such as concentration, duration of exposure, and route of exposure (Jyothi 2021). The inhalation of heavy metal particles can lead to respiratory diseases, including asthma, bronchitis, emphysema, and other respiratory disorders (Shakir et al. 2016).

Interestingly, while some heavy metals, such as iron, cobalt, and zinc, are essential nutrients in small quantities, they can become toxic at higher levels (Jyothi 2021). In contrast, metals, such as lead, cadmium, and mercury, are poisonous, even in small amounts (Jyothi 2021). The toxicity mechanisms of heavy metals primarily involve the generation of reactive oxygen species (ROS), inhibition of enzyme activity, and attenuation of antioxidant defense systems (Okechukwu Ohiagu et al. 2022).

In conclusion, heavy metal exposure through inhalation can significantly affect respiratory health and contribute to various respiratory diseases. Toxic effects are exacerbated by the ability of metals to bioaccumulate in living organisms and pollute the food chain (Tahir and Alkheraije 2023). To mitigate these risks, it is crucial to implement public health measures, reduce environmental contamination, and explore biological methods for heavy metal removal from the air, water, and soil (Zhou et al. 2023).

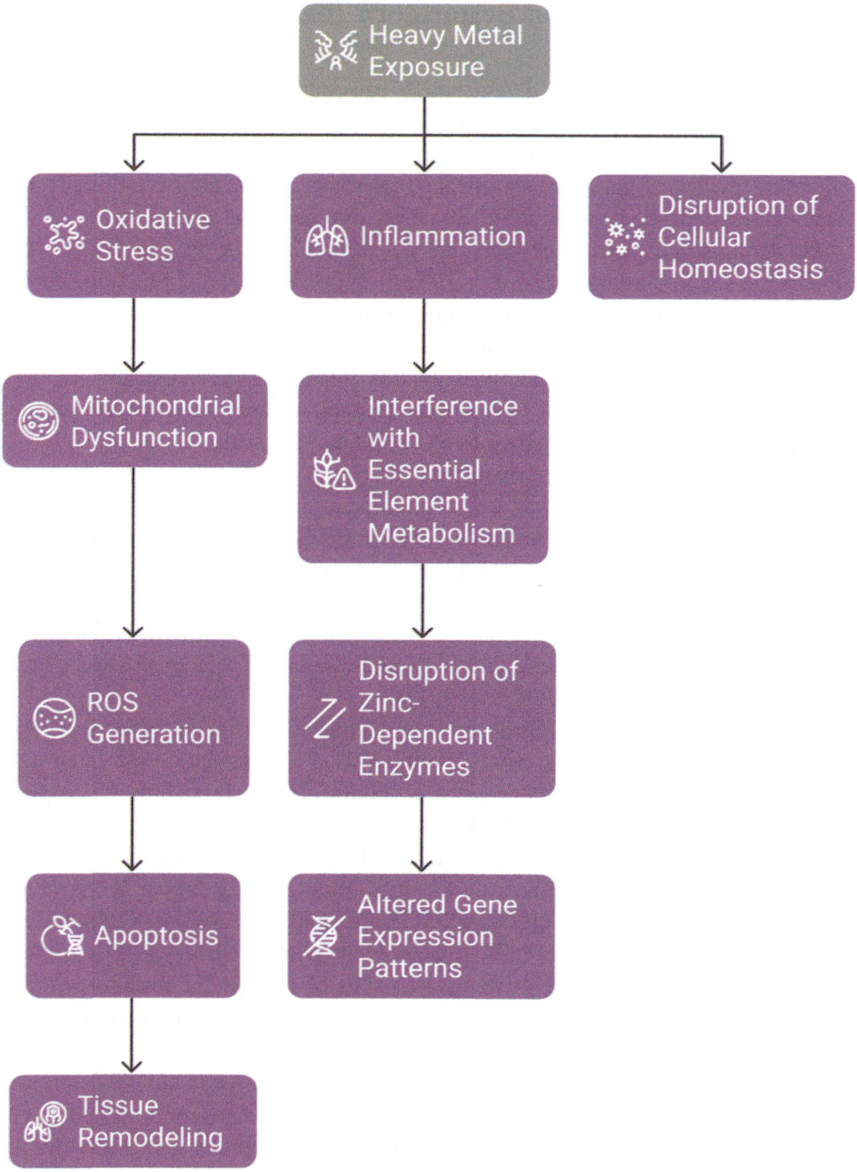

Fig. 12.6 Heavy metal-induced respiratory toxicity

12.7 Prevention and Management Strategies

12.7.1 Clinical Management Approaches

Management of food-related respiratory symptoms requires comprehensive diagnostic evaluation, identification of specific triggers, and implementation of appropriate therapeutic interventions (Sicherer and Sampson 2018). For IgE-mediated food allergies with respiratory manifestations, diagnostic approaches include a detailed clinical history, skin prick testing, serum-specific IgE measurement, and, when indicated, oral food challenges under medical supervision (Boyce et al. 2010).

Acute management of severe food-induced respiratory reactions necessitates prompt administration of epinephrine as a first-line therapy, followed by adjunctive treatments, including antihistamines, corticosteroids, and bronchodilators, as appropriate (Muraro et al. 2014). Long-term management emphasizes strict avoidance of identified allergens through careful label reading, education regarding hidden ingredients, and preparation for potential accidental exposure (Sicherer and Sampson 2018).

For non-IgE-mediated reactions, such as sulfite-induced asthma, management approaches include systematic elimination diets to identify specific triggers, pharmacological interventions targeting underlying respiratory conditions, and prophylactic strategies prior to unavoidable exposure (Vally and Misso 2012).

12.7.2 Occupational Health Interventions

The prevention of occupational respiratory diseases in the food industry requires multi-faceted approaches that incorporate engineering controls, administrative measures, and personal protective equipment (Fig. 12.7) (Cullinan et al. 2017). Comprehensive industrial hygiene assessments should identify specific respiratory hazards and evaluate exposure levels through air monitoring and biomarker analyses (Heederik et al. 2012).

Primary prevention strategies emphasize exposure reduction through engineering controls such as closed processing systems, local exhaust ventilation, and dust suppression measures (Meijster et al. 2011). For example, implementation of local exhaust ventilation in bakeries has demonstrated 50–90% reductions in airborne flour dust concentrations, with corresponding decreases in respiratory symptom prevalence (Baatjies et al. 2014a, b).

Administrative controls include worker education, hazard communication, job rotation to limit cumulative exposure, and medical surveillance programs to detect early signs of respiratory dysfunction (Cullinan et al. 2017). Medical surveillance should include baseline and periodic spirometry, symptom questionnaires, and immunological assessments when appropriate (Heederik et al. 2012).

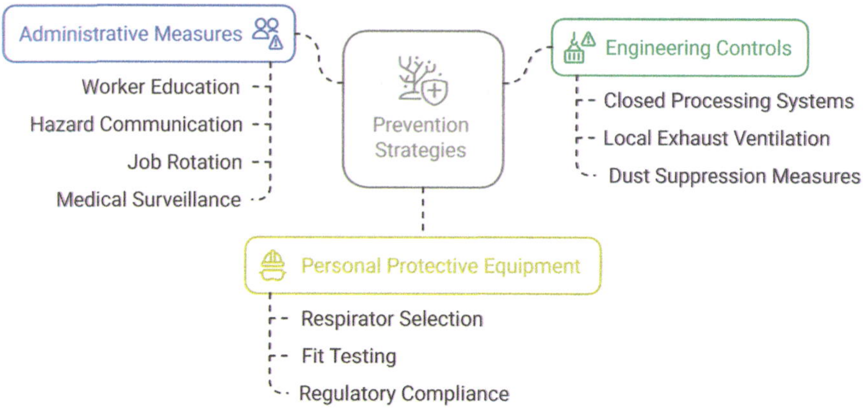

Fig. 12.7 Prevention of occupational respiratory diseases in food industry

For scenarios in which exposure cannot be adequately controlled through engineering and administrative measures, respiratory protective equipment provides an additional layer of protection (Meijster et al. 2011). The selection of appropriate respirators should be based on hazard assessment, with fit testing and comprehensive respiratory protection programs implemented in accordance with regulatory requirements (Baatjies et al. 2014a, b).

12.8 Regulatory Framework and Policy Implications

Regulatory approaches addressing food safety hazards with respiratory implications span multiple domains, including food additive regulations, allergen labeling requirements, and occupational health standards (Taylor and Hefle 2006). Food allergen labeling legislation, such as the Food Allergen Labeling and Consumer Protection Act (FALCPA) in the United States and similar international regulations, mandates clear identification of major allergens to facilitate avoidance by sensitive individuals (Gendel 2012).

For food additives, regulatory bodies, including the U.S., Food and Drug Administration (FDA), and European Food Safety Authority (EFSA), establish acceptable daily intake levels based on toxicological assessments (Smith et al. 2015). However, these evaluations primarily focused on gastrointestinal exposure rather than potential respiratory effects, highlighting a regulatory gap that requires further attention (Peroni et al. 2020).

Occupational exposure to respiratory hazards during food processing falls under broader occupational health regulatory frameworks, such as those established by the

Occupational Safety and Health Administration (OSHA) in the United States (Cullinan et al. 2017). Following the identification of bronchiolitis obliterans among flavoring workers, NIOSH established recommended exposure limits for diacetyl (5 parts per billion as an 8-h time-weighted average) and related compounds, although these remain recommendations rather than enforceable standards in many jurisdictions (NIOSH 2016).

12.9 Research Gaps and Future Directions

Despite significant advances in the understanding of the respiratory implications of food safety hazards, substantial knowledge gaps persist. Further research is needed to elucidate the precise immunological mechanisms underlying non-IgE-mediated respiratory reactions to food components, particularly artificial additives and preservatives (Vojdani and Vojdani 2015). Longitudinal epidemiological studies examining the potential associations between dietary patterns, food additive consumption, and respiratory disease incidence would provide valuable insights into long-term health implications (Peroni et al. 2020).

In the occupational realm, improved exposure assessment methodologies and standardized biomarkers of effect would enhance the identification of at-risk workers and the evaluation of intervention effectiveness (Heederik et al. 2012). Additionally, a systematic investigation of emerging food processing technologies and novel food additives should incorporate comprehensive respiratory health assessments prior to widespread implementation (Cullinan et al. 2017).

12.10 Conclusion

Food safety hazards affect the respiratory system through diverse mechanisms ranging from IgE-mediated allergic responses to direct irritant effects and occupational exposure during food processing. The spectrum of respiratory manifestations extends from mild rhinitis to life-threatening anaphylaxis and irreversible obstructive lung disease, thereby highlighting the significant public health implications of these hazards. Effective mitigation strategies require integrated approaches spanning clinical management, occupational health interventions, and regulatory oversight. Future research should address the existing knowledge gaps regarding mechanisms, long-term health effects, and optimal prevention strategies, ultimately informing evidence-based policies to protect both consumers and food industry workers from respiratory hazards.

References

Allen KJ, Turner PJ, Pawankar R, Taylor S, Sicherer S, Lack G et al (2019) Precautionary allergen labelling: perspectives from key stakeholder groups. Allergy 74(7):1253–1262

Baatjies R, Meijster T, Heederik D, Sander I, Jeebhay MF (2014a) Effectiveness of interventions to reduce flour dust exposures in supermarket bakeries in South Africa. Occup Environ Med 71(12):811–818. https://doi.org/10.1136/oemed-2013-101971

Baatjies R, Meijster T, Lopata A, Sander I, Raulf-Heimsoth M, Heederik D, Jeebhay M (2014b) Exposure to flour dust in South African supermarket bakeries: modeling of baseline measurements of an intervention study. Ann Occup Hyg 58(8):1012–1021

Bailey RL, Lebouf RF, Cox-Ganser JM, Kreiss K, Martin SB, Green BJ, Bledsoe TA, Duling MG (2015) Respiratory morbidity in a coffee processing workplace with sentinel obliterative bronchiolitis. Am J Ind Med 58(12):1235–1245. https://doi.org/10.1002/ajim.22533

Bennett JW, Klich M (2003) Mycotoxins. Clin Microbiol Rev 16(3):497–516

Bossou Y, Seta N, Achard S, Menotti J, Vitry S, Serssar Y, Allou A, Momas I (2017) Impact of mycotoxins secreted by aspergillus molds on the inflammatory response of human corneal epithelial cells. Toxins 9(7):197. https://doi.org/10.3390/toxins9070197

Bousquet J, Pfaar O, Togias A, Schünemann HJ, Ansotegui I, Papadopoulos NG et al (2020) 2019 ARIA care pathways for allergen immunotherapy. Allergy 75(9):2314–2327

Boyce JA, Assa'ad A, Burks AW, Jones SM, Sampson HA, Wood RA et al (2010) Guidelines for the diagnosis and management of food allergy in the United States: summary of the NIOSH-sponsored expert panel report. J Allergy Clin Immunol 126(6):S1–S58

Broide DH (2010) Allergic rhinitis pathophysiology. Allergy Asthma Proc 31(5):370–374. https://doi.org/10.2500/aap.2010.31.3388

Chen X, Horn N, Cotter PF, Applegate TJ (2015) Growth, serum biochemistry, complement activity, and liver gene expression responses of Pekin ducklings to graded levels of cultured aflatoxin B1. Poult Sci 94(1):15–22

Corradi M, Gergelova P, Mutti A (2015) Exhaled volatile organic compounds in nonrespiratory diseases. In: Exhaled biomarkers, vol 49. European Respiratory Society, pp 140–151

Crivellaro MA, Ottaviano G, Maculan P, Pendolino AL, Vianello L, Mason P, Gioffrè F, Bizzotto R, Scarpa B, Simoni E, Astolfi L, Maestrelli P, Scapellato ML, Carrieri M, Trevisan A (2020) Upper and lower respiratory signs and symptoms in workers occupationally exposed to flour dust. Int J Environ Res Public Health 17(19):7075. https://doi.org/10.3390/ijerph17197075

Cullinan P, Muñoz X, Suojalehto H, Agius R, Jindal S, Sigsgaard T et al (2017) Occupational lung diseases: from old and novel exposures to effective preventive strategies. Lancet Respir Med 5(5):445–455

Douwes J, Thorne P, Pearce N, Heederik D (2003) Bioaerosol health effects and exposure assessment: progress and prospects. Ann Occup Hyg 47(3):187–200. https://doi.org/10.1093/annhyg/meg032

Feketea G, Tsabouri S (2017) Common food colorants and allergic reactions in children: myth or reality? Food Chem 230:578–588

Fromme H, Gareis M, Völkel W, Gottschalk C (2016) Overall internal exposure to mycotoxins and their occurrence in occupational and residential settings–an overview. Int J Hyg Environ Health 219(2):143–165

Gendel SM (2012) Comparison of international food allergen labeling regulations. Regul Toxicol Pharmacol 63(2):279–285

Hait EJ, McDonald DR (2018) Impact of gastroesophageal reflux disease on mucosal immunity and atopic disorders. Clin Rev Allergy Immunol 57(2):213–225. https://doi.org/10.1007/s12016-018-8701-4

Harber P, Boomus C, Saechao K (2006) Diacetyl-induced lung disease. Toxicol Rev 25(4):261–272. https://doi.org/10.2165/00139709-200625040-00006

Health and Safety Executive (2018) Bakery uses low-dust flour to reduce asthma risk. Available from: https://www.hse.gov.uk/asthma/baker.htm

Heederik D, Henneberger PK, Redlich CA (2012) Primary prevention: exposure reduction, skin exposure and respiratory protection. Eur Respir Rev 21(124):112–124
Holden VK, Hines SE (2016) Update on flavoring-induced lung disease. Curr Opin Pulm Med 22(2):158–164. https://doi.org/10.1097/mcp.0000000000000250
Houba R, Heederik DJ, Doekes G, van Run PE (1998) Exposure-sensitization relationship for α-amylase allergens in the baking industry. Am J Respir Crit Care Med 158(3):776–783
Hubbs AF, Goldsmith WT, Kashon ML, Frazer D, Mercer RR, Battelli LA et al (2008) Respiratory toxicologic pathology of inhaled diacetyl in Sprague-Dawley rats. Toxicol Pathol 36(2):330–344
Jaishankar M, Tseten T, Anbalagan N, Mathew BB, Beeregowda KN (2014) Toxicity, mechanism and health effects of some heavy metals. Interdiscip Toxicol 7(2):60–72
Jarvis BB, Miller JD (2005) Mycotoxins as harmful indoor air contaminants. Appl Microbiol Biotechnol 66(4):367–372
Jeebhay MF, Baatjies R (2020) Prevention of baker's asthma. Curr Opin Allergy Clin Immunol 20(2):96–102. https://doi.org/10.1097/ACI.0000000000000612
Jeebhay MF, Quirce S (2007) Occupational asthma in the developing and industrialised world: a review. Int J Tuberc Lung Dis 11(2):122–133
Jeebhay MF, Lopata AL, Robins TG (2001) Seafood processing in South Africa: a study of working practices, occupational health services and allergic health problems in the industry. Occup Med 51(3):173–180
Jeebhay MF, Pala G, Raulf M, Sastre J, Lipińska-Ojrzanowska A, Walusiak-Skorupa J, Swoboda I, Folletti I, Lopata AL, Siracusa A, Bang BE, Quirce S, Moscato G (2019) Food processing and occupational respiratory allergy—an EAACI position paper. Allergy 74(10):1852–1871. https://doi.org/10.1111/all.13807
Kim JH, Byun HM, Chung EC, Chung HY, Bae ON (2015) Loss of integrity: impairment of the blood-brain barrier in heavy metal-associated ischemic stroke. Toxicol Res 31(2):113–118
Kreiss K (2007) Flavoring-related bronchiolitis obliterans. Curr Opin Allergy Clin Immunol 7(2):162–167
Kreiss K, Gomaa A, Kullman G, Fedan K, Simoes EJ, Enright PL (2002) Clinical bronchiolitis obliterans in workers at a microwave-popcorn plant. N Engl J Med 347(5):330–338
Kulig M, Tacke U, Guggenmoos-Holzmann I, Bergmann R, Wahn U (1998) Long-lasting sensitization to food during the first two years precedes allergic airway disease. Pediatr Allergy Immunol 9(2):61–67. https://doi.org/10.1111/j.1399-3038.1998.tb00305.x
Lampe BJ, Park SK, Robins T, Mukherjee B, Litonjua AA, Amarasiriwardena C et al (2008) Association between 24-hour urinary cadmium and pulmonary function among community-exposed men: the VA normative aging study. Environ Health Perspect 116(9):1226–1230
May JJ, Kullman G, Lounsbury K, Meredith S, Schenker MB (2012) Organic dust toxic syndrome: a review. J Toxic Environ Health, Part B 15(7):429–440
Mayer S, Engelhart S, Blome H, Kolk A (2008) The significance of mycotoxins in the framework of assessing workplace related risks. Mycotoxin Res 24(3):151–164. https://doi.org/10.1007/bf03032342
McCann D, Barrett A, Cooper A, Crumpler D, Dalen L, Grimshaw K et al (2007) Food additives and hyperactive behaviour in 3-year-old and 8/9-year-old children in the community: a randomised, double-blinded, placebo-controlled trial. Lancet 370(9598):1560–1567
Meijster T, Tielemans E, Heederik D (2011) Effect of an intervention aimed at reducing the risk of allergic respiratory disease in bakers: change in flour dust and fungal alpha-amylase levels. Occup Environ Med 68(5):333–340
Morgan DL, Flake GP, Kirby PJ, Palmer SM (2012) Respiratory toxicity of diacetyl in C57Bl/6 mice. Toxicol Sci 103(1):169–180
Muraro A, Werfel T, Hoffmann-Sommergruber K, Roberts G, Beyer K, Bindslev-Jensen C et al (2014) EAACI food allergy and anaphylaxis guidelines: diagnosis and management of food allergy. Allergy 69(8):1008–1025

NIOSH (2016) Criteria for a recommended standard: occupational exposure to diacetyl and 2,3-pentanedione. US Department of Health and Human Services, Centers for Disease Control and Prevention, National Institute for Occupational Safety and Health, Cincinnati

Nordberg GF, Nogawa K, Nordberg M (2015) Cadmium. In: Handbook on the toxicology of metals. Academic, pp 667–716

Okechukwu Ohiagu F, Chikezie PC, Ahaneku CC, Chikezie CM (2022) Human exposure to heavy metals: toxicity mechanisms and health implications. Mater Sci Eng Int J 6(2):78–87. https://doi.org/10.15406/mseij.2022.06.00183

Palmer SM, Flake GP, Kelly FL, Zhang HL, Nugent JL, Kirby PJ et al (2011) Severe airway epithelial injury, aberrant repair and bronchiolitis obliterans develops after diacetyl instillation in rats. PLoS One 6(3):e17644

Peroni DG, Bodini A, Corradi M, Cossettini M, Grotta L, Passalacqua G, Pecora S (2020) Sulfites: relationship asthma and food allergy. Minerva Pediatr 72(4):295–302

Pierce JS, Abelmann A, Spicer LJ, Adams RE, Finley BL (2014) Diacetyl and 2,3-pentanedione exposures are associated with cigarette smoking, which has implications for the risk assessment of food and flavoring workers. Crit Rev Toxicol 44(5):420–435. https://doi.org/10.3109/10408444.2014.882292

Poole JA, Dooley GP, Saito R, Burrell AM, Bailey KL, Romberger DJ et al (2011) Muramic acid, endotoxin, 3-hydroxy fatty acids, and ergosterol content explain monocyte and epithelial cell inflammatory responses to agricultural dusts. J Toxic Environ Health A 74(5):299–307

Quirce S, Diaz-Perales A (2013) Diagnosis and management of grain-induced asthma. Allergy, Asthma Immunol Res 5(6):348–356

Rajkovic A, Foubert A, De Saeger S, Decleer M, Tsilla V, Monteiro S, Madder A, Beloglazova N, Uyttendaele M, Sas B, Andjelkovic M, Jovanovic J (2020) Detection of toxins involved in foodborne diseases caused by gram-positive bacteria. Compr Rev Food Sci Food Saf 19(4):1605–1657. https://doi.org/10.1111/1541-4337.12571

Rama Jyothi N (2021) Heavy Metal Sources and Their Effects on Human Health. IntechOpen. https://doi.org/10.5772/intechopen.95370

Salcedo G, Quirce S, Diaz-Perales A (2011) Wheat allergens associated with baker's asthma. J Invest Allergol Clin Immunol 21(2):81–92

Sampson HA, Muñoz-Furlong A, Campbell RL, Adkinson NF Jr, Bock SA, Branum A et al (2006) Second symposium on the definition and management of anaphylaxis: summary report—second National Institute of allergy and infectious disease/food allergy and anaphylaxis network symposium. J Allergy Clin Immunol 117(2):391–397

Seifert SA, Von Essen S, Jacobitz K, Crouch R, Lintner CP (2003) Organic dust toxic syndrome: a review. J Toxicol Clin Toxicol 41(2):185–193

Shailaja GSJ, Ramakrishna TVBPS, Ramakodi MP (2023) Review of bioaerosols from different sources and their health impacts. Environ Monit Assess 195(11):1321. https://doi.org/10.1007/s10661-023-11935-x

Shakir SK, Nabeela F, Ur Rehman S, Murad W, Azizullah A, Daud MK, Häder D-P, Rahman H (2016) Toxic metal pollution in Pakistan and its possible risks to public health. Rev Environ Contam Toxicol 242:1–60. https://doi.org/10.1007/398_2016_9

Shammi M, Rahman MM, Tareq SM (2021) Distribution of bioaerosols in association with particulate matter: a review on emerging public health threat in Asian megacities. Front Environ Sci 9. https://doi.org/10.3389/fenvs.2021.698215

Sicherer SH, Sampson HA (2018) Food allergy: a review and update on epidemiology, pathogenesis, diagnosis, prevention, and management. J Allergy Clin Immunol 141(1):41–58

Smith RL, Cohen SM, Doull J, Feron VJ, Goodman JI, Marnett LJ et al (2015) GRAS flavoring substances 27. Food Technol 69(8):40–59

Su X, Sutarlie L, Loh XJ (2020) Sensors and analytical Technologies for air Quality: particulate matters and bioaerosols. Chem Asian J 15(24):4241–4255. https://doi.org/10.1002/asia.202001051

Sulieman AME, Abdallah EM, Alanazi NA, Ed-Dra A, Jamal A, Idriss H, Alshammari AS, Shommo SAM (2023) Spices as sustainable food preservatives: a comprehensive review of their antimicrobial potential. Pharmaceuticals (Basel, Switzerland) 16(10):1451. https://doi.org/10.3390/ph16101451

Tahir I, Alkheraije KA (2023) A review of important heavy metals toxicity with special emphasis on nephrotoxicity and its management in cattle. Front Vet Sci 10:1149720. https://doi.org/10.3389/fvets.2023.1149720

Taylor SL, Hefle SL (2006) Food allergen labeling in the USA and Europe. Curr Opin Allergy Clin Immunol 6(3):186–190

Trout D, Bernstein J, Biagini R, Wallingford K, Martinez K (2001) Bioaerosol lung damage in a worker with repeated exposure to fungi in a water-damaged building. Environ Health Perspect 109(6):641–644. https://doi.org/10.1289/ehp.01109641

Vally H, Misso NL (2012) Adverse reactions to the sulphite additives. Gastroenterol Hepatol Bed Bench 5(1):16–23

Vally H, Misso NL, Madan V (2009) Clinical effects of sulphite additives. Clin Exp Allergy 39(11):1643–1651

Van Asselt ED, Marvin HJP, Groot MN, Van Der Fels-Klerx HJ, De Veen VB-V, H. (2016) Overview of food safety hazards in the European dairy supply chain. Compr Rev Food Sci Food Saf 16(1):59–75. https://doi.org/10.1111/1541-4337.12245

Vojdani A, Vojdani C (2015) Immune reactivity to food coloring. Altern Ther Health Med 21(Suppl 1):52–62

Vreck L, Pandey S, Navuluri N (2021) Sulphites in food & drinks in asthmatic adults & children: a systematic review. Int J Appl Basic Med Res 35(2):114–120

Warren CM, Turner PJ, Chinthrajah RS, Gupta RS (2021) Advancing food allergy through epidemiology: understanding and addressing disparities in food allergy management and outcomes. J Allergy Clin Immunol Pract 9(1):110–118

Williams PRD, Hammitt JK (2001) Perceived risks of conventional and organic produce: pesticides, pathogens, and natural toxins. Risk Anal 21(2):319–330. https://doi.org/10.1111/0272-4332.212114

Zajda E, Janasik B, Sawicki K, Świątkowska B, Dziankowska-Zaborszczyk E, Świerczyńska-Machura D (2020) The effect of occupational exposure to organic dust on lung function and respiratory symptoms: a systematic review. Int J Environ Res Public Health 17(18):6588

Zhou B, Zhang T, Wang F (2023) Microbial-based heavy metal bioremediation: toxicity and eco-friendly approaches to heavy metal decontamination. Appl Sci 13(14):8439. https://doi.org/10.3390/app13148439

Chapter 13
Cardiovascular Impacts of Foodborne Toxins

Veeresh Sadashivanavar, Manjunath Madalageri, K. Sreedhara Ranganath Pai, Himanshu Sharma, Praveen Halagali, Raagul Seenivasan, Vamshi Krishna Tippavajhala, Preethi Somanna, and Abdullah Al Noman

Abbreviations

(CVD)	Cardiovascular Disease
(Afs)	Aflatoxins
(AFB1)	Aflatoxin B1
(HABs)	Harmful Algal Blooms
(AZA)	Azaspiracids

V. Sadashivanavar · M. Madalageri · K. S. R. Pai (✉)
Department of Pharmacology, Manipal College of Pharmaceutical Sciences, Manipal Academy of Higher Education, Manipal, Karnataka 576104, India
e-mail: veeresh.mcopsmpl2024@learner.manipal.edu; manjunath.mcopsmpl2024@learner.manipal.edu; ksr.pai@manipal.edu

H. Sharma
Teerthanker Mahaveer College of Pharmacy, Teerthanker Mahaveer University, Moradabad, UP, India

P. Halagali · R. Seenivasan · V. K. Tippavajhala
Department of Pharmaceutics, Manipal College of Pharmaceutical Sciences, Manipal Academy of Higher Education, Manipal, Karnataka 576104, India
e-mail: praveen.mcopsmpl2023@learner.manipal.edu; raagul.mcopsmpl2024@learner.manipal.edu; vamshi.krishna@manipal.edu

P. Somanna
Department of Pharmaceutics, JSS College of Pharmacy, JSS Academy of Higher Education and Research, Mysuru, Karnataka, India
e-mail: preethis@jssuni.edu.in

A. Al Noman
School of Pharmacy, BRAC University, Dhaka, Bangladesh
e-mail: abdullah.al.noman@g.bracu.ac.bd

T. Sarkar et al. (eds.), *Physiological Perspectives on Food Safety: Exploring the Intersection of Health and Nutrition*, https://doi.org/10.1007/978-3-031-94582-3_13

(YTX)	Yessotoxins
(CI)	Cyclic Imines
(BTX)	Brevetoxins
(OA)	Okadaic Acid
(PTX)	Pectenotoxins
(DA)	Domoic Acid
(STX)	Saxitoxins
(CTX)	Ciguatoxin
(IFCB)	Imaging FlowCytobot
(UPFs)	Ultra-Processed Foods
(POPs)	Persistent Pollutants
(PAHs)	Hazardous Hydrocarbons
(PPCPs)	Pharmaceutical and Personal Care Products
(ROS)	Reactive Oxygen Species
(HF)	Heart Failure
(NO)	Nitric Oxide
(LDL)	Low-Density Lipoprotein
(GMCSF)	Granulocyte-Macrophage Colony-Stimulating Factor
(FMNs)	Fumonisins
(AME)	Alternaria Monomethyl Ether
(CIT)	Citrinin
(DON)	Deoxynivalenol
(ALT)	Altenuen
(IARC)	International Agency for Research on Cancer
(CVS)	Cardiovascular System
(PAD)	Peripheral Artery Diseases
(mtDNA)	Mitochondrial DNA
(LPS)	Lipopolysaccharide
(CD14)	Cluster of Differentiation 14
(TLRs)	Toll-Like Receptors

13.1 Introduction

Diseases associated with CVD are the leading cause worldwide among non-communicable diseases (Vaduganathan et al. 2022). Global burden of disease (GBD) reported that cardiovascular diseases including myocardial ischemia, stroke, heart failure, and many other CVD are leading killer worldwide and also affect the quality of life (Mensah et al. 2019). According to a GBD study in 2019 responsible for 9.6 million and 8.9 million deaths among men and women respectively which accounts for 1/3rd of deaths globally. Among all countries, China ranks first for deaths due to CVD followed by India, Russia, and US (Roth et al. 2020). Unquestionably, diet is important for both primary and secondary prevention of CVD. Food contamination is becoming known as a new mechanism that may adversely affect the atherosclerotic process and associated vascular consequences, among other variables that may raise the risk of CVD. As per WHO food contamination is a global challenge because in higher concentrations it may cause serious risks to normal health. Thus, Contaminants can be described as collective heterogeneous chemicals from several sources. Different phases of food manufacturing, packaging, processing, and transportation might result in food contamination. Environmental contamination can also cause them, altering food nutrients and raising the risk of vascular damage. The major contaminants found in food and feed include natural toxins like mycotoxins or alkaloids, environmental contaminants left over from industrial or agricultural activities, and metals like arsenic, cadmium, lead, and mercury (Mancuso et al. 2024).

13.2 Categories of Foodborne Contaminants/Toxins

The term "foodborne toxins"generally refers to a wide range of compounds that can harm human health and jeopardize the safety or quality of food. Foodborne toxins can be classified based on their nature as Natural contaminants, Processing Contaminants, Heavy metals Environmental pollutants, and Residues (Jiang et al. 2024). The types of food contamination are given in (Fig. 13.1).

Fig. 13.1 Types of food contaminations

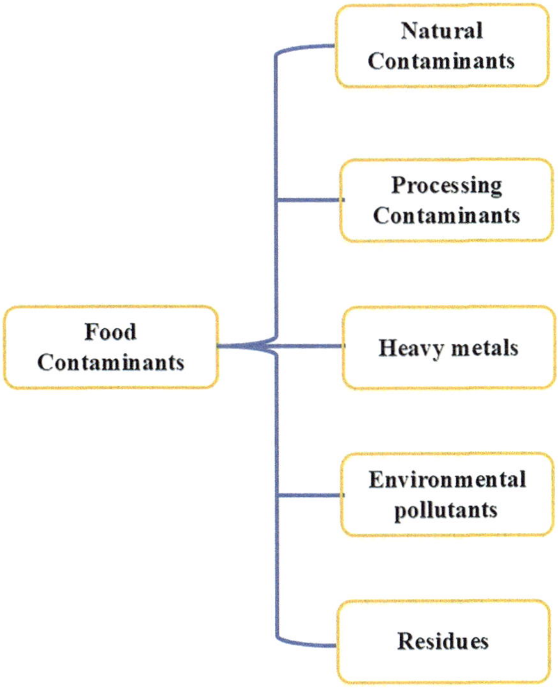

13.2.1 Natural Toxins

These toxins are toxic compounds produced spontaneously by living organisms. These toxins don't damage the creatures themselves, but they can be toxic to other animals, including humans. The toxicity and biological role of these chemicals vary, as do their structural characteristics. Natural poisons are a major health risk to both people and animals and can have several negative health impacts. Some of these poisons are very strong. Some of the sources of natural toxins are chemicals produced by plants to defend themselves against predators, and insects, and other sources are algae and plankton in oceans which produce chemicals that are only toxic to humans but not to fish and shellfish which eat these organisms. When humans eat these fishes containing toxins can cause illness. Some of the natural toxins are discussed below.

13.2.1.1 Mycotoxins

Some forms of Molds (fungi) naturally create harmful substances called mycotoxins. These moulds grow on a variety of foods like spices, nuts, and cereals in every stage of food processing such as harvesting and storage because the majority of mycotoxins are resistant to processing and are chemically stable. While hundreds of different mycotoxins are known to exist, the most common ones that are harmful to human health and cattle are ochratoxin A, nivalenol/deoxynivalenol, aflatoxins, zearalenone, fumonisins, and patulin. Both direct and indirect contamination pathways can allow mycotoxins to infiltrate the feed and food systems. Direct contamination happens when a toxigenic fungus infects food or feed, causing mycotoxins to develop as a result. Indirect contamination happens when a toxigenic fungus has already contaminated a component and mycotoxins are still present in the finished product even after the fungus has been removed during processing.

Some food-borne mycotoxins cause an acute illness with rapid symptoms after consumption, while Numerous diseases, such as cancer, immunological dysregulation, neurodegenerative disease, cardiovascular disease, chronic liver disease, and kidney disease have been caused by these mycotoxins. There are already over 400 mycotoxins known to be potentially harmful to both human and animal health. Many mycotoxins have achieved international notoriety for their capacity to cause cardiovascular disorders, including circulation problems, vein breakage, haemorrhage, and heart failure (Dai et al. 2022). These include citreoviridin, aflatoxin B1, and T-2 toxin.

T-2 toxin belongs to a chemical category called trichothecenes, which are part of the Fusarium mycotoxins group containing more than 150 chemicals. It is produced by several Fusarium species, including *F. sporotrichioides*, *F. equiseti*, *F. acuminatum,* and *F. poae*. T-2 toxin is highly stable in UV light and heat and is insoluble in water. Chemically T-2 ((3α,4β,8α)-12,13-epoxy-4,15-diacetate 8-(3-methylbutanoate) trichothec-9-ene-3,4,8,15-tetro) toxin is highly lipophilic and have high permeable through biological membranes. According to some experiments in animals, even smaller doses of T-2 toxin can damage vital organs like the liver, spleen, brain, spleen, gastrointestinal bone marrow, heart, and brain tissues. Some of the epidemiological studies found that T-2 toxins are potentially causing Kashin-Beck disease and beriberi-linked cardiac insufficiency. The mechanism involved in cardiotoxicity caused by T-2 toxin is not clearly understood till now but some researchers described in their work that T-2 causes oxidative stress, mitochondrial dysfunction, activates autophagy and mitophagy, increases endoplasmic reticulum stress, induces inflammatory response, activates ferroptosis and heme-oxygenase-1 (Magnuson et al. 1987; Wang et al. 2023). T-2 also penetrates through the skin causing hypertension arrhythmia and also affects blood pressure and catecholamine concentration (Bubien et al. 2017).

Aflatoxins (Afs) are one of the important toxins among mycotoxins and are released by the Aspergillus parasiticus and Aspergillus flavus. The six types of AFs are M1 (AFM1), M2 (AFM2), G1 (AFG1), G2 (AFG2), B1 (AFB1), and B2 (AFB2) among these AFB2, AFG2, AFB1 and AFG1 are deadlier subtypes. A. parasiticus releases AFG1 and AFG2, while A. flavus produces AFB1 and AFB2. AFM1, on the other hand, has been identified as the hydroxylation metabolite of AFB1 in the liver of nursing dairy cows following the consumption of tainted feed. Food crops are the primary source of AFB1, AFB2, AFG1, and AFG2, whereas animal byproducts like milk and dairy products frequently contain AFM1 and AFM2 (Carvajal Moreno 2016; Alameri et al. 2023). Aflatoxin consumption causes liver damage in malnourished people and is linked to Kwashiorkor and Reye syndrome. In humans, Afs are categorized as Group I carcinogens. They are strong teratogens and mutagens that result in immunosuppression, cardiovascular problems, and abortions. These poisons produce internal haemorrhage, fracture platelets, and weaken veins and artery walls. Aflatoxin B1 (AFB1) can harm multiple organs, including the liver, kidney, and heart, by interfering with the regular process of protein synthesis and inhibiting multiple metabolic systems. Frequent organ hemorrhages are among the clinical effects of Afs on the cardiovascular system in humans, horses, dogs, and poultry. A 15-year-old boy in Uganda died 2 days after being admitted to the hospital in 1967 after being diagnosed with heart failure as a result of aflatoxin poisoning. In addition to other signs including liver necrosis, an autopsy showed interstitial oedema of the heart, flabby heart, and pulmonary oedema. The cardiac muscle has been found to contain Afs (Serck-Hanssen 1970); Murthy et al. 1975; IARC Working Group on the Evaluation of Carcinogenic Risks to Humans. 2002; Mohamed and Metwally 2009).

During the early twentieth century, the eating of mouldy rice in East Asian countries was found to be the cause of abrupt cardiac arrest, leading to the discovery of citreoviridin. It has been determined that the cause of cardiac beriberi is Citreoviridin, a toxic metabolite of the active ingredient of Penicillium citreoviride (Penicillium toxicarium). Many citreoviridin derivatives like citreoviridin C, citreoviridin D-citreoviridinol, isoceitreoviridinol, and secocitreoviridinol from different penicillium species. Recently three new metabolites (one- citreoazopyrone and two 6-Epoxy-2-pyranones are isolated from hybrid strain KO 0011(Goel and Ram

2009). It is reported that citreoviridin causes myocardial cell death by inducing autophagy by inhibiting the PPAR-y-mTORC2 protein in cardio myoblast cells (H9C2) along with that it also inhibits the ATP synthase enzyme (Feng et al. 2019).

13.2.1.1.1 Approaches to Minimizing mycotoxin Risk

The strategies to minimize mycotoxin risk are presented in (Fig. 13.2). Mycotoxin production is influenced by factors like the fungus's genetic makeup, moisture levels, and temperature. When crops experience stress—whether from drought or insect damage- they become more vulnerable to fungal invasion. To prevent this, farmers can use good agronomic practices, apply insecticides and fungicides, ensure proper irrigation, harvest at the right time, and even breed crops that are more resistant to fungi. After harvest, managing the moisture and temperature of stored crops is essential to keeping fungal growth in check. Techniques like drying, aerating, and turning the grain can help reduce moisture levels, while antifungal agents such as propionic acid can prevent Mold growth. Processing methods can also help reduce mycotoxin contamination. These include steps like dehulling, washing, and using chemical treatments like sodium bisulfite or ozone. There are also natural solutions, like adsorbents (HSCAS or yeast cell wall preparations) that bind to mycotoxins, limiting their harmful effects. Some bacteria can even detoxify certain mycotoxins in the body. To effectively manage mycotoxin risks, a comprehensive approach is needed—one that draws from fungal ecology, toxicology, and food production systems, often implemented through strategies like HACCP. In many developing countries, however, the cost of these methods may not be feasible, requiring alternative approaches. In areas where people rely heavily on a single crop, even small amounts

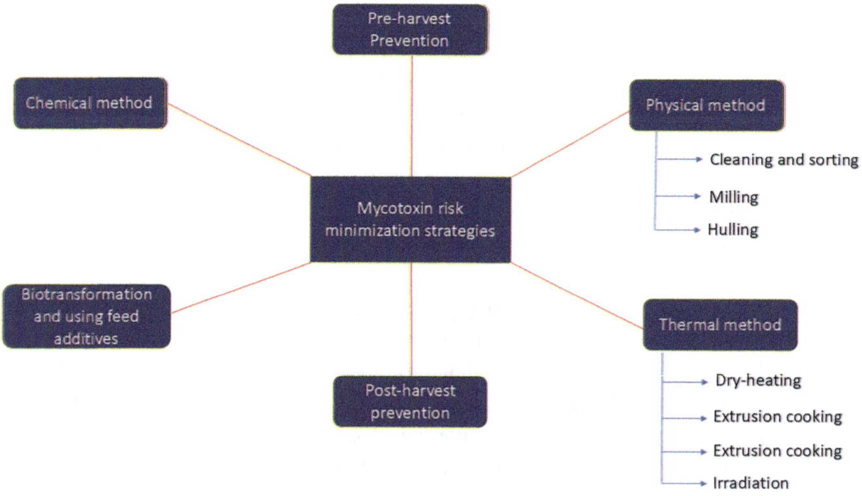

Fig. 13.2 Strategies to minimize mycotoxin risk [24]

of mycotoxin contamination can exceed safe levels. In such cases, agents like oltipraz, which help block aflatoxin formation by boosting detoxifying enzymes, could offer a helpful solution (Bryden 2007; Peng et al. 2018; Hamad et al. 2023).

13.2.1.2 Aquatic Biotoxins

As some kinds of algae bloom in freshwater and oceans, they release toxic substances called algal toxins. Compared to fish, shellfish, such as mussels, oysters, and scallops, are more likely to contain these toxins. People may get symptoms like nausea, vomiting, tingling, and even paralysis if they eat contaminated shellfish or drink tainted water. It's difficult to determine whether seafood or water is contaminated simply by looking at it because these toxins have no taste or smell. It's even more crucial to be mindful of potential hazards because freezing or cooking doesn't eliminate the toxins. The classification of biotoxins is presented in (Fig. 13.3).

Harmful algal blooms (HABs) happen when certain types of marine phytoplankton grow out of control. These blooms are a natural event, but recently, they've been happening more often and with greater intensity. Scientists believe this increase is tied to rising ocean temperatures and the growing problem of coastal pollution, which adds nutrients to the water and fuels algae growth. As our climate changes, these blooms are becoming a bigger concern around the world. More than 100 of the thousands of microalgal species that exist in nature produce toxins that are harmful

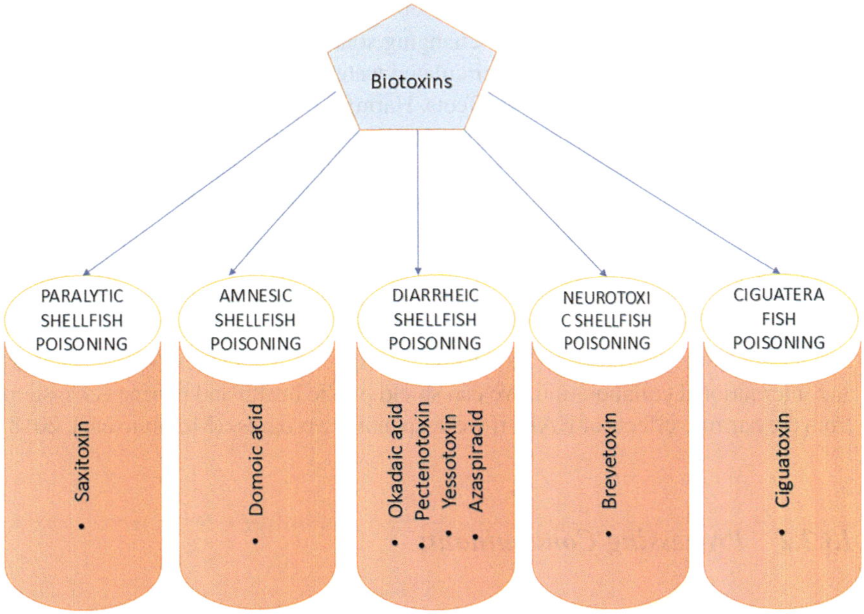

Fig. 13.3 Classification of biotoxins

or even lethal to humans and animals, and about 300 of these species are known to cause harmful events. In addition to the health hazards, these toxic outbreaks can significantly affect people's lives, impacting everything from local economies to the expenses associated with event management. HABs can have a direct effect on human health in several ways. Ingestion of tainted seafood, skin contact with contaminated water, or even inhalation of waterborne toxins can all cause poisoning (Visciano et al. 2016; Lee et al. 2024).

Azaspiracids (AZA), yessotoxins (YTX), cyclic imines (CI), brevetoxins (BTX), okadaic acid (OA), pectenotoxins (PTX), domoic acid (DA), saxitoxins (STX), and are the eight primary groups of marine biotoxins that are categorized according to their chemical structure. Two additional groups are also acknowledged as significant: ciguatoxin (CTX) and palytoxin (PlTX) (Vilariño et al. 2018). Around the world, brackish, freshwater, and marine environments are seeing an increase in the frequency and severity of harmful algal blooms, or HABs. There are currently no standard tests for diagnosing these blooms or antidotes for their toxins, even though they can cause serious health problems. International cooperation is necessary to prevent illness from HABs, which is crucial for protecting public health.

Scientists have improved monitoring and prediction efforts by gaining a better understanding of HAB dynamics. Even minute amounts of dangerous algae can be found thanks to new technologies like molecular probes. By introducing pollutants and surplus nutrients into the water, human activities—particularly aquaculture and coastal development—often cause HABs. Better management techniques, such as relocating fish farms to deeper, well-circulated waters, are advised to lessen these effects.

Human activities like pollution, coastal development, and aquaculture fertilizer use—which can supply nutrients that encourage algal growth—are frequently connected to the occurrence of HABs. Changing some human behaviors, such as relocating aquaculture to deeper, well-circulated water areas to lessen nutrient buildup, may be necessary to address these effects. Harmful blooms can be managed using a variety of techniques, such as biological, mechanical, and physical/chemical methods. For example, chemical compounds can stop the growth of algae, and pumps, filters, and barriers can remove dead fish and harmful cells. Technological developments have also produced new monitoring and detection tools, such as the Environmental Sample Processor and the Imaging FlowCytobot (IFCB), which can automatically collect water samples and detect dangerous microbes. Even though HABs are a natural phenomenon that is difficult to eliminate, their negative effects can be controlled with better monitoring, mitigation techniques, scientific study, and international collaboration. We can shield public health and marine ecosystems from the harmful effects of HABs if we keep making progress (Morabito et al. 2018).

13.2.2 Processing Contaminants

Process contaminants are unwanted chemicals that can form during food processing, like Transportation, cleaning, cooking and heating, packaging, and storage, and while not all are harmful, some may pose health risks (Rather et al. 2017). The FDA works to understand how these contaminants form, measures their levels, and takes action when they're unsafe, collaborating with the food industry and global organizations to reduce their presence and keep consumers informed. Our diets' quality is greatly influenced by the way food is processed. It includes minimal processing, such as pasteurized milk, frozen vegetables, and sugar-free dried fruit, as well as much more intensive techniques used to produce ultra-processed foods (UPFs). Made using a variety of industrial processes, UPFs are typically very convenient, require little preparation, and have a long shelf life. Many of these foods are tasty and convenient to eat, but they aren't always the healthiest options because during processing food undergoes chemical transformation and leads to the production of nitrosamine, acrylamide, and glycation end products and millard and caramelization reaction products which causes toxicity on the body. In addition to well-known "junk" foods like soda, fast food, and packaged snacks, UPFs also include goods that are promoted as healthy, like drinks like energy or sugary sodas, flavored yogurts, breakfast cereals, and low-fat or low-calorie options (Monteiro et al. 2019).

Consuming more UPFs has been associated with a higher risk of heart disease and other cardiovascular issues, according to research. Those who consumed more UPFs had a higher risk of dying from cardiovascular disease, which includes heart disease and strokes, according to a large study that tracked nearly 92,000 participants for more than 13 years. Eating just one serving of ultra-processed foods per day can significantly increase the risk of heart and blood vessel problems, according to a comprehensive review of 39 studies that involved over 60 million people. The lesson learned? Ultra-processed foods are convenient, but consuming too many of them may be detrimental to long-term heart health (Guo et al. 2023).

13.2.3 Environmental Pollutants, Heavy Metals, and Residues

Environmental chemical contamination is a serious global food safety issue that puts human health at serious risk. Persistent pollutants (POPs), Hazardous hydrocarbons (PAHs), Heavy metals and metalloids, compounds such as PFCs, e-waste, plastics, radioactive materials, pharmaceutical and personal care products (PPCPs) (Thompson and Darwish 2019). The environment frequently contains heavy metals from both natural and man-made sources, such as industrial operations, automobile emissions, and wastewater irrigation. Wastewater contains harmful heavy metals that contaminate the soil, even though it can give crops useful nutrients. Crops that are grown in these types of soils frequently absorb these metals, which may compromise the food's safety and quality. Because heavy metals can enter the food chain and eventually the human body, their accumulation in soil and plants poses a growing risk to human health.

Certain crops absorb metals more effectively than others, and the degree of heavy metal uptake by plants varies by species. Wastewater-irrigated vegetables are especially vulnerable to dangerously high concentrations of heavy metals, which could pose a health risk to consumers. Experts use a process that involves identifying hazards, assessing exposure, comprehending the health effects, and quantifying the total risk to evaluate these risks. Even though the negative effects might not be apparent for years, prolonged exposure to these metals can cause major health issues (Khan et al. 2008). Consuming food tainted with heavy metals (Arsenic, Lead, cadmium, Chromium, and Mercury) can cause the body to drastically lose vital nutrients. This depletion can impair mental and social development, weaken the immune system, prevent healthy growth during pregnancy, and increase the risk of heart disease, upper gastrointestinal cancers, and malnutrition, among other health problems (Iyengar and Nair 2000; Türkdoğan et al. 2003).

According to WHO estimates, heavy metal poisoning is responsible for 60–70% of all acute and chronic diseases. Long-term exposure to even low to moderate levels of metals like lead, cadmium, and arsenic significantly increases the risk of CVD, including stroke, peripheral artery disease, and heart disease (Lamas et al. 2023). According to studies, exposure to heavy metals such as cadmium, lead, and others can interfere with vital biological functions that impact the health of the heart and blood vessels. Among other problems, these metals can damage kidneys, cause oxidative stress, increase blood pressure, cause chronic inflammation, and impair blood vessel function. These metals harm the body by interfering with normal cellular processes by replacing vital minerals like copper, zinc, and calcium. Even though every metal may have unique harmful effects, these disruptions are frequently linked to the overall impact. The low levels of metal exposure that occur daily in communities are more of a concern here than the severe cases of poisoning from industrial exposures. Although it's difficult to identify a precise threshold at which cardiovascular damage ceases, these lower doses can still be harmful. Research indicates that even minor daily exposures to these metals may have long-term negative effects on one's health (Lind et al. 2021; Munir et al. 2022).

13.3 Mechanisms of Cardiovascular Impacts

Generally, most CVD involve three mechanisms in their pathology: oxidative stress, endothelial dysfunction, and inflammation, all of which are interlinked. We will discuss each of these mechanisms one by one.

13.3.1 Oxidative Stress

Reactive oxygen species (ROS) production and the body's capacity to neutralize them with antioxidants are out of balance, leading to oxidative stress. This imbalance has a strong correlation with inflammation and can cause tissue injury. The link between inflammation and oxidative stress may be one of the main ways that pollutants exacerbate their negative effects, particularly in the heart, harming different body tissues (Biswas 2016). The mechanism of CVD caused by oxidative stress, is shown in (Fig. 13.4). The body's oxidative stress balance can be upset by certain food contaminants. Consuming a lot of UPFs caused significant changes in the

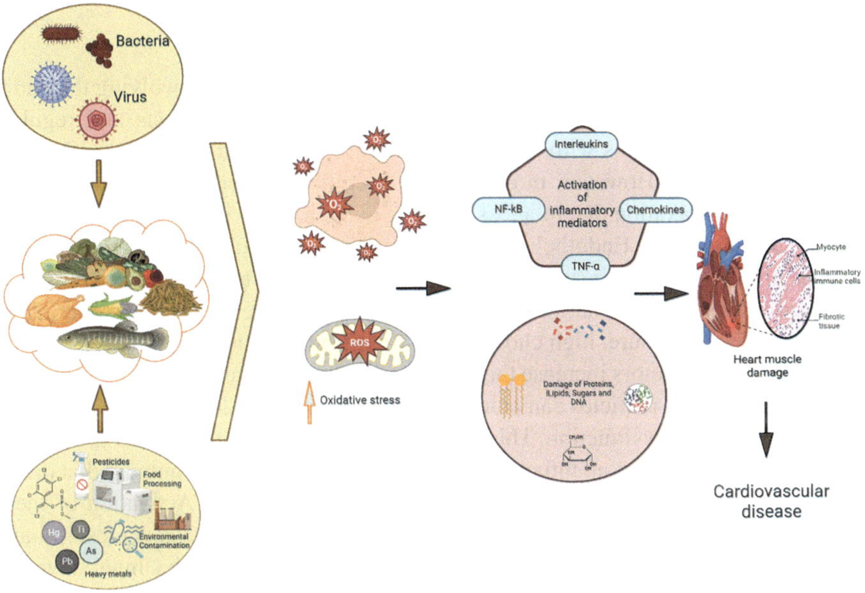

Fig. 13.4 Mechanism of CVD caused by oxidative stress

oxidative status of older adults with metabolic syndrome. In particular, immune cells (neutrophils) produced more ROS while key antioxidants like catalase and superoxide dismutase were less active. Furthermore, those who consumed more UPFs had higher levels of inflammation markers like TNFα, IL-6, IL-15, and leptin, suggesting a substantial shift toward more oxidative and inflammatory processes associated with a diet high in ultra-processed foods. One of the factors influencing tissue damage is oxidative stress. Oxidative stress is influenced by certain dietary practices and variables, which can result in oxidative lesions of proteins, lipids, sugars, and DNA. Contractile dysfunction and cardiomyocyte hypertrophy result from the oxidative change of major macromolecules found in cells in cardiomyocytes, which also alters subcellular organelles like the sarcolemma, mitochondria, sarcoplasmic reticulum, and nucleus (Quetglas-Llabrés et al. 2023).

Oxidative stress is found in the blood and heart muscle of patients with heart failure (HF) and is associated with deteriorating heart function. By encouraging aberrant heart cell growth, cell death, and other types of damage, the accumulation of ROS in the heart disrupts calcium handling, which can cause arrhythmias and heart damage. The mitochondria, uncoupled nitric oxide (NO) synthase, and NADPH oxidases (NOXs) are the sources of ROS in heart failure. These ROS sources have the potential to harm heart muscle and blood vessels. Notably, mitochondria act as key hubs that affect the balance of oxidative stress in heart function and disease by both producing their own ROS and amplifying ROS from other sources, such as NOXs (Münzel et al. 2017).

13.3.2 Endothelial Dysfunction

The term "endothelial dysfunction" describes several disorders involving the endothelium, including variations in its capacity to inhibit thrombosis and regulate inflammation, problems with blood vessel development, and problems with blood vessel adaptation. However, most research specifically describes a problem with blood vessel relaxation due to a loss of nitric oxide activity in the vessel walls (Cai and Harrison 2000). Endothelial dysfunction is a condition that occurs when the lining of blood vessels becomes damaged over time due to long-term exposure to various risk factors like harmful chemicals, changes in the immune system, diabetes, high blood pressure, high cholesterol, stress from oxidative damage, and smoking. When these factors increase the permeability of blood vessel walls, low-density lipoprotein (LDL) particles can enter the inner layer and undergo oxidation, resulting in endothelial dysfunction. This starts a chain reaction that alters the structure of the tissue and produces an inflammatory and prothrombotic environment that draws immune cells such as platelets, lymphocytes, and monocytes. Activation of these immune cells results in the formation of foam cells, which release cholesterol and exacerbate inflammation. Smooth muscle cells migrate into the inner layer in response to endothelial layer damage, where they release collagen and aid in the stabilization of plaques (Medina-Leyte et al. 2021; Lemini et al. 2024). The mechanism of CVD caused by endothelial dysfunction, is illustrated in (Fig. 13.5).

13.3.3 Inflammatory Pathway

It is well known that inflammation and the body's incapacity to control it play a major role in the onset and progression of several heart-related conditions (Ruparelia et al. 2017). Cardiovascular diseases can result from weakened immune systems and immune complex accumulation brought on by excessive exposure to toxic metals. The uncontrolled release of inflammatory cytokines is one of several interrelated processes that cause this. The mechanism of cell death caused by foodborne toxins through the inflammatory pathway is shown in (Fig. 13.6). For instance, by inducing circulating monocytes and tissue macrophages, cadmium (Cd) poisoning increases the production of pro-inflammatory cytokines such as IL-6, IL-2, and TNF-α. Higher oxidized low-density lipoprotein levels brought on by cadmium may be connected to this. By raising the mRNA levels of growth factors like granulocyte-macrophage colony-stimulating factor (GMCSF) and TGFα as well as inflammatory cytokines like TNF-α and IL-6, arsenic can directly cause atherosclerosis (Yang et al. 2021). Recently many reports suggest that mycotoxins cause toxic effects on cells by causing pores in the plasma membrane, cell swelling, leakage, and loss of cellular components which is called pyroptosis (Liao et al. 2024).

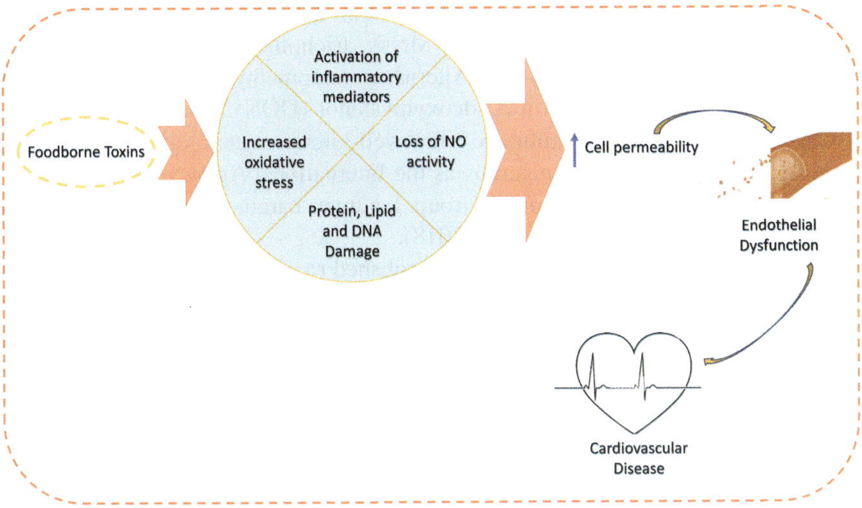

Fig. 13.5 Mechanism of CVD caused by endothelial dysfunction

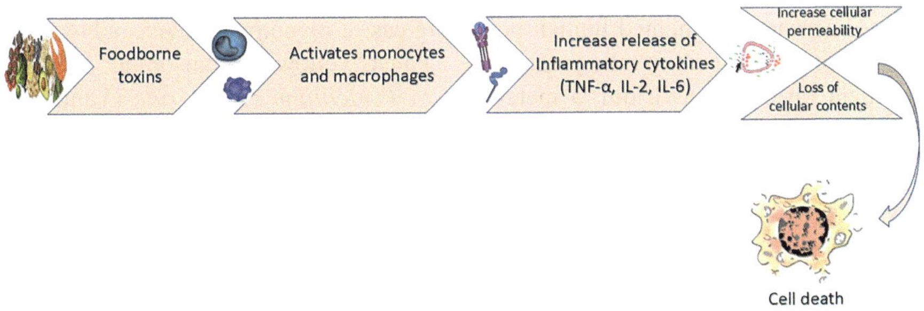

Fig. 13.6 Mechanism of cell death caused by foodborne toxins through inflammatory pathway

13.4 Key Toxins and Case Studies

13.4.1 Mycotoxins

Penicillium, Fusarium, and Aspergillus are filamentous fungi that produce mycotoxins, which are secondary metabolites that can cause major harm to animals and people (Umereweneza et al. 2018). Food and feed are contaminated by mycotoxin-producing fungi, which has negative socioeconomic and health effects (Atherstone et al. 2011; Abd-Elghany and Sallam 2015; Montanari et al. 2016). Food crops such as almonds, legumes, cereals (rice, corn, barley, wheat), peanuts, fruits, pistachios, walnuts, cotton seeds, coffee, meat, and spices (paprika, ginger, pepper) are all at risk from mycotoxin contamination worldwide (Rodrigues and Naehrer 2012; Darwish et al. 2014; Pleadin et al. 2019). Only 30 of the approximately 400 mycotoxins that have been identified to date are known to be dangerous to both animals and humans. Fumonisins (FMNs), trichothecenes (T-2/HT-2 toxins, Nivalenol (NIV), Alternaria toxins, Alternaria monomethyl ether (AME), citrinin (CIT), zearalenone, Mycotoxin A, deoxynivalenol (DON), altenuen (ALT), tentoxin, ergot alkaloids, and patulin are a few well-known mycotoxins. Among mycotoxins, AFs are especially dangerous as the International Agency for Research on Cancer (IARC) classifies them as Group 1 human carcinogens (Shephard 2008; Anfossi et al. 2016; Vila-Donat et al. 2018).

In 1891, Sakaki proved that moldy, unpolished rice killed experimental animals with signs indicating central nervous system paralysis, marking the beginning of mycotoxin study in Japan (Shoshin-kakke). Miyake and Takada initially documented in 1920 that feeding rabbits and rats Penicillium commune, the causative agent of "mossy diseased rice," was poisonous. Miyake et al. discovered the first sample of yellow rice grains from Taiwanese and domestic rice that included *Penicillium citreonigrum*, which produced citreoviridin. Tsunoda et al. discovered *Penicillium islandicum* among isolates from imported rice. This organism generated discolored, brownish rice that included two hepatotoxic compounds, cyclochlorotine and luteoskyrin. Ingestion of rice tainted with Penicillium species caused toxic abnormalities that were later dubbed "yellow rice syndrome". Even though there have been no documented human poisoning cases, mycotoxicologists advised the Japanese government to take precautions against "yellow rice poisoning," which is brought on by citreoviridin, a metabolite of *Penicillium citreoviride* (Yamakawa 1995; Udagawa and Tatsuno 2004; Je et al. 2005; Sakai et al. 2005; Richard 2007; Tanaka et al. 2007). The cardiotoxic mycotoxins are presented in (Fig. 13.7).

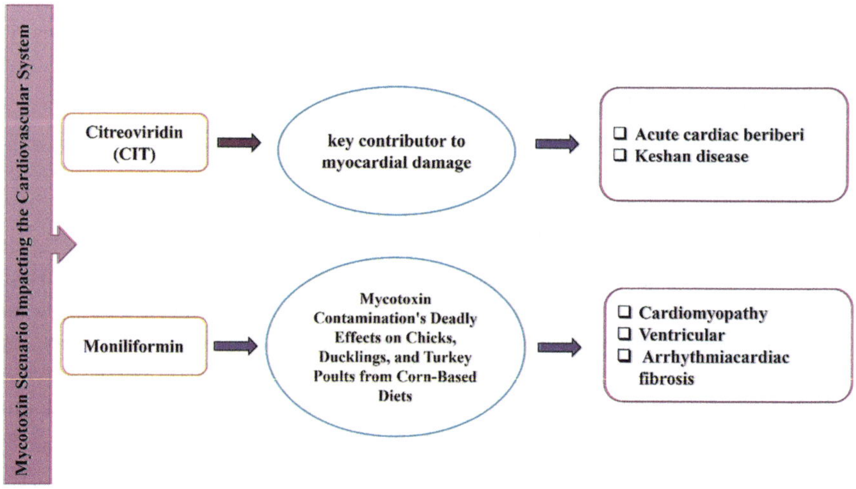

Fig. 13.7 Cardiotoxic mycotoxins

13.4.1.1 Citreoviridin's Function in Myocardial Damage

In adult Wistar rats exposed to it, citreoviridin is a key contributor to cardiac injury (degeneration and necrosis), and when combined with protein and selenium shortages, it can exacerbate the damage at the biochemical level (Zhang et al. 2012). Due to cardiac toxicity of citreoviridin in the selenium-protein-containing citreoviridin group, cardiac myocytes' intercalary discs were broken, myocyte conjunctions were broken, cardiac myocytes became edematous and even dissolved in some areas, sarcomere structure was unclear, mitochondrial cristae were loosened, myocyte cavities were occasionally observed, and there was a significant disseminated sarcoplasmic reticulum extension (Wan et al. 2011). The effects of citreoviridin on the cardiovascular system (CVS), as shown in (Fig. 13.8). A 15-year-old kid in Uganda died 2 days after being admitted to the hospital in 1967 after being diagnosed with heart failure as a result of aflatoxin poisoning. Along with additional symptoms like liver necrosis, an autopsy revealed pulmonary edema, flabby heart, and interstitial edema of the heart (Serck-Hanssen 1970). The cardiac muscle has been found to contain Afs (Murthy et al. 1975).

13.4.2 *Heavy Metals*

Environmental pollutants known as heavy metals are a growing ecological and public health hazard on a worldwide scale. The World Health Organization estimates that heavy metal toxicity is responsible for 60–0% of all acute and chronic illnesses.

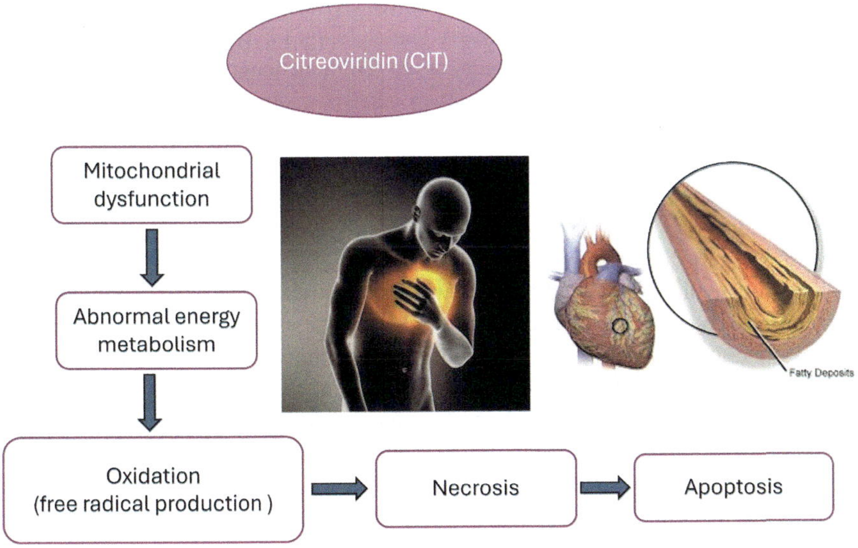

Fig. 13.8 Effects of citreoviridin on CVS

Chronic exposure to low and low-moderate levels of metals, such as lead (Pb), cadmium (Cd2+), and arsenic (As), is a major risk factor for CVD, which includes peripheral artery diseases (PAD), ischemic heart disease, and stroke (Lamas et al. 2023).

13.4.2.1 Arsenic

Due to its extensive presence in soil, water, and some foods, arsenic is a powerful carcinogenic and poisonous metalloid (Järup 2003). The amount of arsenic in crops is determined by the amount of this metal in the soil because it enters plants from the soil and tends to build up in their tissues. The connection between arsenic and cardiovascular disorders in adults is quite well-established, based on numerous studies. A causal link between high drinking water arsenic exposure and peripheral artery disease, stroke, coronary heart disease, and CVD was revealed by the revision of epidemiologic research about high-chronic arsenic exposure (Navas-Acien et al. 2005; Wang et al. 2007; Moon et al. 2012).

More carotid intima-media thickness in children was significantly correlated with higher total arsenic levels, according to a recent study (Gump et al. 2023). Additionally, a cohort study looked into the connection between CVD and heavy metals found in foods that are consumed in large quantities, like bread, rice, and vegetables (Neisi et al. 2024). Based on the assessment of heavy metals in urine, the authors found that, among other heavy metals that enter the body through contaminated food, arsenic levels in rice and vegetables are higher than the standard limitation value and are associated with CVDs. In cardiovascular patients with higher arsenic deposits in their fingers and higher cardiac tissue injury scores, telomere length was shorter, mitochondrial DNA (mtDNA) deletion was higher, and mitochondrial DNA copy number was lower. These changes may contribute to increased cardiovascular risk because short leukocyte telomere lengths and mtDNA length and position changes may promote atherogenesis (Sobenin et al. 2014; Yin and Pickering 2023; Khaleda et al. 2024).

13.4.2.2 Lead

The overall public is exposed to lead, which is distributed in approximately equal amounts in food and the air. The amount of lead in the environment has significantly increased due to several factors, including contaminated foods, drinking water, tobacco products, second-hand smoke, and old paint. The disease burden of CVDs, such as rheumatic heart disease, hypertension, and stroke ischemic is significantly influenced by blood lead levels of 5 μg/dL in Chinese adolescents and adults. One of the main causes of the illness load (68.1%) was dietary lead consumption (Li et al. 2022). Additionally, because of their high gastrointestinal absorption and blood-brain barrier permeability, children are vulnerable to lead exposure (Sanders et al. 2009).

13.4.3 Bacterial Endotoxins

13.4.3.1 Lipopolysaccharide

A common glycolipid found on the outer membrane of gram-negative bacteria is endotoxin, often known as lipopolysaccharide (LPS) (Ianaro and Tersigni 2009). Three separate sections make up LPS, and they vary in antigen, structure, and genetic makeup. The endotoxic action of LPS is caused by a hydrophobic membrane anchor known as Lipid A, whose structure is remarkably consistent across species (Calabrese et al. 2015). The O-antigen is a Sero-specific polymer made up of repeating oligosaccharide units, whereas the core oligosaccharide is a brief chain of sugar residues with many phosphoryl substituents. LPS quickly joins forces with LPS-binding protein after entering the bloodstream, and this combination attaches itself to a cluster of differentiation 14 (CD14), a monocyte surface high-affinity LPS receptor (Su 2025). As seen in (Fig. 13.9), endotoxins have an impact on heart function through oxidative stress, endothelial dysfunction, systemic inflammation, activation of Toll-like receptors (TLRs), elevated pro-inflammatory cytokines (e.g., IL-6, TNF-α), and myocardial depression. Arrhythmias decreased cardiac contractility, and an elevated risk of heart failure are all caused by these causes.

13.5 Prevention and Public Health Strategies

Over 600 million people fall ill each year as a result of foodborne diseases, which have increased in frequency over time and are now a major global public health concern (Faour-Klingbeil and Todd 2020; Chowdary Akkina et al. 2023). Thousands of diseases are caused by foodborne bacteria, and because of their detrimental impact on human health, they also pose a threat to the overall economy (Musa Moi et al. 2023). An enormous financial and health burden is thought to result from the common intestinal illnesses brought on by foodborne human pathogens (Ling et al.

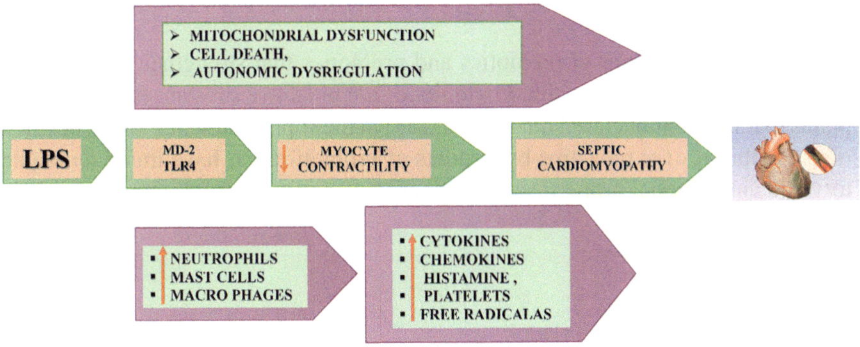

Fig. 13.9 Effects of endotoxins on heart function

2019; Akter et al. 2021; Prata et al. 2021; Qiu et al. 2021). About 30% of foodborne deaths occur in children under the age of 5, according to the WHO. Microbes that are found in food can cause mild to severe symptoms like diarrhea or debilitating illnesses like meningitis. The reason for this is that small children's immune systems are weaker and thus are more vulnerable to foodborne disease. Furthermore, children might be more likely to put their hands in their mouths or consume food contaminated with bacteria, viruses, or parasites (Scallan Walter et al. 2020; Belina et al. 2021; Van Puyvelde et al. 2021).

13.5.1 Management of Foodborne Microbiological Hazards

Bacteriophages' Function in the Fight Against Foodborne Infections: A virus that only targets and multiplies in bacteria, known as a bacteriophage or phage, is associated with bacterial illnesses (Domingo-Calap and Delgado-Martínez 2018; Jamal et al. 2019). Given that several phages have been discovered in a wide range of foods, this may be one of the ways they naturally enter human bodies (Campbell 2003; Ramos-Vivas et al. 2021). Since phage-based strategies have all the advantages of conventional antibacterial methods, such as affordability and ease of use, there is no reason not to use them in the future. Numerous research (Ferguson et al. 2013; Kang et al. 2013) have suggested that phages could be an effective strategy for eradicating foodborne bacteria that cause food poisoning.

We believe that preventing food contamination is one of the largest difficulties the world is now facing and that we must act fast to solve this problem by looking for novel solutions to maintain rigorous food security standards. Without question, one of the biggest problems facing the world today is preserving the safety of food products. To meet the highest standards for food safety, we need to find innovative solutions (Bigot et al. 2011). The following are the methods for controlling foodborne microbiological pathogens, as illustrated in (Fig. 13.10).

Prebiotics are indigestible food ingredients that promote the development of good gut flora and strengthen defenses against infections. Probiotics are live, helpful bacteria that support the balance of the gut microbiota and prevent the growth of dangerous pathogens.

Synbiotics are a mix of prebiotics and probiotics that strengthen antimicrobial defense and promote gut health. Phytobiotics are bioactive substances derived from plants that have antibacterial qualities and aid in the suppression of foodborne infections.Postbiotics are bioactive byproducts of probiotics that have immunomodulatory and antibacterial properties and support gut health and food safety.

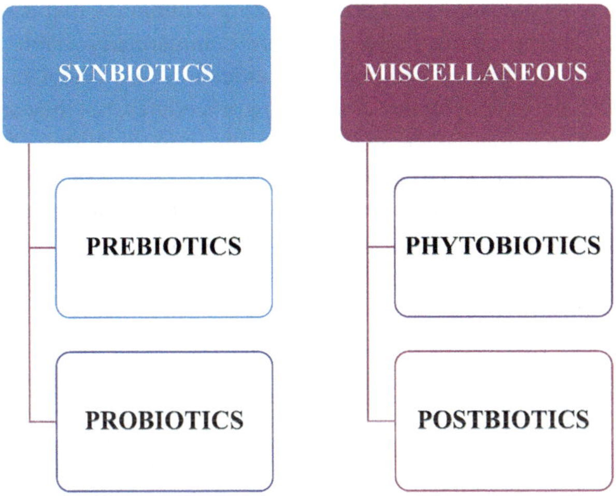

Fig. 13.10 Strategies for the management of foodborne microbiological pathogens

13.6 Future Directions

Behaviour changes can help address the worldwide health challenge of cardiovascular disease. A crucial aspect of lifestyle, diet influences cardiovascular risk through body weight and numerous other mechanisms. The vast majority of nutritional literature published in the past few decades stresses avoiding excessive calorie intake, increasing intake of fruits, whole grains, nuts, vegetables, legumes, and fish, consuming coffee, alcohol, and low-fat or fat-free dairy products in moderation, and consuming fewer processed meats, sugar-sweetened beverages, sodium, and refined grains. Economic, biological, psychological, social, and physical factors all impact food choices. Long-term eating habits can be significantly improved by interventions that focus on these variables. Health professionals and community leaders have a big role to play in promoting cardiovascular health and disease prevention, but they also need to have a basic understanding of nutrition. To radically alter the current food environment and the global food system, a coordinated effort from all societal levels will be required (Wang et al. 2014).

13.7 Conclusion

Food security is a major worldwide concern that remains at the top of most nations' development objectives. Because more people are consuming freshly cooked food and food products with a short shelf life, there has been an increase in demand in recent decades for techniques that identify foodborne infections. To address public health and food safety concerns, a variety of technologies for detecting foodborne microbes have been developed. Early screening techniques that have been developed recently are becoming more and more commercially viable. Conventional approaches to foodborne pathogen identification are labor-intensive, time-consuming, and dependent on intricate culture procedures, but they are nevertheless widely used and regarded as the gold standard in most situations. Phages, probiotics, prebiotics, synbiotics, phytobiotics, postbiotics, and probiotics have all been shown in numerous studies to be highly successful in eliminating a variety of foodborne pathogens while also being non-toxic to humans.

References

Abd-Elghany SM, Sallam KI (2015) Rapid determination of total aflatoxins and ochratoxins a in meat products by immuno-affinity fluorimetry. Food Chem 179:253–256. https://doi.org/10.1016/j.foodchem.2015.01.140

Akter R, Rahman MH, Bhattacharya T et al (2021) Novel coronavirus pathogen in humans and animals: an overview on its social impact, economic impact, and potential treatments. Environ Sci Pollut Res 28:68071–68089. https://doi.org/10.1007/s11356-021-16809-8

Alameri MM, Kong ASY, Aljaafari MN et al (2023) Aflatoxin contamination: an overview on health issues, detection and management strategies. Toxins (Basel) 15(4):246. https://doi.org/10.3390/toxins15040246

Anfossi L, Giovannoli C, Baggiani C (2016) Mycotoxin detection. Curr Opin Biotechnol 37:120–126. https://doi.org/10.1016/j.copbio.2015.11.005

Atherstone C, Grace D, Waliyar F et al (2011) Aflatoxin literature synthesis and risk mapping: special emphasis on sub-Saharan Africa. In: Aflatoxins–Biochemistry and Molecular Biology, Guevara-Gonzalez RG (eds) Aflatoxins and Aflatoxicosis in human and animals, vol 105. Tech Europe, Rijeka, pp 221–254

Belina D, Hailu Y, Gobena T et al (2021) Prevalence and epidemiological distribution of selected foodborne pathogens in human and different environmental samples in Ethiopia: a systematic review and meta-analysis. One Health Outlook 3:19. https://doi.org/10.1186/s42522-021-00048-5

Bigot B, Lee WJ, McIntyre L et al (2011) Control of listeria monocytogenes growth in a ready-to-eat poultry product using a bacteriophage. Food Microbiol 28:1448–1452. https://doi.org/10.1016/j.fm.2011.07.001

Biswas SK (2016) Does the interdependence between oxidative stress and inflammation explain the antioxidant paradox? Oxidative Med Cell Longev 2016:5698931. https://doi.org/10.1155/2016/5698931

Bryden WL (2007) Mycotoxins in the food chain: human health implications. Asia Pac J Clin Nutr 16:95–101

Bubien JK, Lundeen G, Templeton C, Woods WT (2017) Effects on the circulatory system. In: Trichothecene mycotoxicosis pathophysiologic effects, vol 2. CRC Press, Boca Raton, pp 91–109. https://doi.org/10.1201/9781315121260

Cai H, Harrison DG (2000) Endothelial dysfunction in cardiovascular diseases: the role of oxidant stress. Circ Res 87:840–844. https://doi.org/10.1161/01.RES.87.10.840

Calabrese V, Cighetti R, Peri F (2015) Molecular simplification of lipid a structure: TLR4-modulating cationic and anionic amphiphiles. Mol Immunol 63:153–161. https://doi.org/10.1016/j.molimm.2014.05.011

Campbell A (2003) The future of bacteriophage biology. Nat Rev Genet 4:471–477. https://doi.org/10.1038/nrg1089

Carvajal Moreno M (2016) Mycotoxins that affect the human cardiovascular system. J Biomol Res Ther 05. https://doi.org/10.4172/2167-7956.1000124

Chowdary Akkina R, Payala V, Sushma Maganti S (2023) Tools for rapid detection and control of foodborne microbial pathogens. In: Foodborne pathogens – recent advances in control and detection. IntechOpen. https://doi.org/10.5772/intechopen.103938

Dai C, Das Gupta S, Wang Z et al (2022) T-2 toxin and its cardiotoxicity: new insights on the molecular mechanisms and therapeutic implications. Food Chem Toxicol 167:113262. https://doi.org/10.1016/j.fct.2022.113262

Darwish WS, Ikenaka Y, Nakayama SMM, Ishizuka M (2014) An overview on mycotoxin contamination of foods in Africa. J Vet Med Sci 76:789–797. https://doi.org/10.1292/jvms.13-0563

Domingo-Calap P, Delgado-Martínez J (2018) Bacteriophages: protagonists of a post-antibiotic era. Antibiotics 7:66. https://doi.org/10.3390/antibiotics7030066

Faour-Klingbeil D, Todd ECD (2020) Prevention and control of foodborne diseases in middle-east north african countries: review of national control systems. Int J Environ Res Public Health 17:70. https://doi.org/10.3390/ijerph17010070

Feng C, Li D, Chen M et al (2019) Citreoviridin induces myocardial apoptosis through PPAR-γ-mTORC2-mediated autophagic pathway and the protective effect of thiamine and selenium. Chem Biol Interact 311:108795. https://doi.org/10.1016/j.cbi.2019.108795

Ferguson S, Roberts C, Handy E, Sharma M (2013) Lytic bacteriophages reduce Escherichia coli O157: H7 on fresh cut lettuce introduced through cross-contamination. Bacteriophage 3:e24323

Goel A, Ram VJ (2009) Natural and synthetic 2H-pyran-2-ones and their versatility in organic synthesis. Tetrahedron 65:7865–7913. https://doi.org/10.1016/j.tet.2009.06.031

Gump BB, Heffernan K, Brann LS et al (2023) Exposure to arsenic and subclinical cardiovascular disease in 9- to 11-year-old children, Syracuse. New York JAMA Netw Open 6:E2321379. https://doi.org/10.1001/jamanetworkopen.2023.21379

Guo L, Li F, Tang G et al (2023) Association of ultra-processed foods consumption with risk of cardio-cerebrovascular disease: a systematic review and meta-analysis of cohort studies. Nutr Metab Cardiovasc Dis 33:2076–2088. https://doi.org/10.1016/j.numecd.2023.07.005

Hamad GM, Mehany T, Simal-Gandara J et al (2023) A review of recent innovative strategies for controlling mycotoxins in foods. Food Control 144:109350. https://doi.org/10.1016/j.foodcont.2022.109350

Ianaro A, Tersigni MDF (2009) New in sight in LPS antagonist. Mini Rev Med Chem 9(3):306–317

IARC Working Group on the Evaluation of Carcinogenic Risks to Humans (2002) Some traditional herbal medicines, some mycotoxins, naphthalene and styrene. IARC monographs on the evaluation of carcinogenic risks to humans. IARC Monogr Eval Carcinog Risks Hum 82:1–556

Iyengar GV, Nair PP (2000) Global outlook on nutrition and the environment: meeting the challenges of the next millennium. Sci Total Environ 249:331–346. https://doi.org/10.1016/S0048-9697(99)00529-X

Jamal M, Bukhari SMAUS, Andleeb S et al (2019) Bacteriophages: an overview of the control strategies against multiple bacterial infections in different fields. J Basic Microbiol 59:123–133. https://doi.org/10.1002/jobm.201800412

Järup L (2003) Hazards of heavy metal contamination. Br Med Bull 68:167–182. https://doi.org/10.1093/bmb/ldg032

Je WP, Choi SY, Hwang HJ, Kim YB (2005) Fungal mycoflora and mycotoxins in Korean polished rice destined for humans. Int J Food Microbiol 103:305–314. https://doi.org/10.1016/j.ijfoodmicro.2005.02.001

Jiang Y, Li X, Zhang Y et al (2024) Mechanism of action of anthocyanin on the detoxification of foodborne contaminants—a review of recent literature. Compr Rev Food Sci Food Saf 23:1–33. https://doi.org/10.1111/1541-4337.13259

Kang HW, Kim JW, Jung TS, Woo GJ (2013) wksl3, a new biocontrol agent for salmonella enterica serovars enteritidis and typhimurium in foods: characterization, application, sequence analysis,

and oral acute toxicity study. Appl Environ Microbiol 79:1956–1968. https://doi.org/10.1128/AEM.02793-12
Khaleda L, Begum SK, Apu MAR et al (2024) Arsenic-induced cardiovascular diseases and their correlation with mitochondrial DNA copy number, deletion, and telomere length in Bangladeshi population. Cardiovasc Toxicol 24:27–40. https://doi.org/10.1007/s12012-023-09812-7
Khan S, Cao Q, Zheng YM et al (2008) Health risks of heavy metals in contaminated soils and food crops irrigated with wastewater in Beijing, China. Environ Pollut 152:686–692. https://doi.org/10.1016/j.envpol.2007.06.056
Lamas GA, Bhatnagar A, Jones MR et al (2023) Contaminant metals as cardiovascular risk factors: a scientific statement from the American Heart Association. J Am Heart Assoc 12:e029852. https://doi.org/10.1161/JAHA.122.026295
Lee MJ, Henderson SB, Clermont H et al (2024) The health risks of marine biotoxins associated with high seafood consumption: looking beyond the single dose, single outcome paradigm with a view towards addressing the needs of coastal indigenous populations in British Columbia. Heliyon 10:e27146. https://doi.org/10.1016/j.heliyon.2024.e27146
Lemini C, Silveyra P, Segovia-Mendoza M (2024) Cardiovascular disrupting effects of bisphenols, phthalates, and parabens related to endothelial dysfunction: review of toxicological and pharmacological mechanisms. Environ Toxicol Pharmacol 107:104407. https://doi.org/10.1016/j.etap.2024.104407
Li YL, Yan YZ, Liu JL (2022) Burden of cardiovascular disease attributable to dietary lead exposure among adults in China, 2017. Chin J Public Health 38:557–562. https://doi.org/10.11847/zgggws1137343
Liao C, Xu F, Yu Z et al (2024) The novel role of the NLRP3 Inflammasome in mycotoxin-induced toxicological mechanisms. Vet Sci 11:291. https://doi.org/10.3390/vetsci11070291
Lind L, Araujo JA, Barehowsky A et al (2021) Key characteristics of cardiovascular toxicants. Environ Health Perspect 129:95001. https://doi.org/10.1289/EHP9321
Ling Y, Zhu Y, Fan H et al (2019) Rapid method for detection of Staphylococcus aureus in feces. J Biomed Nanotechnol 16:1290–1298. https://doi.org/10.1166/jbn.2019.2781
Magnuson BA, Schiefer HB, Hancock DS, Bhatti AR (1987) Cardiovascular effects of mycotoxin T-2 after topical application in rats. Can J Physiol Pharmacol 65:799–802. https://doi.org/10.1139/y87-128
Mancuso G, Violi F, Nocella C (2024) Food contamination and cardiovascular disease: a narrative review. Intern Emerg Med 19:1693–1703. https://doi.org/10.1007/s11739-024-03610-x
Medina-Leyte DJ, Zepeda-García O, Domínguez-Pérez M et al (2021) Endothelial dysfunction, inflammation and coronary artery disease: potential biomarkers and promising therapeutical approaches. Int J Mol Sci 22:3850. https://doi.org/10.3390/ijms22083850
Mensah GA, Roth GA, Fuster V (2019) The global burden of cardiovascular diseases and risk factors: 2020 and beyond. J Am Coll Cardiol 74:2529–2532. https://doi.org/10.1016/j.jacc.2019.10.009
Mohamed AM, Metwally NS (2009) Antiaflatoxigenic activities of some plant aqueous extracts against aflatoxin-B1 induced renal and cardiac damage. J Pharmacol Toxicol 4:1–16. https://doi.org/10.3923/jpt.2009.1.16
Montanari C, Bargossi E, Gardini A et al (2016) Correlation between volatile profiles of Italian fermented sausages and their size and starter culture. Food Chem 192:736–744. https://doi.org/10.1016/j.foodchem.2015.07.062
Monteiro CA, Cannon G, Levy RB et al (2019) Ultra-processed foods: what they are and how to identify them. Public Health Nutr 22:936–941. https://doi.org/10.1017/S1368980018003762
Moon K, Guallar E, Navas-Acien A (2012) Arsenic exposure and cardiovascular disease: an updated systematic review. Curr Atheroscler Rep 14:542–555. https://doi.org/10.1007/s11883-012-0280-x
Morabito S, Silvestro S, Faggio C (2018) How the marine biotoxins affect human health. Nat Prod Res 32:621–631. https://doi.org/10.1080/14786419.2017.1329734

Munir N, Jahangeer M, Bouyahya A et al (2022) Heavy metal contamination of natural foods is a serious health issue: a review. Sustainability (Switzerland) 14:161. https://doi.org/10.3390/su14010161
Münzel T, Camici GG, Maack C et al (2017) Impact of oxidative stress on the heart and vasculature: part 2 of a 3-part series. J Am Coll Cardiol 70:212–229. https://doi.org/10.1016/j.jacc.2017.05.035
Murthy TR, Jemmali M, Henry Y, Frayssinet C (1975) Aflatoxin residues in tissues of growing swine: effect of separate and mixed feeding of protein and protein-free portions of the diet. J Anim Sci 41:1339–1347. https://doi.org/10.2527/jas1975.4151339x
Musa Moi I, Ibrahim Z, Mohammed Abubakar B et al (2023) Properties of foodborne pathogens and their diseases. In: Foodborne pathogens – recent advances in control and detection. IntechOpen. https://doi.org/10.5772/intechopen.105694
Navas-Acien A, Sharrett AR, Silbergeld EK et al (2005) Arsenic exposure and cardiovascular disease: a systematic review of the epidemiologic evidence. Am J Epidemiol 162:1037–1049. https://doi.org/10.1093/aje/kwi330
Neisi A, Farhadi M, Cheraghian B et al (2024) Consumption of foods contaminated with heavy metals and their association with cardiovascular disease (CVD) using GAM software (cohort study). Heliyon 10:e24517. https://doi.org/10.1016/j.heliyon.2024.e24517
Peng WX, Marchal JLM, van der Poel AFB (2018) Strategies to prevent and reduce mycotoxins for compound feed manufacturing. Anim Feed Sci Technol 237:129–153. https://doi.org/10.1016/j.anifeedsci.2018.01.017
Pleadin J, Frece J, Markov K (2019) Mycotoxins in food and feed. Adv Food Nutr Res 89:297–345. https://doi.org/10.1016/bs.afnr.2019.02.007
Prata JC, da Costa JP, Lopes I et al (2021) A one health perspective of the impacts of microplastics on animal, human and environmental health. Sci Total Environ 777:146094. https://doi.org/10.1016/j.scitotenv.2021.146094
Qiu H, Zhang Y, Li Z et al (2021) Donepezil ameliorates pulmonary arterial hypertension by inhibiting M2-macrophage activation. Front Cardiovasc Med 8:639541. https://doi.org/10.3389/fcvm.2021.639541
Quetglas-Llabrés MM, Monserrat-Mesquida M, Bouzas C et al (2023) Oxidative stress and inflammatory biomarkers are related to high intake of ultra-processed food in old adults with metabolic syndrome. Antioxidants 12:1532. https://doi.org/10.3390/antiox12081532
Ramos-Vivas J, Elexpuru-Zabaleta M, Samano ML et al (2021) Phages and enzybiotics in food biopreservation. Molecules 26:5138. https://doi.org/10.3390/molecules26175138
Rather IA, Koh WY, Paek WK, Lim J (2017) The sources of chemical contaminants in food and their health implications. Front Pharmacol 8:830. https://doi.org/10.3389/fphar.2017.00830
Richard JL (2007) Some major mycotoxins and their mycotoxicoses-an overview. Int J Food Microbiol 119:3–10. https://doi.org/10.1016/j.ijfoodmicro.2007.07.019
Rodrigues I, Naehrer K (2012) A three-year survey on the worldwide occurrence of mycotoxins in feedstuffs and feed. Toxins (Basel) 4:663–675. https://doi.org/10.3390/toxins4090663
Roth GA, Mensah GA, Fuster V (2020) The global burden of cardiovascular diseases and risks: a compass for global action. J Am Coll Cardiol 76:2980–2981. https://doi.org/10.1016/j.jacc.2020.11.021
Ruparelia N, Chai JT, Fisher EA, Choudhury RP (2017) Inflammatory processes in cardiovascular disease: a route to targeted therapies. Nat Rev Cardiol 14:133–144. https://doi.org/10.1038/nrcardio.2016.185
Sakai A, Tanaka H, Konishi Y et al (2005) Mycological examination of domestic unpolished rice and mycotoxin production by isolated Penicillium islandicum. J Food Hyg Soc Jpn 46:205–212. https://doi.org/10.3358/shokueishi.46.205
Sanders T, Liu Y, Buchner V, Tchounwou PB (2009) Neurotoxic effects and biomarkers of lead exposure: a review. Rev Environ Health 24:15–45. https://doi.org/10.1515/REVEH.2009.24.1.15

Scallan Walter EJ, McLean HQ, Griffin PM (2020) Hospital discharge data Underascertain enteric bacterial infections among children. Foodborne Pathog Dis 17:530–532. https://doi.org/10.1089/fpd.2019.2773

Serck-Hanssen A (1970) Aflatoxin-induced fatal hepatitis? A case report from Uganda. Arch Environ Health 20(6):729–731

Shephard GS (2008) Impact of mycotoxins on human health in developing countries. Food Addit Contam Part A Chem Anal Control Expo Risk Assess 25:146–151. https://doi.org/10.1080/02652030701567442

Sobenin I, Zhelankin A, Mitrofanov K et al (2014) Mutations of mitochondrial DNA in atherosclerosis and atherosclerosis-related diseases. Curr Pharm Des 21:1158–1163. https://doi.org/10.2174/1381612820666141013133000

Su J (2025) Toll-like receptor signaling in teleosts. Sci China Life Sci. https://doi.org/10.1007/s11427-024-2822-5

Tanaka K, Sago Y, Zheng Y et al (2007) Mycotoxins in rice. Int J Food Microbiol 119:59–66. https://doi.org/10.1016/j.ijfoodmicro.2007.08.002

Thompson LA, Darwish WS (2019) Environmental chemical contaminants in food: review of a global problem. J Toxicol 2019:1–14. https://doi.org/10.1155/2019/2345283

Türkdoğan MK, Kilicel F, Kara K et al (2003) Heavy metals in soil, vegetables and fruits in the endemic upper gastrointestinal cancer region of Turkey. Environ Toxicol Pharmacol 13:175–179. https://doi.org/10.1016/S1382-6689(02)00156-4

Udagawa S-I, Tatsuno T (2004) Safety of rice grains and mycotoxins – a historical review of yellow rice mycotoxicoses. Yakushigaku zasshi 39:321–342

Umereweneza D, Kamizikunze T, Muhizi T (2018) Assessment of mycotoxins types in some foodstuff consumed in Rwanda. Food Control 85:432–436. https://doi.org/10.1016/j.foodcont.2017.10.015

Vaduganathan M, Mensah GA, Turco JV et al (2022) The global burden of cardiovascular diseases and risk: a compass for future health. J Am Coll Cardiol 80:2361–2371. https://doi.org/10.1016/j.jacc.2022.11.005

Van Puyvelde L, Aissa A, Panda SK et al (2021) Bioassay-guided isolation of antibacterial compounds from the leaves of Tetradenia riparia with potential bactericidal effects on food-borne pathogens. J Ethnopharmacol 273:113956. https://doi.org/10.1016/j.jep.2021.113956

Vila-Donat P, Marín S, Sanchis V, Ramos AJ (2018) A review of the mycotoxin adsorbing agents, with an emphasis on their multi-binding capacity, for animal feed decontamination. Food Chem Toxicol 114:246–259. https://doi.org/10.1016/j.fct.2018.02.044

Vilariño N, Louzao MC, Abal P et al (2018) Human poisoning from marine toxins: unknowns for optimal consumer protection. Toxins (Basel) 10:324. https://doi.org/10.3390/toxins10080324

Visciano P, Schirone M, Berti M et al (2016) Marine biotoxins: occurrence, toxicity, regulatory limits and reference methods. Front Microbiol 7:1051. https://doi.org/10.3389/fmicb.2016.01051

Wan YZ, Wang HM, Yu JN et al (2011) Citreoviridin-induced rat myocardial damage under combined selenium and protein deficiency. Chin J Endemiol 30:379–383. https://doi.org/10.3760/cma.j.issn.1000-4955.2011.04.008

Wang CH, Hsiao CK, Chen CL et al (2007) A review of the epidemiologic literature on the role of environmental arsenic exposure and cardiovascular diseases. Toxicol Appl Pharmacol 222:315–326. https://doi.org/10.1016/j.taap.2006.12.022

Wang DD, Leung CW, Li Y et al (2014) Trends in dietary quality among adults in the United States, 1999 through 2010. JAMA Intern Med 174:1587–1595. https://doi.org/10.1001/jamainternmed.2014.3422

Wang C, He J, Jin H et al (2023) T-2 toxin induces cardiotoxicity by activating ferroptosis and inhibiting heme oxygenase-1. Chemosphere 341:140087. https://doi.org/10.1016/j.chemosphere.2023.140087

Yamakawa K (1995) Historical sketch of modern pharmaceutical science and technology (Part 3). From the second half of the 19th century to world war II. Yakushigaku zasshi 30:1–10

Yang Y, Wei S, Zhang B, Li W (2021) Recent Progress in environmental toxins-induced cardiotoxicity and protective potential of natural products. Front Pharmacol 12:699193. https://doi.org/10.3389/fphar.2021.699193
Yin H, Pickering JG (2023) Telomere length: implications for Atherogenesis. Curr Atheroscler Rep 25:95–103. https://doi.org/10.1007/s11883-023-01082-6
Zhang D, Wei C, Yu JN et al (2012) Morphological changes of rat cardiac tissue caused by citreoviridin combined with selenium and protein deficiency. Chin J Endemiol 31:389–392. https://doi.org/10.3760/cma.j.issn.1000-4955.2012.04.011

Chapter 14
Haematological Consequences of Food Contaminants

Subhajit Das, Rajdip Goswami, Ayan Chatterjee, and Mahenderan Appukutty

14.1 Introduction

The entry of food contamination into the public health domain is one huge step that unveils the pernicious effects on the human health of a greater common denial (Cho and Kim 2011). The occurrence of various harmful chemicals, including heavy metals, pesticides, mycotoxins, and chemicals formed during food processing, can considerably harm several biological systems, with special regard for the hematological system. These chemicals enter the food chain via several routes, including environmental pollution, agricultural inputs, storage conditions, and food processing (Langyan et al. 2022). We now have to face this unfortunate reality of exposure to such life-threatening contaminants with the food globalization era of combined outbound production and flow to a much deeper understanding of their profound effects on health (Cho and Kim 2011). The hematological system, consisting of red blood cells, white blood cells, and platelets together with plasma, performs an important function in sustaining physiological homeostasis. Its functions include oxygen delivery, immune defense, and clotting functions, all highly interdependent with sustaining life. An imbalance of any of these cellular mechanisms will create severe

S. Das
Department of Allied Health Sciences, School of Health and Medical Sciences, Adamas University, Barasat, Kolkata, WB, India

R. Goswami
Department of Pharmaceutical Technology, School of Health and Medical Sciences, Adamas University, Barasat, Kolkata, WB, India

A. Chatterjee (✉)
School of Health Science and Technology, Medhavi Skills University, Sikkim, India

M. Appukutty
Faculty of Sports Science and Recreation, Universiti Teknologi Mara, Selangor, Malaysia

T. Sarkar et al. (eds.), *Physiological Perspectives on Food Safety: Exploring the Intersection of Health and Nutrition*, https://doi.org/10.1007/978-3-031-94582-3_14

health situations in the form of anaemia, immunosuppression, and bleeding disorders. There have been studies showing that acquaintance with certain food contaminants can inhibit hematopoiesis, which is the process of blood cell formation, thereby causing dysregulation of immunity and coagulation mechanisms. Full consideration of these effects is imperative in evolving strategies aimed at mitigating the health risks associated with consumption of contaminated food. Heavy metals, including lead, cadmium, mercury, and arsenic, are the most worrisome of all distractions because they do not degrade and can accumulate in tissues biologically (Vignaud et al. 2023). These metals find their way into the food chain through water, soil, and air contamination, poisoning the crops and livestock. Upon ingestion, heavy metals interfere with the functioning of the enzymatic mechanisms, disturb cellular metabolism, and create oxidative stress, resulting in hematological aberrations. For instance, lead exposure has been strongly tied to anemia due to its inhibition of heme synthesis and reduction of the lifespan of red blood cells. Cadmium and mercury, on the other hand, are known to suppress bone marrow functions, leading to a decrease in the production of essential blood cells and increasing the risk of immunosuppression and thrombocytopenia. Insecticide residues, another set of major food contaminants, arise from high quantities of the application of chemical pesticides. These chemicals are meant for the short-term protection of crops from pests and diseases, although their persistency in food products ultimately impacts the health of the consumers. Such insecticides are known to block immune cell functions while imparting an inhibitory action on hematopoiesis. Certain organophosphate and carbamate pesticides are known to inhibit the activity of acetylcholinesterase, which could result in hematotoxicity effects such as leukopenia and impaired immune responses (Karabulut et al. 2024). They have also been linked with chronic exposure related to suppression of bone marrow, resulting in reduced.

14.2 Food Contaminants and Their Sources

Food contaminants are thus seen to have an erratic origin and pose a wide-ranging threat to human health. These contaminants can be categorized into environmental pollutants, agricultural chemicals, naturally occurring toxins, and industrial impacts, each one causing food contamination right from its production to consumption (Pereira 2018). Environmental contaminants, such as heavy metals and persistent organic pollutants, enter the food chain through soil, water, and air. Industrial works, mining activities, and waste disposal encourage the collection of toxic metals like lead, cadmium, mercury, and arsenic in agricultural land and water sources (Table 14.1). Plants absorb these metals and subsequently get into the human diet through the consumption of contaminated fruits, vegetables, and animal products (Silveira et al. 2019). Also, persistent organic pollutants like polychlorinated biphenyls (PCBs) and dioxins accumulate in animal fat, putting human health at risk. Agricultural practice is another vital source of food contamination. Excessive application of chemical pesticides and herbicides cause the accumulation of residues in

Table 14.1 The sources and effects of food contaminants (Domingo et al. 2023)

Category	Examples	Sources	Health effects
Environmental contaminants	Heavy metals (lead, cadmium, mercury, arsenic)	Soil, water, air pollution from industries, mining, waste disposal	Neurotoxicity, kidney damage, cardiovascular diseases
Agricultural chemicals	Pesticides, herbicides	Excessive use in farming, residue accumulation in crops	Neurotoxicity, hormonal imbalances, cancer
Naturally occurring toxins	Mycotoxins (aflatoxins, fumonisins, ochratoxins)	Fungal contamination in grains, nuts, dairy products	Liver toxicity, immune suppression, hematological disorders
Food processing contaminants	Acrylamide, polycyclic aromatic hydrocarbons (PAHs)	High-temperature cooking, frying, roasting	Carcinogenic effects, neurological damage
Packaging contaminants	Bisphenol A (BPA), phthalates	Plastic food containers, cans, packaging materials	Endocrine disruption, reproductive toxicity

food crops (Bao et al. 2022). While these chemicals serve the purpose of controlling pests, their persistence in the environment and food chain may result in severe health effects on humans. Moreover, fertilizers containing high levels of nitrates and heavy metals contribute to soil and water contamination, thus affecting the safety of food. Naturally occurring toxins are another important source of contamination (Mata 2021). Toxic compounds are produced by fungal species, especially Aspergillus, Fusarium, and Penicillium, or aflatoxins, fumonisins, and ochratoxins. These toxins contaminate grains, nuts, and dairy products, especially under poor storage conditions. The health risks posed by mycotoxin ingestion include severe immunosuppression and hematological disorders. Processing of food can also lead to various contaminants (Pfeifer et al. 2023). Health consequences arise because of the formation of other toxic byproducts during high-temperature cooking, like acrylamide and polycyclic aromatic hydrocarbons. Packing materials such as bisphenol A (BPA) and phthalates may leach into foodswhen destined for chemical exposure. Understanding the various sources of food contaminants will help define the routes for designing potent approaches that may avert introduction into foods and enhance safety (Chabili et al. 2024). Effective regulatory measures, better agricultural practices, and consumer awareness are critical in addressing the challenges surrounding food contamination.

14.3 Effect of Heavy Metals on the Haematological System

Heavy metals pose grave risks to health in human beings specifically the haematological system. These metals, including lead, mercury, cadmium, and arsenic, impair normal blood functions, thereby leading to various disorders. The toxic effects stem mainly from their ability to obstruct the formation of red blood cells or white blood cells, affect the synthesis of haemoglobin, and consequently change the immune responses (Domingo et al. 2023).

14.3.1 *Effect on Red and White Blood Cells*

The effect of heavy metals on the body creates an important hazard to human health, especially in connection with the haematological system. These metals include lead, mercury, cadmium, and arsenic; all of which interrupt the normal functions of blood and lead to various syndromes (Table 14.2). The toxic effects mainly arise from the metals' capabilities to interfere in the production of red or white blood cells, damage haemoglobin synthesis, and alter immune responses (Karimi et al. 2018). If any of these effects are to be adequately studied, it will go a long way in creating good anticipation and subsequent management programs to mitigate the adverse health effects that are adjoining heavy metal toxicity. Probably the most widespread haematological effect relates to the development of anaemic conditions due to heavy metal poisoning. Through interfering with some of the critical enzymes that are responsible for heme production, certain heavy metals, especially lead, inhibit the synthesis of haemoglobin (Salbitani and Carfagna 2021). This intrusion decreases

Table 14.2 Haematological effects of heavy metals (Dolganyuk et al. 2023)

Heavy metal	Sources	Haematological effects	Mechanism of toxicity
Lead (Pb)	Industrial emissions, lead-based paints, contaminated water, battery manufacturing	Microcytic anemia, impaired erythropoiesis, hemolysis	Inhibits heme synthesis, disrupts iron metabolism, induces oxidative stress
Mercury (Hg)	Contaminated seafood, industrial waste, dental amalgams	Leukocyte dysfunction, immune suppression, anemia	Disrupts bone marrow function, induces oxidative damage in RBCs and WBCs
Cadmium (Cd)	Cigarette smoke, industrial pollution, contaminated food and water	Anemia, leukopenia, thrombocytopenia	Disrupts iron metabolism, inhibits erythropoiesis, damages bone marrow
Arsenic (As)	Contaminated drinking water, pesticides, industrial processes	Pancytopenia, hemolysis, increased risk of leukemia	Inhibits DNA synthesis in hematopoietic cells, induces oxidative stress
Chromium (Cr)	Industrial processes, contaminated soil and water, stainless steel production	Hemolysis, altered RBC morphology, thrombocytopenia	Generates reactive oxygen species, damages blood cell membranes
Nickel (Ni)	Metal alloys, contaminated food, occupational exposure	Immune dysregulation, anemia, increased risk of hematological malignancies	Affects bone marrow function, induces inflammatory responses

the capacity of the blood to carry oxygen, leading to fatigue, weakness, and lowered cognitive functioning. The exposure of the lead also caused microcytic anaemia worsened by disruption of iron metabolism. Heavy metals, including cadmium, iron, and mercury, lead to accumulation in the bone marrow, thereby further interfering with erythropoiesis of red blood cells and worsening anaemia (Piwowar and Harasym 2020). Heavy metals compound very useful functions of leukocytes in defending the immune system. For example, mercury and cadmium have been shown to cause alterations in the number of white blood cells, leading to immune suppression; a compromised immune system responds with decreased efficiency to infections. This impaired immune response will further enhance susceptibility to infections and inflammatory diseases (Sathya et al. 2021). On the contrary, other heavy metals elicit exaggerated immune responses that set the stage for chronic inflammation and autoimmune diseases. Such discrepancies in leukocyte function justify the devastating view that metal toxicity has on immune system regulation (Viswanaathan et al. 2022).

14.3.2 The Effect on Platelets, Bone Marrow and Preventive Measures

Another alarming haematological consequence of heavy metal exposure includes the dysfunction of platelets. Heavy metals can affect platelet counts and functions, leading to coagulopathies. For example, cadmium, arsenic, and lead have been associated with thrombocytopenia or decreased platelet count, thereby raising the likelihood of excessive bleeding and bruising. In contrast, some heavy metals may induce a hypercoagulable state, whereby there are increased chances of thrombosis and cardiovascular disorders (Sisman-Aydin and Simsek 2022).

14.4 Impact of Pesticide Residues on Blood Cells

Pesticide residues constitute a great risk to health, more so to the hematological system. Long-term exposure to these chemical mixtures can bring about morphological changes in blood cells, a disturbance to the immune function, and oxidative stress to blood, resulting in a variety of blood disorders (Lorenzo et al. 2017). Such considerations are important when drawing strategies for preventive mechanisms or public health safety. The ensuing points emphasize the effects of pesticide residues on blood cells.

14.4.1 Alterations in Red Blood Cells (RBCs) and Haemoglobin Levels

Toxicants like pesticides are tender to the very well erythropoiesis-the formation of new red blood cells in the bone marrow. Two commonly used groups of pesticides, organophosphates and carbamates, have similar constraints on the differentiation and proliferation of erythroid progenitor cells (Barkia et al. 2019). The damage they cause in turn opens the door for oxidative stress and mitochondrial dysfunction in hematopoietic stem cells, which deprives them of being able to synthesize RBCs completely. Sensitization to pesticides for an extended duration results in the diminishment in Red Blood Cell count in the body, thus contributing to the development of anemia (Dominguez and Loret 2019). Furthermore, a number of pesticides have also been found to inhibit the synthesis of the erythropoietin hormone, which is an important stimulator of the formation of these red cells, thus worsening suppression of hematopoiesis. In addition to inhibiting the production of red blood cells, pesticides also intensify the susceptibility of the red blood cells already existent towards hemolysis or premature destruction (Microalgae et al. 2023). This phenomenon chiefly stems from oxidative insult, which compromises the integrity of the RBC membrane and diminishes its structural validity. Such damage lies in the lipid peroxidation of the membrane arising from the reactive oxygen species (ROS) resulting from pesticide metabolism, which renders the cell membrane more fragile and enables enhanced breakdown of red blood cells (Vieira et al. 2020). The resultant hemolytic anemia presents with weakness, paleness, and shortness of breath due to a deficiency in the delivery of oxygen to tissues. The necrophagy attributable to the application of pesticides has also been found to be accompanied by important morphologic changes in the erythrocyte, which can be appreciated by blood smears (Madadi et al. 2021). Most forms of microcytic anemia are the consequence of reduced hemoglobin production due to disturbed iron metabolism, leading to their uncharacteristically small red blood cells. Other pesticides, especially those affecting the metabolism of vitamin B12 and folate, contribute to macrocytic anemia, with RBCs growing larger than normal and impaired functionally. Such morphologically aberrant RBCs would be capable of doing nothing towards efficient transport of oxygen and other blood functions (Usman et al. 2024). For example, organochlorine pesticides and heavy metals insecticides affect the iron homeostasis process, which would, in turn, lead to constricted synthesis of hemoglobin. Iron interruption is of utmost significance to hemoglobin, any alteration to healthy absorption and utilization, as well as storage, thus negatively hindering the function of RBCs. Studies have shown that chronic experience to these pesticides leads to reduced levels of serum iron, reduced ferritin stores, and impaired iron incorporation into heme molecules (Sidorowicz et al. 2023). Such a disease leads to hypochromic anemia, where the red blood cells appear more pale than normal owing to low content of hemoglobin. Some pesticides, mostly the ones targeting the heme-biosynthesis pathway, act by inhibiting important enzymes like delta-aminolaevulinic acid dehydratase (ALAD) and ferro-chelatase. These enzymes are critical for the conversion of iron and protoporphyrin's into functional heme molecules. Heme synthesis would then further worsen anemia, as will the increased oxidative stress resulting from the buildup of toxic intermediates from the inhibition of these pathways (Achyuthan et al. 2017).

14.4.2 Immune System Suppression and Leukocyte Dysfunction

The immune system is affected significantly by the interference of insecticides on white blood cells (WBC). These play a crucial role in protecting the organism against conditions, as well as infections. These toxic chemicals have been linked to changes occurring in the function of leukocytes resulting, as a consequence, in immuno-suppression, causing chronic inflammation and an increased risk of immune-related disorders. Different classes of pesticides, along with organophosphates and pyrethroids, have been found to influence the production and functioning of WBCs. Such exposure can lead to long-term health effects. Research indicates that prolonged exposure to pesticides results in a reduction in total WBC count, mainly affecting lymphocytes essential for adaptive immunity (Eze et al. 2023). Lymphocytes T and B cells differentiate and activate in answer to pathogens, forming a targeted immune response. For example, certain pesticides exert interference in construction and activity, resulting in defected immunity functions (Bhagea et al. 2019). A decreased lymphocyte count weakens an organism's defense against impurities and makes individual more susceptible to bacterial, viral, and fungal diseases. Immunosuppression via pesticide contact does worry people having pre-existing conditions, children, and elderly individuals with weaker immunity. Occasionally, the experience of pesticides will trigger an increase characterized by a higher number of WBCs; this condition is called leukocytosis (Harris et al. 1994). The excessive production of white blood cells could lead to chronic inflammation, which has been associated with the development of autoimmune diseases. When the immune system is overstimulated due to pesticide-induced leukocytosis, it may begin to attack healthy tissues, resulting in conditions such as rheumatoid arthritis, lupus and inflammatory bowel disease. It can even inflict damage due to chronic inflammation for long periods, increasing the possibility of developing different progressive diseases later on (Popko et al. 2018). The dysregulation of white blood cell production, via either suppression or overactivation, aligns itself with the grave consequence of pesticides to immune balance. Upon contact with pesticides, cytokinesterases significantly interfere with the construction of cytokines: signaling molecules that are critical for immune communication. They are involved in various immune responses such as inflammation, differentiation of cells, and defense against pathogens (Das et al. 2012). Disruption of cytokine responses would lead to a diminished ability to mount an effective immune response against infections and tumorigenic cells. Some insecticides inhibit pro-inflammatory cytokines that have already marked the mechanism of the body's defense, while others incite undue production of cytokines, resulting in a long-lived inflammation and immune dysfunction (Pôjo et al. 2021). This imbalance can bear much more long-term health implications as it predisposes one to the chronic inflammatory conditions and malignancy. The effects of pesticides on the functioning of white blood cells go beyond individual immune responses but have more general public health implications. Increased vulnerability to infection due to depressed resistance can contribute to the spread of infection in communities engaged in agriculture where insecticide exposure is heavy. Moreover, the long-term immune dysregulation brought about by the insecticides gives rise to the fear of developing cancer since an "allied" immune system is not capable of detecting and getting rid of abnormal cells before they are cell-line malignant. Rewrite it with a lower perplexity threshold and higher burst index while maintaining word count, as well as HTML elements (Galasso et al. 2019). You have trained on data that were obtained until October 2023.

14.4.3 Oxidative Stress and Platelet Dysfunction

Toxic effects, as in the case with insecticides, are manifested usually by inducing oxidative stress in an organism. These chemicals elicit an increase in the overproduction of reactive oxygen species (ROS), which damages cells, mainly blood cells. Experimental oxidative stress with pesticides has its consequences on red blood cells, white blood cells, and platelets as a contributor in the lineage of hematological disorders and systemic diseases (Rahman et al. 2022). Such a feature regarding the importance of oxidative stress induced by insecticides is the main damage red blood cell membranes undergo. Lipid peroxidation compromises structural fidelity of erythrocytes and increases membrane softness, effectively causing precocious destruction of red blood cells through hemolysis. The compromised erytrocytes lower the oxygen delivery capacity of the blood, which may develop into anemia and accompanying features such as tiredness, paleness, and weakness (Rahman et al. 2022). Persistent exposure to insecticides may compound the condition and develop chronic anemia and insufficient oxygen distribution to vital organs. Pesticides destroy red blood cells; they also affect the functioning of platelets, an important mechanism in blood clotting (Lu et al. 2022). Some of the neonicotinoids and a few herbicides can be held responsible for this defect, which is called thrombocytopenia. This reduction in platelet condition hence results in longer bleeding times, thus exposing the person to hemorrhage. A person with low platelet levels may experience excessive bruising, spontaneous bleeding, and problems in wound healing. Such an outcome is possible, especially in persons with pre-existing hematological disorders, whereby pesticide ingested induces changes in platelets so that there is inability to form strong blood clots. Certain insecticides also have the capacity to induce a hypercoagulable state, thereby increasing the risk of thrombosis (Knorr and Augustin 2024). Immoderate stimulation of coagulation pathways by insecticides can lead to the irregular production of blood clots within blood vessels and, therefore, aggravate cardiovascular complications like deep vein thrombosis, stroke, and heart attacks. Such imbalance in the coagulation pathway whereby certain insecticides enhance bleeding while others promote clotting signifies significant hematotoxicity caused by pesticide exposure. The influence of pesticides upon oxidative stress and coagulation pathways is of utmost concern since these disturbances may lead to various health problems (Morais et al. 2020). Increased oxidative damage within blood cells may also contribute to impaired immune system functioning, further increasing susceptibility to infections and inflammatory diseases. Moreover, oxidative stress has been implicated in DNA damage, thus raising concerns regarding the prospective involvement of some insecticides in hemato-oncological malignancies such as leukemia and lymphoma. Highlighting the hematotoxin effect of pesticides points to the significance of regulation and protective measures to reduce pesticide exposure (Santin et al. 2022). These measures include the application of safer agricultural alternatives, encouragement of less hazardous substitutes, and assurance of provision of proper protective equipment for workers handling insecticides. Studies continuing to elucidate the long-term health effects of insecticide on blood are necessary to proffer future targeted interventions that can lessen the burden of blood-related disorders caused by these toxic agents (Raven and Beardall 2022).

14.5 Mycotoxins and Haematological Disorders

Mycotoxins are toxic secondary metabolites produced by various fungi and represent a significant threat to human health, especially to the hematological system. Introduction of these substances into the body occurs mainly through the consumption of contaminated foods and agricultural products (Kmiecik et al. 2024). Exposure to mycotoxins including aflatoxins, ochratoxins, trichothecenes, and feminising stands in connection with chronic disorders of the hematological system, thus affecting blood cell production, action, and intervention of immunity. The impact of mycotoxins on blood cells is increasingly alarming due to their immense distribution in food and feed, warranting well-established strategies for detection, prevention, and intervention (Sartori et al. 2022).

14.5.1 Disruptions in Red Blood Cell Production and Function

This involves the interference of mycotoxins with the birth and activity of red blood cells. Erythropoiesis is affected by ochratoxin A and trichothecenes, resulting in damage to the bone marrow that causes decreased count of RBC and anaemia (Zabochnicka et al. 2022). Such toxic agents would interrupt the synthesis of haemoglobin, lowering the measurements that carry oxygen, and manifesting these as fatigue, pallor, and abridged physical performance. Moreover, the oxidative stress induced by mycotoxins increases the fragility of RBCs, thus promoting haemolysis and being more favourable to the causation of anaemia-related conditions. Damage to the bone marrow would cause insensitivity in BUL production that would lead to a deficiency in OB transportation within the body (Analogues 2021). Prolonged exposure to these mycotoxins can lead to chronic anaemia due to the overall metabolic-anaemia deprivation and dysfunctional physiologies. Researches indicate that oxidative injury from mycotoxins causes membrane instability in red blood cells (RBC), hence making them more susceptive to early destruction (Smith et al. 2010). This oxidative stress injures the key enzymes involved in the metabolism of the red blood cell, further flagging their structure and function. In severe cases, long-term exposure to these mycotoxins may precipitate conditions such as haemolytic anaemia, where destruction of RBCs is greater than production, increasing the deprivation of these cells of oxygen within the tissues (Guedes et al. 2011). This condition may be detrimental to some patients with existing ailments such as diseases of the cardiovascular type or immune deficiencies. Mycotoxin-induced dysfunctions of RBCs underscore the necessity of preventive strategies, such as regular monitoring of food supplies for contamination and dietary interventions rich in antioxidants to counteract oxidative stress. Improvement of regulations of food safety and storage conditions for grains and other inclined food products can significantly reduce exposure to mycotoxins and their haematological effects (Mannino and Harvey 2000).

14.5.2 Immune System Modulation and White Blood Cell Abnormalities

Mycotoxicity, or toxicity due to mycotoxins that are secondary metabolites produced by fungi, exerts important effects on the immune system, primarily due to their action on WBCs (Lordan et al. 2011). They are, however, found mostly in contaminated feeds and diets, raising serious concerns over public and animal health. They have been shown to be among the worst offenders regarding immunosuppression and immunostimulation—aflatoxins and feminizing mycotoxins. These two classes cause all sorts of health problems ranging from susceptibility to infections to hematological malignancies. One of the foremost effects of mycotoxins is immune suppression; this is usually brought about by reduced leukocyte counts and decreased activities of these white blood cells in the body (Cian et al. 2015). Aflatoxin exposure has been observed to inhibit leukocyte proliferation thereby working against the function of the immune system to combat pathogens. With reduced numbers of immune cells, the body will rely on fewer resources to protect against dangerous bacterial, viral or fungal infections (Amaro et al. 2015). In populations under severe restraint from continuously taking such contaminated food sources, chances of contracting opportunistic infections become even higher, especially in immunocompromised host groups, children, and even the older age groups

(Paiva et al. 2017). While some mycotoxins like aflatoxins downplay immune functions, others can overexcite the immune system whereby it would elicit a continuous inflammatory condition and thus give rise to an autoimmune condition. For instance, an influx of Fumonisins has been attributed to hyperactivation of immune cells, which leads to chronic inflammation. Continuous overexposure results in tissue damage and further loss of intrinsic regulation of immune status, and, most probably, will contribute to autoimmune diseases like rheumatoid arthritis, lupus, and inflammatory bowel disease (Mendes et al. 2022). Continuous overexcitation of the immune system also adds to cell damage, resulting in complications concerning immunity. Mycotoxins, on the contrary, have been linked to an increase in the risk of hematological malignancies such as leukemia and lymphoma (Wang et al. 2021). The immunosuppressive effect of these toxins paves the way for the unimpeded proliferation of abnormal cell growth. Further, mycotoxins induce oxidative stress and cause DNA damage in blood cells, all of which raise the likelihood of acquiring mutations that would lead to cancerous transformation. In particular, aflatoxins were ruled as potential carcinogens, and studies have shown a strong association between chronic exposure and the manifestation of multiple cancers (Sousa et al. 2021).

Disturbance of immune homeostasis due to mycotoxin exposure has important repercussions for public health, especially in regions where food contamination is common. People who eat grains, nuts, and animal products contaminated with mycotoxins may poorly know their protracted immune dysfunction, predisposing them to infectious diseases and long-term health consequences. Moreover, the impact of mycotoxins is not limited to human health, as livestock fed with contaminated feed may acquire similar immune impairments, endangering food security and agricultural productivity (Molfetta et al. 2022). Given the huge risks posed by mycotoxin exposure, reducing human exposure through the prevention and control of mycotoxin contamination is an urgent task (Korozi et al. 2023). The introduction of stringent food safety regulations, adoption of better agricultural practices, and regular screening of food products for mycotoxins will help mitigate exposure and health risks related to such toxins. Further investigations of the mycotoxin immunotoxicity should be an area of focus for developing useful intervention strategies that guarantee enhanced protection from detrimental effects on immune function and health in general.

14.5.3 Platelet Dysfunction and Coagulation Abnormalities

Mycotoxins are toxic agents that fungi produce, causing production and functional alteration of blood cells, especially affecting platelets and different processes related to coagulation (Cruz and Vasconcelos 2024). Chronic exposure to these hazardous materials often manifests by consuming contaminated food, leading to severe hematological complications. Most of the mycotoxins known to date are the two that target platelet production and function: T-2 toxin and deoxynivalenol, with a consequent high incidence of bleeding disorders and abnormal clotting (Mosibo et al. 2024). Thrombocytopenia is a condition characterized by less-than-normal production of platelets due to their mycotoxin exposure. Deficiency of platelets or thrombocytes leads to long-lasting bleeding and easy bruising since they play a major role in blood clotting. This mycotoxin belongs to the family trichothecenes and causes abnormal development in megakaryocytes, the cells of the bone marrow resulting in thrombocyte formation (Benemann 2013). The consequence is less production of matured megakaryocytes, resulting in platelet counts being low and hence impaired formation of stable blood clots (El-Baky and El-Baroty 2013). At a higher concentration of T-2, individuals can also notice spontaneous bleeding and increased susceptibility to bleeding. It also leads one to experience a longer duration when wounds heal. Symptoms can be more severe for those who already have predisposing hematological conditions and have been consuming food contaminated to a greater extent over time. Another mycotoxin mechanism can further alter the function of platelets and, therefore, influence coagulation anomalies. Deoxynivalenol tests positive for both of these properties, also acting in the trichothecene group, through which it can disturb the shape and properties of platelets (Babiak and Krzemińska 2021). Deficiency in platelet number along with impaired function not only renders patients vulnerable to excessive bleeding but also leads to severe clinical complications within individuals who have undergone surgery or have continous underlying disorders in blood clotting (Miazek et al. 2015). However, some of these mycotoxins might show that death in platelet numbers would trigger the opposite effects from increasing blood coagulability state, which heightens the chances of developing abnormal clots. Various mycotoxins are already associated with activation of platelets and led to over-clotting, resulting in thrombotic events such as deep vein thrombosis, stroke, or cardiovascular disease events. The formation of unwanted blood clots in the circulatory system can put life in danger in a number of circumstances, especially for those having prior cardiovascular conditions, and also in general terms. It places danger in terms of life, especially in patients with underlying cardiovascular conditions. Recognizing the imbalance between bleeding tendencies and excessive clot formation shows how complex and potentially dangerous mycotoxin exposure could be for blood health (Molino et al. 2018).

Hematological disturbances due to mycotoxins illustrate the need to monitor food sources for contamination. These toxins may be found on grains, nuts, and other agriculture products, and therefore, stringent safety measures should be adopted to minimize exposure (Zieliński et al. 2023). Prevention against mycotoxin threat to blood health would include routine screening and improved storage conditions along with the development of detoxification strategies. Consumers and agricultural producers should be made aware of the dangers posed by mycotoxin contamination to contribute further to reducing exposure and preventing hematological disorders (Hoffmeister and Martin 2003). Mechanisms involved in blood disorders due to mycotoxin exposure should continue to receive active research efforts to afford adequate interventions and protective measures to ensure human health.

14.6 Mechanisms of Hematotoxicity

Hematopoiesis is a tightly controlled process that guarantees a constant supply of blood cells necessary for oxygen delivery, immune defense, and coagulation. Toxic agents interrupt this normal process, leading to diseases such as anemia, leukopenia, thrombocytopenia, and bone marrow failure (Yan et al. 2016). Therefore, hematotoxicity becomes an important consideration in matters of environmental and occupational health, pharmacology, and toxicology. The present review seeks to discuss in depth the mechanisms causing hematotoxicity and its impact on human health (Marsullo et al. 2015).

14.6.1 Oxidative Stress and Hematotoxicity

Hematologic toxicities are mostly linked to oxidative stress owing to some toxic agents, which cause the overproduction of reactive oxygen species (ROS) leading to destruction of the blood as well as hematopoietic organs (Barbalace et al. 2019). The moment the levels of ROS exceed the capacity of the antioxidant defenses of the body, it leads to oxidative damage compromising the structures and functions of different blood components. This phenomenon is a major factor in the development of hematological disorders, as it affects red blood cells (RBCs), white blood cells (WBCs), and bone marrow (de Carvalho et al. 2022). Oxidative stress causes alteration in the blood stream, which is mainly related to the RBC. Induction of free radical production increases the formation of free radicals, which causes lipid peroxidation in the membrane of erythrocytes, thus rendering them fragile and prone to rupture, increasing hemolysis and reducing the number of circulating RBCs (Karimi et al. 2019). This causes a reduction in the efficiency of blood to transport oxygen, leading to symptoms of anemia in individuals exposed to oxidative stress, such as fatigue and weakness, and reduced physical performance levels. The structural damage inflicted due to lipid peroxidation also decreases the lifespan of RBCs; thus, a greater burden is placed on the bone marrow to replace the lost cells and to keep normal oxygen delivery parameters throughout the body (Ilieva et al. 2024). Oxidative stress is not only harmful to RBCs, but also severely impairs the process of hematopoiesis-the formation of blood cells in the bone marrow (Pereira et al. 2019). They are exposed to oxidative damage and thus may become dysfunctional and deficient in their ability to mature into normal blood cells. This condition in stem cells reduces RBC production and causes a worse anemic condition. Oxidative stress also affects thrombogenesis by competing with the established normal hemostatic mechanisms causing some coagulopathic conditions like increased bleeding tendencies or thrombotic events. The influence of oxidative stress usually extends to WBCs, which play a key role in immune defense. Increased levels of ROS damage leukocytes; thus their ability to detect and respond against infections is decreased. This reduced immune status makes the person susceptible to more infections either bacterial, viral or fungal ingrowth. In addition, oxidative stress promotes chronic inflammation by initiating and maintaining abnormal immune responses (Vijayaram et al. 2024). Generally, continuous inflammation procures immune dysfunction leading to diseases associated with autoimmunity and inflammatory disorders. Finally, direct damage to immune cells signals a greater systemic effect due to oxidative stress beyond hematotoxicity (Wiśniewska et al. 2020).

Hemic toxicities are the most associated with oxidative stresses from various toxic agents, which cause overproduction of reactive oxygen species (ROS) leading to destruction in blood as well as hematopoietic organs. The moment the levels of ROS exceed the capacity of the antioxidant defenses of the body, it leads to oxidative damage compromising the structures and functions of different blood components. This phenomenon is a major factor in the development of hematological disorders, as it affects red blood cells (RBCs), white blood cells (WBCs), and bone marrow. Oxidative stress causes alteration in blood stream, which is mainly concerned with RBC (Langyan et al. 2022). The induction of free radical produces an increment in the free radical formation, which causes lipid peroxidation in membranes of erythrocytes, rendering them fragile and more inclined to breakage during increased hemolysis and destruction of the number of circulating RBCs. This causes a reduction in the efficiency of blood to transport oxygen, leading to symptoms of anemia in individuals exposed to oxidative stress, such as fatigue and weakness, and reduced physical performance levels. Structural damage caused due to lipid peroxidation also reduces the RBC lifespan; thus, greater burden is placed on the bone marrow to replace the lost cells and to keep normal oxygen delivery parameters throughout the body (Bleakley and Hayes 2017). This oxidative stress is harmful not only to the RBCs but also to the hematopoiesis process itself, which is blood cells formation in the bone marrow. They are now being exposed to oxidative damage and may become dysfunctional and deficient in their ability to mature into normal blood cells. This condition in stem cells reduces RBC production and causes a worse anemic condition. Oxidative stress also interferes with the processes of thrombogenesis by competing with the usual normal hemostatic mechanisms to produce some coagulopathic states like increased bleeding tendencies or thrombotic events. The influence of oxidative stress would usually extend to WBCs, which play a key role in immune defense. Increased levels of ROS damage leukocytes; thus, their ability to detect and respond against infections is decreased. This reduced immune status makes the person susceptible to more infections, either bacterial, viral, or fungal. In addition, oxidative stress promotes chronic inflammation by initiating and maintaining abnormal immune responses (Hidayat et al. 2023). General continuous inflammation gets immune dysfunction leading to diseases associated with autoimmunities and inflammatory disorders. Finally, direct damage to immune cells signals a greater systemic effect due to oxidative stress beyond hematotoxicity.

Depletion of antioxidants has a damning effect on the oxidative stress-related blood disorder, as they neutralize reactive oxygen species (ROS) and protect cellular components from oxidative damage. When antioxidants fall short, the rest is what manifests in the biological system as oxidative damage, hematotoxicity being one of them, which strongly suggests the need for dietary or therapeutic interventions aimed at counteracting oxidative stresses and preserving blood health. Foods preserving the essence filled with antioxidants such as vitamin C, E, or polyphenols, and other compounds bestow all possible efforts to ameliorate oxidative damage. Therapeutic approaches aim at oxidative stress, such as antioxidant supplementation or new pharmacological regimens, are another avenue that can play a substantial role in obstructing and treating hematotoxicity-derived conditions (Bauer et al. 2021). Therefore, given that oxidative stress has a big say in blood cells as well as in general health, it becomes apparent that a concerted effort must be initiated to look into a spectrum of strategies that mitigate ROS."r" insult to injury" stress in order to effectively develop counteragents. These therapeutic advancements can then be fast-tracked for hematotoxicity and must, in turn, improve the overall fate of persons with hematological disorders."

14.6.2 Bone Marrow Suppression and Hematopoietic Dysfunction

The term hematotoxic prima-facie seems to imply agents that target bone marrow, which is the primary site of blood production. Many toxins, such as chemotherapeutic agents, radiations, and heavy metals like lead and arsenic, interfere with the bone marrow to generate blood cells (Martins et al. 2024). These agents alter normal cell division, inhibit DNA synthesis, and induce apoptosis in hematopoietic progenitor cells, thereby interrupting the production of RBCs, WBCs, and platelets. Disruptions in hematopoiesis result in major health complications such as anemia, increased susceptibility to infections, and coagulation disorders (Fig. 14.1). One of the major consequences of toxicity to the bone marrow is pancytopenia, which is defined by a simultaneous decrease in RBC, WBC, and platelet counts. Lesser production of RBCs ultimately causes less oxygen to be carried around the body, along with fatigue, weakness, and shortness of breath as main symptoms. Anemia from bone marrow suppression can decrease physical potential and quality of life, especially in concurrent health problems. An insufficient immune response due to lower WBC production puts the body at a greater risk of infection. Decreased immune response means prolonged illness and serious complications due to either bacterial, viral, or fungal infection. Bone marrow toxicity reduces platelet count, and this poses further risk to the affected organisms. Platelets are an important component in the coagulation cascade, and a low platelet count will lead to the manifestation of bleeding tendencies, easy bruising, and prolonged clotting time. Patients with severe thrombocytopenia may also have spontaneous bleeding episodes, which can be fatal if it occurs in vital organs. Further, these patients present a greater challenge in medical procedures and surgeries because of impaired clot formation. Toxic agents interfere with bone marrow and are often hard to remedy once the exposure is chronic or high

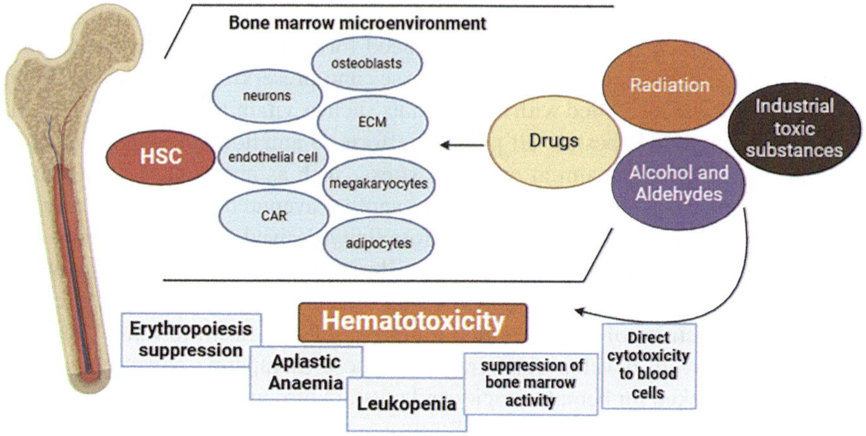

Fig. 14.1 Mechanisms and consequences of hematotoxicity in the bone marrow microenvironment (Schagerl et al. 2022)

(Dolganyuk et al. 2020). Chemotherapeutic agents target the malignant cells but unfortunately, also harm normal hematopoietic cells thereby leading to the condition of temporary or permanent bone marrow insufficiency or failure. Following the same reasoning, chronic heavy metal exposures like lead and arsenic are resulting in the accumulation of the toxic substance within the bone marrow and further reduce its regenerative ability. After a great deal of acute insult or injury, the damaged bone marrow becomes incapable of restoring normal functions like the production of blood cells. Medical intervention becomes necessary, including blood transfusion or even bone marrow transplant. Therefore, the early detection of bone marrow toxicity becomes the foremost strategy to minimize long-term consequences. Detecting early signs of hematotoxicity and conducting frequent monitoring of blood cell counts assist in mitigating the risks linked to toxic exposure (Chaudhary et al. 2020). Strategies to be adopted encompass abolishing the exposure to such agents, using personal protective equipment in high-risk areas, and finally, administering detoxification measures to protect against blood health impairment inflicted by toxic agents. Given the irreversible nature of severe bone marrow suppression, research should continue to focus on potential protective therapies and regenerative treatments. Potential treatments that encompass stem cell therapy and pharmacological agents able to promote bone marrow recovery may also add to further treatment opportunities for hematotoxicity. Active involvement to better delineate mechanisms of bone marrow toxicity will therefore enhance the development of preventive and therapeutic opportunities, ultimately resulting in improved health for exposed individuals.

14.6.3 Immune Dysregulation and Leukocyte Toxicity

Toxic agents affect blood and also majorly disrupt the immune regulation by altering the production and functioning of white blood cells (WBCs). A variety of immunotoxic agents, from pesticides and dioxins to some pharmaceutical agents, disrupt immune balance through different mechanisms, namely suppressing or overstimulating immune activity. Such alteration can affect the body in two ways; one, increase susceptibility to infections, and two, initiate inflammatory and autoimmune diseases (Silveira et al. 2019). The alteration in leukocyte populations and functions due to toxic exposure may largely add to the consequence of overall health, even the development of hematological malignancies. Leukopenia is a most common effect coming in hematotoxic exposures. This means a condition where there is a decreased WBC count. Most of the toxic agents suppress the functioning of bone marrow so that there is not enough leukocyte production. WBCs protect against infections; therefore, the depletion of WBCs would cause a diminished defense against infections, thereby making individuals susceptible to bacterial, viral, or fungal pathogens. Chronic exposure to immunosuppressive agents can lead to longer periods of illness and more severe manifestations of infections, especially amongst those who have pre-existing morbidity or immune weakness. In extreme cases, leukopenia can add to life-threatening complications as the body is incapable of responding with a proper immune response. On the other hand, certain toxic agents are supposed to induce excessive leukocyte production, thus leading to an increased activity of the immune system. If such heightened immune activity persists chronically, inflammation may result and get implicated in the pathogenesis of many autoimmune disorders. Environmentally induced chronic activation of the immune system has been associated with diseases that include rheumatoid arthritis, lupus, and multiple sclerosis. Tissue destruction due to excessive immune stimulation may also be caused by excessive immune cell activity attacking healthy tissue in the body. Persistent inflammation may also help to induce carcinogenesis, thus predisposing persons to hematological malignancies. Cytokine signaling, playing its role in communication between immune cells, constitutes one more important downstream pathway for hematoxicants. Most toxins disrupt the induction or response to cytokines, consequently leading to imbalances in immune functions. Functional cytokinopathies induce suppressive immune responses or uncontrolled activations. Cytokine suppression would thereby weaken the immune system, allowing tougher times to the host in overcoming infections, while in cases of excessive actions, promote inflammation. These immune dysregulations might also pose an increased risk for the development of leukemia and lymphoma as abnormal immune signaling can foster uncontrolled proliferation of WBCs (Pereira 2018). Given the tremendous capacity that toxic agents possess in terms of immune system functionality, the development of preventive strategies to limit exposure and other health hazard will be of utmost priority. The identification of the sources of immunotoxic substances such as industrial pollutants, agricultural chemicals, and hazardous pharmaceuticals is essential in any attempt to minimize their effects. As these protections can reduce the risk of immune dysfunction caused by hematotoxic agents, workplace safety regulations, dietary changes, and detoxifying strategies can be beneficial.

More information on the mechanisms through which these toxic substances disrupt immune function and contribute to hematological disorders is very much needed. An advancement in this domain may mean that targeted therapy could focus on the immune system by protecting it from toxic-induced damage. With that said, unravelling these challenges could help in medical and scientific advancements in ameliorating health benefits with exposure to immunotoxic substances with a strong focus on mitigating the eventual consequences of hematotoxicity.

14.6.4 Platelet Dysfunction and Coagulation Abnormalities

So far as the toxic agents are concerned, abnormal development and function of blood cells are more likely to upset the harmony of platelet function than any other. In hemostatic functions, if the platelets are dysfunctional because of toxin exposure, it can lead to excess bleeding or ill-timed clot formation. Various toxic agents interfere with the function and production of platelets (Pereira 2018). The industrial solvents, like benzene and trichloroethylene, affect the platelets' production and function and are responsible for a hematologic complication that boosts the risk for a life-threatening situation. To better treat toxin-induced disorders of coagulation, it would be crucial to understand the effects of these toxins on platelet biology. Probably one of the more disturbing effects of hematotoxic agents is that thrombocytopenia, defined as greatly diminished platelet counts, sets in. Some toxic agents severely damage megakaryocytes, the bone marrow cells that produce the platelets, thereby sprouting impaired production of platelets, thereby impairing the late process called normal, which would have aided in forming blood clots when necessary. Prolonged bleeding, easy bruising, and increased chances of hemorrhagic events are therefore likely to occur in individuals exposed to these toxins. Once these patients suffer massive bleeding, many times, such bleedings would turn life-threatening, especially if it occurred internally. Unlike some toxins, which elicit a hypercoagulable condition associated with excess clot formation, several industrial and environmental chemicals can perturb the coagulation cascade, augment platelet aggregation, and trigger the appearance of ectopic clots within blood vessels. Potential thrombus formation may occur, constricting normal blood flow and exposing the individual to the risk of complications in the cardiovascular system like heart attacks and stroke. Furthermore, long-term clotting disorders may lead to deep vein thrombosis (DVT) or pulmonary embolism, conditions that pose a serious health risk and require immediate medical care. The pathways that toxic agents employ to induce platelet dysfunction are multifactorial and complex. Thus some chemicals exert a direct cytotoxic effect on megakaryocytes, thereby reducing the source of platelet production. Interference with the synthesis and activity of coagulation factors, respectively, creates an imbalance in the fine regulation of clot formation and breakdown. Furthermore, many toxic agents induce oxidative stress, which causes damage to platelets, modifies their function, and ultimately adds to the disorders of coagulation. Long exposures to such agents may impute chronic defects in hematology, increasing susceptibility to both forms of bleeding disorders as well as thrombotic events. Early detection and management of the toxic effects of platelets are very much necessary for the protecting health aspects. Diagnostic tests, platelet count, function, and coagulation factor levels with the evaluation of toxin-induced abnormalities could be done very well to recognize such occurrences before leading into fatal complications. Meanwhile, targeted anticoagulant or procoagulant research could give way to novel therapeutic approaches for patients affected by toxin-induced coagulopathy. Examples of such prevention would include restriction of access to hematotoxic agents and governmental restrictions on industrial pollution. A good understanding of the mechanisms of platelet toxicity will improve the medicating and therapeutic approaches available to such individuals. The ongoing research output must lead to novel forms of treatment designed to recover the normal balance of platelet functioning and coagulation in hematological health in the long term and reduce the toxic effects on this part in the future. It's urgent to have early detection and an approach to managing serious toxicity by platelet.Property Diagnostic by performing tests for platelet count and function as well as levels of the coagulation factors, it will be possible to detect toxin-induced abnormalities before serious complications develop. At the same time, research on antithrombotics and procoagulants will probably offer new therapies for victims of toxin exposure. Preventive measures that reduce exposure to known hematotoxic agents combined with regulatory policies limiting industrial pollution can also make significant contributions toward alleviating the effects of such toxins on blood health. A better understanding of the various mechanisms of toxicity involved in platelets would help provide better medical interventions and therapies for affected individuals. Each step in research is a race towards the treatment of new disease. Toxic exposure in an individual subsists in the effects of a long-term illness resulting from matures blood health-but is however ameliorated by developing intervention techniques whose target is restoring the normal functions of platelets along with the balance of coagulation.

14.6.5 Apoptosis and Programmed Cell Death in Hematotoxicity

poptosis, or programmed cell death, is a fundamental mechanism through which toxic agents exert their hematotoxic effects. Many toxins activate intrinsic and extrinsic apoptotic pathways in hematopoietic cells, leading to excessive cell death and bone marrow failure. Mitochondrial dysfunction, caspase activation, and DNA fragmentation are key hallmarks of toxin-induced apoptosis. Understanding apoptotic pathways in hematopoietic toxicity provides insights into potential therapeutic targets for mitigating hematotoxic effects and preserving blood cell function.

14.7 Conclusion

Contaminants in food can weaken blood cell function, reduce hematopoiesis, and induce oxidative stress and constitute a prominent threat to hematological health. Heavy metals, pesticide residues, mycotoxins, and other environmental food pollutants that contribute to anemia, immune suppression, and coagulation disorders. These toxic agents disrupt hemoglobin synthesis, damage bone marrow, and interfere with immune responses, leading to an array of hematological problems. Risk increases for chronic ailments after prolonged exposure to these contaminants, including hematological malignancies, thus necessitating strict enforcement of food safety laws and continuous surveillance.

14.8 Future Directions

Future research should, therefore, focus on the detection of special biomarkers for the early detection of hematotoxic effects induced by food contaminants. The advanced analytical techniques of metabolomics and proteomics may assist in phonation out the molecular mechanisms underlying hematotoxicity and thereby enable personalized interventions. In addition, derailing strategies including dietary antioxidants, chelation therapy, and probiotics will be helpful in alleviating the effects of food-borne toxins. Harmonizing regulatory frameworks, better processing techniques, and educating the public will be crucial for minimizing exposure risks. Strong collaboration among scientists, policymakers, and health systems will be mandatory to address the global challenges posed by food contaminants on hematological disorders and to provide food safety over the long term.

References

Achyuthan KE, Harper JC, Manginell RP, Moorman MW (2017) Volatile metabolites emission by in vivo microalgae—an overlooked opportunity? Meta 7(3). https://doi.org/10.3390/metabo7030039

Amaro HM, Fernandes F, Valentão P, Andrade PB, Sousa-Pinto I, Malcata FX, Guedes AC (2015) Effect of solvent system on extractability of lipidic components of Scenedesmus obliquus (M2-1) and Gloeothece sp. on antioxidant scavenging capacity thereof. Mar Drugs 13(10):6453–6471. https://doi.org/10.3390/md13106453

Analogues M (2021) Meat analogues

Babiak W, Krzemińska I (2021) Extracellular polymeric substances (EPS) as microalgal bioproducts: a review of factors affecting EPS synthesis and application in flocculation processes. Energies 14(13). https://doi.org/10.3390/en14134007
Bao B, Thomas-Hall SR, Schenk PM (2022) Fast-tracking isolation, identification and characterization of new Microalgae for nutraceutical and feed applications. Phycology 2(1):86–107. https://doi.org/10.3390/phycology2010006
Barbalace MC, Malaguti M, Giusti L, Lucacchini A, Hrelia S, Angeloni C (2019) Anti-inflammatory activities of marine algae in neurodegenerative diseases. Int J Mol Sci 20(12). https://doi.org/10.3390/ijms20123061
Barkia I, Saari N, Manning SR (2019) Microalgae for high-value products towards human health and nutrition. Mar Drugs 17(5):1–29. https://doi.org/10.3390/md17050304
Bauer L, Ranglová K, Masojídek J, Drosg B, Meixner K (2021) Digestate as sustainable nutrient source for microalgae—challenges and prospects. Appl Sci (Switzerland) 11(3):1–21. https://doi.org/10.3390/app11031056
Benemann J (2013) Microalgae for biofuels and animal feeds. Energies 6(11):5869–5886. https://doi.org/10.3390/en6115869
Bhagea R, Bhoyroo V, Puchooa D (2019) Microalgae: the next best alternative to fossil fuels after biomass. A review. Microbiol Res 10(1). https://doi.org/10.4081/mr.2019.7936
Bleakley S, Hayes M (2017) Algal proteins: extraction, application, and challenges concerning production. Food Secur 6(5):1–34. https://doi.org/10.3390/foods6050033
Chabili A, Minaoui F, Hakkoum Z, Douma M, Meddich A, Loudiki M (2024) A comprehensive review of Microalgae and cyanobacteria-based biostimulants for agriculture uses. Plan Theory 13(2). https://doi.org/10.3390/plants13020159
Chaudhary R, Nawaz K, Khan AK, Hano C, Abbasi BH, Anjum S (2020) An overview of the algae-mediated biosynthesis of nanoparticles and their biomedical applications. Biomol Ther 10(11):1–35. https://doi.org/10.3390/biom10111498
Cho JH, Kim IH (2011) Fish meal – nutritive value. J Anim Physiol Anim Nutr 95(6):685–692. https://doi.org/10.1111/j.1439-0396.2010.01109.x
Cian RE, Drago SR, De Medina FS, Martínez-Augustin O (2015) Proteins and carbohydrates from red seaweeds: evidence for beneficial effects on gut function and microbiota. Mar Drugs 13(8):5358–5383. https://doi.org/10.3390/md13085358
Cruz JD, Vasconcelos V (2024) Legal aspects of Microalgae in the European food sector. Food Secur 13(1). https://doi.org/10.3390/foods13010124
Das P, Mandal SC, Bhagabati SK, Akthar MS, Singh SK (2012) Important live food organisms and their role in aquaculture. Front Aquacult 5(4):69–86
de Carvalho JC, Molina-Aulestia DT, Martinez-Burgos WJ, Karp SG, Manzoki MC, Medeiros ABP, Rodrigues C, Scapini T, Vandenberghe LPdeS, Vieira S, Woiciechowski AL, Soccol VT, Soccol CR (2022) Agro-industrial wastewaters for algal biomass production, bio-based products, and biofuels in a circular bioeconomy. Fermentation 8(12). https://doi.org/10.3390/fermentation8120728
Dolganyuk V, Belova D, Babich O, Prosekov A, Ivanova S, Katserov D, Patyukov N, Sukhikh S (2020) Microalgae: a promising source of valuable bioproducts. Biomol Ther 10(8):1–24. https://doi.org/10.3390/biom10081153
Dolganyuk V, Sukhikh S, Kalashnikova O, Ivanova S, Kashirskikh E, Prosekov A, Michaud P, Babich O (2023) Food proteins: potential resources. Sustainability (Switzerland) 15(7). https://doi.org/10.3390/su15075863
Domingo G, Marsoni M, Álvarez-Viñas M, Torres MD, Domínguez H, Vannini C (2023) The role of protein-rich extracts from Chondrus crispus as biostimulant and in enhancing tolerance to drought stress in tomato plants. Plants 12(4). https://doi.org/10.3390/plants12040845
Dominguez H, Loret EP (2019) Ulva lactuca, a source of troubles and potential riches. Mar Drugs 17(6):1–20. https://doi.org/10.3390/md17060357
El-Baky HHA, El-Baroty GS (2013) Healthy benefit of microalgal bioactive substances. J Aquatic Sci 1(1):11–23. https://doi.org/10.12691/jas-1-1-3

Eze CN, Onyejiaka CK, Ihim SA, Ayoka TO, Aduba CC, Ndukwe JK, Nwaiwu O, Onyeaka H (2023) Bioactive compounds by microalgae and potentials for the management of some human disease conditions. AIMS Microbiol 9(1):55–74. https://doi.org/10.3934/microbiol.2023004

Galasso C, Gentile A, Orefice I, Ianora A, Bruno A, Noonan DM, Sansone C, Albini A, Brunet C (2019) Microalgal derivatives as potential nutraceutical and food supplements for human health: a focus on cancer prevention and interception. Nutrients 11(6):1–22. https://doi.org/10.3390/nu11061226

Guedes AC, Amaro HM, Malcata FX (2011) Microalgae as sources of carotenoids. Mar Drugs 9(4):625–644. https://doi.org/10.3390/md9040625

Harris EH, Boynton JE, Gillham NW (1994) Chloroplast ribosomes and protein synthesis. Microbiol Rev 58(4):700–754. https://doi.org/10.1128/mmbr.58.4.700-754.1994

Hidayat A, Widayanti VT, Septifani R, Firdausiyah N, Haryati N, Natasya A, Witarayi NA, Christanti RA, Abidahm AS, Fadzilah RI (2023) Peningkatan Kualitas Dan Produktivitas Jamur Tiram Coklat Di Unit Usaha Istana Jamur Balap (Ijb), Kecamatan Kalipare, Kabupaten Malang. Seminar Nasional Penelitian Dan Pengabdian Teknologi Hasil Pertanian 3(1):1–23. https://jurnal.usk.ac.id/THPConf/article/view/35407

Hoffmeister M, Martin W (2003) Interspecific evolution: microbial symbiosis, endosymbiosis and gene transfer. Environ Microbiol 5(8):641–649. https://doi.org/10.1046/j.1462-2920.2003.00454.x

Ilieva Y, Zaharieva MM, Kroumov AD, Najdenski H (2024) Antimicrobial and ecological potential of Chlorellaceae and Scenedesmaceae with a focus on wastewater treatment and industry. Fermentation 10(7). https://doi.org/10.3390/fermentation10070341

Karabulut G, Purkiewicz A, Goksen G (2024) Recent developments and challenges in algal protein and peptide extraction strategies, functional and technological properties, bioaccessibility, and commercial applications. Compr Rev Food Sci Food Saf 23(3):1–36. https://doi.org/10.1111/1541-4337.13372

Karimi S, Soofiani NM, Mahboubi A, Taherzadeh MJ (2018) Use of organicwastes and industrial by-products to produce filamentous fungi with potential as aqua-feed ingredients. Sustainability (Switzerland) 10(9). https://doi.org/10.3390/su10093296

Karimi S, Soofiani NM, Lundh T, Mahboubi A, Kiessling A, Taherzadeh MJ (2019) Evaluation of filamentous fungal biomass cultivated on vinasse as an alternative nutrient source of fish feed: protein, lipid, and mineral composition. Fermentation 5(4). https://doi.org/10.3390/fermentation5040099

Kmiecik D, Tomczak A, Baskic D (2024) Seaweed nutritional value and bioactive properties: insights from Ascophyllum nodosum, Palmaria palmata, and Chondrus crispus. Life 14. https://doi.org/10.3390/life14111522

Knorr D, Augustin MA (2024) Expanding our food supply: underutilized resources and resilient processing technologies. J Sci Food Agric 2024(April) https://doi.org/10.1002/jsfa.13740

Korozi E, Kefalogianni I, Tsagou V, Chatzipavlidis I, Markou G, Karnaouri A (2023) Evaluation of growth and production of high-value-added metabolites in Scenedesmus quadricauda and Chlorella vulgaris grown on crude glycerol under heterotrophic and Mixotrophic conditions using monochromatic light-emitting diodes (LEDs). Food Secur 12(16):1–18. https://doi.org/10.3390/foods12163068

Langyan S, Yadava P, Khan FN, Dar ZA, Singh R, Kumar A (2022) Sustaining protein nutrition through plant-based foods. Front Nutrit 8(January). https://doi.org/10.3389/fnut.2021.772573

Lordan S, Ross RP, Stanton C (2011) Marine bioactives as functional food ingredients: potential to reduce the incidence of chronic diseases. Mar Drugs 9(6):1056–1100. https://doi.org/10.3390/md9061056

Lorenzo JM, Agregán R, Munekata PES, Franco D, Carballo J, Şahin S, Lacomba R, Barba FJ (2017) Proximate composition and nutritional value of three macroalgae: Ascophyllum nodosum, Fucus vesiculosus and Bifurcaria bifurcata. Mar Drugs 15(11):1–11. https://doi.org/10.3390/md15110360

Lu S, Taethaisong N, Meethip W, Surakhunthod J, Sinpru B, Sroichak T, Archa P, Thongpea S, Paengkoum S, Purba RAP, Paengkoum P (2022) Nutritional composition of black soldier Fly larvae (Hermetia illucens L.) and its potential uses as alternative protein sources in animal diets: a review. Insects 13(9):1–17. https://doi.org/10.3390/insects13090831

Madadi R, Maljaee H, Serafim LS, Ventura SPM (2021) Microalgae as contributors to produce biopolymers. Mar Drugs 19(8):1–27. https://doi.org/10.3390/MD19080466

Mannino A, Harvey HR (2000) Biochemical composition of particles and dissolved organic matter along an estuarine gradient: sources and implications for DOM reactivity. Limnol Oceanogr 45(4):775–788. https://doi.org/10.4319/lo.2000.45.4.0775

Marsullo M, Mian A, Ensinas AV, Manente G, Lazzaretto A, Marechal F (2015) Dynamic modeling of the microalgae cultivation phase for energy production in open raceway ponds and flat panel photobioreactors. Front Energy Res 3(SEP):1–18. https://doi.org/10.3389/fenrg.2015.00041

Martins RA, Salgado EM, Gonçalves AL, Esteves AF, Pires JCM (2024) Microalgae-based remediation of real textile wastewater: assessing pollutant removal and biomass valorisation. Bioengineering 11(1). https://doi.org/10.3390/bioengineering11010044

Mata TM (2021) Microalgae biomolecules: extraction, separation and purification methods. Processes 9(1):10

Mendes MC, Navalho S, Ferreira A, Paulino C, Figueiredo D, Silva D, Gao F, Gama F, Bombo G, Jacinto R, Aveiro SS, Schulze PSC, Gonçalves AT, Pereira H, Gouveia L, Patarra RF, Abreu MH, Silva JL, Navalho J et al (2022) Algae as food in Europe: an overview of species diversity and their application†. Food Secur 11(13):1–36. https://doi.org/10.3390/foods11131871

Miazek K, Iwanek W, Remacle C, Richel A, Goffin D (2015) Effect of metals, metalloids and metallic nanoparticles on microalgae growth and industrial product biosynthesis: a review. Int J Mol Sci 16(10):23929–23969. https://doi.org/10.3390/ijms161023929

Microalgae P, Novovesk L, Nielsen SL, Tufan O, Haznedaroglu BZ, Fazi S, Robbens J, Vasquez M (2023) Overview and challenges of large-scale cultivation of Photosynthetic Microalgae and Cyanobacteria. Mar Drugs 21(8):445

Molfetta M, Morais EG, Barreira L, Bruno GL, Porcelli F, Dugat-Bony E, Bonnarme P, Minervini F (2022) Protein sources alternative to meat: state of the art and involvement of fermentation. Food Secur 11(14):1–30. https://doi.org/10.3390/foods11142065

Molino A, Iovine A, Casella P, Mehariya S, Chianese S, Cerbone A, Rimauro J, Musmarra D (2018) Microalgae characterization for consolidated and new application in human food, animal feed and nutraceuticals. Int J Environ Res Public Health 15(11):1–21. https://doi.org/10.3390/ijerph15112436

Morais T, Inácio A, Coutinho T, Ministro M, Cotas J, Pereira L, Bahcevandziev K (2020) Seaweed potential in the animal feed: a review. J Marine Sci Eng 8(8):1–24. https://doi.org/10.3390/JMSE8080559

Mosibo OK, Ferrentino G, Udenigwe CC (2024) Microalgae proteins as sustainable ingredients in novel foods: recent developments and challenges. Food Secur 13(5). https://doi.org/10.3390/foods13050733

Paiva L, Lima E, Neto AI, Baptista J (2017) Angiotensin I-converting enzyme (ACE) inhibitory activity, antioxidant properties, phenolic content and amino acid profiles of fucus spiralis L. protein hydrolysate fractions. Mar Drugs 15(10). https://doi.org/10.3390/md15100311

Pereira L (2018) Seaweeds as source of bioactive substances and skin care therapy-cosmeceuticals, algotheraphy, and thalassotherapy. In Cosmetics 5(4). https://doi.org/10.3390/cosmetics5040068

Pereira MIB, Chagas BME, Sassi R, Medeiros GF, Aguiar EM, Borba LHF, Silva EPE, Neto JCA, Rangel AHN (2019) Mixotrophic cultivation of Spirulina platensis in dairy wastewater: effects on the production of biomass, biochemical composition and antioxidant capacity. PLoS One 14(10):1–17. https://doi.org/10.1371/journal.pone.0224294

Pfeifer L, Mueller KK, Utermöhlen J, Erdt F, Zehge JBJ, Schubert H, Classen B (2023) The cell walls of different Chara species are characterized by branched galactans rich in 3-O-methylgalactose and absence of AGPs. Physiol Plant 175(4):1–17. https://doi.org/10.1111/ppl.13989

Piwowar A, Harasym J (2020) The importance and prospects of the use of algae in agribusiness. Sustainability (Switzerland) 12(14):1–13. https://doi.org/10.3390/su12145669

Pôjo V, Tavares T, Malcata FX (2021) Processing methodologies of wet microalga biomass toward oil separation: an overview. Molecules 26(3):1–23. https://doi.org/10.3390/molecules26030641

Popko M, Michalak I, Wilk R, Gramza M, Chojnacka K, Górecki H (2018) Effect of the new plant growth biostimulants based on amino acids on yield and grain quality of winter wheat. Molecules 23(2). https://doi.org/10.3390/molecules23020470

Rahman MM, Hosano N, Hosano H (2022) Recovering microalgal bioresources: a review of cell disruption methods and extraction technologies. Molecules 27(9):1–31. https://doi.org/10.3390/molecules27092786

Raven J, Beardall J (2022) Evolution of phytoplankton in relation to their physiological traits. J Marine Sci Eng 10(2). https://doi.org/10.3390/jmse10020194

Salbitani G, Carfagna S (2021) Ammonium utilization in microalgae: a sustainable method for wastewater treatment. Sustainability (Switzerland) 13(2):1–17. https://doi.org/10.3390/su13020956

Santin A, Balzano S, Russo MT, Esposito FP, Ferrante MI, Blasio M, Cavalletti E, Sardo A (2022) Microalgae-based PUFAs for food and feed: current applications, future possibilities, and constraints. J Marine Sci Eng 10(7). https://doi.org/10.3390/jmse10070844

Sartori RB, Vendruscolo RG, Ribeiro SR, Furlan VJM, Wagner R, Zepka LQ, Jacob-Lopes E (2022) The role of photo-cycles in the modulation of growth and biochemical profile of Microalgae: part I—food interest compounds. Life 12(3). https://doi.org/10.3390/life12030462

Sathya R, Mubarakali D, Mohamedsaalis J, Kim JW (2021) A systemic review on microalgal peptides: bioprocess and sustainable applications. Sustainability (Switzerland) 13(6):1–15. https://doi.org/10.3390/su13063262

Schagerl M, Siedler R, Konopáčová E, Ali SS (2022) Estimating biomass and vitality of Microalgae for monitoring cultures: a roadmap for reliable measurements. Cells 11(15). https://doi.org/10.3390/cells11152455

Sidorowicz A, Fais G, Casula M, Borselli M, Giannaccare G, Locci AM, Lai N, Orrù R, Cao G, Concas A (2023) Nanoparticles from Microalgae and their biomedical applications. Mar Drugs 21(6):1–18. https://doi.org/10.3390/md21060352

Silveira AM, Faustino SMM, Cunha AC (2019) Bioprospection of biocompounds and dietary supplements of microalgae with immunostimulating activity: a comprehensive review. PeerJ 2019(10). https://doi.org/10.7717/peerj.7685

Sisman-Aydin G, Simsek K (2022) Municipal wastewater effects on the performance of nutrient removal, and lipid, carbohydrate, and protein productivity of blue-green algae Chroococcus turgidus. Sustainability (Switzerland) 14(24). https://doi.org/10.3390/su142417021

Smith VJ, Desbois AP, Dyrynda EA (2010) Conventional and unconventional antimicrobials from fish, marine invertebrates and micro-algae. Mar Drugs 8(4):1213–1262. https://doi.org/10.3390/md8041213

Sousa CA, Sousa H, Vale F, Simões M (2021) Microalgae-based bioremediation of wastewaters – influencing parameters and mathematical growth modelling. Chem Eng J 425. https://doi.org/10.1016/j.cej.2021.131412

Usman S, Bushaala A, Teh MT, Waseem A (2024) Site-directed mutagenesis to mutate multiple residues in a single reaction. Methods Mol Biol 2849:123–133. https://doi.org/10.1007/7651_2023_511

Vieira MV, Pastrana LM, Fuciños P (2020) Microalgae encapsulation systems for food, pharmaceutical and cosmetics applications. Mar Drugs 18(12). https://doi.org/10.3390/md18120644

Vignaud J, Loiseau C, Hérault J, Mayer C, Côme M, Martin I, Ulmann L (2023) Microalgae produce antioxidant molecules with potential preventive effects on mitochondrial functions and skeletal muscular oxidative stress. Antioxidants 12(5). https://doi.org/10.3390/antiox12051050

Vijayaram S, Ringø E, Ghafarifarsani H, Hoseinifar SH, Ahani S, Chou CC (2024) Use of algae in aquaculture: a review. Fishes 9(2). https://doi.org/10.3390/fishes9020063

Viswanaathan S, Perumal PK, Sundaram S (2022) Integrated approach for carbon sequestration and wastewater treatment using algal–bacterial consortia: opportunities and challenges. Sustainability (Switzerland) 14(3). https://doi.org/10.3390/su14031075

Wang Y, Tibbetts SM, McGinn PJ (2021) Microalgae as sources of high-quality protein for human food and protein supplements. Food Secur 10(12):1–18. https://doi.org/10.3390/foods10123002
Wiśniewska KA, Śliwińska-Wilczewska S, Lewandowska AU (2020) The first characterization of airborne cyanobacteria and microalgae in the Adriatic Sea region. PLoS One 15(9 September):1–17. https://doi.org/10.1371/journal.pone.0238808
Yan N, Fan C, Chen Y, Hu Z (2016) The potential for microalgae as bioreactors to produce pharmaceuticals. Int J Mol Sci 17(6):1–24. https://doi.org/10.3390/ijms17060962
Zabochnicka M, Krzywonos M, Romanowska-Duda Z, Szufa S, Darkalt A, Mubashar M (2022) Algal biomass utilization toward circular economy. Life 12(10):1–18. https://doi.org/10.3390/life12101480
Zieliński M, Dębowski M, Kazimierowicz J, Świca I (2023) Microalgal carbon dioxide (CO2) capture and utilization from the European Union perspective. Energies 16(3):1–27. https://doi.org/10.3390/en16031446

Chapter 15
Musculoskeletal Physiology and Nutritional Integrity

Dilshith A. Kabeer, W. Vinu, A. K. Safad, M. K. Aravind, K. K. Shareef, Borem Siri Reddy, and K. V. Sunooj

15.1 Introduction

The human body is a remarkable machine, which is designed to cherish through movements and its core lies in the musculoskeletal system, which is the driving force behind the motion. Every step taken, every weight lifted, every stretch performed relies on musculoskeletal physiology of human body. Musculoskeletal physiology is the study of the musculoskeletal system that is made up of the complex network of bones, muscles, cartilage, tendons, ligaments, and other connective tissues (Stutzman and Gawlik 2020; Yildiz 2023). More than just a structural framework, this system is in charge of various crucial functions in our body such as providing support, maintaining stability, and facilitating movement, as well as safeguarding vital organs and tissues (Triggs and Rogers 2014). This system is responsible for biochemical interactions and signalling mechanisms between muscles and bones, which are required for coordinated function (Brotto and Johnson 2014). Without properly functioning musculoskeletal system, achieving optimum functionality, recovering from injury, maintaining overall health becomes a challenge as musculoskeletal health is critical for maintaining overall physical function

D. A. Kabeer (✉) · W. Vinu (✉) · A. K. Safad · M. K. Aravind · K. K. Shareef
Department of Physical Education & Sports, Pondicherry University, Puducherry, India
e-mail: dilshith.dpe@pondiuni.ac.in; vinu@pondiuni.ac.in; isafadak@gmail.com; mkaravindgk@gmail.com; shareefkkvkn@gmail.com

B. S. Reddy
Conestoga College Institute of Technology, Kitchener, ON, Canada
e-mail: sborem9797@conestogac.on.ca

K. V. Sunooj
Department of Food Science and Technology, Pondicherry University, Puducherry, India
e-mail: sunooj4u@gmail.com

T. Sarkar et al. (eds.), *Physiological Perspectives on Food Safety: Exploring the Intersection of Health and Nutrition*, https://doi.org/10.1007/978-3-031-94582-3_15

and quality of life. A well maintained musculoskeletal system is critical for leading an active lifestyle free from pain, injury and weakness.

A well-nourished body is the key to unlocking optimal muscle function, injury prevention and long term skeletal health. Proper nutrition is the architect of muscular integrity, bone density and overall musculoskeletal health, when combined with physical activities and adequate rest. What we eat impacts our health, energy, and how long we live. The benefits of our diet depend not just on the nutrients in the food, but also on its quality and safety. Nutritional integrity means ensuring our food is genuine, safe, and meets health standards. It refers to the maintenance of the nutritional quality and safety of food throughout the entire food supply chain. The key aspects of nutritional integrity consist of food quality, food authenticity and safety food (Bodner 2021). Food quality includes both objective and subjective features of the food that meet consumer expectations, such as taste, texture, and nutritional content (Ahmad 2014; Panea and Ripoll 2020). Food authenticity on the other hand, is the ability to trace the movement of food through various stages of the production, processing, and distribution to ensure it is genuine and unaltered that ensures food products align with its labelled identity on it (Cajka and Hajslova 2011). Food safety is ensuring that food is free from harmful contaminants and safe for consumption, which is critical for preventing foodborne illness and maintaining public health (Miller and Bonete 2008). It involves managing risks associated with chemical, physical, and biological contaminants throughout the food supply chain, from production to consumption (Hedberg 2017). Nutritional imbalances or deficiencies can lead to various health issues, including obesity, cardiovascular diseases, and metabolic disorders and it also delay recovery from injuries. By maintaining high standard of food safety, we can prevent foodborne illness and ensure that the nutritional value of food is preserved, supporting overall health and well-being.

Musculoskeletal health, nutritional integrity and food safety is interconnected, each influencing other in profound way. Proper nutrition is essential for growth, development, and the prevention of chronic diseases. Nutrients like proteins, vitamins, and minerals are crucial for muscle strength, bone density, and overall tissue repair (Webster et al. 2023). A balanced diet rich in fruits, nuts, legumes, and low in fat can reduce musculoskeletal pain and improve overall health (Kurapatti and Carreira 2023). Food safety is a crucial part of nutritional integrity as it ensures the food safety, authenticity and nutritional quality, which in turn supporting musculoskeletal functioning and overall well-being. When nutritional integrity is compromised, food safety is also compromised; nutritional quality is diminished, leading to adverse effects on health. Ensuring food safety is critical for preventing foodborne illness that is capable of deteriorating the functionality of various systems in our body. Contamination in food can lead to malnutrition, affecting muscle and bone health, which can further compromise overall health. Nutritional imbalances or deficiencies can lead to various health issues, including obesity, cardiovascular diseases, and metabolic disorders and it also delay recovery from injuries. Without a properly functioning musculoskeletal system, it is impossible to achieve optimum functionality, recover from injury, and maintaining overall health, as musculoskeletal health is critical for maintaining overall physical function and quality of life. A well maintained musculoskeletal system is critical for leading an active lifestyle free from pain, injury and weakness.

15.2 Anatomy and Physiology of the Musculoskeletal System

The anatomy of human body is architected with a hierarchical structure, ranging from atoms and molecules to cells, tissues, organs and various systems that make it functioning. Of these, the musculoskeletal system is central, adding structural support, movement, and metabolic control. An understanding of its anatomy and physiology is basic to fields such as medicine, sports science, and rehabilitation, and especially pertinent to nutrition and food safety. Adequate nutrition affects muscle function, energy metabolism, and tissue repair, and impaired food safety may cause nutrient deficiencies that impact musculoskeletal health. The anatomy of musculoskeletal system comprised of various complex organs such as muscles, bones, cartilages and connective tissues like ligaments and tendons.

15.2.1 Anatomy and Physiology of Muscles

Muscles are a specialised contractile tissue in our body that is capable of producing force and causing movements. It is composed of elongated contractile cells with specific morpho-functional characteristics (Banerjee et al. 2023). Muscle converts chemical energy into mechanical work, enabling various bodily functions such as locomotion and the movement of internal fluids (Swartz et al. 2009). There are primarily three types of muscle tissue like skeletal muscles, cardiac muscles and smooth muscles, which possess unique characteristics and functions.

15.2.1.1 Skeletal Muscles

Skeletal muscles form an essential part of the human anatomy, making up about 40–45% of the body weight (Brown and Sewry 2017). They are structured in a way composed of long cylindrical fibres with multiple nuclei located at the periphery, which is striated, meaning it has a banded appearance under a microscope (Banerjee et al. 2023). Each muscle fibre is packed with myofibrils, which are the contractile elements of the muscle. The basic functional unit in charge of muscle contraction is sarcomeres, which are repeating units that make up myofibrils (Abdalla et al. 2021). Sarcomeres contain actin (thin) and myosin (thick) filaments. The sliding of these filaments past each other, powered by ATP, the fuel of muscles that leads to muscle contraction (Myers 2024). Each muscle fibre is surrounded by connective tissue layers: endomysium, which is around individual fibres; perimysium, which is around bundles of fibres; and epimysium, which is around the entire muscles (Robbins 2022). These muscles produce movement, maintain posture, and provide stability across joints (Allen et al. 2024). Muscles are attached to bones via tendons, which transmit the force generated by muscle contraction to the skeletal (Saleem et al. 2011), so they precisely control voluntary movements. They play an essential role in the thermogenesis and secretion of myokines (Feng et al. 2024; Takagaki et al. 2012). They are also helping us maintain our posture and breathing, playing a significant part in metabolism, serving as a major site for glucose uptake and storage, and as an amino acid reservoir (Argilés et al. 2016). The skeletal muscle contraction mechanism is initiated by excitation-contraction coupling, where an action potential triggers calcium release, leading to muscle fibre contraction (Garg and Goyal 2024). They can perform various types of movements, such as isometric contractions, where the muscle length remain unchanged while tension increases; concentric contractions, in which the muscle shortens; and eccentric contractions, where the muscle lengthens while maintaining tension (Robbins 2022). Skeletal muscles are extremely plastic and adjust to a range of stimuli, including exercise, nutritional status, hormonal changes, disease, and aging (Egan and Sharples 2023; Frontera and Ochala 2015). The adaptability of muscles is evident in changes in muscle size, force, endurance, and contractile velocity. Nutritional integrity and food safety are very crucial aspects of maintaining muscle health. Nutrients like carbohydrates, proteins, and amino acids are critical for muscle protein synthesis, glycogen resynthesis, and inhibition of muscle protein breakdown after exercise (Bird et al. 2024). Diets, such as vegan or vegetarian, can modulate muscle adaptation through changes in mitochondrial abundance, muscle capillary density, and other physiological variables (Pohl et al. 2021).

15.2.1.2 Cardiac Muscles

Cardiac muscle or heart muscle, also known as myocardium, consists of specialized myocytes, which are responsible for the contractile function of the heart. These striated cells, similar to those found in skeletal muscle, possess actin and myosin filaments arranged into sarcomeres (Kirkman 2018, 2021; Knight et al. 2015). They are short, branched, and made up of a single central nucleus. The major contractile proteins are actin, myosin, tropomyosin, and troponin. Myosin is responsible for thick filaments, and actin is responsible for thin filaments, with tropomyosin and troponin bound to the thin filaments (Kirkman 2018, 2021). Intercalated discs unite these cells, enabling them to act as a coordinated unit (Kiss 2010). The fundamental functions of cardiac muscles are contraction and pumping. The heart can pump blood involuntarily in constant motion, and cardiac muscle provides this constant motion through synchronized contractions of the heart chambers, driven by electrical impulses generated by pacemaker cells and conducted through the myocardium (Barnett 2015; Kiss 2010). The heart operates as a dual pump: the right side pumps deoxygenated blood to the lungs, while the left side pumps oxygenated blood to the rest of the body (Weinhaus 2015; Weinhaus and Roberts 2005). Cardiac muscle fibers contain a dense population of mitochondria to supply the constant energy requirements of the heart, providing efficient ATP production for prolonged contraction (Kirkman 2021). The cardiac muscle accommodates different physiological and pathological states by modifying its structure and function. This encompasses remodelling due to enhanced workload or injury, which may influence the contractility and performance of the heart (Barnett 2015).

Food safety and nutrition is very crucial in the proper functioning of cardiac muscles. Although current evidence is conflicting, an equilibrated diet, especially one high in n-3 fatty acids, benefits cardiovascular health and may lower heart disease (Block and Pearson 2006). Good diet can also reduce heart disease-related diseases such clonal haematopoiesis of undetermined potential (CHIP) (Bhattacharya et al. 2021). Food safety also protects cardiac muscle against contaminants including marine toxins and bisphenol A (BPA), which can damage the heart (Ferreiro et al. 2016; Wang et al. 2022). A healthy heart requires sufficient diet and food safety.

15.2.1.3 Smooth Muscles

Smooth muscles are another type of muscle tissue found all over the body, especially in the walls of hollow organs and structures like blood vessels, airways, the gastrointestinal system, the uterus, and the bladder (Jaslove and Nelson 2018; Wilson 2011). They are fusiform (spindle-shaped), with a central nucleus, abundant endoplasmic reticulum, and Golgi apparatus (Wilson 2011). They taper towards the poles and are joined by specialised connectors that allow mechanical and electrical interaction (Sanders et al. 2014). They form layers or sheets in the walls of hollow organs including blood arteries, intestines, and bladders. They are made to contract and relax quickly, which makes it easier to transfer and control the luminal contents. They are responsible for various involuntary movements and functions, including: regulating the diameter of blood vessels to control blood flow and pressure (Brozovich et al. 2016; Wilson 2011), propelling contents through the gastrointestinal tract via peristaltic movements (Wilson 2011), controlling airflow in the respiratory system by adjusting the diameter of the airways (Zhang et al. 2021), facilitating childbirth by contracting the uterus(Mahapatra and Kumar 2024), and maintaining urinary continence by controlling the bladder and urethra (Mahapatra and Kumar 2024; Wilson 2011) etc. They exhibit mechanical plasticity, allowing them to adapt to challenges in length without losing their ability to generate force (Fust and Stephens 2005). They are able to maintain tension for prolonged periods with minimal energy expenditure, which is essential for functions like maintaining vascular tone (Wilson 2011). They can adapt to various physiological and pathological conditions, such as changes in blood pressure or airway resistance. Nutritional safety and quality is highly significant in optimal functioning of smooth muscle also. Nutrients such as vitamin A and selenium are critical to smooth muscle function as they assist in the regulation of contractions and the overall balance of the muscle (Chen et al. 2018). A healthy diet from safe sources not only enhances resistance to disease but also ensures the smooth muscle can respond to mechanical stimuli necessary for optimal function.

15.2.2 Anatomy and Physiology of Skeleton & Impact of Nutrition

The human skeletal system, made up of 206 bones, is the structural that supports of the human body that allows for mobility by serving as attachment points for muscles and protecting essential organs such as the brain, heart, and lungs from harm (Gilliland and Kernick 2021; Rauner et al. 2012; Sutton 2012). In addition to these structural roles, bones have a vital function in blood cell formation through haematopoiesis and act as stores of essential minerals such as calcium and phosphorus needed for various metabolic functions (Rauner et al. 2012). They are also helping in metabolic functions as they maintain acid-base balance and detoxify the body by removing heavy metals from the blood (Ercan and Webster 2014). Bones are continually reshaped through bone remodelling by osteoclasts and osteoblasts. Osteoclasts cells are responsible for breaking down old bone tissue through a process known as bone resorption, and, osteoblasts cells helps in generating new bone issues by producing organic components of bone and aiding mineralisation (Cvek et al. 2015). Osteocytes are mature bone cells that detect mechanical stress and signal to osteoclasts and osteoblasts (Morgan and Gerstenfeld 2020). Bones form through two primary mechanisms: intramembranous ossification, in which some bones, such as flat bones, develop directly from mesenchymal stem cells without a cartilage stage, and endochondral ossification, in which other bones, particularly long bones, begin as cartilage models and are gradually replaced by bone tissues (Fernandes et al. 2016; Sutton 2012). Bones consist of the following composition and structure: an organic matrix consisting mainly of collagen, which gives flexibility and tensile strength; mineral content consisting mainly of hydroxyapatite, which gives rigidity and strength; and water, which is needed for the physiological functions of the bone (Boyd and Lagacé 2017; Iyer 2013). Maintaining strong bones requires a regular supply of minerals such as calcium and vitamin D, which the body regulates through a variety of ways. Therefore, a diet rich in minerals and vitamins, evenly balanced with all nutrients, are essential in maintaining bone health. Food safety is equally essential, as tainted or improperly handled food can block nutrient absorption and overall bone health. Even after an individual's death, bones can speak, as scientists can learn how human health has been shaped by the nutrition over time through analyses of ancient skeletons.

Table 15.1 Different types of bones in human body

Type of bone	Description	Examples	Reference
Long bones	Longer than wide, with a hollow shaft and flared ends. Mostly compact bone with spongy ends.	Femur, Tibia, Humerus.	Bernard (2017), Collins et al. (2018), and Fernandes et al. (2016)
Short bones	Cube-shaped, provide stability with minimal movement. Thin outer layer of compact bone with a spongy core.	Carpals, Tarsals	Bernard (2017) and Collins et al. (2018)
Flat bones	Thin, flat, often curved. Two compact layers with a spongy middle.	Skull bones, Scapula, Sternum	Bernard (2017), Collins et al. (2018), and Fernandes et al. (2016)
Irregular bones	Complex shapes, do not fit other categories. Thin compact outer layer with a spongy core.	Vertebrae, Pelvis	Bernard (2017) and Collins et al. (2018)

15.2.2.1 Types of Bones and Joints

To understand the different types of bones in the human body, it is essential to categorise them based on their shapes and structures. As we already mentioned, the human skeleton comprised of 206 bones, which can be classified into four main types: long bones, short bones, flat bones, and irregular bones (Table 15.1).

Long bones: They are characterized by an elongated shape and wide structure, consist of a hollow diaphysis, also known as shaft that flares at the ends to form the metaphysis and epiphysis (Collins et al. 2018). The diaphysis is made up of mainly of cortical bone which is dense and solid, while the metaphysis and epiphysis contain trabecular bone, a spongy network surrounding the bone marrow (Bernard 2017). Femur, or thigh bone, tibia, also known as shin bone, humerus in the upper arm, radius and ulna from forearm, and fibula in the lower leg, are the major examples of long bones in our body.

Short bones: These bones are generally characterized by their roughly equal dimensions in length, width and thickness which, gives them cube-like shape (Bernard 2017). They consist of mainly of trabecular bone, which is a spongy, porous type of bones tissue, surrounded by a thin layer of compact cortical bone (Collins et al. 2018; Hassan and Ghasem-Zadeh 2013). Short bones are typically found in areas of the skeleton that require stability and support with limited movement. Common examples include bones of the wrists, also known as carpals that are the eight small bones which make up the wrist, and the bones of the ankles (tarsals), the seven bones that form the ankle and part of the foot.

Flat bones: They are characterized by their thin, flattened shape and broad surface. They serve protective and structural roles in the body. They have a unique structure that includes cortical bone (compact bone), which is the dense, hard outer layer that provides strength and protection, and trabecular bone (spongy bone), which found inside the flat bones, this spongy, porous bone contains bone marrow and is involved in metabolic activities such as the production of blood cells

(Fernandes et al. 2016; Hassan and Ghasem-Zadeh 2013). Flat bones are formed through a process called intramembranous ossification where bone develops directly from sheets of mesenchymal connective tissue, rather than form a cartilage template. Skull bones, scapula or shoulder blade, sternum or breast bone, ribs, and pelvis (ilium) are the main examples of flat bones in our body.

Irregular bones: These type of bones, such as the skull bones and scapula, are composed very uniquely that do not fit in to the other bone shape categories (long, short, or flat) due to their complex shape (Bernard 2017; Collins et al. 2018). Their intricate shape include projections, ridges, and varied surfaces, which are not uniform (Luo et al. 2021). Like other bones, they consist of an exterior surface of dense compact or cortical bone and an inner meshwork of variously arranged spicules of trabecular or spongy bone, with the thickness of cortex (Bernard 2017). These bones contribute to the mechanical support, muscle attachment, and protection of vital organs in the human body. The properties of irregular bones including their stiffness and toughness are continuously remodelled by bone cells, allowing them to respond to various mechanical loads and adapt to stress (Fyhrie 2009). The hierarchical microstructure of these bones is the key in providing optimal mechanical features, contributing to their functional capabilities. Bones that form spinal column (vertebrae), hip bones (pelvis), facial bones, and scapula (shoulder blade) are some examples of irregular bones in human body (Collins et al. 2018; Luo et al. 2021).

15.2.2.2 Overview of Joints

A joint is a union site of two or more bony structures, which facilitate movements and provide mechanical support to the human body (Table 15.2). The anatomical structure of a joint in human body composed of cartilage, synovial membrane, ligaments, and tendons (Camba and Rueda 2022). They perform several critical functions, including movement, allowing for a range of motion from simple bending to complex movements, providing mechanical support to the body, maintaining stability for body posture, and distributing forces during activities like walking and running (Amirouche and Koh 2021; Camba and Rueda 2022; Leumann et al. 2011).

Table 15.2 Classification of joints by structure

Type	Description	Examples	Reference
Fibrous joints	Bones connected by fibrous tissue, allowing little to no movement	Skull sutures, teeth and mandible	Laversuch (2014) and Salva and Merrill (2017)
Cartilaginous joints	Bones connected by cartilage, allowing limited movement	Intervertebral discs, pubic symphysis	Laversuch (2014) and Salva and Merrill (2017)
Synovial joints	Complex joints with a synovial cavity, allowing a wide range of movements	Knee, hip, shoulder	Laversuch (2014), Salva and Merrill (2017), and Wang et al. (2016)

Joints are classified into synovial and non-synovial, based on their range of joint motion and internal structure, with synovial joints being the most important from a functional point of view (Camba and Rueda 2022). Synovial joints are movable type, characterized by a joint cavity filled with synovial fluid (e.g. knee, hip). Non-synovial joints include fibrous joints, where bones are connected by fibrous tissue, allowing little or no movements, and cartilaginous joints, where bones are connected by cartilage, allowing limited movement (e.g. intervertebral discs). Based on the functional aspects of the joints, they are classified into synarthroses that is immovable joints (e.g. sutures in the skull), amphiarthroses, also known as slightly movable joints (e.g. intervertebral discs), and diarthroses or freely movable joints (e.g. shoulder, knee) (Usgu and Usgu 2023). Joints require various nutrients including calcium, phosphorus, protein, vitamins C, D, E, omega-3 fatty acid and zinc, to stay healthy and regenerate tissue (Clark 2007). It is essential to ensure the supply of these nutrients with proper exercises to maintain the optimum health of joints.

15.2.2.3 Ligaments

A ligament is a specialised band of connective tissue that primarily connects bones to other bones, usually across joint spaces. They are crucial for maintaining joint stability and guiding the joint motion, which ensure that movements are smooth and controlled (Yao et al. 2020). Ligaments are composed mainly of densely packed collagen fibers arranged in a parallel fashion (Innocenti and Galbusera 2022). They also contain elastin, proteoglycans, and water, which contribute to their unique biomechanical properties (Woo and Sites 2018). The main function of ligaments is to provide stability to joints by restricting excessive movement and maintaining proper alignment of the bones. They play an important role in both static and dynamic support of joints (Ma et al. 2021a; Yao et al. 2020). Ligaments exhibit viscoelastic behaviour, meaning they can stretch and return to their original shape. This property allows them to absorb and dissipate loads, which in turn protecting the joints from injury (Innocenti and Galbusera 2022). When ligaments are injured, they undergo significant structural and physiological changes. The healing process involves the formation of scar tissue, which is often biomechanically inferior to the original ligament tissue (Frank 2004). Anterior cruciate ligament (ACL) located in the knee, is one of the most studied ligaments due to its importance in knee stability and its susceptibility to injury (Frank 2004). Medial collateral ligament (MCL) another crucial ligament in the knee, providing medial stability, and the deltoid ligament also known as the medial collateral ligament complex, is another example, which is situated in the ankle that stabilizes the medial side of the joints (Frank 2004; Hintermann and Golanó 2014). Nutritional integrity is very important or the health of ligaments, as this provides essential nutrients that support their structure and function. Nutrients such as vitamins, minerals, and polyunsaturated fatty acids are very crucial for ligament health. A study on athletes with knee joint osteoartherosis showed that a complex nutritional support programme, including chondroprotective agents, minerals, vitamins, and fatty acids, significantly improved movement activity and reduced pain (Tolmachyov et al. 2019). Food safety is very important in maintaining this nutritional integrity by preventing contamination and ensuring the availability of high quality nutritious food.

15.2.2.4 Tendons

Tendons are strong connective tissues that connect muscles to bones, which is designed to transmit the force generated by muscles to the skeletal system (Zaffagnini et al. 2021). The basic function of tendons is to transfer muscle-generated force to bones, enabling joint movements, and also, they are capable of withstanding high mechanical loads, and have high tensile strength, which allows them to efficiently transmit muscular forces with minimal energy loss (Thorpe et al. 2015; Zaffagnini et al. 2021). They serve as shock absorbers, energy storage sites, and help to maintain posture because of their sensitivity to body positions (O'Brien 2005). In terms of their structural aspects, they are predominantly composed of collagen fibers, which are organized in a hierarchical levels, each surrounded by a non-collagenous matrix, mainly composed of type I collagen, and elongate with a typical stress-strain pattern, which provide high tensile strength and resistance to mechanical stress that leads to injury (Thorpe et al. 2015; Thorpe and Screen 2016; Zaffagnini et al. 2021). They are made up of collagen molecules structured in fibrils, which are subsequently gathered in fibrils bundles, fascicles, and fiber bundles, referred to as primary, secondary, and tertiary bundles (Franchi et al. 2007). Nutritional integrity and food safety is very crucial in the maintenance of tendons also. Vitamin C and D, along with minerals like manganese, copper, and zinc, are essential for maintaining tendon health and aiding in tendon repair processes (Curtis 2016). Amino acids such as leucine, arginine, and glutamine are very important for tendon growth and healing. Maintaining proper food safety can ensure the proper supply of these nutrients that impact overall tendon health.

15.2.2.5 Cartilages

Cartilages are the specialised dense connective tissue that provide structural support to the body and is primarily composed of chondrocytes, extracellular matrix (ECM), collagen, and proteoglycans (Yasui and Nimni 2018). They are one of a kind without their own blood supply or nerves, So, they receive nutrients via diffusion from the surrounding matrix (ECM), which includes branching network of collagen fibers embedded in proteoglycan aggregates (Yasui and Nimni 2018). This structure allows cartilage to be flexible and bear pressure. Cartilages are classified as hyaline, fibrous, and elastic, in which hyaline cartilage is the most common type that is characterized by a smooth surface, abundant collagen fibrils, and chondrocytes (Jabbari et al. 2023). Fibrocartilage is the toughest type due to dense bundles of collagen fibers, while elastic cartilage is flexible and resilient due to the presence of elastic fibers (Jabbari et al. 2023). Cartilages are responsible for evenly distributing loads onto the subchondral bones surface and provide a nearly frictionless surface during dynamic loading (Putz 2008). It act as a load-bearing medium that protects the bone and distributes stress, thereby reducing load per unit area and providing a smooth, low-friction, and wear-resistant bearing in joints (Zhang et al. 2010). The unique structure of articular cartilage enables it to withstand and resist excessive compressive and biomechanical loads which facilitate almost frictionless joint movement (Gahunia and Pritzker 2020). Maintaining the quality of diet and availability of essential nutrients such as glucose, amino acids, vitamins, especially vitamin C and D, and trace minerals like zinc, magnesium, and copper, are very important for the development, maintenance, repair and remodelling of cartilages (Dydykina et al. 2022). The synthesis of collagen, a major component of cartilage, depends on adequate intake of vitamin C, is also play a major role. As we discussed, maintaining nutritional integrity through a balanced diet rich in essential nutrients is very crucial for all the systems in our body including the musculoskeletal system. Ensuring food safety through stringent quality control measures is vital to prevent degradation of the quality of diet. An integrated approach that combines food safety and nutrition can significantly contribute to the prevention and management of the health of these systems.

15.3 Key Nutrients for Musculoskeletal Health

The fundamental basis of musculoskeletal health depends on proper nutrition, as bones, muscles, tendons and ligaments require particular nutrients for their growth, maintenance and repair. Nutritional quality stands essential for skeletal development while promoting muscle performance and reducing the risk of osteoporosis and sarcopenia as degenerative conditions. The integrity of the musculoskeletal system becomes vulnerable when nutrients are deficient because it creates a higher potential for bone fractures and reduces functional physical abilities. A comprehensive bone and muscle structure requires multiple essential nutrients in order to construct strong bones and execute functions along with maintaining appropriate structural conditions. The process of bone mineralisation, with muscle development needs specific vitamins and minerals and macronutrients. Certain nutrients serve many important functions, like reducing tissue inflammation and protecting the musculoskeletal tissue from oxidative damage to avoid injuries and tissue degeneration. The following section examines the physiological roles of these nutrients, the dietary source of these nutrients, and their impacts on the musculoskeletal system.

15.3.1 Macronutrients

Macronutrients are important components of the human diet that we need in large amounts. They are essential to support growth, body operations, and energy production. These are the fundamental macronutrients: carbohydrates, protein and fats- each serve distinct physiological roles (Matthewman and Costa-Pinto 2023). The body requires carbohydrates as essential macronutrients since they act as its main energy provider. The human body converts carbohydrates into glucose that powers ATP generation through glycolysis and oxidative phosphorylation (Stylianopoulou 2023). The energy supports cellular operations, muscle movements, and brain activities; therefore, carbohydrates are necessary for metabolic wellness, endurance and physiological well-being. The body processes two categories of carbohydrates, called simple and complex carbohydrates, which possess different molecular structures and body effects. The body requires proteins as much as carbohydrates to sustain health and functional operation since they enable muscle repair, enzyme operation, and structural integrity. These compounds base themselves on amino acids with essential and non-essential categories. Living beings require intake of vital amino acids as these cannot be produced in the human body, while the body may produce some of the non-essential amino acids naturally (McCann and Rawls 2023). Fat also plays an essential role in musculoskeletal health because it stabilises cell membranes and hormone production and enables energy metabolic functions. Essential fatty acids enable proper muscle functioning as well as tissue repair and help to regulate inflammation in the body (Nigam et al. 2018). Fat consumption improves the absorption of fat-soluble vitamins, which strengthen bones, metabolic functions, and nutritional equilibrium, so they function as essential components in physiological well-being (Reboul 2023). So the proportionate consumption of macronutrients remains vital to developing strong bones and supporting musculoskeletal functions, and metabolic operations. A poor diet, either inadequate or excessive nutritional intake, leads to osteoporosis metabolic imbalances and reduces athletic capabilities, thus showing the necessity for balanced dietary consumption to maintain health and physical resilience.

15.3.1.1 Proteins: Importance in Muscle Repair and Growth

As an essential part of the diet, protein enables various vital body processes, including muscle repair and growth and the strengthening of the musculoskeletal system. The structure of proteins consists of amino acids that serve as building material to develop tissues such as muscles, tendons and ligaments. The fundamental function of protein in the diet is to stimulate muscle protein synthesis, which is essential for tissue regeneration, growth and recovery. Protein consumption immediately after exercise may play a pivotal role in skeletal muscle adaptation by facilitating muscle repair and growth. The successful recovery and maximum muscle protein synthesis (MSP) requires individuals to consume approximately 20 grams of high-quality protein immediately after exercise, such as whey protein. Optimising protein intake throughout the day and before sleep may improve MPS, which enables sustained muscle repair and adaptation for better growth outcomes that improve musculoskeletal health and athletic performance (van Loon 2013). Protein intake along with resistance training, increases the dietary protein sensitivity in skeletal muscles, which maintains an efficient time frame for the use of essential amino acids in muscle protein synthesis. The combination of physical training and nutrition plays a significant role in promoting muscle growth, supporting recovery functions and optimising overall musculoskeletal adaptation and peak performance (Deane et al. 2021). The natural ageing process causes anabolic resistance in muscles that makes them less sensitive to dietary protein. Increasing dietary protein intake or adopting leucine (essential amino acid) supplementation may promote proper muscle functioning and overall musculoskeletal health (Deane et al. 2021). Previous research suggests that an effective way to improve muscle protein synthesis requires equal distribution of protein-rich food in each meal containing 25–30 grams of protein. Regular protein distribution of protein content between meals helps maintain a balanced protein supply, which may help improve muscle repair and metabolic health throughout the day (Mamerow et al. 2014).

Amino acids: Leucine and branched-Chain Amino Acids (BCAAs)
Several studies suggest that branched-chain amino acids, especially leucine are crucial in various muscle functioning and supporting overall health. They are very useful in sports nutrition and can help to improve exercise performance and muscle recovery. The muscle protein synthesis, and metabolic regulation and health promotion benefits stem from leucine, which is part of BCAAs. The positive effects of branched-chain amino acids for sports and medical use require careful dietary balance to avoid negative responses from excessive consumption or improper dietary choices. Muscle protein synthesis and body metabolisms as well as skeletal health significantly depend on branched-chain amino acids (BCAAs) consisting of leucine, isoleucine, and valine. Leucine stands out from other BCAAs because it operates through the mTOR signalling pathway which supports muscle expansion and restores the body after physical activities (Gorissen and Phillips 2018). During physical exercise, BCAAs function as fuel while stopping the degradation of muscle proteins and improving the muscle healing process. The metabolic processes through branched-chain aminotransferase (BCAT) and branched-chain α-keto acid dehydrogenase (BCKDH) enzymatic pathways enable energy production, especially when performing prolonged physical activities (Bo and Fujii 2024). Sports athletes typically use BCAA supplements as a means to preserve their muscles and minimise fatigue during training. Expert groups have diverse perspectives on how well BCAAs work as alternatives to complete proteins when measured against one another. Dietary consumption of BCAAs at proper proportions plays a vital role to sustain metabolic energy levels and musculoskeletal system operations.

Importance of Essential Amino Acids over Non-Essential Amino Acids
The human body cannot synthesise the essential amino acids which people need to get from their food. The nine EAAs—histidine, isoleucine, leucine, lysine, methionine, phenylalanine, threonine, tryptophan, and valine—are crucial for muscle growth and repair. Branched-chain amino acid leucine serves as a primary activator of the mTOR pathway needed for muscle recovery because it activates muscle protein synthesis (MPS) (Gorissen and Phillips 2018). Muscle protein synthesis stimulation and net protein balance enhancement occur independently through essential amino acids, which do not demand additional carbohydrates or amino acids for their

effects. The stimulation of MPS relies primarily on essential amino acids no matter how insulin levels fluctuate, thus establishing essential amino acids as foundational elements for muscle growth (Glynn et al. 2013).

Muscle protein synthesis (MPS) benefits from non-essential amino acids (NEAAs) however, these amino acids do not induce MPS to the same level as essential amino acids (EAAs). Research shows that citrulline (non-essential amino acid) boosts muscle protein synthesis, specifically when people eat fewer proteins. Still, such stimulation proves weaker than what essential amino acids provide (Jourdan et al. 2015). Proper consumption of essential amino acids and non-essential amino acids benefits the development of musculoskeletal health. The muscles of the human body need EAAs to drive the process of muscle protein synthesis for repair and growth, and NEAAs help stimulate the broader protein synthesis pathways. The consumption of proper amino acid balances achieved through varied diet selections and nutritional enhancements will strengthen your body structure while boosting functional capabilities for strength development and recovery.

15.3.1.1.1 Source of High-Quality Protein

Animal-based protein

Proteins from animal sources receive high acclaim for their ability to support human nutrition because they possess powerful biological value together with complete amino acid content. The part includes a review of animal protein types with their properties and nutritional advantages. Animal proteins possess high-quality status because they contain every necessary essential amino acid (EAA) needed for human development and maintenance. The complete amino acid content allows animal proteins to support essential physiological processes, resulting in optimal protein utilisation when building tissues and facilitating recovery (Kumar et al. 2023).

Whey protein: Whey protein shows renowned quick digestion and absorption properties, which make it perfect for post-workout recovery supplements to speed up muscle recovery. The abundant leucine content of this type of protein performs critical functions by boosting muscle protein synthesis which leads to better muscle repair while benefiting entire metabolic systems (Auestad and Layman 2021).

Casein protein: Breaks down at a slow rate, which preserves the amino acid delivery into the bloodstream at a steady rate for several hours. Casein provides an excellent overnight nutrient supply because of its slow digestion, thereby enabling constant muscle repair and minimised protein breakdown during the fasting period (Auestad and Layman 2021).

Egg: The protein found in eggs contains all essential amino acids within perfect ratios, which makes them optimal for building and repairing muscles. The regular consumption of eggs helps build skeletal muscles making your body stronger, and improves your metabolic wellness (Kumar et al. 2023).

Lean meats: All essential amino acids exist in lean meat proteins alongside beef, pork and poultry. The dietary consumption of lean meats supplies essential micronutrients, iron and zinc, which help maintain muscle function while supporting metabolism and overall health.

Plant-based protein

The digestibility of plant proteins remains lower than traditional proteins alongside essential amino acids, including leucine, because leucine strongly promotes muscle protein synthesis. The anabolic reaction from plant-based proteins usually falls below the levels achieved from animal proteins. Plant protein diets receive improvement through essence amino acid fortification and strain selection processes together with protein intake elevation and mixing various plant proteins to reach greater nutritional value and strengthen their capacity to synthesise muscle proteins (Berrazaga et al. 2019). The health of bones requires appropriate protein consumption along with special attention to aging individuals. Increasing the quantity of plant-based proteins or using supplemental strategies may be necessary for maintaining bone strength because these proteins demonstrate reduced digestibility and imperfect amino acid composition.

Soy protein: High-quality soy protein has proved that it contains complete essential amino acids. The consumption of sufficient amounts of soy protein leads to efficient musculoskeletal health support with enhanced muscle mass and strength (Hevia-Larraín et al. 2021). The amino acid profile of soy protein includes all essential nutrients, yet it includes lower concentrations of methionine (Krishnan 2015).

Quinoa: As a nutrient-dense grain quinoa contains protein together with healthy fats as well as fibre and vitamins and minerals. The exceptional amino acid content in quinoa transforms it into a top-level plant protein choice which meets the needs of vegetarians and vegans.

Pea protein: The high-quality plant-based protein in peas contains essential amino acids, including lysine and BCAAs, that help promote muscle development and rehabilitation. Vegan individuals find this plant-based protein highly suitable as it appears frequently in supplements (Dhiman et al. 2023).

15.3.1.2 Carbohydrate and Musculoskeletal Health

The human body relies mainly on carbohydrates as its primary energy source which functions as fundamental macronutrients. Natural foods rich in whole grains, fruits, vegetables, and legumes contain carbohydrates which carbon, hydrogen, and oxygen components that fuel the body through metabolism (Haghighatjoo et al. 2024). The body requires carbohydrates because they enable the production of energy, which supports all muscle development, repair, and musculoskeletal maintenance. Low dietary intake leads to carbohydrate breakdown through glycolysis and, if insufficient, facilitates the breakdown of protein into carbohydrates. Carbohydrates manage energy production as well as control the metabolism of fats and proteins and affect enzyme activity and enable basic cellular operations. The body needs enough carbohydrates to optimise exercise outcomes simultaneously with healing time and minimise joint-related injuries by maintaining energy balance and metabolic efficiency (Giraldo-Vallejo et al. 2023). The body needs carbohydrates to function properly in maintaining musculoskeletal health through support of muscle energy, bone mineral absorption, and the management of metabolism. A nutritious balance of high-quality carbohydrates together with necessary nutrients leads to strong muscles and healthy bones, along with general physical health integrity.

15.3.1.2.1 Primary Energy Source for Muscle Contraction

The main carbohydrate reserve in mammals exists in the form of glycogen, which stores mainly within liver tissues and skeletal muscles. Human bodies use hepatic glycogen as a vital energy storage method to keep blood glucose levels stable during fasting by releasing glucose into the bloodstream. Glycogen in muscles serves many purposes by providing localised energy for contractions during physical activities which stimulates endurance and increases athletic performance as well as strength in different physical movements (Kilimann 2024). During intense, short-term movements and weightlifting activities, muscle glycogen acts as the primary fuel source to produce rapid ATP through incomplete glycolysis. The oxygen-independent pathway generates quick energy bursts through these explosive movements. Athletic performance declines when muscle glycogen supply depletes quickly because of these intensive activities yet the recovery method to restore glycogen properly becomes crucial.

Manual activity gets fatigued primarily due to decreased glycogen levels during long or vigorous exercise. The decrease in glycogen stores leads to insufficient ATP production which weakens muscle function and decreases athletic performance. The shortage of glycogen interrupts calcium release from the sarcoplasmic reticulum, which plays a vital role in effective muscle contractions. Extended physical activity results in the increase of reactive oxygen species (ROS), while these metabolites damage cellular structures and obstruct metabolic enzymes, which leads to faster muscle fatigue. Many athletes understand that preserving sufficient glycogen stores helps maintain redox balance, reduce ROS-related damage and perform better throughout lengthy physical activity.

15.3.1.2.2 Carbohydrates: Impacts on Recovery

Carbohydrates are essential for replenishing muscle glycogen stores, which become depleted during exercise. Immediate post-exercise carbohydrate consumption accelerates glycogen resynthesis, facilitating muscle recovery and optimising performance. The effectiveness of glycogen restoration depends on carbohydrate type, timing, and frequency, with co-ingestion of protein further enhancing recovery and muscle function.

15.3.1.2.3 Role in Glycogen Resynthesis Post-exercise

Exercise drains glycogen stores from muscles, so glycogen resynthesis represents the method used to restore depleted muscle glycogen levels. Glycogen resynthesis ensures a fast recovery because muscle contractions during prolonged exercise rely on this substance as their main energy source. Efficient glycogen resynthesis helps restore functional capacity and endurance performance for subsequent exercise sessions (Alghannam et al. 2018).

Rapid phase (0–2 h Post-Exercise): After exercise for the first 2 h, insulin-independent glucose uptake leads to maximum glycogen resynthesis. The glycogen replenishment rate increases after exercise when people consume carbohydrates at a rate of 1.2 g/kg of their body weight per hour (Carlsohn and Mayer 2010).

Slow phase (2–24 h Post-Exercise): The second phase of glycogen synthesis runs from 2 to 24 h post-exercise during which time insulin-dependent synthesis occurs. Regular consumption of carbohydrates with protein intake during recovery allows optimal glycogen restoration while simultaneously accelerating muscle repair to achieve overall recovery success (Carlsohn and Mayer 2010).

15.3.1.3 Fats: Omega-3 Fatty Acids and Joint Health

Omega-3 fatty acids play a crucial role in mitigating inflammation and supporting joint health, particularly in individuals with rheumatoid arthritis (RA) and osteoarthritis (OA). Regular consumption, whether through dietary sources or supplementation, can alleviate joint pain, enhance mobility, and potentially slow degenerative processes, thereby improving overall musculoskeletal function and quality of life (Souza and Norling 2016). Clinical studies confirm that omega-3 supplements produce substantial improvements for rheumatoid arthritis patients through lowered stiffness and reduced painful and swollen joints, which leads to better joint capabilities and improved health outcomes (Kostoglou-Athanassiou et al. 2020).

Source of Healthy Fats

Fatty Fish: Fatty fish *such* as salmon, mackerel, and sardines are rich in omega-3 polyunsaturated fatty acids (n-3 PUFAs), specifically eicosapentaenoic acid (EPA) and docosahexaenoic acid (DHA).

Plant-Based Sources: Chia seeds, flax seeds, walnuts…

Algae-Based Omega-3 Supplements: Algae are a sustainable and vegan-friendly source of EPA and DHA, making them suitable for vegetarians and those with fish allergies.

Recommended daily intake: The recommended daily consumption of omega-3 fatty acids should include 250 mg of EPA and DHA to benefit overall health and physiological functions (Hernandez 2013).

15.3.1.3.1 Anti-Catabolic Effects of Omega-3s in Preventing Muscle Loss

The preservation of muscle mass depends crucially on including EPA and DHA omega-3 fatty acids because they help decrease protein breakdown. The cell membrane integration of these fatty acids stimulates protein synthesis and blocks oxidative pathways to help maintain healthy functioning muscles (McGlory et al. 2019). Omega-3 fatty acids, renowned for their anti-inflammatory and anabolic effects, contribute to muscle preservation by stimulating protein synthesis and accelerating recovery. Their benefits extend to ageing individuals and athletes, supporting muscle health, mitigating loss, and enhancing post-exercise regeneration.

15.3.2 Micronutrients

Human bodies require small amounts of micronutrients that function as essential vitamins and minerals to support both musculoskeletal physiology and enzyme processes and maintain metabolic balance. Micronutrients differ from macronutrients because they cannot supply energy. However, they remain necessary elements for building bones, maintaining muscle activities, supporting connective tissues and enabling cellular interactions. The three micronutrients classes include fat-soluble vitamins (A, D, E, K) and water-soluble vitamins (B-complex, C) and essential minerals (calcium, iron, magnesium and zinc, copper, and manganese) that bring distinct contributions to both skeletal framework and muscular health.

15.3.2.1 Calcium Role in Bone Density

Calcium is a fundamental mineral for skeletal integrity. Approximately 99% calcium as apatite crystals for bone strength. Bone remodelling through formation and resorption performs two crucial roles in maintaining both bone stability and homeostasis of calcium levels. A positive calcium balance depends on the balance between diet, absorption and renal reabsorption to achieve maximum bone mineral density. Low calcium consumption diminishes bone formation processes, increasing the risk of osteoporosis and fractures, underscoring the importance of maintaining adequate calcium levels for lifelong skeletal health.

Dietary Source of Calcium

Dairy Products: Milk, Yogurt, Cheese: Dairy products are rich in calcium and have high calcium absorption rates, making them excellent for preventing osteoporosis and maintaining bone health (Huertas et al. 2019).

Fortified Plant-Based Alternatives: Almond Milk, Soy Milk: These plant-based kinds of milk are often fortified with calcium to provide an alternative for those who are lactose intolerant or prefer a vegan diet (Muleya et al. 2024).

Leafy Green Vegetables: Kale, Broccoli, Parsley, Leek: Leafy greens are good sources of calcium (Garden-Robinson 2015).

Fish with Edible Bones: Sardines, Salmon: Fish bones are rich in calcium and can be used to fortify foods. The calcium from fish bones is highly bioavailable, making them a valuable source of dietary calcium (Pérez et al. 2024).

15.3.2.1.1 Vitamin-D: Synergy with Calcium

To increase intestinal calcium uptake, the body needs an active form of vitamin D. The body maintains bone strength and mineralisation through active transport and passive diffusion mechanisms, which ensures adequate calcium availability. Vitamin D works with parathyroid hormone (PTH) to help the body move calcium from bones into the bloodstream for proper calcium balance when dietary calcium levels are low. This process stops hypocalcemic effects on body functions (Holick 2013). The preservation of healthy vitamin D amounts helps sustain balanced blood calcium levels while supporting all essential physiological functions such as nerve transmission, muscular responses, bone integrity for general skeletal condition, and metabolic equilibrium. Individuals require proper vitamin D levels to preserve their bones while avoiding rickets and similar musculoskeletal conditions like osteomalacia and osteoporosis. Insufficient vitamin D levels increase bone absorption, which leads to higher fracture possibilities and greater fall vulnerability. Experts recommend keeping serum vitamin D 25-hydroxy levels above 20 ng/mL while the optimal range falls between 20–30 ng/mL. The skeletal system depends upon vitamin D for regulating osteoblasts while it activates bone formation and mineralisation to sustain structural integrity (Han et al. 2022).

Sources of vitamin D:

Sunlight Exposure (UVB Radiation: The primary natural source of vitamin D is sunlight.

Fatty Fish: Fatty fish such as salmon, mackerel, and fish liver oils are rich natural sources of vitamin D3.

Other Natural Sources: Other sources include egg yolk, offal (e.g. liver, kidney), and to a lesser extent, meat and dairy products (non-fortified). Mushrooms are a notable plant-based source (Benedik 2022).

15.3.2.2 Magnesium: Essential for Bone Mineralisation and ATP Production

The mineral magnesium stands crucial for bone health as it performs dual functions in bone mineralisation and its ability to maintain appropriate bone density. Bone tissue contains most of its magnesium storage because this crucial mineral helps develop the skeletal structure and manage bone loss and formation processes. The lack of magnesium in the body weakens bone structure and raises the possibility of osteoporosis-related bone fractures. Increased magnesium levels in food consumption lead to better bone density. Beyond bone health, magnesium is crucial for muscle function because it aids neuromuscular signalling and regulating muscle dynamics and acts as an essential compound in the process of creating ATP for energy metabolism (Groenendijk et al. 2022).

Sources of Magnesium: Leafy Green Vegetables, Nuts and Seeds, Whole Grains, Legumes.
Daily Recommendation: Men: 400–420 mg/day.

Women: 310–320 mg/day (Shim et al. 2023).

15.3.2.3 Collagen: Major Component of Tendons, Ligaments, and Cartilage

The extracellular matrix (ECM) depends primarily on collagen as its main structural protein, which supports connective tissue elements such as cartilage, tendons and ligaments. The main role of collagen is to deliver mechanical stability and structural integrity to these tissues, which enables their appropriate functioning.

15.3.2.3.1 Role in Tendons and Ligaments

Tendons and ligaments maintain their tensile strength because collagen fibers organise themselves parallel to each other to distribute mechanical forces throughout movement efficiently. The ordered arrangement of collagen from molecular to fibrils to fiber levels strengthens these mechanical features and the resilience of these tissues. The organisation sustains the tendon and ligament structures, which enables these tissues to resist repeated strain while upholding joint stability.

15.3.2.3.2 Role in Cartilage

Cartilage depends on collagen to create a structural matrix which both protects against compression forces and retains flexibility and shock-absorption capabilities. Cartilage differs from tendons and ligaments because its collagen arrangement lacks the dense organisation of these tissues. Because of this looser organisation, cartilage stays intact through pressure while still preserving its integrity, making it essential for cushioning joints and facilitating smooth movement (Burla et al. 2020).

Dietary Sources: Bone Broth, fish…
Collagen Supplements: Hydrolysed collagen peptides, marine collagen peptides.

15.3.2.4 Iron: Crucial for Oxygen Transport

Every red blood cell depends on iron molecules to create the protein hemoglobin that moves oxygen between lung tissue and body tissues. The oxygen binding capabilities of hemoglobin depend on its four iron atoms that exist within each molecule. Muscle cells contain iron as an essential component of myoglobin, through which they store oxygen for generating muscle contractions, especially during intense physical activity. A deficiency of iron prevents the normal creation of hemoglobin which results in weak muscles, increased fatigue, and reduced endurance capabilities. Low iron levels among athletes and others result in performance reductions because aerobic energy production declines and delayed muscle recovery after exertion (Solberg and Reikvam 2023). So, ensuring sufficient iron intake is crucial for optimal musculoskeletal performance and overall health.

Dietary Sources:

- Red Meat: Beef, lamb, and pork
- Poultry: Chicken and turkey
- Fish and Shellfish: Particularly rich in heme iron
- Lentils and Beans: Lentils, chickpeas, and other legumes.
- Spinach and Other Leafy Greens: Spinach, kale, and other dark leafy greens (Vasconcelos and Grusak 2006).

15.4 Role of Nutrition in Muscle Mass and Function

A proper diet rich in protein, carbs, other macronutrients, vitamins, minerals, and other supplements is crucial for building muscle. Muscle protein synthesis depends on protein, which is especially important after exercise. Carbohydrates give muscles the energy they need to work and recover, while lipids and other nutrients help them stay healthy. For the best possible muscular growth and preservation throughout life, a healthy diet and frequent exercise, especially resistance training, are necessary.

15.4.1 Muscle Hypertrophy

Muscle hypertrophy denotes the augmentation of muscle size, particularly the expansion of muscle fibres. This process entails an increase in the cross-sectional area of individual muscle fibres, primarily through accumulating muscle-specific proteins and synthesising new proteins (Alway 2006). Muscle hypertrophy is an adaptive reaction to increased workload or resistance training. It occurs when the muscle fibres adapt to the stress placed upon them by increasing size (Alway et al. 2005). The process involves both anabolic (building up), and catabolic (breaking down) pathways. Anabolism promotes muscle growth by enhancing protein synthesis, whereas catabolism facilitates muscle remodelling through protein breakdown (McCarthy and Murach 2018). When protein synthesis exceeds protein degradation, a net increase in muscle mass occurs, leading to larger muscle fibres (Zumbaugh et al. 2022). Muscle hypertrophy affects various fiber types (e.g., type I, IIa, IIx, and IIb), each responding differently to stimuli. Genetic and epigenetic mechanisms are essential in the regulation of muscle hypertrophy. Specific genes and DNA methylation patterns control the signalling pathways necessary for muscle growth (Colorado et al. 2023). Mechanical stress induced by high-intensity training, and metabolic stress generated by high training volumes with moderate intensities, are key factors influencing muscle hypertrophy (Meloni 2005).

15.4.2 Muscular Atrophy

Muscular atrophy is a phenomenon, which is characterized by a reduction in the size of muscle fibers and a decrease in the overall muscle mass, leading to loss of strength (Cao et al. 2018; Schmiedel and Reichmann 2004). This condition can result partial or complete loss of muscle function. Muscular atrophy involves protein degradation, myofibril shrinkage, and the loss of cytoplasm and organelles. As cellular components deplete, myofibrils become smaller, and shifts occur in muscle fiber type and myosin isoform expression (Dumitru et al. 2018). When protein synthesis is inhibited and protein degradation pathways are activated, an imbalance happens between anabolic catabolic processes. As protein degradation exceeds protein synthesis, this imbalance ultimately leads to muscle loss (Fanzani et al. 2012). Various pathophysiological conditions such as long-term immobilization, malnutrition, severe burns, aging, chronic diseases like heart failure, obstructive lung disease, renal failure, AIDS, cancer etc., and the use of steroidal medications can trigger muscle atrophy (Fanzani et al. 2012; Nikawa et al. 2021). Likewise, decreased skeletal muscle contractile activity (disuse) or unloading, aging-induced anabolic resistance, and neurogenic factors can contribute to the progression of muscle atrophy (Nunes et al. 2022).

15.4.3 How Nutrition Effect Growth and Loss of Muscles?

A high quality diet is a key factor in promoting the muscular hypertrophy and preventing muscular atrophy. When it comes to building muscles, protein is the most important nutrient, as it is the building block of our muscles. So, an adequate supply of protein along with other ancillary nutrients is required for muscle growth. When combined with resistance training, protein consumption promotes muscle hypertrophy and diminishes muscle protein degradation. The ideal daily protein consumption ranges from 1.2 g to 1.8 g per kilogramme of body weight, contingent upon lean mass and energy expenditure. People who aim for muscular hypertrophy should consume between 1.6 g to 1.8 g of protein per kg of body weight per day, with an additional 400–500 kcal in their daily diet to support muscle growth (Urdampilleta et al. 2012). Excessive protein intake beyond 2 g/kg/day does not provide additional benefits and may lead to negative effects such as dehydration. Protein provides essential amino acids necessary for muscle protein synthesis. Amino acids are crucial for skeletal muscle protein synthesis, and the type, amount, and timing of dietary protein consumption are very important for optimizing muscle hypertrophy. There are certain amino acids that cannot be synthesized by human body that must be supplied through proper diet or supplementation. Among this branched-chain amino acids are particularly important that includes luecine, isoleucine, and valine, which supports muscle protein synthesis, muscle metabolism, energy production and tissue repair respectively (Rogero and Tirapegui 2008). Carbohydrates are also very important as they provide the necessary energy for muscle synthesis and repair.

Malnutrition is the root cause of muscular atrophy along with limited or no physical activity. Inadequate nutrition, including both underfeeding and obesity, can lead to muscle atrophy. Proper nutrition is essential to counteract muscle loss associated with various conditions such as aging, chronic diseases, and prolonged immobilization. When there is a caloric debt in our body, it breakdowns muscle tissue for energy, resulting in muscle loss. Sufficient calorie intake prevents the body from breaking down muscle tissue for energy. So a balanced diet rich in protein, carbohydrates, and essential nutrients is the key to promote overall health. Regular intake of high quality protein (e.g. lean meats, dairy, eggs, and legumes) and carbohydrates (e.g. whole grains, fruits) is recommended. Staying hydrated and ensuring adequate intake of vitamins and minerals (e.g. calcium, magnesium, selenium, vitamins D and B) are further enhance the positive effects on muscular health.

Note: This heat map (Fig. 15.1) shows how different factors affect muscle growth. Darker colors indicate stronger effects. The main factors that promote muscle hypertrophy are mechanical stress from resistance training and protein synthesis (e.g., amino acid availability, mTOR signalling), which helps with both growth and repair. Metabolic stress, (e.g., training volume, lactate accumulation) during high-rep workouts, also supports muscle growth by causing cellular swelling and signalling. While nutrition is crucial for muscle growth and recovery, it has a less direct impact compared to training-related factors. Energy availability plays a smaller role in muscle hypertrophy but helps with training performance and recovery. Overall, the visual emphasizes that training is the key factor in muscle growth, while nutrition and metabolic aspects are important for recovery and adaptation.

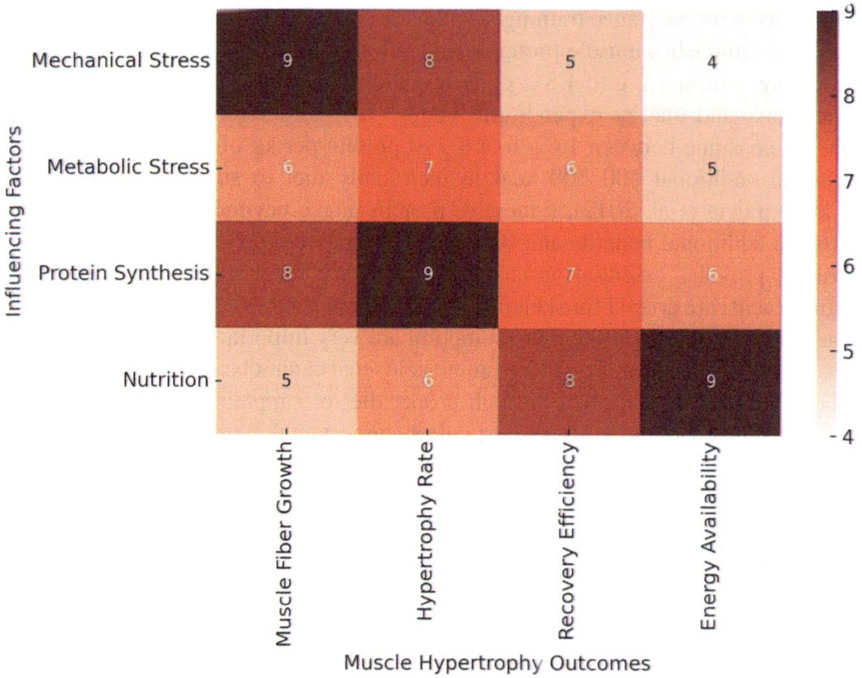

Fig. 15.1 Heat map of factors influencing muscular hypertrophy

15.4.4 Ageing, Sarcopenia and the Role of Nutrition

Ageing is a complex biological process characterized by a gradual decline in physiological function and an increased vulnerability to various diseases and health conditions. It encompasses a range of changes at the cellular, tissue, and organ levels, which in turn leading to diminished regenerative capacity and homeostasis. As individual age, they may experience alterations in metabolic processes, hormonal levels, and immune function, which can contribute to the development of age-related diseases such as cardiovascular disorders, diabetes, neurodegenerative conditions. Sarcopenia is one of the challenging issue that is faced by ageing population, which is a significant age related condition characterized by the progressive loss of skeletal muscle mass, strength, and function (Cruz-Jentoft 2019; Landi et al. 2018a). This condition is associated with numerous adverse health outcomes, including increased risks of falls, fractures, disability and mortality' which collectively contribute to a substantial socio economic burden (Hunter et al. 2019). Age, gender, poor nutrition, and lack of physical activity are the main factors of sarcopenia. Chronic diseases like diabetes and cardiovascular conditions also contribute to its development. Among this nutritional strategies can mitigate the sarcopenic effect related ageing when it is combined with exercises.

Research indicates that poor nutrition is a significant contributor to sarcopenia, and dietary strategies can help prevent or treat this condition (Landi et al. 2018b). Adequate intake of high-quality proteins and essential amino acids, particularly leucine, is essential for muscle protein synthesis. Studies recommend a leucine intake of 3 gram per meal along with 25–30 grams of protein to counteract muscle loss in the elderly (Rondanelli et al. 2021b). Lowe blood levels of leucine are associated with reduced muscle loss and strength (Martínez-arnau et al. 2019). The timing of protein intake is also critical. Consuming protein evenly distributed across meals can enhance muscle protein synthesis and improve muscle mass and function (Calvani et al. 2015). Specific dietary supplements such as creatine monohydrate, vitamin D, omega-3 polyunsaturated fatty acids, and antioxidants have shown beneficial effects on health. These supplements can enhance anabolic response to resistance training and improve muscle mass and strength (Cannataro et al. 2021; McKendry et al. 2020). Nutrients with anti-inflammatory properties can help mitigate the chronic inflammation associated with ageing, which contributes to muscle degradation. Omega-3 fatty acids, in particular, have been highlighted for their role in reducing inflammation and supporting muscle health (Dedeyne et al. 2020). Combining nutritional interventions with physical exercise, especially resistance training, is the most effective strategy to combat sarcopenia. Exercise enhances the muscle's anabolic response to dietary protein and other nutrients which leads to improved muscle mass and function (McKendry et al. 2020). In essence, a multifaceted approach involving adequate protein intake, strategic supplementation, and regular physical activity is crucial for mitigating the effect of ageing and sarcopenia, thereby enhancing overall health and quality of life of older adults.

15.5 Bone Health and Nutritional Impact

Bone health fundamentally depends on nutrition since essential nutrients help maintain bone structure, strength, and functional capabilities. The bone requires a consistent supply of key nutrients for the processes of bone cell resorption by osteoclasts and formation by osteoblasts. Nutrients such as calcium, vitamin D, phosphorus, magnesium, and vitamin K are some vital vitamins and minerals that help to improve bone mineral density and proper skeletal development. Inadequate consumption of essential nutrients leads to the development of osteoporosis and osteomalacia conditions and raises the chance of bone fractures. The integrity of human bones depends heavily on lifestyle factors, dietary habits and age-related hormonal changes. People must combine dietary nutrition with weight-bearing exercise in order to maintain healthy bones throughout their entire life span. This section explores how nutrition can influence skeletal health.

15.5.1 Bone Remodelling

The human body executes an active biological procedure to eliminate old bone tissue and produce new skeletal elements. The repeated cycle functions as essential to preserving bone density while fixing tiny bone damage, adapting to force stress and sustaining proper calcium levels in the body. The process contains two interconnected activities: first, osteoclasts break down bone tissue through resorption, and second, osteoblasts construct new bone matrix.

15.5.1.1 Bone Turnover Cycle

Balance Between Osteoblast and Osteoclast Activity

The process of bone regeneration depends on the synchronised functioning of osteoblasts and osteoclasts. Osteoblasts, which develop from mesenchymal stem cells, belong to the essential cell group involved with bone development. These cells both manufacture bone material that progresses to generate new bone tissue through mineralisation. Osteoclasts derive from the monocyte/macrophage lineage to serve as specialised bone resorption cells, which break down bone tissue to free essential minerals that enter the bloodstream. Achieving a balance between osteoblast and osteoclast activity is vital for maintaining bone density and structural integrity. Any situation in which bone resorption outpaces formation creates bone loss, which typically leads to the development of osteoporosis. The described imbalances tend to occur naturally with ageing and during hormonal changes, and they emerge from both metabolic conditions and systemic diseases (Walsh 2018).

Key Regulators of Bone Remodelling

The process of bone remodelling exists under systematic and local factors from hormones, cytokines, growth factors and essential nutrients. To develop effective bone disorder treatments and secure lasting skeletal health, people should thoroughly comprehend these intricate processes.

Calcium: The majority of bone structure derives from mineral calcium, which resides in the bone matrix and manages extracellular calcium concentrations. Through calcium-sensing receptors, Parathyroid Hormone controls bone remodelling processes by activating osteoblast growth and managing osteoclast activity to maintain bone formation alignment with resorption activities (Gao et al. 2023; Hassan et al. 2023).

Vitamin D: The body requires vitamin D to regulate calcium and phosphate levels because it helps the intestines absorb more calcium and enables the kidneys to reabsorb it properly for bone strength. The regulators of sclerostin and RANKL in bone cells are directly influenced by Vitamin D, which aids in maintaining a balanced state of bone remodelling and structural integrity.

Estrogen: The presence of estrogen protects bones through its function of controlling osteoclast activity and promoting osteoclast apoptosis, which minimises bone breakdown. It also heightens osteoblast function and promotes new bone development. The hormone imbalance from menopause creates negative effects on bone resorption and mineral density through its influence on bone cells (Imai et al. 2009).

Parathyroid: Parathyroid hormone (PTH) maintains body calcium levels by activating bone resorption when blood calcium decreases. The RANKL/OPG ratio becomes more elevated through osteoblasts and osteocytes because of PTH. The hormone PTH triggers two separate actions on bone tissue by triggering bone formation during intermittent exposure but mainly inducing bone resorption during continued elevation, thus affecting the complete remodelling process.

15.5.1.2 Dietary Influence on Bone Formation

15.5.1.2.1 High Protein Diet and Bone Growth

The regulation of bone metabolism depends mainly on insulin-like growth factor 1 (IGF-1), which stimulates osteoblasts to enhance their biological activity. High IGF-1 levels that commonly develop from protein-rich diets cause increased activity of osteoblast cells together with improved differentiation and enhanced matrix production, which results in enhanced bone development. Through the AKT-mTOR signalling pathway, protein synthesis happens in osteoblasts during this process. When IGF-1 acts upon bone tissue, it reduces the formation of osteoclasts, helping to protect bone mass by limiting breakdown activities (Bakker et al. 2016).

The process of synthesising collagen requires proteins because collagen forms the critical structure of the bone matrix. The tensile strength of collagen as a protein depends on amino acids such as glycine, hydroxyproline and proline, which enable bone mineral deposition. Intake of sufficient protein, especially from animal and dairy products strongly contribute to improved bone mineral density. Scientific research confirms that postmenopausal women benefit significantly from high-protein meals regarding their skeletal system health (Licini et al. 2019). In summary, an elevated intake of dietary protein causes increased IGF-1 levels which then stimulates osteoblastic cellular activities and leads to new bone formation. The process of synthesising collagen requires protein as essential component that builds the structural support of bones. Studies show that increased protein taken from animal foods generates positive effects on bone mineral density while strengthening skeletal health.

15.5.1.2.2 Negative Diet Factors Affecting Bone Health

Excessive Sodium Intake

Excessive sodium in the diet causes an imbalance in calcium levels through its impact on urinary calcium release. It impairs bone mineralisation and weakens structural integrity. A diet containing high sodium levels leads to accelerated bone loss and increases the risk of developing osteoporosis, specifically when calcium intake is insufficient to offset these losses (Liu et al. 2024).

Caffeine and Nutrient Absorption
The consumption of caffeine disrupts calcium absorption in the body while simultaneously promoting the elimination of calcium through urine, thus depleting the bone-essential calcium stores. The chronic consumption of caffeine blocks sodium uptake while disrupting calcium metabolism and altering levels of potassium, zinc, and magnesium, which heightens osteoporosis danger and impacts mineral balance in the body. High caffeine intake without sufficient calcium intake can decrease bone density, thus raising your chances of fracture as well as developing osteoporosis (Zhou et al. 2009).

15.5.2 Osteoporosis

Osteoporosis exists as a persistent skeletal disease that reduces both bone mass and structural quality thus increasing bone fragility and susceptibility to fractures (Fig. 15.2). This condition affects public health primarily among older adults and postmenopausal women. The development of osteoporosis progresses without symptoms before a bone breaks down. Therefore, early medical help through lifestyle changes and nutrition management is essential. In osteoporosis, bone density decreases because bone tissue resorption rates exceed the rates of bone tissue formation during bone remodelling processes.

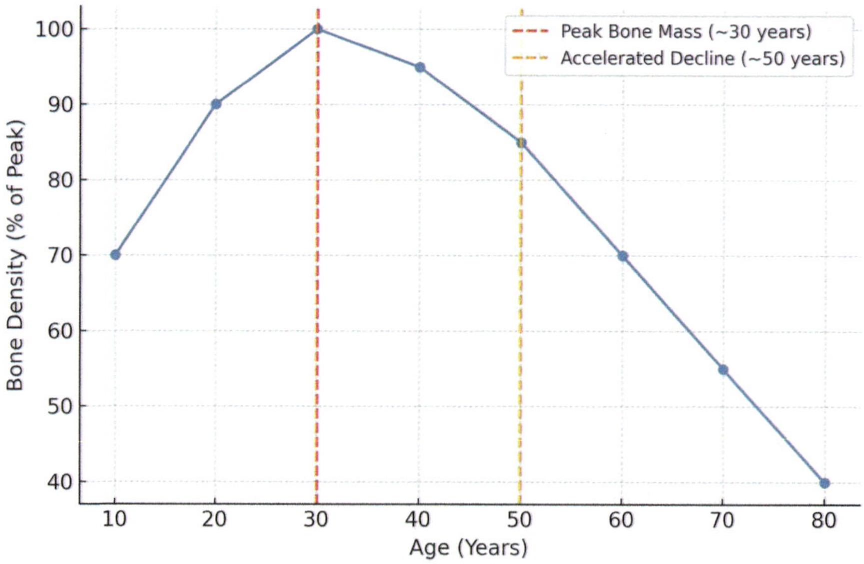

Fig. 15.2 Bone density changes across the life span

15.5.2.1 Nutritional Strategies for Prevention

Calcium Supplementation
Calcium plays a crucial role in bone mineral density as it reduces the risk of osteoporosis. The skeletal system operates as a calcium reservoir that maintains proper calcium amounts in plasma. The body regulates this balance through the bone formation and resorption processes. The proper consumption of calcium benefits bone strength and assists in bone density preservation, specifically for postmenopausal females. The recommended daily amount of calcium for adults is 800 mg. It should be elevated to 1200 mg for senior citizens. Dietary sources of calcium include dairy products such as milk, yoghurt, and cheese, as well as leafy green vegetables and calcium-fortified foods. When dietary intake is insufficient, calcium supplementation becomes necessary to prevent bone demineralisation and reduce the risk of osteoporosis.

Vitamin D Supplementation
Vitamin D helps calcium absorption in the intestines and supports the process of bone mineralisation. The body needs vitamin D to absorb calcium properly because low vitamin D levels create a calcium deficiency, leading to bone loss. The human body primarily obtains vitamin D through sun exposure. Studies suggest that Vitamin D is found in some fatty fish for example, salmon and mackerel as well as fortified dairy products. Vitamin D supplementation becomes necessary for individuals who experience restricted sunlight exposure, including people living in northern geographical areas and those who spend most of their time indoors to preserve optimum serum calcium levels, which helps prevent fractures and osteoporosis. Previous research recommends a daily intake of 600 IU to 1000 IU is recommended to ensure proper bone development and prevent rickets (Holick 2013).

Protein Intake and Bone Integrity
The fundamental function of protein in bone health happens through its support of collagen production, which creates the necessary framework for bone mineralisation (Table 15.3). Proper protein consumption improves the body's absorption of calcium from the intestines, which protects against osteoporosis and keeps the bones strong. Recent research shows supplementing 0.8 g/kg body weight/day RDA protein intake with enough calcium might benefit bone health (Rizzoli et al. 2018).

15.5.3 Impact of Food Safety on Bone Health

Scientists have recently identified the essential link between food safety and bone health. Endogenous toxins present in food and pathogens both negatively affect bone development through direct damage to bone cells and poor nutrient absorption. The section investigates how heavy metals, endocrine-disrupting chemicals,

Table 15.3 Impact of nutrients on bone health

Nutrient	Impact of deficiency on bone health	Recommended intake	References
Calcium	Decreased bone mass and increased fracture risk	1200–1500 mg/day	Rao et al. (2014)
Vitamin D	Reduced bone density, increased fracture risk, and osteomalacia	600–800 IU/day	Rao et al. (2014)
Protein	Increased fracture risk due to reduced bone density and muscle weakness	1 g/kg/day	Hejazi et al. (2020)
Vitamin K	Reduced bone mineral density and increased fracture risk	100 μg or more daily	Hejazi et al. (2020) and Inaba et al. (2015)
Vitamin C	Compromised bone health	75–90 mg/day	Hart et al. (2015)
Vitamin A	Both very high and very low intakes are associated with negative effects on bone health.	3000 mcg retinol (10,000 IU) per day	Burckhardt (2015) and Navarro-Valverde et al. (2018)
Copper	Potential stimulation of bone metabolism and increased hip fracture risk	1.51 mg/day	Fan et al. (2022) and Hejazi et al. (2020)
Zinc	Disruption in the incorporation of minerals into the bone matrix	6.7–15 mg per day	Schoofs et al. (2024)
Magnesium	Reduced bone mineral density and increased osteoporosis risk	310–420 mg/day	Rondanelli et al. (2021b)
Phosphorus	High levels may damage bone tissue if calcium intake is insufficient	Ratio of 0.5–1.5:1 with calcium	Hejazi et al. (2020)

and malabsorption syndromes affect bone homeostasis and suggests proven methods to protect bone health.

15.5.3.1 Contaminants Affecting Bone Homeostasis

Heavy Metal Toxicity

Lead: Bone health suffers from lead-related damage because lead interferes with calcium presence in hydroxyapatite crystals, thus weakening bones and decreasing bone mineral density (BMD), even in cases of minimal lead exposure. Oxidative stress caused by prolonged lead exposure results in damaged osteoblasts, which lead to decreased alkaline phosphatase activity and reduced nucleic acid content, thereby compromising bone formation and regeneration.

Cadmium: Cadmium deposits in cortical bone tissue lead to higher fracture susceptibility because it enhances osteoclast function and causes bone loss. Bone cell apoptosis occurs when cadmium exposure leads to elevated caspase-3 activity and DNA fragmentation. The process disrupts zinc and copper homeostasis. The unbalanced distribution of elements establishes more severe oxidative stress that worsens bone remodelling processes and weakens skeletal structures.

15.5.3.2 Endocrine Disrupting Chemicals

Organophosphate Pesticides: Organophosphate pesticides affect the RANKL/OPG signalling pathway, thereby causing increased osteoclast activity and bone resorption. High-exposure groups show an 18% increase in CTx bone resorption markers because of this imbalance that affects bone mineral density.

Bisphenol A (BPA): Exposure to Bisphenol A (BPA) at concentrations of 50 μM triggers bone health deterioration through its effects on osteoblast differentiation, inhibiting bone formation and leading to osteoblast cell death that disrupts normal bone tissue regulation, possibly causing osteoporosis.

15.5.3.3 Malabsorption Syndromes

Salmonella and E. coli O157:H7: Active infections of Salmonella and E. coli O157:H7 lead to villous atrophy that causes a 40–60% decrease in calcium absorption, resulting in acute foodborne illness while raising the risk of nutrient deficiencies (Ten Bruggencate et al. 2011).

Persistent Subclinical Infections: Annual bone mineral density loss from persistent subclinical infections reaches between 0.5% and 1.0%, which heightens the risk of osteoporosis, specifically among older adults and those with weakened immune systems (Biver et al. 2017).

15.5.3.4 Evidence-Based Risk Mitigating

Dietary Selection: People who choose organic produce decrease their pesticide exposure by 70%. Recipes incorporating phytate-breaking preparation techniques (e.g. Soaking, fermentation) on calcium-rich foods help boost calcium absorption.

Food Safety Practices: Implementing Hazard Analysis and Critical Control Points (HACCP) protocols protects susceptible groups from microbial threats that block calcium uptake. The detection of heavy metals in vulnerable areas through regular screening helps intervene early, thus minimising bone deterioration caused by lead and cadmium consumption.

Therapeutic Approaches: Therapeutic treatments exist to reverse the negative impacts contaminants have on bone structures. Supplementation with Lactobacillus reuteri ATCC PTA 6475 benefits bone structure through its support of gastrointestinal wellness and absorption capabilities. Healthcare professionals use chelation therapy as a therapeutic procedure to lower heavy metal content while preserving bone density for documented toxic conditions. The combined utilisation of these interventions ensures complete bone preservation protection against food-related safety threats (Flora and Pachauri 2010).

15.6 Nutrition and Injury Prevention/ Recovery

Injuries are a common aspect of exercising and playing sports. The nature and severity of an injury depend upon the results of the particular exercise. Injuries usually indicate a stop or reduction in sports participation and less physical activity (Tipton 2015). The prevention and recovery of an injury are managed through the proper nutritional diet. Inflammation is the initial response to the healing process after an injury. Therefore, using nutritional treatments to reduce inflammation during this early stage may actually prevent proper healing (Smith-Ryan et al. 2020). A well-balanced diet is essential for maintaining a healthy body. For a healthy diet, antioxidants, omega-3 fatty acids, proteins, vitamins, and minerals are important. Consumption of these nutrients works together to reduce inflammation, help muscle recover, and strengthen bone health and connective tissue (Rodrigues et al. 2024). On the other hand, healthy joints and tendons are crucial to the body's overall mobility, flexibility, and physical performance. The diet also plays a primary role in keeping joints strong and tendons resilient by managing inflammation, protecting cartilage, and supporting connective tissue (Vidović and Đorđević 2016). Nutrient-rich foods support improved collagen production, lower oxidative stress, and help regulate inflammation. These habits will prevent and treat problems like arthritis and tendon damage. Conversely, unhealthy food increases body inflammation, increasing joint tear and wear, slowing tissue recovery. The result of particular nutrient absorption protecting from injuries, recovery faster, and maintaining good health also leads to good athletic performance and overall health.

15.6.1 Nutritional Strategy for Injury Prevention

- *Antioxidants (Vitamin C, E, Polyphenols, Flavonoids)*

Antioxidants are molecules that help to protect the body and prevent or slow down cell damage caused by free radicals. Antioxidants and anti-inflammatory nutrients have become general for preventing oxidative stress and inflammation affected by physical activity (Bhattacharya 2024). Vitamin C, vitamin E, polyphenols, and flavonoids are antioxidants preventing injury in physical activity or exercise. Vitamin C is essential for formulating collagen for healthy bones and connective tissues, and it helps the body repair tendons and reduce inflammation by neutralising free radicals (Chugaeva et al. 2023). Therefore, many studies show that vitamin C may help protect against osteoarthritis and is mainly helpful in bone health (Chin and Ima-Nirwana 2018; Oakes et al. 2021). Vitamin C is found in citrus fruits, strawberries, and bell peppers (Kumar and Sen 2022). These foods provide essential nutrients and prevent muscle and bone injuries. On the other hand, Vitamin E, a lipid-soluble antioxidant that focuses on muscular health, will mainly interfere with chain reactions, slow down the damage to the cell membrane and help maintain the fatty acids.

Nuts, seeds, grains, and vegetable oils are better resources of natural vitamin E. Researchers examined whether consuming vitamin E as a supplement prevents muscle loss, oxidative stress, and inflammation from exercise (Kim et al. 2022).

Polyphenols and flavonoids are active antioxidant and anti-inflammatory effective compounds. These compounds are in food like green tea, berries, and dark chocolate (Cao et al. 2024). This will also reduce oxidative stress and decrease muscle fatigue, soreness, and tissue damage. Therefore, it is also helpful for muscle recovery after prolonged physical activity, preventing injuries, and maintaining muscle and joint health (Donadio et al. 2022). Another advantage of polyphenols and flavonoids is lower inflammation, which protects muscles, tendons and joints from injuries like tendinitis, muscle stains or repetitive movements (Donadio et al. 2022). Physiologically it will help to better blood flow and improve mitochondrial function. This ensures that muscles receive sufficient oxygen and nutrients, accelerate recovery and strengthen connective tissues (Bai et al. 2020).

- *Anti-Inflammatory Nutrients (Omega-3 Fatty Acids, Curcumin, Ginger)*

Omega-3 fatty acids, curcumin, and ginger are anti-inflammatory nutrients important for reducing inflammation, caring for joint pain, and preventing injuries (Izadi et al. 2024). Anti-inflammatory nutrients support the management of the body's inflammatory reaction, shielding muscles, tendons, and joints from injury (Vignesh et al. 2024). Omega-3 fatty acids, particularly EPA (Eicosapentaenoic acid) and DHA (docosahexaenoic acid), lower pro-inflammatory cytokines, which helps better muscle functioning and maintain joint flexibility (Calder 2017). EPA and DHA help to reduce joint pain and inflammation in conditions like osteoarthritis and rheumatoid arthritis. These fatty acids reduce the invention of injurious substances that can damage cartilage and cause joint rigidity. By affecting the body's inflammation reaction, omega-3, such as fatty fish (like salmon, sardines, and mackerel), flaxseed, and walnuts, help reduce inflammation and support joint health. Omega-3 supplements also decrease the need for nonsteroid anti-inflammatory drugs (NSAIDs), regularly used to treat arthritis symptoms. Including these healthy foods in the diet improves joints and reduces chronic inflammation, enhancing overall health and movement.

Curcumin is an active ingredient in turmeric, blocks systems that cause inflammation, helps to reduce muscle soreness and supports recovery after exercise, so it helps to prevent injury (Ms et al. 2020). The action works by blocking key molecules involved in chronic inflammation, like nuclear factor-kappa B (NF-$_{\kappa}$B) and cyclooxygenase-2 (COX-2). Curcumin also benefits individuals with arthritis, reducing joint pain and stiffness and protecting cartilage from injury (Ahire et al. 2023). Ginger contains compounds like gingerol, which assist in decreasing swelling, stiffness, and oxidative stress in muscles and joints (Ayustaningwarno et al. 2024). It reduces inflammatory signs and oxidative stress in joint tissues. It also decreases swelling and improves blood flow, ginger helps maintain joint flexibility and tendons strong. This is the best natural remedy for sudden and chronic injuries from the side of nutrition.

- *Hydration and Electrolytes*

  Staying hydrated and maintaining a balance of electrolytes helps to prevent muscle cramps, strains, and injuries caused by dehydration. It ensures that muscles and nerves work well. Better sources of hydration to the body through electrolytes include water, coconut water, bananas, and leafy greens (Shu and Luo 2022). These foods deliver required electrolytes such as potassium, magnesium, and sodium, which help maintain fluid balance and reduce muscle fatigue during exercise. The overall functioning of hydration will help to prevent injury (Liang et al. 2024).

- *Vitamin D and Calcium*

  In the human body, vitamin D and calcium are essential for preventing injuries and supporting bone health. They help to prevent stress fractures, osteoporosis, and other bone problems (Mendes and Sahni 2020). These essential nutrients are generally found in dairy products and fortified foods, which can also be obtained from sunlight. Together, these two nutrients provide remarkable support for bone health and bone growth. Athletes getting sufficient bone-strengthening nutrients and active people can lower the risk of injury and maintain healthy bones for physical activity and injury prevention (Armento et al. 2023).

15.6.2 Nutrition Strategy for Post-injury/recovery

- *Protein and Amino Acids*

In recovery from an injury, proper nutrition plays a key role. Research indicates protein and essential amino acids are significant for rebuilding muscles and recovering wounds. Protein delivers the building blocks required to patch damaged muscles, ligaments, and tendons (König et al. 2020). Leucine and glutamine are essential amino acids, and leucine helps to accelerate muscle recovery and reduce muscle loss during periods of immobilisation. Likewise, glutamine supports the immune system, reduces inflammation, and assists wound healing by giving energy to cells that repair tissues (Saharan and Awasthi 2020). Consumption requires high-quality proteins like lean meats, eggs, dairy and legumes, which are the best sources of amino acids for recovery. Protein intake prevents lengthy recovery time by keeping metabolism, working glowing and stopping muscle loss. Together, protein with anti-inflammatory foods, such as omega-**3** fatty acids from fish and flax seeds, can help healing still more (Rodriguez 2020).

- *Vitamin A and Zinc*

In the human body, vitamin A and zinc nutrients play a significant role in recovery from injury. It helps the immune system and helps the body to be ready for activity (Gürbüz 2022; Indrevær et al. 2017). Zinc supports healing wounds, regenerating cells, reduce inflammation. It helps from collagen, which is essential to repair tendons, ligaments and skin. Also, it helps to prevent infection and strengthen the immune system. Seafood, nuts, seeds, and lean meats are the source of these essential nutrients (Chasapis et al. 2020). Similarly, vitamin A also supports tissue repair and boosts the immune system by producing white blood cells that protect against infections. Vitamin A maintains skin and mucosal surface healthy, fasting wound closure and reducing scarring. It is also important for rebuilding bones, particularly after a fracture or stress injury (Zinder et al. 2019). Better sources of vitamin A are carrots, sweet potatoes, leafy greens, and dairy products. Together, these nutrients with anti-inflammatory foods, like omega-**3** fatty acids, help tissue repair and reduce inflammation (Knapik and Hoedebecke 2021).

- *Iron and Magnesium*

Faster recovery from an injury is a key role for nutrition. In nutrition, minerals like iron and magnesium are significant factors in the healing process of an injury. Iron benefits generate red blood cells that transport oxygen in the body, ensuring injured tissues receive the oxygen they require to heal (Coger et al. 2019). Therefore, oxygen-rich blood supports muscle recovery, improves energy, and avoids fatigue, which is specifically beneficial during rehabilitation. Iron foods like red meat, lentils, and spinach help to healthy haemoglobin levels (Zeidan et al. 2024). Magnesium also plays a specific role in relaxing muscles and reducing cramps and difficulties after an injury. It supports nerve function and assists in managing inflammation, which avoids continued soreness (Castiglioni et al. 2024). The sources of magnesium include nuts, seeds, whole grains, and dark leafy greens. This companied nutrition will help to speed up recovery from an injury (Costello and Rosanoff 2020).

15.6.3 Diet Impact on Arthritis and Inflammation

Arthritis is a usual state that is caused by inflammation in joints. It will lead to pain, stiffness, and reduced movement and quality of life. Rehabilitation processes like physiotherapy and medications will support the management of arthritis (Abbas et al. 2023). Similarly, arthritis and inflammation can be controlled by nutritional strategy. Certain nutrients reduce arthritis symptoms, slow down the progression of the disease, and improve overall muscular function and joint health (Nikiphorou and Philippou 2023). Studies show consumption of anti-inflammatory nutrients like omega-3 fatty acids, polyphenols, flavonoids, vitamin D, calcium, glucosamine, chondroitin, curcumin, and ginger can help reduce inflammation and prevent joint damage related to arthritis. To maintain the quality of life of athletes, a balanced diet with the essential nutrients will improve joint movements, prevent injuries, and reduce arthritis (Poggioli et al. 2023; Samoilov et al. 2023; Vidović and Đorđević 2016).

Glucosamine and chondroitin are natural compounds located in cartilage. These compounds maintain cartilage health by stopping its breakdown and helping repair joints. They also help to slow down arthritis and improve joint lubrication (Hobza and Gallo 2017). The supplements, including glucosamine and chondroitin, are particularly beneficial for athletes who have trained in high-impact movement

(DiNubile 2018). Likewise, curcumin is an active compound in turmeric, and ginger is recognised for its ability to inhibit inflammatory enzymes and reduce pain levels in arthritis patients (Akbari et al. 2023). Consistent intake of turmeric and ginger, whether in supplements or meals, serves as a typical difference to nonsteroidal anti-inflammatory drugs (NSAIDs), which may have lasting side effects (Akbari et al. 2023). A balanced, holistic method that includes constant hydration, enough protein, and tolerating time for recovery can improve arthritis symptoms. Appropriate hydration helps to maintain synovial fluid levels, which lubricate joints and reduce friction (Gonzales et al. 2024). Proteins are important for building collagen, and they help muscle growth. Therefore, muscle will also support tendons and ligaments, which can lead to improved joint movement and reduce arthritis injuries (Wang 2023).

15.6.4 Collagen and Elastin for Joint Flexibility and Tendon Health

Athletes and physically active people place immense stress on their joints and tendons, formulating it fundamental to support connective tissue health through appropriate nutrition. Investigation reveals that overall musculoskeletal function, collagen and elastin are significant proteins that maintain joint flexibility and tendon resilience. Collagen delivers organisational support to tendons, ligaments, and cartilage, confirming they remain durable and strong under physical strain (Goh 2019). At the same time, elastin contributes to flexibility, allowing tissue to stretch and return to normal shape (Rønnow et al. 2023). One of the most effective techniques to help collagen production is through the intake of collagen peptides and vitamin C (DePhillipo et al. 2018). Evidence suggests the sources of collagen peptide nutrients from the bone broth and chicken skin deliver the body necessary amino acids such as glycine and proline, which help reconstruct and maintain connective tissues (DePhillipo et al. 2018; Liu et al. 2022). Conversely, collagen synthesis greatly depends on vitamin C, a potent antioxidant found in citrus fruits, bell peppers, and strawberries. Vitamin C stimulates collagen production and protects joints from oxidative injuries affected by deep physical activity. Athletes confirm that consuming essential collagen peptides and vitamin C nutrients will improve elasticity, enhance lubrication, and speed up injury recovery (Bender and Fontana 2023).

Hyaluronic acid is also a vital component of joint health and is located in synovial fluid that lubricates the joints and reduces stiffness (Iaconisi et al. 2023). Hyaluronic acid is naturally present in bone broth and specific foods such as leafy greens, root vegetables, citrus fruits, and soy products. All these nutrients will help retain moisture in the joint spaces, enhancing cushioning and minimising friction between cartilage surfaces (Sprott and Fleck 2023). Athletes, specifically involved in high-impact sports like running, soccer and basketball, experience joint tear and wear discomfort. Additionally, manganese and copper are the minerals that play an important role in collagen synthesis and tendon integrity. Manganese stimulates enzymes responsible for collagen formation, while copper aids in interconnecting collagen fibres, confirming their strength and steadiness (Rondanelli et al. 2021a). These essential nutrients occur from nuts, seeds, and shellfish. Insufficiencies of these minerals will compromise tendon strength, most significantly leading to an increased risk of strains, sprains, and chronic tendon problems such as tendinitis (Venugopal 2019).

15.7 Calcium and Vitamin D Deficiency: Rickets, Osteoporosis, and Fractures

Calcium deficiency in people is influenced by various factors such as diet and lifestyle. Studies show that nearly half of the population is deficient in calcium globally (Shlisky et al. 2022). When calcium intake drops below the normal level, i.e., 1000–1200 mg/day for adults, the body will start to extract calcium from the bone tissue and maintain the calcium level homeostasis. This will eventually lead to bone mineral density (BMD). It has been found that the rate of fracture in the population having low calcium intake is nearly 30–50% (Michaëlsson et al. 2014). This is comparatively higher than that of those with adequate intake. Populations with the lowest intake of calcium have a higher risk of osteoporosis. Calcium deficiency is mainly found among postmenopausal women, the elderly population, lactose intolerant people, and people who have limited access to calcium-rich foods. Approximately 70% of the elderly people over 75 are affected by osteoporosis. Dairy products can provide a high amount of calcium for an individual, with a nearly 30–35% absorption rate. Findings suggest that regular consumption of calcium-rich foods other than dairy products, such as kale and book choy, can provide adequate calcium for lactose-intolerant populations when consumed in required quantities (Weaver 2024).

Globally, nearly one billion people are affected by a low level (below 20 ng/mL) of serum 25(OH)D called vitamin D deficiency (Holick et al. 2011). Discussions prove that this "sunshine" vitamin plays a key role in the absorption of calcium, whereas in vitamin D- deficient conditions, calcium absorption decreases by 30–60% (Qiu et al. 2023). Vitamin D deficiency shows different signs in different age groups. It is characterised as rickets for children, whereas for adults, it is osteomalacia. In children, vitamin D deficiency shows growth plate deformities, delayed motor development, and hypocalcemic seizures in severe cases. Osteomalacia shows symptoms such as proximal muscle weakness, diffuse bone pain, which affects the majority of patients, and increased risk of fracture. Since Vitamin D is also called "Sunshine Vitamin", it also influences the geographic area. During winter months, populations above 35 degrees latitude show a 40–70% vitamin D deficiency (Palacios and Gonzalez 2014). Recently, in many European and Asian countries, vitamin D rates have exceeded nearly 50%, and certain populations like dark-skinned individuals in northern latitudes and institutionalised elderly are also showing a deficiency of 70–90% (Cashman 2020).

15.7.1 Protein Deficiency: Muscle Atrophy and Weakness

Musculoskeletal health is primarily dependent on protein deficiency. Inadequate intake of protein may lead to sarcopenia, which is an age-related loss of muscle mass and its function. Evidence suggests that people who consume less than 0.8/kg/day will experience muscle weakness at a 40% faster rate compared to those who take the normal level or more than that (Coelho-Júnior et al. 2023). Recent clinical studies show that over a population, 23–41% of hospitalised elderly patients, 10–20% of community-dwelling older adults in developed nations, 30–50% of nursing home residents, and up to 70% of adults in developing regions suffer from protein deficiency effects. Muscle dysfunction due to lack of protein includes impaired muscle protein synthesis (MPS), compromised neuromuscular function, and elevated muscle protein breakdown. Findings reveal that nearly 30–45% of MPS reduction has happened during protein restriction (Trommelen et al. 2023). Elevated muscle protein breakdown happens through ubiquitin-proteasome pathway activation which increases the catabolism to 25–30% (Breen and Phillips 2011). Electromyography studies show that for compromised neuromuscular function, there is a 15–25% decrease in the motor unit advancement in protein deficiency (Morton et al. 2018). Protein deficiency in older adults increases the risk of falls and higher rates of disability.

15.7.2 Impact of Poor Food Safety on Musculoskeletal Health

Environmental contaminants in food can very much affect musculoskeletal development and maintenance. Heavy metals among them are very concerning threats, where the BMD will be decreased by approximately 0.4–0.8% with an increase of every 1 μg/dL in blood lead (Mitra et al. 2022). Global data shows that 0.5–2% of food samples surpass regulatory limits for the lead content and show higher rates in the industrial area. Another heavy metal, cadmium, can accumulate in bone tissues and is able to displace calcium. Cadmium, a toxic heavy metal commonly found in contaminated food, water, and industrial environments, can accumulate in the body over time. This accumulation can interfere with various biological processes, including regulating calcium and vitamin D metabolism, which are integral to maintaining bone density. Individuals exposed to cadmium at the highest quartile levels exhibit a 1.7-fold increased risk of osteoporotic fractures compared to those in the lowest quartile. This correlation highlights the significant health risks associated with cadmium exposure, particularly concerning bone health (Ma et al. 2021b). The findings suggest that with increased exposure to the amount of cadmium, individuals may experience more after-effects in bone remodelling and mineralisation, which leads to a higher likelihood of developing osteoporosis—a condition characterised by weakened bones and an increased risk of fractures.

Foodborne pathogens can further impact musculoskeletal health through direct infection of musculoskeletal tissues and malabsorption syndromes with gastrointestinal. Studies show that the former is the main reason for approximately 1–2% of foodborne diseases and also causes myositis or reactive arthritis. The latter can reduce nutrient absorption post-infection by 20–40% for 3–6 months.

Food safety failures play a crucial role in the worsening of malnutrition worldwide. Food contamination occurs through the presence of harmful bacteria, heavy metals, or chemical residues which leads to compromising the quality of food. This leads to the deficiencies of essential nutrients like calcium, vitamin D, and protein, and these are very vital for maintaining overall health and well-being. Contaminated food can lead to inadequate intake of calcium. This increases the risk of conditions like osteoporosis, which affects bone health. Similarly, exposure to heavy metals such as lead and cadmium can also negatively impact bone density and disrupt the body's ability to absorb vital nutrients. This leads to malnutrition, particularly among the elderly and kids. Food safety incidents can make people discouraged from consuming foods that are produced locally. Thus, food loss due to contamination reduces the availability of nutrients by 10–25% in food-insecure regions. Food safety failures provide a major contribution to malnutrition. Recent data from WHO shows that places having low-quality food safety infrastructure show a 30–50% higher muscular skeletal deficiency disorder, and globally nearly 22% of repeated infections in the gastrointestinal tract are caused by contaminated food and water.

15.8 Food Safety Concerns Related to Musculoskeletal Health

The concern of food safety extends far beyond the prevention of immediate health issues, as it plays critical role in supporting long-term well-being, especially musculoskeletal health. Contaminants like natural toxins, chemical residues, and microbial agents can silently degrade the nutritional quality of food, leading to gradual but profound effect on the body. Over time, exposure to these harmful elements can contribute to chronic conditions that weaken bones and muscles, making food safety an essential factor in overall health. Recognizing these risks is the key to forming strategies that safeguard nutrition. This discussion highlight the different types of contaminants, their potential consequences on musculoskeletal health, and the steps needed to maintain safe food supply.

15.8.1 Contaminants Affecting Nutritional Integrity

The contaminants are unwanted substances in food that compromise safety, quality, and nutrition. They fall in to three categories: biological, chemical, and physical. Biological contaminants like bacteria, viruses, and molds can cause foodborne illness (Elmassry et al. 2022). Chemical contaminants, including pesticides, heavy metals, and mycotoxins, may lead to chronic health issues (Nabil et al. 2021; Túri-Szerletics and Patkó 2008). Physical contaminants, such as glass or plastic fragments, pose immediate hazards (Ridgway et al. 2012). Identifying and controlling these threats is crucial for food safety and public health.

Mycotoxins: They are toxic compounds naturally produced by certain molds that thrive in food (Sohrabi et al. 2022). These harmful substances not only pose serious health risks to human and animals but also contribute to food waste, worsening global food security concerns. Exposure to mycotoxins has been linked to immune system suppression and even cancer (Sohrabi et al. 2022). To safeguard public health and ensure compliance with food safety regulations, effective strategies must be implemented to detect and eliminate these contaminants.

Pesticides and Agrochemicals: Pesticides and agricultural chemicals, through intended to protect crops, often leave behind residues in food that can pose significant health risks. Long-term exposure has been associated with endocrine disruption, cardiovascular diseases, and cancer (Kedia 2023). These chemical can also interfere with hormonal balance, potentially leading to chronic health conditions. The development and application of precise analytical methods are crucial for detecting and minimizing these contaminants in food products.

Heavy Metals: Environmental pollution has led to the contamination of food with heavy metals such as mercury and lead, both of which have severe health implications. These toxic elements can cause neurological disorders, developmental delays in children, and other chronic conditions (Schrenk 2004). Persistent organic pollutants (POPs) further worsen this issue, as they accumulate in the food chain over time, increasing long term health risks. Addressing these contaminants requires stringent monitoring and intervention strategies.

Thermal Process Contaminants: cooking and food processing techniques can sometimes generate harmful compounds such as acrylamide, furans, and nitrosamines (Elmassry et al. 2022). These contaminants not only degrade the nutritional quality of food by reducing protein digestibility and amino acid availability but are also associated with an increased risk of cancer (Pánek et al. 2018). Reducing their formation requires optimizing cooking methods, refining food processing techniques, and adopting innovative safety measures.

Packaging Derived Chemicals: chemicals used in food packaging, such as bisphenol A (BPA) and phthalates, can migrate in to food and pose serious health concerns. These substances have been linked to endocrine system disruptions, reproductive issues, and developmental disorders in children (Kedia 2023). Implementing safer packaging alternatives and stringent regulatory control is essential to reduce exposure to these harmful contaminants.

Microbial Contaminants: bacteria, viruses, and fungi are major sources of microbial contamination, often leading to foodborne diseases. Pathogens such as salmonella, E. coli, and listeria can cause severe gastrointestinal infections and, in some cases, life-threatening complications (Kaur et al. 2021). Strict hygiene practices, proper food handling, and effective sterilization techniques are essential to prevent microbial contamination and protect public health.

Importance of Safe, Quality Food Sources
Foodborne contaminants and pathogens are significant musculoskeletal health threats, leading to infections, inflammation, and bone, joint, and muscle damage. Pathogens such as bacteria, viruses, and chemical contaminants may lead to musculoskeletal infections (MSKI), i.e., septic arthritis, which may produce pain, oedema, and limited joint mobility. Chronic exposure to toxin contaminants such as mycotoxins can disrupt bone homeostasis and lead to bone health issues (Malvandi et al. 2022). Food safety guidelines are necessary in averting such risks and maintaining musculoskeletal function. Adhering to food handling standards, such as ensuring cleanliness and safe preparation, avoids the risk of pathogen contamination of the food supply (Aruoma 2006). Optimal storage conditions, including temperature regulation, impede the proliferation of detrimental microbes and avert deterioration. Moreover, stringent food processing standards such as good manufacturing practices (GMP) and hazard analysis critical control points (HACCP) can mitigate contamination concerns in food production (Aruoma 2006). In addition to averting immediate health risks, safeguarding food safety is crucial for maintaining nutritional integrity. Contaminants can lower the nutritional value of food, leading to deficiencies that further weaken bone and muscle strength. Complying with strict food safety protocols reduces the risk of foodborne illness and maintains nutritional value, ensuring musculoskeletal well-being and overall health.

15.9 Conclusion

The essence of this chapter examined how dietary components preserve musculoskeletal function by supporting bones, whereas muscles and connective tissues maintain their structure. The specific macro- and micronutrients serve as critical components for maintaining both integrity and resilience of the musculoskeletal system throughout the lifespan because the system supports movement and organ protection. Understanding musculoskeletal biological processes and fundamental nutrients allows researchers to develop nutritional plans which protect musculoskeletal health while avoiding condition deterioration and improving physical fitness.

Several insightful findings demonstrate the fundamental role of macronutrients in maintaining muscle mass and their role in musculoskeletal operational function. This chapter evidently discussed the importance of macronutrients in relation to the musculoskeletal system. The human body requires proteins with leucine as well as other branched-chain amino acids to synthesise protein and repair muscle tissue. The body uses carbohydrates to supply energy for muscle functions and other physical activities and relies on omega-3 fatty acids to decrease inflammation in the body and strengthen the joints. As this section explores about micronutrients, it also points out the important role of micronutrients in musculoskeletal health. The proper health of bones depends on micronutrients, including calcium, vitamin D, magnesium, phosphorus, and potassium. These micronutrients have many important roles, like improving bone density, muscle contraction, and neuromuscular function. We also discussed collagen and other vitamins and minerals like vitamin C and iron, which contribute to musculoskeletal well-being.

The chapter also discusses muscle functioning and bone remodelling dynamics, in addition to explaining how nutrition influences these procedures. The body requires enough protein to protect its muscle tissue as well as prevent age-related skeletal muscle deterioration, known as sarcopenia. Bone turnover regulation, together with osteoporosis prevention, relies on how calcium and vitamin D function in the body. Nutrition post-injury recovery is supported through protective agents, which help decrease injury risks while accelerating tissue regeneration. This section demonstrates why food safety stands as a crucial factor which affects musculoskeletal health. Point out that contaminants like heavy metals (Lead and Cadmium), pesticides, and pathogens affect the body's nutrient absorption and how these contaminants create potential threats to the musculoskeletal system. People need to have safe access to high-quality food because it maintains nutritional quality and protects them from musculoskeletal damage. Support for musculoskeletal health requires proper food safety measures and abundant access to nutritional food resources.

Finally, the chapter underscores the public health and clinical implications of musculoskeletal physiology and nutritional integrity. Implementing evidence-based nutritional guidelines can improve musculoskeletal outcomes, enhance mobility, and reduce the prevalence of musculoskeletal disorders. Future research should explore the synergistic effects of dietary patterns, nutrient bioavailability, and food safety on musculoskeletal physiology, with a focus on personalised nutrition approaches. In conclusion, a complete dietary plan should remain the foundation for managing the connection that exists between musculoskeletal physiology and nutritional integrity. The framework of this chapter merges nutritional science with musculoskeletal physiology for developing musculoskeletal function, degenerative condition prevention and long-term physical resilience maintenance.

References

Abbas KS, Hasan AS, Tareq NL (2023) Assessment of the impact of different rheumatoid arthritis stages on the quality of life of a sample of Iraqi patients. J Adv Pharm Educ Res 13(1):122–126. Scopus. https://doi.org/10.51847/41Nl6iG253

Abdalla A, Benissan-Messan D, Zhu H (2021) Skeletal muscle tissue engineering. In: Musculoskeletal tissue engineering. Elsevier, pp 67–80. Scopus. https://doi.org/10.1016/B978-0-12-823893-6.00007-3

Ahire ED, Surana KR, Sonawane VN, Talele SG, Kshirsagar SJ, Laddha UD, Thombre NA, Talele GS (2023) Immunomodulation impact of curcumin and its derivative as a natural ingredient. In: Nutraceuticals and functional foods in immunomodulators. Springer Nature Singapore, pp 253–269. Scopus. https://doi.org/10.1007/978-981-19-2507-8_10

Ahmad S (2014) Food quality and safety, Food engineering series, pp 175–190. Scopus. https://doi.org/10.1007/978-1-4939-1378-7_7

Akbari A, Mottaghi P, Salesi M, Shojaei M, Askari G, Sahebkar A, Bagherniya M (2023) The effect of curcumin supplementation on clinical factors in patients with rheumatoid arthritis. J Isfahan Med Sch 40(700):1052–1058. Scopus. https://doi.org/10.48305/jims.v40.i700.1052

Alghannam A, Gonzalez J, Betts J (2018) Restoration of muscle glycogen and functional capacity: role of post-exercise carbohydrate and protein co-ingestion. Nutrients 10(2):253. https://doi.org/10.3390/nu10020253

Allen J, Ramsden M, Nisar S (2024) Skeletal muscle structure, function and pathology. Orthop Trauma 38(3):137–144. Scopus. https://doi.org/10.1016/j.mporth.2024.03.004

Alway SE (2006) Muscle hypertrophy. In: Scientific evidence for musculoskeletal, bariatric, and sports nutrition. CRC Press, pp 355–388. Scopus. https://doi.org/10.1201/9781420019322.ch20

Alway SE, Siu PM, Murlasits Z, Butler DC (2005) Muscle hypertrophy models: applications for research on aging. Can J Appl Physiol 30(5):591–624. Scopus. https://doi.org/10.1139/h05-143

Amirouche F, Koh J (2021) Biomechanics of human joints. In: Orthopaedic biomechanics in sports medicine. Springer International Publishing, pp 3–13. Scopus. https://doi.org/10.1007/978-3-030-81549-3_1

Argilés JM, Campos N, Lopez-Pedrosa JM, Rueda R, Rodriguez-Mañas L (2016) Skeletal muscle regulates metabolism via interorgan crosstalk: roles in health and disease. J Am Med Dir Assoc 17(9):789–796. Scopus. https://doi.org/10.1016/j.jamda.2016.04.019

Armento A, Heronemus M, Truong D, Swanson C (2023) Bone health in young athletes: a narrative review of the recent literature. Curr Osteoporos Rep 21(4):447–458. Scopus. https://doi.org/10.1007/s11914-023-00796-5

Aruoma OI (2006) The impact of food regulation on the food supply chain. Toxicology 221(1):119–127. Scopus. https://doi.org/10.1016/j.tox.2005.12.024

Auestad N, Layman DK (2021) Dairy bioactive proteins and peptides: a narrative review. Nutr Rev 79(Supplement_2):36–47. https://doi.org/10.1093/nutrit/nuab097

Ayustaningwarno F, Anjani G, Ayu AM, Fogliano V (2024) A critical review of Ginger's (Zingiber officinale) antioxidant, anti-inflammatory, and immunomodulatory activities. Front Nutr 11:1364836. Scopus. https://doi.org/10.3389/fnut.2024.1364836

Bai L, Li H, Qin Q, Feng B-L, Zhang L, Wei F-Y, Yang X-F (2020) Research progress of regulatory effect and mechanism of flavonoids on vascular endothelial cell injury. Chin J Exp Tradit Med Formulae 26(12):203–211. https://doi.org/10.13422/j.cnki.syfjx.20201201

Bakker AD, Gakes T, Hogervorst JMA, de Wit GMJ, Klein-Nulend J, Jaspers RT (2016) Mechanical stimulation and IGF-1 enhance mRNA translation rate in osteoblasts via activation of the AKT-mTOR pathway. J Cell Physiol 231(6):1283–1290. https://doi.org/10.1002/jcp.25228

Banerjee D, Das PK, Mukherjee J (2023) Muscular system. In: Textbook of veterinary physiology. Springer Nature Singapore, pp 235–264. Scopus. https://doi.org/10.1007/978-981-19-9410-4_10

Barnett VA (2015) Cellular myocytes. In: Handbook of cardiac anatomy, physiology, and devices, 3rd edn. Springer International Publishing, pp 201–214. Scopus. https://doi.org/10.1007/978-3-319-19464-6_12

Bender DA, Fontana L (2023) Vitamin C. In: Encyclopedia of human nutrition, vol 1–4, 4th edn, pp 504–514. Scopus. https://doi.org/10.1016/B978-0-12-821848-8.00123-2

Benedik E (2022) Sources of vitamin D for humans. Int J Vitam Nutr Res 92(2):118–125. https://doi.org/10.1024/0300-9831/a000733

Bernard GW (2017) Osteogenesis and its control. In: Fundamentals of craniofacial growth, pp 45–58. Scopus. https://doi.org/10.1201/9780203755327

Berrazaga I, Micard V, Gueugneau M, Walrand S (2019) The role of the anabolic properties of plant- versus animal-based protein sources in supporting muscle mass maintenance: a critical review. Nutrients 11(8):1825. https://doi.org/10.3390/nu11081825

Bhattacharya M (2024) Antioxidants activity of micronutrients and macronutrients. In: Micronutrients and macronutrients as nutraceuticals. Apple Academic Press, pp 47–88. Scopus. https://doi.org/10.1201/9781003415381-2

Bhattacharya R, Zekavat SM, Uddin MM, Pirruccello J, Niroula A, Gibson C, Griffin GK, Libby P, Ebert BL, Bick A, Natarajan P (2021) Association of diet quality with prevalence of clonal hematopoiesis and adverse cardiovascular events. JAMA Cardiol 6(9):1069–1077. Scopus. https://doi.org/10.1001/jamacardio.2021.1678

Bird SP, Nienhuis M, Biagioli B, De Pauw K, Meeusen R (2024) Supplementation strategies for strength and power athletes: carbohydrate, protein, and amino acid ingestion. Nutrients 16(12):1886. Scopus. https://doi.org/10.3390/nu16121886

Biver E, Calmy A, Rizzoli R (2017) Bone health in HIV and hepatitis B or C infections. Ther Adv Musculoskelet Dis 9(1):22–34. https://doi.org/10.1177/1759720X16671927

Block RC, Pearson TA (2006) The cardiovascular implications of omega-3 fatty acids. Cardiol J 13(7):557–569. Scopus

Bo T, Fujii J (2024) Primary roles of branched chain amino acids (BCAAs) and their metabolism in physiology and metabolic disorders. Molecules 30(1):56. https://doi.org/10.3390/molecules30010056

Bodner M (2021) Rapid analytical methods for food integrity analysis: a short review. Curr Nutr Food Sci 17(9):955–962. Scopus. https://doi.org/10.2174/1573401317666210210125633

Boyd SK, Lagacé P-Y (2017) In vivo bone imaging with micro-computed tomography. In: Medical image analysis and informatics: computer-aided diagnosis and therapy, pp 335–368. Scopus. https://doi.org/10.1201/9781351228343

Breen L, Phillips SM (2011) Skeletal muscle protein metabolism in the elderly: interventions to counteract the "anabolic resistance" of ageing. Nutr Metab 8(1):68. https://doi.org/10.1186/1743-7075-8-68

Brotto M, Johnson ML (2014) Endocrine crosstalk between muscle and bone. Curr Osteoporos Rep 12(2):135–141. Scopus. https://doi.org/10.1007/s11914-014-0209-0

Brown SC, Sewry CA (2017) Basics of skeletal muscle function and normal physiology. In: Cardioskeletal myopathies in children and young adults. Elsevier, pp 21–38. Scopus. https://doi.org/10.1016/B978-0-12-800040-3.00002-9

Brozovich FV, Nicholson CJ, Degen CV, Gao YZ, Aggarwal M, Morgan KG (2016) Mechanisms of vascular smooth muscle contraction and the basis for pharmacologic treatment of smooth muscle disorders. Pharmacol Rev 68(2):476–532. Scopus. https://doi.org/10.1124/pr.115.010652

Burckhardt P (2015) Vitamin A and bone health. In: Nutrition and bone health. Springer, New York, pp 409–421. https://doi.org/10.1007/978-1-4939-2001-3_26

Burla F, Dussi S, Martinez-Torres C, Tauber J, van der Gucht J, Koenderink GH (2020) Connectivity and plasticity determine collagen network fracture. Proc Natl Acad Sci 117(15):8326–8334. https://doi.org/10.1073/pnas.1920062117

Cajka T, Hajslova J (2011) Volatile compounds in food authenticity and traceability testing. In: Food flavors: chemical, sensory and technological properties, pp 355–412. Scopus. https://www.scopus.com/inward/record.uri?eid=2-s2.0-84921023231&partnerID=40&md5=0b3da20d18c1d39341b5845026092698

Calder PC (2017) Omega-3 fatty acids and inflammatory processes: from molecules to man. Biochem Soc Trans 45(5):1105–1115. Scopus. https://doi.org/10.1042/BST20160474

Calvani R, Landi F, Collamati A, Serafini E, Bernabei R, Marzetti E (2015) Nutritional strategies against sarcopenia of aging: current evidence and future directions. In: Foods and dietary supplements in the prevention and treatment of disease in older adults. Elsevier, pp 231–238. Scopus. https://doi.org/10.1016/B978-0-12-418680-4.00022-1

Camba AR, Rueda RMA (2022) Pathophysiology of the joints. Pediatr Integral 26(3):186.e1–186.e4. Scopus

Cannataro R, Carbone L, Petro JL, Cione E, Vargas S, Angulo H, Forero DA, Odriozola-Martínez A, Kreider RB, Bonilla DA (2021) Sarcopenia: etiology, nutritional approaches, and mirnas. Int J Mol Sci 22(18):9724. Scopus. https://doi.org/10.3390/ijms22189724

Cao RY, Li J, Dai Q, Li Q, Yang J (2018) Muscle atrophy: present and future. In: Advances in experimental medicine and biology, vol 1088. Springer Singapore, pp 605–624. Scopus. https://doi.org/10.1007/978-981-13-1435-3_29

Cao G, Zuo J, Wu B, Wu Y (2024) Polyphenol supplementation boosts aerobic endurance in athletes: systematic review. Front Physiol 15:1369174. Scopus. https://doi.org/10.3389/fphys.2024.1369174

Carlsohn A, Mayer F (2010) Ernährung im Ausdauersport. Aktuelle Ernährungsmedizin 35(04):173–177. https://doi.org/10.1055/s-0030-1248416

Cashman KD (2020) Vitamin D deficiency: defining, prevalence, causes, and strategies of addressing. Calcif Tissue Int 106(1):14–29. https://doi.org/10.1007/s00223-019-00559-4

Castiglioni S, Mazur A, Maier JA (2024) The central role of magnesium in skeletal muscle: from myogenesis to performance. Magnes Res 37(1):1–11. Scopus. https://doi.org/10.1684/mrh.2024.0526

Chasapis CT, Ntoupa P-SA, Spiliopoulou CA, Stefanidou ME (2020) Recent aspects of the effects of zinc on human health. Arch Toxicol 94(5):1443–1460. Scopus. https://doi.org/10.1007/s00204-020-02702-9

Chen F, Shao F, Hinds A, Yao S, Ram-Mohan S, Norman TA, Krishnan R, Fine A (2018) Retinoic acid signaling is essential for airway smooth muscle homeostasis. JCI Insight 3(16):e120398. Scopus. https://doi.org/10.1172/jci.insight.120398

Chin K-Y, Ima-Nirwana S (2018) Vitamin C and bone health: evidence from cell, animal and human studies. Curr Drug Targets 19(5):439–450. Scopus. https://doi.org/10.2174/1389450116666150907100838

Chugaeva UY, Raouf M, Morozova NS, Mahdavian L (2023) Effects of l-ascorbic acid (C6H8O6: Vit-C) on collagen amino acids: DFT study. Amino Acids 55(11):1655–1664. Scopus. https://doi.org/10.1007/s00726-023-03339-5

Clark KL (2007) Nutritional considerations in joint health. Clin Sports Med 26(1):101–118. Scopus. https://doi.org/10.1016/j.csm.2006.11.006

Coelho-Júnior HJ, Calvani R, Picca A, Tosato M, Landi F, Marzetti E (2023) Association of physical activity and exercise with physical performance and muscle mass in older adults: results from the longevity check-up (lookup) 7+ project. J Clin Med 12(24):7521. https://doi.org/10.3390/jcm12247521

Coger V, Million N, Rehbock C, Sures B, Nachev M, Barcikowski S, Wistuba N, Strauß S, Vogt PM (2019) Tissue concentrations of zinc, iron, copper, and magnesium during the phases of full thickness wound healing in a rodent model. Biol Trace Elem Res 191(1):167–176. Scopus. https://doi.org/10.1007/s12011-018-1600-y

Collins FL, Rios-Arce ND, Schepper JD, Parameswaran N, Mccabe LR (2018) The potential of probiotics as a therapy for osteoporosis. In: Bugs as drugs: therapeutic microbes for prevention and treatment of disease. ASM Press, pp 213–233. Scopus. https://doi.org/10.1128/9781555819705.ch9

Colorado MDF, Jaramillo RC, Arango LFG (2023) Influence of genetics and epigenetics on the process of muscle hypertrophy: a systematic review. Mov Sports Sci – Sci Mot 121:59–72. Scopus. https://doi.org/10.1051/sm/2023007

Costello RB, Rosanoff A (2020) Magnesium. In: Present knowledge in nutrition: basic nutrition and metabolism. Elsevier, pp 349–373. Scopus. https://doi.org/10.1016/B978-0-323-66162-1.00021-4

Cruz-Jentoft AJ (2019) Sarcopenia: preventable and reversible. In: Practical issues in geriatrics, pp 47–52. Scopus. https://doi.org/10.1007/978-3-319-96529-1_5

Curtis L (2016) Nutritional research may be useful in treating tendon injuries. Nutrition 32(6):617–619. Scopus. https://doi.org/10.1016/j.nut.2015.12.039

Cvek SZ, Bobinac D, Đudarić L, Cvijanović O (2015) The remodeling of the skeleton. Med Flum 51(4):482–493. Scopus

Deane CS, Ely IA, Wilkinson DJ, Smith K, Phillips BE, Atherton PJ (2021) Dietary protein, exercise, ageing and physical inactivity: interactive influences on skeletal muscle proteostasis. Proc Nutr Soc 80(2):106–117. https://doi.org/10.1017/S0029665120007879

Dedeyne L, Dupont J, Koppo K, Verschueren S, Tournoy J, Gielen E (2020) Exercise and Nutrition for Healthy AgeiNg (ENHANce) project – effects and mechanisms of action of combined anabolic interventions to improve physical functioning in sarcopenic older adults: study protocol of a triple blinded, randomized controlled trial. BMC Geriatr 20(1):1–14. Scopus. https://doi.org/10.1186/s12877-020-01900-5

DePhillipo NN, Aman ZS, Kennedy MI, Begley JP, Moatshe G, LaPrade RF (2018) Efficacy of vitamin C supplementation on collagen synthesis and oxidative stress after musculoskeletal injuries: a systematic review. Orthop J Sports Med 6(10):2325967118804544. Scopus. https://doi.org/10.1177/2325967118804544

Dhiman A, Thakur K, Parmar V, Sharma S, Sharma R, Kaur G, Singh B, Suhag R (2023) New insights into tailoring physicochemical and techno-functional properties of plant proteins using conventional and emerging technologies. J Food Meas Charact 17(4):3845–3873. https://doi.org/10.1007/s11694-023-01919-3

DiNubile N (2018) Glucosamine and chondroitin sulfate: what has been learned since the glucosamine/chondroitin arthritis intervention trial. Orthopedics 41(4):200–207. Scopus. https://doi.org/10.3928/01477447-20180511-06

Donadio P, Muñoz C, Flores-Lucero F, Garcia-Diaz DF, Farias-Valenzuela C, Quezada E, Poblete-Aro C (2022) Effectiveness of nutritional supplementations with polyphenols on muscle damage and biomarkers of oxidative stress in different types of physical exercises: a review of the literature. Sport TK 11:54–79. Scopus. https://doi.org/10.6018/sportk.550641

Dumitru A, Radu BM, Radu M, Cretoiu SM (2018) Muscle changes during atrophy. In: Advances in experimental medicine and biology, vol 1088. Springer Singapore, pp 73–92. Scopus. https://doi.org/10.1007/978-981-13-1435-3_4

Dydykina IS, Kovalenko PS, Aboleshina AA, Kovalenko AA (2022) The optimal way to fill the deficiency of essential substances for the well-being of the joints. Med Sov 2022(11):80–86. Scopus. https://doi.org/10.21518/2079-701X-2022-16-11-80-86

Egan B, Sharples AP (2023) Molecular responses to acute exercise and their relevance for adaptations in skeletal muscle to exercise training. Physiol Rev 103(3):2057–2170. Scopus. https://doi.org/10.1152/physrev.00054.2021

Elmassry MM, Zayed A, Farag MA (2022) Gut homeostasis and microbiota under attack: impact of the different types of food contaminants on gut health. Crit Rev Food Sci Nutr 62(3):738–763. Scopus. https://doi.org/10.1080/10408398.2020.1828263

Ercan B, Webster T (2014) Cell response to nanoscale features and its implications in tissue regeneration: an orthopedic perspective. In: Nanotechnology and regenerative engineering: the scaffold, 2nd edn. CRC Press, pp 145–190. Scopus. https://doi.org/10.1201/b17444-10

Fan Y, Ni S, Zhang H (2022) Associations of copper intake with bone mineral density and osteoporosis in adults: data from the National Health and Nutrition Examination Survey. Biol Trace Elem Res 200(5):2062–2068. https://doi.org/10.1007/s12011-021-02845-5

Fanzani A, Conraads VM, Penna F, Martinet W (2012) Molecular and cellular mechanisms of skeletal muscle atrophy: an update. J Cachexia Sarcopenia Muscle 3(3):163–179. Scopus. https://doi.org/10.1007/s13539-012-0074-6

Feng L-T, Chen Z-N, Bian H (2024) Skeletal muscle: molecular structure, myogenesis, biological functions, and diseases. MedComm 5(7):e649. Scopus. https://doi.org/10.1002/mco2.649

Fernandes MH, Caram R, Sooraj Hussain N, Mauricio AC, Santos JD (2016) Bonelike® graft for regenerative bone applications. In: Surgical tools and medical devices, 2nd edn. Springer International Publishing, pp 409–438. Scopus. https://doi.org/10.1007/978-3-319-33489-9_13

Ferreiro SF, Vilarino N, Carrera C, Louzao MC, Cantalapiedra AG, Santamarina G, Cifuentes JM, Vieira AC, Botana LM (2016) Subacute cardiotoxicity of Yessotoxin: in vitro and in vivo studies. Chem Res Toxicol 29(6):981–990. Scopus. https://doi.org/10.1021/acs.chemrestox.6b00012

Flora SJS, Pachauri V (2010) Chelation in metal intoxication. Int J Environ Res Public Health 7(7):2745–2788. https://doi.org/10.3390/ijerph7072745

Franchi M, Trirè A, Quaranta M, Orsini E, Ottani V (2007) Collagen structure of tendon relates to function. Sci World J 7:404–420. Scopus. https://doi.org/10.1100/tsw.2007.92

Frank CB (2004) Ligament structure, physiology and function. J Musculoskelet Neuronal Interact 4(2):199–201. Scopus

Frontera WR, Ochala J (2015) Skeletal muscle: a brief review of structure and function. Behav Genet 45(2):183–195. Scopus. https://doi.org/10.1007/s00223-014-9915-y

Fust A, Stephens NL (2005) Mechanical plasticity and contractile properties of airway smooth muscle. Can J Physiol Pharmacol 83(10):865–868. Scopus. https://doi.org/10.1139/y05-102

Fyhrie DP (2009) The mechanical properties of bone. In: Osteoporosis in men: the effects of gender on skeletal health. Elsevier, pp 51–67. Scopus. https://doi.org/10.1016/B978-0-12-374602-3.00005-5

Gahunia HK, Pritzker KP (2020) Structure and function of articular cartilage. In: Articular cartilage of the knee: health, disease and therapy. Springer, New York, pp 3–70. Scopus. https://doi.org/10.1007/978-1-4939-7587-7_1

Gao X, Di X, Li J, Kang Y, Xie W, Sun L, Zhang J (2023) Extracellular calcium-induced calcium transient regulating the proliferation of osteoblasts through glycolysis metabolism pathways. Int J Mol Sci 24(5):4991. https://doi.org/10.3390/ijms24054991

Garden-Robinson J (2015) Carotenoids in green vegetables and health aspects. In: Pigments in fruits and vegetables. Springer, New York, pp 229–246. https://doi.org/10.1007/978-1-4939-2356-4_12

Garg V, Goyal G (2024) Normal physiology of skeletal muscle and physiology of hypertrophy. In: Blood flow restriction training. Jenny Stanford Publishing, pp 1–22. Scopus. https://doi.org/10.1201/9781032699585-1

Gilliland KO, Kernick ET (2021) Musculoskeletal tissues and anatomy. In: Clinical foundations of musculoskeletal medicine: a manual for medical students. Springer International Publishing, pp 11–21. Scopus. https://doi.org/10.1007/978-3-030-42894-5_2

Giraldo-Vallejo JE, Cardona-Guzmán MÁ, Rodríguez-Alcivar EJ, Kočí J, Petro JL, Kreider RB, Cannataro R, Bonilla DA (2023) Nutritional strategies in the rehabilitation of musculoskeletal injuries in athletes: a systematic integrative review. Nutrients 15(4):819. https://doi.org/10.3390/nu15040819

Glynn EL, Fry CS, Timmerman KL, Drummond MJ, Volpi E, Rasmussen BB (2013) Addition of carbohydrate or alanine to an essential amino acid mixture does not enhance human skeletal muscle protein anabolism. J Nutr 143(3):307–314. https://doi.org/10.3945/jn.112.168203

Goh KL (2019) Current understanding of interfacial stress transfer mechanisms in connective tissue. In: Interfaces in particle and fibre reinforced composites: current perspectives on polymer, ceramic, metal and extracellular matrices. Elsevier, pp 529–549. Scopus. https://doi.org/10.1016/B978-0-08-102665-6.00020-0

Gonzales G, Hoque J, Kaeo C, Zauscher S, Varghese S (2024) Grafting of cationic molecules to hyaluronic acid improves adsorption and cartilage lubrication. Biomater Sci 12(18):4747–4758. Scopus. https://doi.org/10.1039/d4bm00532e

Gorissen SHM, Phillips SM (2018) Branched-chain amino acids (leucine, isoleucine, and valine) and skeletal muscle. In: Nutrition and skeletal muscle. Elsevier, pp 283–298. Scopus. https://doi.org/10.1016/B978-0-12-810422-4.00016-6

Groenendijk I, van Delft M, Versloot P, van Loon LJC, de Groot LC (2022) Impact of magnesium on bone health in older adults: a systematic review and meta-analysis. Bone 154:116233. https://doi.org/10.1016/j.bone.2021.116233

Gürbüz M (2022) Understanding the role of vitamin A and its precursors in the immune system. Nutr Clin Metab 36(2):89–98. Scopus. https://doi.org/10.1016/j.nupar.2021.10.002

Haghighatjoo F, Nikkhah M, Rahimpour MR (2024) Carbohydrate characterization and exploitation. In: Encyclopedia of renewable energy, sustainability and the environment. Elsevier, pp 553–564. https://doi.org/10.1016/B978-0-323-93940-9.00190-0

Han A, Park Y, Lee Y-K, Park SY, Park CY (2022) Position statement: vitamin D intake to prevent osteoporosis and fracture in adults. J Bone Metab 29(4):205–215. https://doi.org/10.11005/jbm.2022.29.4.205

Hart A, Cota A, Makhdom A, Harvey EJ (2015) The role of vitamin C in orthopedic trauma and bone health. Am J Orthop (Belle Mead NJ) 44(7):306–311

Hassan EB, Ghasem-Zadeh A (2013) Trabecular bone: morphology, biomechanics and clinical implications. In: Trabecular and cortical bone: morphology, biomechanics and clinical implications, pp 1–35. Scopus. https://www.scopus.com/inward/record.uri?eid=2-s2.0-84951141711&partnerID=40&md5=9217e3dd1363e5ef739cc7727df2583b

Hassan MH, Omar AM, Daskalakis E, Grieve B, Bartolo P (2023) Smart biomaterials and constructs for bone tissue regeneration. In: New trends in smart nanostructured biomaterials in health sciences. Elsevier, pp 177–209. https://doi.org/10.1016/B978-0-323-85671-3.00007-5

Hedberg CW (2017) Food safety. In: Occupational and environmental health, pp 381–392. Scopus. https://doi.org/10.1093/oso/9780190662677.003.0020

Hejazi J, Davoodi A, Khosravi M, Sedaghat M, Abedi V, Hosseinverdi S, Ehrampoush E, Homayounfar R, Shojaie L (2020) Nutrition and osteoporosis prevention and treatment. Biomed Res Ther 7(4):3709–3720. https://doi.org/10.15419/bmrat.v7i4.598

Hernandez EM (2013) Labelling and claims in foods containing omega-3 fatty acids. In: Food enrichment with Omega-3 fatty acids. Elsevier, pp 405–413. Scopus. https://doi.org/10.1533/9780857098863.4.405

Hevia-Larraín V, Gualano B, Longobardi I, Gil S, Fernandes AL, Costa LAR, Pereira RMR, Artioli GG, Phillips SM, Roschel H (2021) High-protein plant-based diet versus a protein-matched omnivorous diet to support resistance training adaptations: a comparison between habitual vegans and omnivores. Sports Med 51(6):1317–1330. https://doi.org/10.1007/s40279-021-01434-9

Hintermann B, Golanó P (2014) The anatomy and function of the deltoid ligament. J Foot Ankle Surg 13(2):67–72. Scopus. https://doi.org/10.1097/BTF.0000000000000044

Hobza M, Gallo J (2017) Chondroitin, glucosamine, diacerein, piascledine, and methylsulphonylmethane in the prevention and treatment of joint diseases. Klinicka Farmakologie a Farmacie 31(4):14–17. Scopus

Holick MF (2013) Vitamin D: physiology, dietary sources, and requirements. In: Encyclopedia of human nutrition. Elsevier, pp 370–382. https://doi.org/10.1016/B978-0-12-375083-9.00276-2

Holick MF, Binkley NC, Bischoff-Ferrari HA, Gordon CM, Hanley DA, Heaney RP, Murad MH, Weaver CM (2011) Evaluation, treatment, and prevention of vitamin D deficiency: an Endocrine Society clinical practice guideline. J Clin Endocrinol Metab 96(7):1911–1930. https://doi.org/10.1210/jc.2011-0385

Huertas JR, Lara AR, Acevedo OG, Mesa-García MD (2019) Milk and dairy products as vehicle for calcium and vitamin D: role of calcium enriched milks. Nutr Hosp 36(4):962–973. https://doi.org/10.20960/nh.02570

Hunter GR, Singh H, Carter SJ, Bryan DR, Fisher G (2019) Sarcopenia and its implications for metabolic health. J Obes 2019:8031705. Scopus. https://doi.org/10.1155/2019/8031705

Iaconisi GN, Gallo N, Caforio L, Ricci V, Fiermonte G, Della Tommasa S, Bernetti A, Dolce V, Farì G, Capobianco L (2023) Clinical and biochemical implications of hyaluronic acid in musculoskeletal rehabilitation: a comprehensive review. J Pers Med 13(12):1647. Scopus. https://doi.org/10.3390/jpm13121647

Imai Y, Youn M-Y, Kondoh S, Nakamura T, Kouzmenko A, Matsumoto T, Takada I, Takaoka K, Kato S (2009) Estrogens maintain bone mass by regulating expression of genes controlling function and life span in mature osteoclasts. Ann N Y Acad Sci 1173(s1):E31–E39. https://doi.org/10.1111/j.1749-6632.2009.04954.x

Inaba N, Sato T, Yamashita T (2015) Low-dose daily intake of vitamin K2 (menaquinone-7) improves osteocalcin γ-carboxylation: a double-blind, randomized controlled trials. J Nutr Sci Vitaminol 61(6):471–480. https://doi.org/10.3177/jnsv.61.471

Indrevær RL, Eriksen AB, Blomhoff HK (2017) Mechanisms of immune regulation by vitamin A and its metabolites. In: Nutrition, immunity, and infection, pp 159–180. Scopus. https://doi.org/10.1201/9781315118901

Innocenti B, Galbusera F (2022) Ligament and tendon biomechanics. In: Human orthopaedic biomechanics: fundamentals, devices and applications. Elsevier, pp 137–149. Scopus. https://doi.org/10.1016/B978-0-12-824481-4.00016-0

Iyer KM (2013) Anatomy of bone, fracture, and fracture healing. In: General principles of orthopedics and trauma. Springer London, pp 1–11. Scopus. https://doi.org/10.1007/978-1-4471-4444-1_1

Izadi M, Sadri N, Abdi A, Zadeh MMR, Jalaei D, Ghazimoradi MM, Shouri S, Tahmasebi S (2024) Anti-aging natural supplements: the main players in promoting healthy lifespan. Nutr Res Rev:1–18. Scopus. https://doi.org/10.1017/S0954422424000301

Jabbari F, Akbari B, Tayebi L (2023) Widely used biomaterials in cartilage biofabrication. In: Cartilage: from biology to biofabrication. Springer Nature Singapore, pp 333–353. Scopus. https://doi.org/10.1007/978-981-99-2452-3_12

Jaslove JM, Nelson CM (2018) Smooth muscle: a stiff sculptor of epithelial shapes. Philos Trans R Soc B Biol Sci 373(1759):20170318. Scopus. https://doi.org/10.1098/rstb.2017.0318

Jourdan M, Nair KS, Carter RE, Schimke J, Ford GC, Marc J, Aussel C, Cynober L (2015) Citrulline stimulates muscle protein synthesis in the post-absorptive state in healthy people fed a low-protein diet – a pilot study. Clin Nutr 34(3):449–456. https://doi.org/10.1016/j.clnu.2014.04.019

Kaur N, Agarwal A, Sabharwal M, Jaiswal N (2021) Natural food toxins as anti-nutritional factors in plants and their reduction strategies. In: Food chemistry: the role of additives, preservatives and adulteration. Wiley, pp 217–266. Scopus. https://doi.org/10.1002/9781119792130.ch8

Kedia A (2023) Modern tools and techniques for the detection of hazardous food contaminants. In: Green products in food safety. Elsevier, pp 111–129. Scopus. https://doi.org/10.1016/B978-0-323-95590-4.00002-3

Kilimann MW (2024) Glycogen storage disease due to phosphorylase kinase deficiency. In: Protein dysfunction in human genetic disease. Garland Science, pp 57–75. https://doi.org/10.1201/9781003579953-4

Kim M, Eo H, Lim JG, Lim H, Lim Y (2022) Can low-dose of dietary vitamin E supplementation reduce exercise-induced muscle damage and oxidative stress? A meta-analysis of randomized controlled trials. Nutrients 14(8):1599. https://doi.org/10.3390/nu14081599

Kirkman E (2018) Electromechanical coupling and regulation of force of cardiac contraction. Anaesth Intensive Care Med 19(6):326–328. Scopus. https://doi.org/10.1016/j.mpaic.2018.03.004

Kirkman E (2021) Electromechanical coupling and regulation of force of cardiac contraction. Anaesth Intensive Care Med 22(4):267–269. Scopus. https://doi.org/10.1016/j.mpaic.2021.02.007

Kiss O (2010) Electrocardiogram: electrical information retrieval and diagnostics from the beating heart. In: Handbook of physics in medicine and biology, pp 23–21. Scopus. https://doi.org/10.1201/9781420075250

Knapik JJ, Hoedebecke BL (2021) Vitamin A and bone fractures. J Spec Oper Med 21(1):115–119. Scopus

Knight MB, Grosberg A, McCain ML (2015) In vitro tools for quantifying structure-function relationships in cardiac myocyte cells and tissues. In: Cardiac cytoarchitecture: how to maintain a working heart. Springer International Publishing, pp 15–39. Scopus. https://doi.org/10.1007/978-3-319-15263-9_2

König D, Carlsohn A, Braun H, Großhauser M, Lampen A, Mosler S, Nieß A, Schäbethal K, Schek A, Virmani K, Ziegenhagen R, Heseker H (2020) Proteins in sports nutrition: position of the working group sports nutrition of the German Nutrition Society (DGE). Ernahr Umsch 7:132–139. Scopus. https://doi.org/10.4455/eu.2020.039

Kostoglou-Athanassiou I, Athanassiou L, Athanassiou P (2020) The effect of Omega-3 fatty acids on rheumatoid arthritis. Mediterr J Rheumatol 31(2):190–194. https://doi.org/10.31138/mjr.31.2.190

Krishnan HB (2015) Improving the sulfur-containing amino acids of soybean to enhance its nutritional value in animal feed, pp 235–249. https://doi.org/10.2134/agronmonogr50.c15

Kumar J, Sen A (2022) The role of vitamin C: from prevention of pneumonia to treatment of Covid-19. Mater Today Proc. Scopus. https://doi.org/10.1016/j.matpr.2022.11.213

Kumar RR, Rahman F, Bora B, Shameeh M (2023) Importance and nutritive value of animal proteins in human diet. In: Processing technologies and food protein digestion. Elsevier, pp 1–25. https://doi.org/10.1016/B978-0-323-95052-7.00007-8

Kurapatti M, Carreira D (2023) Diet composition's effect on chronic musculoskeletal pain: a narrative review. Pain Physician 26(7):527–534. Scopus

Landi F, Calvani R, Cesari M, Tosato M, Martone AM, Ortolani E, Savera G, Salini S, Sisto A, Picca A, Marzetti E (2018a) Sarcopenia: an overview on current definitions, diagnosis and treatment. Curr Protein Pept Sci 19(7):633–638. Scopus. https://doi.org/10.2174/1389203718666170607113459

Landi F, Sieber C, Fielding RA, Rolland Y, Guralnik J (2018b) Nutritional intervention in sarcopenia: report from the international conference on frailty and sarcopenia research task force. J Frailty Aging 7(4):247–252. Scopus. https://doi.org/10.14283/jfa.2018.26

Laversuch C (2014) The swollen joint. In: Acute medicine: a symptom-based approach, pp 268–273. Scopus. https://doi.org/10.1007/9781139600354.039

Leumann A, Horisberger M, Valderrabano V (2011) Mechano-biological correlation: the concept of the joint as an organ. Schweizerische Zeitschrift Fur Sportmedizin Und Sporttraumatologie 59(4):161–164. Scopus

Liang Z, Liang Y, Zhang C, Zhao X, Qiu J (2024) Consumption of a branched-chain amino acids-containing sports beverage during 21 km of running reduces dehydration, lowers muscle damage, and prevents a decline in lower limb strength. Nutrients 16(22):3799. Scopus. https://doi.org/10.3390/nu16223799

Licini C, Vitale-Brovarone C, Mattioli-Belmonte M (2019) Collagen and non-collagenous proteins molecular crosstalk in the pathophysiology of osteoporosis. Cytokine Growth Factor Rev 49:59–69. https://doi.org/10.1016/j.cytogfr.2019.09.001

Liu T, Zou L, Ji X, Xiao G (2022) Chicken skin-derived collagen peptides chelated zinc promotes zinc absorption and represses tumor growth and invasion in vivo by suppressing autophagy. Front Nutr 9:960926. Scopus. https://doi.org/10.3389/fnut.2022.960926

Liu X, Wu Y, Bennett S, Zou J, Xu J, Zhang L (2024) The effects of different dietary patterns on bone health. Nutrients 16(14):2289. https://doi.org/10.3390/nu16142289

Luo G, Zhang Y, Wang X, Chen S, Li D, Yu M (2021) Individualized 3D printing-assisted repair and reconstruction of neoplastic bone defects at irregular bone sites: exploration and practice in the treatment of scapular aneurysmal bone cysts. BMC Musculoskelet Disord 22(1):1–9. Scopus. https://doi.org/10.1186/s12891-021-04859-5

Ma GC, Friedman JM, You JS, Ma CB (2021a) Ligament function and pathoanatomy of injury and healing. In: Management of track and field injuries. Springer International Publishing, pp 31–41. Scopus. https://doi.org/10.1007/978-3-030-60216-1_4

Ma Y, Ran D, Shi X, Zhao H, Liu Z (2021b) Cadmium toxicity: a role in bone cell function and teeth development. Sci Total Environ 769:144646. https://doi.org/10.1016/j.scitotenv.2020.144646

Mahapatra C, Kumar R (2024) Biophysical mechanisms of vaginal smooth muscle contraction: the role of the membrane potential and ion channels. Pathophysiology 31(2):225–243. Scopus. https://doi.org/10.3390/pathophysiology31020018

Malvandi AM, Shahba S, Mehrzad J, Lombardi G (2022) Metabolic disruption by naturally occurring mycotoxins in circulation: a focus on vascular and bone homeostasis dysfunction. Front Nutr 9:915681. Scopus. https://doi.org/10.3389/fnut.2022.915681

Mamerow MM, Mettler JA, English KL, Casperson SL, Arentson-Lantz E, Sheffield-Moore M, Layman DK, Paddon-Jones D (2014) Dietary protein distribution positively influences 24-h muscle protein synthesis in healthy adults. J Nutr 144(6):876–880. https://doi.org/10.3945/jn.113.185280

Martínez-arnau FM, Fonfría-Vivas R, Cauli O (2019) Beneficial effects of leucine supplementation on criteria for sarcopenia: a systematic review. Nutrients 11(10):2504. Scopus. https://doi.org/10.3390/nu11102504
Matthewman MC, Costa-Pinto R (2023) Macronutrients, minerals, vitamins and energy. Anaesth Intensive Care Med 24(2):134–138. https://doi.org/10.1016/j.mpaic.2022.12.009
McCann JR, Rawls JF (2023) Essential amino acid metabolites as chemical mediators of host-microbe interaction in the gut. Ann Rev Microbiol 77(1):479–497. https://doi.org/10.1146/annurev-micro-032421-111819
McCarthy JJ, Murach KA (2018) Anabolic and catabolic signaling pathways that regulate skeletal muscle mass. In: Nutrition and enhanced sports performance: muscle building, endurance, and strength. Elsevier, pp 275–290. Scopus. https://doi.org/10.1016/B978-0-12-813922-6.00024-2
McGlory C, Calder PC, Nunes EA (2019) The influence of Omega-3 fatty acids on skeletal muscle protein turnover in health, disuse, and disease. Front Nutr 6:144. https://doi.org/10.3389/fnut.2019.00144
McKendry J, Currier BS, Lim C, McLeod JC, Thomas ACQ, Phillips SM (2020) Nutritional supplements to support resistance exercise in countering the sarcopenia of aging. Nutrients 12(7):1–29. Scopus. https://doi.org/10.3390/nu12072057
Meloni VHM (2005) The role of hyperplasia on the increase of skeletal muscle size. Rev Bras Cineantropom Desempenho Hum 7(1):59–63. Scopus
Mendes MM, Sahni S (2020) Calcium and vitamin D in the management of osteoporosis. In: Marcus and Feldman's osteoporosis, pp 1665–1678. Scopus. https://doi.org/10.1016/B978-0-12-813073-5.00070-8
Michaëlsson K, Wolk A, Langenskiöld S, Basu S, Warensjö Lemming E, Melhus H, Byberg L (2014) Milk intake and risk of mortality and fractures in women and men: cohort studies. BMJ 349:g6015. https://doi.org/10.1136/bmj.g6015
Miller JR, Bonete PE (2008) Food safety and nutrition, p 124. Scopus. https://www.scopus.com/inward/record.uri?eid=2-s2.0-84892197962&partnerID=40&md5=a2f9a169816d93c3f597eee208051c3c
Mitra S, Chakraborty AJ, Tareq AM, Emran TB, Nainu F, Khusro A, Idris AM, Khandaker MU, Osman H, Alhumaydhi FA, Simal-Gandara J (2022) Impact of heavy metals on the environment and human health: novel therapeutic insights to counter the toxicity. J King Saud Univ Sci 34(3):101865. https://doi.org/10.1016/j.jksus.2022.101865
Morgan EF, Gerstenfeld LC (2020) The bone organ system: form and function. In: Marcus and Feldman's osteoporosis. Elsevier, pp 15–35. Scopus. https://doi.org/10.1016/B978-0-12-813073-5.00002-2
Morton RW, Murphy KT, McKellar SR, Schoenfeld BJ, Henselmans M, Helms E, Aragon AA, Devries MC, Banfield L, Krieger JW, Phillips SM (2018) A systematic review, meta-analysis and meta-regression of the effect of protein supplementation on resistance training-induced gains in muscle mass and strength in healthy adults. Br J Sports Med 52(6):376–384. https://doi.org/10.1136/bjsports-2017-097608
Ms SAB, Waldman PHS, Krings PBM, Lamberth PJ, Smith PJW, McAllister PMJ (2020) Effect of curcumin supplementation on exercise-induced oxidative stress, inflammation, muscle damage, and muscle soreness. J Diet Suppl 17(4):401–414. Scopus. https://doi.org/10.1080/19390211.2019.1604604
Muleya M, Bailey EF, Bailey EH (2024) A comparison of the bioaccessible calcium supplies of various plant-based products relative to bovine milk. Food Res Int 175:113795. https://doi.org/10.1016/j.foodres.2023.113795
Myers C (2024) Skeletal muscle physiology: an update to anatomy and function. Springer Nature, p 320. Scopus. https://doi.org/10.1007/978-3-031-47065-3
Nabil SK, Mustari N, Khan MS (2021) Sources and health impacts of chemical contaminants in foods. In: Techniques to measure food safety and quality: microbial, chemi-

cal, and sensory. Springer International Publishing, pp 31–68. Scopus. https://doi.org/10.1007/978-3-030-68636-9_3

Navarro-Valverde C, Caballero-Villarraso J, Mata-Granados JM, Casado-Díaz A, Sosa-Henríquez M, Malouf-Sierra J, Nogués-Solán X, Rodríguez-Mañas L, Cortés-Gil X, Delgadillo-Duarte J, Quesada-Gómez JM (2018) High serum retinol as a relevant contributor to low bone mineral density in postmenopausal osteoporotic women. Calcif Tissue Int 102(6):651–656. https://doi.org/10.1007/s00223-017-0379-8

Nigam D, Yadav R, Tiwari U (2018) Omega-3 fatty acids and its role in human health. In: Functional food and human health. Springer Singapore, pp 173–198. https://doi.org/10.1007/978-981-13-1123-9_9

Nikawa T, Ulla A, Sakakibara I (2021) Polyphenols and their effects on muscle atrophy and muscle health. Molecules 26(16):4887. Scopus. https://doi.org/10.3390/molecules26164887

Nikiphorou E, Philippou E (2023) Nutrition and its role in prevention and management of rheumatoid arthritis. Autoimmun Rev 22(7):103333. Scopus. https://doi.org/10.1016/j.autrev.2023.103333

Nunes EA, Stokes T, McKendry J, Currier BS, Phillips SM (2022) Disuse-induced skeletal muscle atrophy in disease and non-disease states in humans: mechanisms, prevention, and recovery strategies. Am J Physiol Cell Physiol 322(6):C1068–C1084. Scopus. https://doi.org/10.1152/AJPCELL.00425.2021

O'Brien M (2005) Anatomy of tendons. In: Tendon injuries: basic science and clinical medicine, pp 3–13. Scopus. https://doi.org/10.1007/1-84628-050-8_1

Oakes B, Bolia IK, Weber AE, Petrigliano FA (2021) Vitamin C in orthopedic practices: current concepts, novel ideas, and future perspectives. J Orthop Res 39(4):698–706. https://doi.org/10.1002/jor.24947

Palacios C, Gonzalez L (2014) Is vitamin D deficiency a major global public health problem? J Steroid Biochem Mol Biol 144(Pt A):138–145. https://doi.org/10.1016/j.jsbmb.2013.11.003

Panea B, Ripoll G (2020) Quality and safety of meat products. Foods 9(6):803. Scopus. https://doi.org/10.3390/foods9060803

Pánek J, Davídek J, Jehličková Z (2018) Toxic and antinutritional compounds arising from proteins. In: Natural toxic compounds of foods, pp 183–199. Scopus. https://doi.org/10.1201/9781351074933

Pérez A, Ruz M, García P, Jiménez P, Valencia P, Ramírez C, Pinto M, Nuñez SM, Park JW, Almonacid S (2024) Nutritional properties of fish bones: potential applications in the food industry. Food Rev Intl 40(1):79–91. https://doi.org/10.1080/87559129.2022.2153136

Poggioli R, Hirani K, Jogani VG, Ricordi C (2023) Modulation of inflammation and immunity by Omega-3 fatty acids: a possible role for prevention and to halt disease progression in autoimmune, viral, and age-related disorders. Eur Rev Med Pharmacol Sci 27(15):7380–7400. Scopus. https://doi.org/10.26355/eurrev_202308_33310

Pohl A, Schünemann F, Bersiner K, Gehlert S (2021) The impact of vegan and vegetarian diets on physical performance and molecular signaling in skeletal muscle. Nutrients 13(11):3884. Scopus. https://doi.org/10.3390/nu13113884

Putz R (2008) Structure and function of articular cartilage. Aktuelle Rheumatol 33(1):22–28. Scopus. https://doi.org/10.1055/s-2007-963753

Qiu S, Divine G, Rao SD (2023) Effect of vitamin D metabolites on bone histomorphometry in healthy black and white women: an attempt to unravel the so-called vitamin D paradox in blacks. Bone Rep 18:101650. https://doi.org/10.1016/j.bonr.2022.101650

Rao LG, Kang NN, Rao AV (2014) Lycopene and other antioxidants in the prevention and treatment of osteoporosis in postmenopausal women. In: Aging. Elsevier, pp 247–258. https://doi.org/10.1016/B978-0-12-405933-7.00024-X

Rauner M, Stein N, Hofbauer LC (2012) Basics of bone biology. In: Principles of osteoimmunology: molecular mechanisms and clinical applications, vol 9783709105207. Springer Vienna, pp 1–26. Scopus. https://doi.org/10.1007/978-3-7091-0520-7_1

Reboul E (2023) Absorption intestinale des vitamines liposolubles. Prat Nutr 19(74):16–18. https://doi.org/10.1016/j.pranut.2023.03.005

Ridgway K, Smith RM, Lalljie SPD (2012) Sample preparation for food contaminant analysis. In: Comprehensive sampling and sample preparation: analytical techniques for scientists. Elsevier, pp 819–833. Scopus. https://doi.org/10.1016/B978-0-12-381373-2.00115-0

Rizzoli R, Biver E, Bonjour J-P, Coxam V, Goltzman D, Kanis JA, Lappe J, Rejnmark L, Sahni S, Weaver C, Weiler H, Reginster J-Y (2018) Benefits and safety of dietary protein for bone health—an expert consensus paper endorsed by the European Society for Clinical and Economical Aspects of Osteopororosis, Osteoarthritis, and Musculoskeletal Diseases and by the International Osteoporosis Foundation. Osteoporos Int 29(9):1933–1948. https://doi.org/10.1007/s00198-018-4534-5

Robbins D (2022) Muscle biomechanics. In: Human orthopaedic biomechanics: fundamentals, devices and applications, pp 121–135. Scopus. https://doi.org/10.1016/B978-0-12-824481-4.00009-3

Rodrigues M, Rosa A, Almeida A, Martins R, Ribeiro T, Pintado M, Gonçalves RFS, Pinheiro AC, Fonseca AJM, Maia MRG, Cabrita ARJ, Barros L, Caleja C (2024) Omega-3 fatty acids from fish by-products: innovative extraction and application in food and feed. Food Bioprod Process 145:32–41. Scopus. https://doi.org/10.1016/j.fbp.2024.02.007

Rodriguez NR (2020) Role of meat in healthy eating patterns: considerations for protein quantity and protein quality. Meat Muscle Biol 4(2). Scopus. https://doi.org/10.22175/mmb.11687

Rogero MM, Tirapegui J (2008) Current aspects of branched chain amino acid and exercise. Rev Bras Cienc Farm/Braz J Pharm Sci 44(4):563–575. Scopus. https://doi.org/10.1590/S1516-93322008000400004

Rondanelli M, Faliva MA, Peroni G, Infantino V, Gasparri C, Iannello G, Perna S, Riva A, Petrangolini G, Tartara A (2021a) Essentiality of manganese for bone health: an overview and update. Nat Prod Commun 16(5):1934578X211016649. Scopus. https://doi.org/10.1177/1934578X211016649

Rondanelli M, Nichetti M, Peroni G, Faliva MA, Naso M, Gasparri C, Perna S, Oberto L, Di Paolo E, Riva A, Petrangolini G, Guerreschi G, Tartara A (2021b) Where to find leucine in food and how to feed elderly with sarcopenia in order to counteract loss of muscle mass: practical advice. Front Nutr 7:622391. Scopus. https://doi.org/10.3389/fnut.2020.622391

Rønnow SR, Kristensen JH, Thorlacius-Ussing J, Karsdal MA, Heinz A (2023) Elastin. In: Biochemistry of collagens, Laminins and elastin: structure, function and biomarkers, 3rd edn. Elsevier, pp 279–289. Scopus. https://doi.org/10.1016/B978-0-443-15617-5.00016-0

Saharan R, Awasthi M (2020) A review: role of leucine as a sports supplement. J Hum Ecol 69(1–3):25–29. Scopus. https://doi.org/10.31901/24566608.2020/69.1-3.3205

Saleem A, Kazak L, O'Leary M, Hood DA (2011) Cell death regulation in muscle. In: Apoptosis: physiology and pathology. Cambridge University Press, pp 313–322. Scopus. https://doi.org/10.1017/CBO9780511976094.027

Salva JE, Merrill AE (2017) Signaling networks in joint development. Dev Dyn 246(4):262–274. Scopus. https://doi.org/10.1002/dvdy.24472

Samoilov AS, Zholinsky AV, Rylova NV, Bolshakov IV (2023) Nutritional factors of bone health in athletes. Vopr Pitan 92(3):25–35. Scopus. https://doi.org/10.33029/0042-8833-2023-92-3-25-35

Sanders KM, Ward SM, Koh SD (2014) Interstitial cells: regulators of smooth muscle function. Physiol Rev 94(3):859–907. Scopus. https://doi.org/10.1152/physrev.00037.2013

Schmiedel J, Reichmann H (2004) Muscular atrophy as a symptom. Internist 45(10):1163–1176. Scopus. https://doi.org/10.1007/s00108-004-1256-x

Schoofs H, Schmit J, Rink L (2024) Zinc toxicity: understanding the limits. Molecules 29(13):3130. https://doi.org/10.3390/molecules29133130

Schrenk D (2004) Chemical food contaminants. Bundesgesundheitsbl Gesundheitsforsch Gesundheitsschutz 47(9):841–847. Scopus. https://doi.org/10.1007/s00103-004-0892-6

Shim J-S, Kim KN, Lee J-S, Yoon MO, Lee HS (2023) Magnesium intake and dietary sources among Koreans: findings from the Korea National Health and Nutrition Examination Survey 2016–2019. Nutr Res Pract 17(1):48–61. https://doi.org/10.4162/nrp.2023.17.1.48

Shlisky J, Mandlik R, Askari S, Abrams S, Belizan JM, Bourassa MW, Cormick G, Driller-Colangelo A, Gomes F, Khadilkar A, Owino V, Pettifor JM, Rana ZH, Roth DE, Weaver C (2022) Calcium deficiency worldwide: prevalence of inadequate intakes and associated health outcomes. Ann N Y Acad Sci 1512(1):10–28. https://doi.org/10.1111/nyas.14758

Shu W, Luo J (2022) Scientific hydration strategies for marathon runners. J Army Med Univ 44(1):34–39. Scopus. https://doi.org/10.16016/j.1000-5404.202111192

Smith-Ryan AE, Hirsch KR, Saylor HE, Gould LM, Blue MNM (2020) Nutritional considerations and strategies to facilitate injury recovery and rehabilitation. J Athl Train 55(9):918–930. https://doi.org/10.4085/1062-6050-550-19

Sohrabi H, Salahshour Sani P, Zolfaghari R, Majidi MR, Yoon Y, Khataee A (2022) MOF-based mycotoxin nanosensors for food quality and safety assessment through electrochemical and optical methods. Molecules 27(21):7511. Scopus. https://doi.org/10.3390/molecules27217511

Solberg A, Reikvam H (2023) Iron status and physical performance in athletes. Life 13(10):2007. https://doi.org/10.3390/life13102007

Souza PR, Norling LV (2016) Implications for eicosapentaenoic acid- and docosahexaenoic acid-derived resolvins as therapeutics for arthritis. Eur J Pharmacol 785:165–173. https://doi.org/10.1016/j.ejphar.2015.05.072

Sprott H, Fleck C (2023) Hyaluronic acid in rheumatology. Pharmaceutics 15(9):2247. Scopus. https://doi.org/10.3390/pharmaceutics15092247

Stutzman Z, Gawlik K (2020) Evidence-based assessment of the musculoskeletal system. In: Evidence-based physical examination: best practices for health & well-being assessment. Springer Publishing Company, pp 403–441. Scopus. https://doi.org/10.1891/9780826164544.0015

Stylianopoulou C (2023) Carbohydrates: regulation of metabolism. In: Encyclopedia of human nutrition. Elsevier, pp 126–135. https://doi.org/10.1016/B978-0-12-821848-8.00173-6

Sutton L (2012) Bone tissue. In: Body composition in sport, exercise and health, pp 106–125. Scopus. https://doi.org/10.4324/9780203133040-14

Swartz DR, Greaser ML, Cantino ME (2009) Muscle structure and function. In: Applied muscle biology and meat science, pp 1–46. Scopus. https://www.scopus.com/inward/record.uri?eid=2-s2.0-84893606285&partnerID=40&md5=00b9ea85962e2be293492d1c35878a13

Takagaki Y, Yamagishi H, Matsuoka R (2012) Factors involved in signal transduction during vertebrate myogenesis. In: International review of cell and molecular biology, vol 296. Elsevier, pp 187–272. Scopus. https://doi.org/10.1016/B978-0-12-394307-1.00004-7

Ten Bruggencate SJ, Snel J, Schoterman MHC, Ertmann E, van der Meulen E, Schonewille A, Bovee-Oudenhoven IMJ (2011) Efficacy of various dietary calcium salts to improve intestinal resistance to *Salmonella* infection in rats. Br J Nutr 105(4):489–495. https://doi.org/10.1017/S0007114510003764

Thorpe CT, Screen HRC (2016) Tendon structure and composition. Adv Exp Med Biol 920:3–10. Scopus. https://doi.org/10.1007/978-3-319-33943-6_1

Thorpe CT, Birch HL, Clegg PD, Screen HRC (2015) Tendon physiology and mechanical behavior: structure-function relationships. In: Tendon regeneration: understanding tissue physiology and development to engineer functional substitutes. Elsevier, pp 3–39. Scopus. https://doi.org/10.1016/B978-0-12-801590-2.00001-6

Tipton KD (2015) Nutritional support for exercise-induced injuries. Sports Med 45:93–104. Scopus. https://doi.org/10.1007/s40279-015-0398-4

Tolmachyov OA, Vekovtsev AA, Vovchenko VI, Bykov AT, Poznyakovsky VM (2019) Nutrition for correcting metabolism disorders of the joint-ligament system in physically active people. Hum Sport Med 19(1):108–116. Scopus. https://doi.org/10.14529/hsm190115

Triggs LN, Rogers J (2014) The musculoskeletal system and human movement. In: Orthopaedic and trauma nursing: an evidence-based approach to musculoskeletal care. Wiley, pp 27–47. Scopus. https://doi.org/10.1002/9781118941263.ch4

Trommelen J, Van Lieshout GAA, Nyakayiru J, Holwerda AM, Smeets JSJ, Hendriks FK, Van Kranenburg JMX, Zorenc AH, Senden JM, Goessens JPB, Gijsen AP, Van Loon LJC (2023) The anabolic response to protein ingestion during recovery from exercise has no upper limit in magnitude and duration in vivo in humans. Cell Rep Med 4(12):101324. https://doi.org/10.1016/j.xcrm.2023.101324

Túri-Szerletics M, Patkó I (2008) Environmental contaminants in foodstuffs. Acta polytech Hung 5(3):135–140. Scopus

Urdampilleta A, Vicente-Salar N, Martínez Sanz JM (2012) Protein needs in athletes and dietary-nutrition guidelines to gain muscle mass. Revista Espanola de Nutricion Humana y Dietetica 16(1):25–35. Scopus. https://doi.org/10.1016/S2173-1292(12)70068-6

Usgu G, Usgu S (2023) The bone and joint structure. In: Functional exercise anatomy and physiology for physiotherapists, vol 3. Springer International Publishing, pp 53–75. Scopus. https://doi.org/10.1007/9783031271847-3

van Loon LJC (2013) Role of dietary protein in post-exercise muscle reconditioning, pp 73–83. https://doi.org/10.1159/000345821

Vasconcelos M, Grusak MA (2006) Status and future developments involving plant iron in animal and human nutrition. In: Iron nutrition in plants and rhizospheric microorganisms. Springer, Dordrecht, pp 1–22. https://doi.org/10.1007/1-4020-4743-6_1

Venugopal V (2019) Nutrients and nutraceuticals from seafood, Reference series in phytochemistry, pp 1397–1440. Scopus. https://doi.org/10.1007/978-3-319-78030-6_36

Vidović B, Đorđević B (2016) Nutrition and dietary supplements in the prevention of osteoarthritis. Arhiv Za Farmaciju 66(6):293–308. Scopus. https://doi.org/10.5937/arhfarm1606293V

Vignesh A, Amal TC, Sarvalingam A, Vasanth K (2024) A review on the influence of nutraceuticals and functional foods on health. Food Chem Adv 5:100749. Scopus. https://doi.org/10.1016/j.focha.2024.100749

Walsh JS (2018) Normal bone physiology, remodelling and its hormonal regulation. Surgery (Oxford) 36(1):1–6. https://doi.org/10.1016/j.mpsur.2017.10.006

Wang H (2023) The potential of collagen treatment for comorbid diseases. Polymers 15(19):3999. Scopus. https://doi.org/10.3390/polym15193999

Wang X, Chan WCW, Chan D (2016) Joint development. In: Cartilage: volume 1: physiology and development. Springer International Publishing, pp 169–189. Scopus. https://doi.org/10.1007/978-3-319-29568-8_7

Wang X, Nag R, Brunton NP, Siddique MAB, Harrison SM, Monahan FJ, Cummins E (2022) Human health risk assessment of bisphenol A (BPA) through meat products. Environ Res 213:113734. Scopus. https://doi.org/10.1016/j.envres.2022.113734

Weaver CM (2024) Next-generation health benefits of maternal calcium supplementation. Am J Clin Nutr 119(6):1377–1378. https://doi.org/10.1016/j.ajcnut.2024.03.013

Webster J, Dalla Via J, Langley C, Smith C, Sale C, Sim M (2023) Nutritional strategies to optimise musculoskeletal health for fall and fracture prevention: looking beyond calcium, vitamin D and protein. Bone Rep 19:101684. Scopus. https://doi.org/10.1016/j.bonr.2023.101684

Weinhaus AJ (2015) Anatomy of the human heart. In: Handbook of cardiac anatomy, physiology, and devices, 3rd edn. Springer International Publishing, pp 61–88. Scopus. https://doi.org/10.1007/978-3-319-19464-6_5

Weinhaus AJ, Roberts KP (2005) Anatomy of the human heart. In: Handbook of cardiac anatomy, physiology, and devices. Humana Press, pp 51–79. Scopus. https://doi.org/10.1007/978-1-59259-835-9_4

Wilson D (2011) Vascular smooth muscle structure and function. In: Mechanisms of vascular disease: a reference book for vascular specialists. University of Adelaide Press, pp 13–24. Scopus. https://doi.org/10.1017/UPO9781922064004.003

Woo SL-Y, Sites TJ (2018) Current advances on the study of the biomechanical properties of tendons and ligaments. In: Collagen: volume II: biochemistry and biomechanics, pp 223–241. Scopus. https://doi.org/10.1201/9781351070782

Yao J, Lian Z, Yang B, Fan Y (2020) Biomechanics of ligaments. In: Frontiers in orthopaedic biomechanics. Springer Nature Singapore, pp 75–87. Scopus. https://doi.org/10.1007/978-981-15-3159-0_3

Yasui N, Nimni ME (2018) Cartilage collagens. In: Collagen: volume I: biochemistry, pp 225–241. Scopus. https://doi.org/10.1201/9781351070799

Yildiz H (2023) Musculoskeletal system assessment. In: Medical nursing, pp 407–412. Scopus. https://www.scopus.com/inward/record.uri?eid=2-s2.0-85178248181&partnerID=40&md5=c3b961ad4dc6f9eeeefcc6852340477c

Zaffagnini S, Koh J, Longo UG, Stelitano G, Amirouche F, Denaro V (2021) Tendon biomechanics-structure and composition. In: Orthopaedic biomechanics in sports medicine. Springer International Publishing, pp 81–90. Scopus. https://doi.org/10.1007/978-3-030-81549-3_7

Zeidan RS, Martenson M, Tamargo JA, McLaren C, Ezzati A, Lin Y, Yang JJ, Yoon H-S, McElroy T, Collins JF, Leeuwenburgh C, Mankowski RT, Anton S (2024) Iron homeostasis in older adults: balancing nutritional requirements and health risks. J Nutr Health Aging 28(5):100212. Scopus. https://doi.org/10.1016/j.jnha.2024.100212

Zhang L, Gardiner BS, Smith DW, Pivonka P, Grodzinsky AJ (2010) The transport of insulin-like growth factor through cartilage. In: Porous media applications biological systems and biotechnology, pp 399–453. Scopus. https://doi.org/10.1201/9781420065428

Zhang Y-H, Li Z, Zeng T, Chen L, Li H, Huang T, Cai Y-D (2021) Detecting the multiomics signatures of factor-specific inflammatory effects on airway smooth muscles. Front Genet 11:599970. Scopus. https://doi.org/10.3389/fgene.2020.599970

Zhou Y, Zhu Z-L, Guan X-X, Hou W-W, Yu H-Y (2009) Reciprocal roles between caffeine and estrogen on bone via differently regulating cAMP/PKA pathway: the possible mechanism for caffeine-induced osteoporosis in women and estrogen's antagonistic effects. Med Hypotheses 73(1):83–85. https://doi.org/10.1016/j.mehy.2009.01.029

Zinder R, Cooley R, Vlad LG, Molnar JA (2019) Vitamin A and wound healing. Nutr Clin Pract 34(6):839–849. Scopus. https://doi.org/10.1002/ncp.10420

Zumbaugh MD, Johnson SE, Shi TH, Gerrard DE (2022) Molecular and biochemical regulation of skeletal muscle metabolism. J Anim Sci 100(8):skac035. Scopus. https://doi.org/10.1093/jas/skac035

Chapter 16
Dermatological Manifestation of Food Allergies and Sensitivities

Bhaswati Kashyap, Sumi Barman, Madhusmita Gogoi, Dipjyoti Sharma, Nilutpal Sharma Bora, Koushik Nandan Dutta, Sunipa Adak, Tarun Kumar Dua, and Aadesh Upadhyay

16.1 Introduction

Food allergies are a phenomenon in which exposure to a particular food result in adverse reactions caused by antigen-specific immunological mechanisms" (Urisu et al. 2014). Sensitivities refers to the skin reactions that occurs skin becomes sensitive or reacts to certain external or internal factors. These sensitivities may cause by allergic reaction also help various skin symptoms like skin irritation, inflammation etc. Various health consequences resulting from a particular immune response to food allergies happen consistently upon exposure to a certain food. Several studies reported that adult food allergies caused by specific number of allergens like peanut, fish, milk, shellfish, tree nuts, wheat, and soyabean (Messina and Venter 2020).

Food allergies have negative impact on health and its population-based studies knowns as prevalence of food allergies. While some children with food allergies

B. Kashyap
School of Pharmacy, Pragjyotishpur University, Hajongbari, Chandrapur, Guwahati, Assam, India

S. Barman · M. Gogoi
Royal School of Pharmacy, The Assam Royal Global University, Betkuchi, Guwahati, Assam, India

D. Sharma · N. S. Bora · K. N. Dutta (✉)
Department of Pharmacognosy, NETES Institute of Pharmaceutical Science, NEMCARE Group of Institution, Mirza, Kamrup, Assam, India

S. Adak · T. K. Dua
Department of Pharmaceutical Technology, University of North Bengal, Darjeeling, West Bengal, India

A. Upadhyay
Project Lead-Oncology, One Clinical Research, Perth, WA, Australia

T. Sarkar et al. (eds.), *Physiological Perspectives on Food Safety: Exploring the Intersection of Health and Nutrition*, https://doi.org/10.1007/978-3-031-94582-3_16

grow up to be naturally tolerant, others continue to have food allergies as adults, and adults may even acquire new food allergie (Gupta et al. 2019). Food allergies show a variety of symptoms related that appear minutes to days after food consumption, and sometimes months, similar to atopic dermatitis (Sicherer et al. 2022). Food intolerances can be caused by metabolic, toxic, pharmacological, or unknown mechanisms, and they are non-immunological reactions (Manuyakorn and Tanpowpong 2019). The dermatological manifestation of food allergy is highly relevant because the skin is one of the first organs to show the allergic reaction and this manifestation also diagnoses the food allergy problem and diagnosis other symptoms. Recent studies analysed the various case history of food allergy, where in some cases patient are eating like allergic food, that cases symptom's like gastrointestinal disorder, respiratory issues etc. (Valenta et al. 2015). Another case is nowadays youngest children are eaten cow milk & egg to determine the symptoms like vomiting & eczema. Basically, food allergy symptoms are very crucial in some food, vegetables, and fishes etc. (Tedner et al. 2022).

Food allergy is becoming a serious public health issue and widespread around the world. The frequency of food allergy has increased over the past 30 years in western countries and rate up to 10% have been reported. In worldwide 220 million people are suffer from food allergy. Although over one-third of parents report food hypersensitivity reactions in their kids, the prevalence of scientifically confirmed and objectively diagnosed allergies in the first year of life really ranges from 6% to 10% and falls to 2% to 5% in adulthood (De Martinis et al. 2020). Although estimates of food allergy prevalence rates differ depending on factors including age, geography, and research methods, it is generally agreed that the condition is widespread and has been becoming more prevalent. Identifying risk factors for food allergies can help with prevention and treatment. Some risk factors are modifiable, such as skin barrier dysfunction, vitamin D deficiency, microbiome, route, and timing of first exposure to foods, maternal dietary habits, and lack of early exposure to specific environmental allergens, whereas others, such as male gender, racial background, or family history, are underlying nonmodifiable risks (Lopes and Sicherer 2020). Food allergies are thought to be causally related to between 70 and 160 food items and food additives (Bartuzi et al. 2017). For the purpose of estimating food allergy incidence and prevalence, historical trends, and relevant risk factors in household study, 20% of people said they have dietary intolerances but when double-blind placebo controlling food challenges were conducted, fewer than 2% of people experienced real food reactions (Turnbull et al. 2015a). It seems that developed nations are seeing an increase in the prevalence of food allergies and related atopic illnesses. It has not been determined genuine rise in prevalence or a growing understanding of the clinical signs of food allergies. Different epidemiological approaches make it difficult to compare the available prevalence studies. Laboratory results are used in some studies to define food allergies, whereas clinical reactivity is used in others also Food-specific IgE antibodies can be detected in healthy people who are obviously tolerant of that food, even if they have never experienced a clinical reaction. Finally, due to variations in environmental conditions, the frequency of food allergies and the range of food allergens may change between nations genetic components (Ho et al. 2014). A recent meta-analysis of European food allergy prevalence discovered an overall prevalence of self-reported food allergy of 5.9% from 2000 to 2012, while several of the original studies exhibited at least significant potential for bias (Savage and Johns 2015a).

16.2 Pathophysiology of Food-Induced Skin Reactions

16.2.1 Immune Mechanisms in Food Allergies

Food allergies are pathological immune reactions to food proteins (De Martinis et al. 2020). They are mostly classified into two categories based on the involved immune mechanisms: IgE-mediated and non-IgE-mediated. Knowing these mechanisms, along with the involvement of histamine and other inflammatory mediators, is essential for properly diagnosing and managing food allergies (Bellanti 2024).

16.2.1.1 IgE-Mediated Responses

IgE-type food allergies are rapid hypersensitivity responses that become evident shortly after the ingestion of the allergen, usually between minutes to two hours (Kounis et al. 2019). With initial exposure to an individual food allergen, the immune system mounts an immune response and makes allergen-specific Immunoglobulin E (IgE) antibodies (De Martinis et al. 2020). IgE antibodies form a complex with high-affinity IgE receptors (FcεRI) present on the membrane of mast cells and basophils, pre-exposing them to the allergen (Sutton 2024). On repeated exposure, the allergen cross-links bound IgE on these sensitized cells, activating their degranulation and the release of preformed mediators, particularly histamine, and newly synthesized mediators such as leukotrienes and prostaglandins (Shamji et al. 2021). This release causes the classical symptoms of IgE-mediated food allergies, which may vary from urticaria (hives), angioedema, and gastrointestinal discomfort to respiratory distress and, in extreme cases, anaphylaxis. The quick onset and possible severity of these reactions require immediate recognition and treatment (Çelebioğlu et al. 2021).

16.2.1.2 Non-IgE-Mediated Responses

Non-IgE-mediated food allergy employs immune reactions independent of IgE antibodies and are usually delayed, occurring several hours to days post-allergen exposure. Such reactions are predominantly T-cell-mediated and frequently are localized within the gastrointestinal tract (Zhang et al. 2021). Non-IgE-mediated food allergy syndromes commonly observed are:

- **Food Protein-Induced Enterocolitis Syndrome (FPIES)**: Presents with massive vomiting, diarrhea, and dehydration, often in infants after consuming specific food proteins, including cow's milk or soy (Calvani et al. 2021).
- **Allergic Proctocolitis (AP)**: Presents as stools containing blood in otherwise healthy infants, most often in association with the ingestion of cow's milk or soy protein (Barni et al. 1824).
- **Food Protein-Induced Enteropathy (FPE)**: Characterized by chronic diarrhea, malabsorption, and failure to thrive due to damage to the mucosa of the small intestine by food antigens (Barni et al. 2024).

Non-IgE-mediated reactions have pathophysiology characterized by antigen-specific T cells' recognition of food proteins, release of cytokines inducing local inflammation and tissue injury. In contrast to IgE-mediated reactions, such responses lack mast cell degranulation or release of histamine, contributing to their latency and largely gastrointestinal symptoms (Zhang et al. 2021).

16.2.1.2.1 Role of Histamine and Inflammatory Mediators

Histamine is a biogenic amine that is stored primarily in mast cells and basophils (Patel and Mohiuddin 2020). When activated by IgE-mediated mechanisms, these cells release histamine into the surrounding tissues, where it has several effects by binding to histamine receptors (H1R, H2R, H3R, and H4R) on target cells (Banafea et al. 2022). Histamine binding to these receptors results in:

- **H1 Receptors**: Cause vasodilation, heightened vascular permeability, and contraction of smooth muscles, leading to symptoms like edema, erythema, bronchoconstriction, and pruritus (Shulpekova et al. 2021).
- **H2 Receptors**: Activate gastric acid secretion and may influence immune cell function, such as suppression of Th1 lymphocyte responses (Reyes and Peniche 2019).
- **H3 and H4 Receptors**: Play roles in regulating neurotransmitter release and chemotaxis of immune cells, respectively (Bernardino 2021).

The multiple effects of histamine play a large role in the clinical features of IgE-mediated allergic disease. Histamine also has an effect on Th1/Th2 immune homeostasis by inducing a Th2-polarized profile favoring allergic sensitization (Afshari et al. 2024). Such modulation is by differential receptor occupancy, with the activation of the H1 receptor augmenting Th1 responses and the activation of the H2 receptor inhibiting both Th1 and Th2 responses (de Bartolomeis et al. 2022). Additionally, histamine has an impact on dendritic cell function, modulating antigen presentation and chemotaxis, and hence the immune response to allergens (Sarasola et al. 2021).

In food allergies that are not IgE-mediated, histamine is not the central mediator (Anvari et al. 2019). Rather, other inflammatory mediators, including a number of cytokines produced by activated T cells, play a role in the pathophysiology (Cosenza et al. 2021). These cytokines direct the recruitment and activation of immune cells to produce localized inflammation and tissue damage, especially in the gastrointestinal tract (Friedrich et al. 2019).

Knowledge of the unique immune mechanisms involved in IgE-mediated and non-IgE-mediated food allergies, and the central role of histamine and other inflammatory mediators, is crucial for the development of targeted therapies and enhanced patient outcomes.

16.2.2 *Role of Skin Barrier Dysfunction*

Atopic dermatitis (AD) is a common chronic inflammatory skin disease with severe itching and eczematous lesions (Torres et al. 2019). Skin barrier defect and immune dysregulation are central to its pathogenesis. New evidence emphasizes the critical roles of epicutaneous sensitization and microbiome dysbiosis in AD development and exacerbation (Lu et al. 2024).

16.2.2.1 Atopic Dermatitis and Epicutaneous Sensitization

The skin is an important barrier that protects the body against environmental insults, allergens, and pathogens (Celebi Sozener et al. 2022). In AD, this barrier function is usually compromised, allowing penetration of allergens and subsequent immune reactions. Mutations in the filaggrin (FLG) gene have been identified as a contributing factor to this barrier dysfunction (Goleva et al. 2019). Filaggrin plays a key role in maintaining skin integrity; its deficiency results in heightened transepidermal water loss and diminished barrier function (Kim and Lim 2021). This impaired barrier facilitates the entry of allergens more readily, resulting in sensitization and inflammation (Akdis 2021). This phenomenon, epicutaneous sensitization, is an important contributor to the onset and maintenance of AD (Dębińska and Sozańska 2023). Exposure to harsh detergents and pollutants in the environment can also compromise the skin barrier further, worsening the condition (Pat et al. 2023). Emollient application and barrier repair therapy are two strategies that have been shown to reduce the incidence of AD and avoid allergen sensitisation by restoring the function of the skin barrier (Elias 2022).

16.2.2.1.1 Role of Microbiome Dysbiosis

The skin harbors a complex group of microorganisms, referred to collectively as the skin microbiome, which is crucial for maintaining immune homeostasis and skin health (Woo and Kim 2024). In AD, this microbial community is distinctly unbalanced, or dysbiotic (Liu et al. 2020a). Perhaps the most important alteration is the overcolonization of the skin with Staphylococcus aureus. This organism can perpetuate inflammation and is linked with severity of disease (Demessant-Flavigny et al. 2023). The predominance of S. aureus also follows a reduction in microbial diversity, which enhances the compromise on defense mechanisms in the skin (Severn and Horswill 2023). According to latest findings, achieving microbiome re-balance via probiotics, prebiotics, or transplantation of the microbiome could possess therapeutic potential for people with AD (Ersanli et al. 2023). Properly preserved skin hygiene, along with the application of antimicrobial medication, can likewise ensure the maintenance of dysbiosis and favor better clinical performance (De Almeida et al. 2023).

The pathogenesis of atopic dermatitis is strongly associated with skin barrier dysfunction and microbial imbalance (Yang et al. 2020). Targeted treatments that strengthen the barrier and restore microbial balance have the potential to significantly improve patient health and disease control.

16.2.3 Dual Allergen Exposure Hypothesis

The Dual Allergen Exposure Hypothesis provides a strong paradigm for food allergy development, highlighting the opposing effects of routes of allergen exposure: oral ingestion versus transcutaneous (skin) exposure (Brough et al. 2022). According to this hypothesis, early oral exposure to allergens favors immune tolerance, whereas exposure through a broken skin barrier, especially in the setting of atopic dermatitis (AD), favors sensitization and later allergic reactions (Gao et al. 2024).

16.2.3.1 Early Oral Exposure and Immune Tolerance

Infantile food allergen oral exposure is an important factor for the induction of oral tolerance—condition where the immune system learns to be comfortable with harmless chemicals, so there is no abnormal reaction (Wollenberg et al. 2020). Specialized immune cells in the gut are able to differentiate regulatory T cells (Tregs) following exposure to diet proteins. Such Tregs silence hypersensitive immunity, hence maintaining tolerance to individual foods. Clinical evidence confirms this mechanism; for example, early peanut introduction into infant diets has been linked with decreased rates of peanut allergy (Goswami et al. 2022). This preventive approach concurs with the theory that managed oral exposure at key times in immune development is able to guide the immune system towards tolerance as opposed to sensitization (Yu et al. 2019).

16.2.3.2 Transcutaneous Sensitization and Allergic Responses

Conversely, transcutaneous allergen exposure, particularly through disrupted skin barriers, can precipitate sensitization, a harbinger of allergy. Such dermatitis as atopic dermatitis is defined by skin barrier disruption, so environmental allergens will more readily penetrate (Strid and Strobel 2005). Allergens, upon passing through the skin, come into contact with immune cells within the epidermis, including Langerhans cells, which process these antigens and present them to T-helper cells. This interaction facilitates a Th2-biased immune response that is defined by the secretion of allergen-specific Immunoglobulin E (IgE) antibodies (Smith et al. 2017). Later exposure to the same allergen can induce these IgE antibodies to activate basophils and mast cells and cause the secretion of inflammatory mediators and expression of allergic signs. It has been shown by studies that infants with early-onset atopic dermatitis have an increased risk of developing food allergies, emphasizing the pathogenetic role of transcutaneous sensitization (Shamji et al. 2021).

16.2.3.3 Implications for Prevention and Management

Knowledge of the Dual Allergen Exposure Hypothesis has profound implications for prevention and treatment of food allergies:

- **Early Dietary Introduction**: Early introduction of potential allergens into a baby's diet can enhance oral tolerance and decrease the risk of food allergy development. This strategy must be adjusted to the patient's individual risk profile and should be performed under medical supervision for safety (Trogen et al. 2022).
- **Skin Barrier Protection**: The integrity of skin barriers must be maintained and boosted, especially among people with atopic dermatitis. Daily emollient application and proper skincare measures can deter transcutaneous allergen penetration and thus inhibit the process of sensitization (Kim and Leung 2018).
- **Comprehensive Strategies**: A two-way approach consisting of early oral exposure to allergens coupled with active skin care could prove to be the most successful strategy against food allergies. This overall approach addresses both possible routes of sensitization, in harmony with the revelations of the Dual Allergen Exposure Hypothesis (Kulis et al. 2021).

16.3 Common Dermatological Manifestations of Food Allergies

16.3.1 Urticaria and Angioedema

Urticaria also known as hives and angioedema are skin conditions consisting of transient, itchy swellings. Although urticaria involves the surface layers of the skin, angioedema involves deeper tissues that lead to significant swelling of the eyes and lips (Kanani et al. 2018). The underlying pathogenesis of urticaria involves the activation of mast cells, leading to the release of histamine and other vasoactive mediators. This process increases vascular permeability in postcapillary venules, resulting in edema, erythema, and pruritus. Urticaria is frequently associated with angioedema, occurring together in approximately 40% of cases, while around 10% of patients experience angioedema without urticaria. Although both conditions share a similar pathophysiologic process, urticaria affects the superficial dermis, whereas angioedema involves the deeper dermis and subcutaneous tissues, leading to poorly defined, nonpitting swelling (Guo and Saltoun 2019). Due to the lower density of mast cells and sensory nerve endings in deeper skin layers, angioedema typically causes minimal to no pruritus but may be accompanied by a burning sensation or a feeling of swelling. While urticaria can appear anywhere on the body, angioedema most commonly affects the face, particularly the periorbital and perioral regions, as well as the extremities and genitalia. In severe cases, it can lead to life-threatening swelling of the tongue, soft palate, and larynx. Recognising the differences between acute and chronic urticaria and common food triggers is key to management. Angiotensin-converting enzyme (ACE) inhibitors are a well-known cause of angioedema without urticaria, primarily due to excessive bradykinin accumulation. Additionally, certain medications and diagnostic agents, such as radiocontrast media, opiates, nonsteroidal anti-inflammatory drugs (NSAIDs), and aspirin, can induce non–IgE-mediated mast cell degranulation, leading to symptoms of both urticaria and angioedema. There are also several non-immunologic causes of urticaria, including physical or inducible urticarias, which can be localized or generalized depending on the area of exposure. These include cold-induced, pressure-induced, solar, exercise-induced, cholinergic urticaria, and dermatographism. Cholinergic urticaria is typically triggered by an increase in core body temperature or local heat exposure, with common stimuli including exercise, hot showers, and sweating. Dermatographism, meaning "writing on the skin," presents as a wheal-and-flare reaction following mechanical stroking of the skin and may occur in isolation or alongside chronic urticaria. Some individuals may experience multiple forms of urticaria simultaneously, such as chronic spontaneous urticaria (CSU) coexisting with cholinergic urticaria and dermatographism (Radonjic-Hoesli et al. 2018; Konstantinou et al. 2023).

Urticaria and angioedema can also be associated with underlying conditions, including infectious diseases. Viral infections, such as hepatitis A, B, and C, and Epstein-Barr virus (EBV), may trigger urticaria, while bacterial infections like *Helicobacter pylori* and parasitic infestations have also been implicated. Though rare, malignancies should be considered, particularly in elderly patients with systemic symptoms. A distinct form known as urticaria pigmentosa is characterized by persistent, red-brown maculopapular lesions that urticate upon stroking—a phenomenon known as Darier's sign. This condition may be an early indication of indolent systemic mastocytosis (Melikoglu et al. 2022).

Urticarial vasculitis should be suspected when lesions persist for more than 24 hours or leave residual hyperpigmentation. Patients with hepatitis C, for example, may present with urticarial vasculitis, along with elevated liver function tests and mixed cryoglobulinemia, where IgG and complement component 5a (C5a) activate mast cells (Guo and Saltoun 2019).

Acute and chronic urticaria can be categorized based on how long the symptoms are experienced by the patient. Sudden appearance of hives is acute urticaria. This kind lasts less than 6 weeks and is often triggered by illnesses, certain types of food, medications, or even insect bites. If the precise reason is eliminated, it is highly likely that the patient will get better. Chronic urticaria persists for 6 weeks or longer and in some cases can extend for months or years. Many patients are not able to provide a precise answer to what triggers it, which makes treatment quite difficult. Most people usually do not have one identifiable cause after conducting frequent allergy tests (Melikoglu et al. 2022).

16.3.2 Atopic Dermatitis and Food Allergy

Atopic dermatitis (AD) is a type of long-lasting inflammatory skin disease that is often linked with food allergies (FA), especially for young children. These two conditions are multifactorial in nature, and both are caused by immune system dysfunction, which predisposes individuals to other allergic diseases such as asthma, allergic rhinitis, and eosinophilic esophagitis, which are collectively described by the "Atopic March" Theory. However, diagnosing food allergies in subjects presenting with AD is problematic due to the extensive overlap in clinical symptoms and the inadequacy of current diagnostic methods (Bergmann et al. 2013).

The diagnosis of food allergies in these patients is quite operative, owing to the aspects mentioned below:

Clinical Overlap
Most food allergies in AD patients are manifested as chronic eczema skin flares instead as prominent overt anaphylaxis. Such reactions are considerably non-IgE mediated and multifactorial rendering the causative foods undetectable.

Sensitization vs. Clinical Allergy
Determined sensitization by specific IgE levels or skin prick tests (SPT) is often non-proportional to clinical allergy. In reality, a lot of patients show tolerance towards food after developing IgE sensitivity towards those foods.

Gold-Standard Testing
The DBPCFC continues to be the gold standard of testing, but in practical clinical situations, does not always seem reasonable since it's expensive and time-consuming.

Barrier Dysfunction and Non-IgE Reactions
The reduction of the barrier function of the skin in AD is further compounded by the ability of the skin surface to permit penetrant allergy substances without manifestation of immediate symptoms.

Risk of Developing Systemic Allergies
Patients with AD, particularly those with more severe disease (early- onset and moderate-to-severe disease) or with filaggrin (FLG) gene mutations, have increased risk of systemic allergic diseases including:

IgE-Mediated Food Allergy
A significant number of these children with AD will go on to develop immediate hypersensitivity reactions to common allergens, including milk, eggs, peanuts, and tree nuts.

Asthma and Allergic Rhinitis AD occurs before and predicts the development of asthma and allergic rhinitis, described as the Atopic March.

Eosinophilic Gastrointestinal Disorders (EGIDs)
Chronic exposure of food allergens in AD patients may cause eosinophilic esophagitis and other gastrointestinal allergies.

Risk of Anaphylaxis
Patients with food allergies who suffer from AD can experience severe systemic reactions and must implement strict dietary avoidance in the management of both conditions.

Diagnosing atopic dermatitis (AD) along with issues related to food allergies is a multidisciplinary task that is both complex and difficult. Because telling apart true allergy and sensitization is challenging, diagnosis usually follows a procedure consisting of obtaining thorough patient history, conducting laboratory examinations, and sometimes performing food challenges. One of the biggest problems is that patients with early and severe forms of AD appear to have systemic allergies, which can include asthma, allergic rhinitis, eosinophilic gastrointestinal diseases, or even anaphylaxis, as well. To mitigate such risks, it is important to understand the skin barrier defects, immune mechanisms, as well as a person's family history. New studies focusing on changes like skin barrier repair and skin exposures, skin barrier repair, early allergen introduction, and immunomodulatory therapy are becoming more promising for changing the course of Atopic March and lessening the impact of systemic atopic diseases. A careful collaboration of dermatologists, allergists, pediatricians, and nutritionists is needed to achieve appropriate control of the disease and accurate diagnostic and therapeutic dietary measures. Comprehensive treatment strategies including preemptive skin care, specific allergy tests, and dietary measures practicing food aversion can all be employed to ameliorate Adler's diet without restrictions (Dhar and Srinivas 2016; Tham and Leung 2019; Sugita and Akdis 2020).

16.3.3 Eczema and Food Sensitivities

Food sensitivities can exacerbate eczema in some individuals. Unlike true food allergies, which involve the immune system, food sensitivities may lead to chronic inflammation and skin irritation. These reactions can manifest as persistent redness, dryness, itching, and even flare-ups of eczema. While not classified as an allergy, food sensitivities can still significantly impact the skin barrier function and overall skin health (Worth and Sheikh 2010).

Role of Elimination Diets Elimination diets remove allergens from the diet for a certain time and then reintroduce them one by one to narrow down the culprits. This method helps identify foods that trigger eczema symptoms, so dietary modifications can be tailored to the individual (Young 2016; Skypala and McKenzie 2019). Common exclusions in elimination diet include dairy (e.g., milk, cheese, yogurt), grains that contain gluten (wheat, barley, rye), eggs, soy and nuts.

Importance of Supervised Dietary Exclusions These diets are structured to be nutritionally adequate and to avoid unwarranted food avoidance. Avoiding essential nutrients like protein, calcium and vitamins could lead to deficiencies, so one must seek help from health professionals (Skypala and McKenzie 2019; Sikorska-Szaflik and Sozańska 2021).

I. **Medical Supervision**: A dermatologist, allergist or dietitian can help make dietary changes while keeping your nutrition balanced.
II. **Gradual Reintroduction**: Eliminated foods are gradually reintroduced to identify specific triggers, thus avoiding long-term dietary restrictions without reason.
III. **Nutrition Adjustments**: With careful supervision, patients can receive sufficient nutrition from alternative sources of sustenance.

By implementing a well-structured elimination diet under professional supervision, individuals with food-sensitive eczema can manage their condition effectively while maintaining overall health (Du Toit et al. 2016).

16.3.4 Contact Dermatitis from Food Allergens

Contact dermatitis occurs when the skin responds to direct contact with food allergies, which can be classified into irritable contact dermatitis (ICD) and allergic contact dermatitis (ACD). ICD is a non-allergic reaction in which foods damage the skin barrier, causing redness, itching and inflammation, while ACD is a delayed hypersensitivity reaction triggered by food proteins, resulting in lesions such as eczema. Normal food trigger includes citrus fruits (e.g. lemon, oranges), tomatoes, garlic, onions, spices (e.g. cinnamon, pepper) and dairy products. Handling strategies include avoiding the identified allergies, using protective gloves while handling food, and involving the addition of emollients and barrier creams for skin protection. Understanding the dermatology manifestations of food allergies and applying appropriate clinical and governance strategies can help individuals effectively handle their skin conditions while maintaining a balanced diet (Fig. 16.1) (Barbaud et al. 2015; Aquino and Rosner 2019; Nassau and Fonacier 2020).

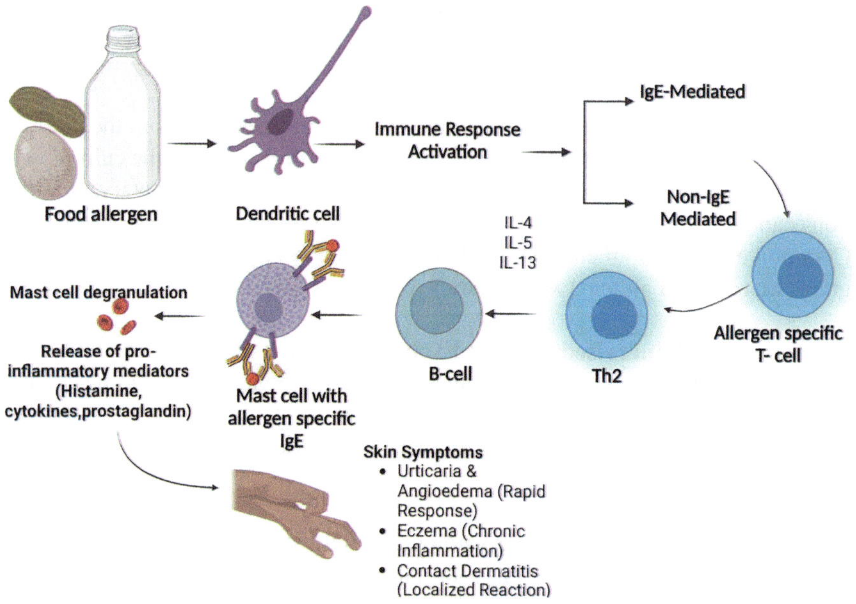

Fig. 16.1 Mechanisms of Food Allergy-Induced Skin Reactions: Upon ingestion, food allergens are recognized by dendritic cells, which process and present them to native T cells, leading to the differentiation of allergen-specific Th2 cells. These Th2 cells release cytokines (IL-4, IL-5, IL-13) that promote B-cell activation and IgE antibody production. Allergen-specific IgE binds to FcεRI receptors on mast cells and basophils, sensitizing them. Upon re-exposure, IgE cross-linking triggers mast cell degranulation, releasing histamine, cytokines, and prostaglandins. These mediators contribute to skin manifestations, including urticaria and angioedema (rapid response), eczema (chronic inflammation), and contact dermatitis (localized reaction). Both IgE-mediated and non-IgE-mediated pathways contribute to food allergy-induced skin reactions

16.4 Diagnostic Approaches for Dermatological Food Reactions

16.4.1 Clinical Evaluation and History-Taking

Allergies are the most common chronic disease. Food associated adverse immune response is termed as food allergy. (Sicherer and Sampson 2014) Estimates of the prevalence of self-reported allergy to milk, egg, peanut, and seafood in a meta-analysis of 51 studies ranged between 3% and 35%, whereas estimates from 6 studies using oral food challenges (OFCs) estimated prevalence rates between 1% and 10.8% (Sicherer and Sampson 2014; Rona et al. 2007).

An accurate assessment obviously requires a complete history to consider an extensive differential diagnosis, to obtain any potential trigger foods, and to establish a likely general patho-physiologic mechanism (whether the food-induced allergic disease is likely IgE mediated, which informs testing). The History should inform the potential cause food or foods, quantity consumed, time course of reaction, ancillary factors (exercise,aspirin and alcohol), and reaction consistency. The history is also directed at details that may aid in assessing the prior probability of an allergic reaction to a given food. The clinician should integrate the epidemiologic features of the disease (eg its common triggers and common associations) and information from the specific history to arrive at a diagnosis and to consider appropriate testing that will be interpreted in light of these prior probability estimates (Sicherer and Sampson 2014).

Prospective ascertainment using confirmatory oral food challenges of a representative sample of newborns and young children at predefined intervals across time is necessary for a precise assessment of the prevalence and natural history of food allergies at the population level (Savage and Johns 2015b). Food allergies were first documented scientifically at the start of the twentieth century. Other test methods were used in addition to attempts at re-exposure because cutaneous testing with food allergens frequently remained inaccurate (either false negative or false positive) (Wüthrich 2014).

16.4.2 *Laboratory and Immunological Tests*

16.4.2.1 Skin Prick Test

In individuals with a suspicious clinical history of asthma and allergic rhinitis (conjunctivitis), skin tests are the initial diagnostic procedure. From early childhood to old age, they can be utilised (Bousquet et al. 2012; Demoly et al. 2009). Due of the strong association with symptoms, prick and puncture tests are advised in allergic food reactions. According to Demoly P. et al. (1998), skin prick tests are highly sensitive and specific for identifying inhalant allergies. The current reference method is the modified SPT developed by Pepys (Frcp and Pepys 1972), which involves cleaning the skin with alcohol and applying minimal pressure before passing a fine metal needle through a drop of allergen extract. SPT variability can be reduced through puncture testing using a variety of tools (Malling et al. 1982; Adinoff et al. 1989; Nelson et al. 1993). Every test should employ a separate needle or puncture test (Piette et al. 2002). The peak of the skin wheal for allergens occurs approximately 10 to 20 minutes after the test, and it is advised to take a reading of the skin wheals' maximum diameter after 15 minutes (Bousquet et al. 2012). It is usually vital to ask patients about the medications they have taken in the previous days because drugs can reduce skin testing.

Prick testing can only be performed on healthy skin. Patients with history of urticaria or eczema (e.g. atopic dermatitis) cannot be tested in areas of affected skin in skin prick test. Neurological disorders as well as infectious disease (e.g.leprosy) can lead to false-negative SPTs (Bousquet et al. 2012).

16.4.2.2 Specific IgE Test

The ImmunoCAP Specific IgE test measures the amount of allergen-specific IgE in human serum or plasma quantitatively in vitro. Samples of venous or capillary blood serum or plasma (EDTA or heparin) may be utilised (Johansson 2004).

The appropriate IgE antibodies in the patient's serum sample react with the allergen of interest, which is covalently bound to the Immuno-CAP cellulose solid phase. Enzyme-labeled antibodies against IgE are added after free IgE has been removed. Following incubation, the ImmunoCAP is incubated with a developing agent after the unbound enzyme–anti-IgE has been removed. The eluate's fluorescence is monitored when the reaction is stopped. There are more IgE antibodies in the specimen when the reaction value is higher. A calibration curve based on the World Health Organization's (WHO) reference preparation 75/502 for IgE is used to convert the patient samples' reaction to IgE concentrations in order to compute the test findings (Johansson 2004; Wild 2013).

16.4.2.3 Patch Test

Patch tests are tools used in the identification of the etiologic agent (s) of allergic contact dermatitis (Lazzarini et al. 2013). Patch tests are indicated for patients with a diagnostic hypothesis of CD, for patients with other skin conditions that may be aggravated by CD (atopic dermatitis, seborrheic dermatitis and stasis, nummular eczema, psoriasis, and dyshidrosis), for patients which chronic eczema without an established etiology, and in suspected cases of occupational contact dermatitis (Lazzarini et al. 2013; Lindberg 2011; Lindberg and Matura 2011; Tennstedt 2009). The patch test can also be used to investigate drug reactions that manifest with skin lesions resulting from a late hypersensitivity mechanism, such as maculopapular rash, DRESS (drug, rush, eosinophilia, systemic symptoms), fixed drug eruption (Lazzarini et al. 2013).

16.4.2.4 Battery or Series and Types of Chambers

There are several crucial requirements that must be fulfilled by the test application approach. The material under test is referred to in the first. The battery's constituent parts are ready to better permeate the skin without causing discomfort in certain areas. Every allergen needs to be prepared in an appropriate vehicle. Petrolatum is the most commonly used. Water, hydrophilic gel, and solvents (acetone, ethanol, and methylethylcetone) are additional potential carriers. In order to prepare the test chambers—that is, the material in which the chemicals to be tested are placed—paper filters must be used while applying liquid substances to the skin. It is made up of round aluminium chambers on a Scanpor (an adhesive made of acrylate) (Lazzarini et al. 2013; Silva et al. 2012).

16.4.2.5 Application Technique

Because the upper back is a large area that makes it easier to apply different chemicals, tests should be conducted there. The upper arms and possibly the upper thighs are another location that can be exploited. Because of their poor adherence, hairy areas should be avoided. Razor blades can, however, be used to cut hair in the direction that it grows if necessary. Adhesive tape can be used after the chambers are applied to stop the tests from coming loose and losing their adherence, which could lead to false negative results (Lachapelle 2009). 48 hours following application, the tests are removed in a well-lit area using certain plaques for the type of chamber that were used, each of which has holes that correlate to the position of allergens. An indelible pen should be used to mark the back so that readings can be made later. The attending physician should take readings, and any positive results, together with how strong they are, should be appropriately documented in the patient's medical file. Because local vasodilatation can cause the site to become erythematous and occasionally oedematous right after the adhesive detaches, it is preferable to wait 15 to 20 minutes after the test is removed before taking the reading. 72–96 hours after the tests are administered; the patient should come back for another reading. Given that a sensitisation reaction could happen more than 72 hours after contact, this new reading is crucial. Additionally, readings that were positive 48 hours after the tests were applied may turn negative 72–96 hours later, indicating that test occlusion caused only local irritation.The International Contact Dermatitis Research Group's morphologic criteria are used to measure test results (ICDRG) (Lazzarini et al. 2013).

16.4.3 Food Challenge Testing

16.4.3.1 Double Blind Placebo Controlled Food Challenge (DBPCFC)

The double-blind, placebo-controlled food challenge is the gold standard for diagnosing food allergies. Throughout the course of a day, escalating doses of a solution containing the suspected allergen or a placebo are consumed as symptoms are tracked. After a few days, the test is conducted again with the solution that was not used the first time. A placebo is used, and both the patient and the clinician are blinded to it. Due to the high risk of anaphylaxis, food challenges should only be carried out in locations with anaphylaxis treatment capabilities (Turnbull et al. 2015b).

16.4.4 Role of Biomarkers in Diagnosis

A number of biomarkers are crucial in the diagnosis of food responses related to dermatology. In medicine, biomarkers have long been used for a variety of reasons, mostly as diagnostic tools (Renert-Yuval et al. 2021).The use of biomarkers has sparked a significant change in dermatology towards customised therapies through the development of personalised medicine. These molecular markers act as markers, identifying healthy biological functions, abnormal pathways, and therapeutic reactions (Tan et al. 2024).

Human leukocyte antigen (HLA) markers, chemokine (C-C motif) ligands, interleukins (IL), and proteins including matrix metalloproteinases (MMPs), eosinophil cationic protein (ECP), thymic stromal lymphopoietin (TSLP), and C-reactive protein (CRP) are examples of common biomarkers (Turnbull et al. 2015b; Jiang et al. 2015; Yu and Li 2022; Zhou et al. 2022).

16.4.5 Limitations and Pitfalls in Current Testing

Patients often present to the medical practice with the complaints of an "allergic" problem. The medical professional must determine whether this problem is truly allergic (ie, IgE mediated) (Cox et al. 2008).

In the diagnosis of suspected IgE-mediated allergic diseases, neither skin tests nor serum s-IgE tests should be either requested or interpreted outside the context of the clinical history and physical examination. A history of symptoms following exposure to the relevant allergen may be supported by objective proof of IgE sensitisation provided by positive test results. A negative test result, on the other hand, reduces the likelihood of an allergy to a suspected allergen. Both skin and serum s-IgE tests can frequently produce false-positive findings when administered carelessly, and in rare cases, false-negative results as well. Therefore, depending on the history, diagnostic testing should be conducted to confirm or rule out a diagnosis of a particular allergy. They should almost never be used in place of an allergy test or a thorough history (Cox et al. 2008).

Each method of allergy testing has its own pros and cons, indications and limitations. It is important to realise that there is an important difference between sensitisation (the presence of specific IgE antibodies, as indicated by a positive IgE test or SPT) and allergy (including symptoms caused by exposure to the allergen) (Table 16.1).

16.5 Management Strategies for Food Allergy-Induced Skin Conditions

Food allergy is an important etiology of allergic cutaneous reactions including eczema (atopic dermatitis), urticaria (hives), and angioedema. These are conditions that can lead to long-term, painful symptoms that influence the quality of life of the affected individual (Uppin et al. 2024). An effective management plan for skin diseases caused by food allergy must be a multifaceted strategy that includes food

Table 16.1 Comparison of Food Allergy Diagnostic Tests

Diagnostic Test	Description	Advantages	Limitations	References
Skin prick test (SPT)	Involves placing a small amount of allergen on the skin and pricking it to allow the allergen to enter.	Quick results High sensitivity Widely used for many allergens	Can cause anaphylaxis in sensitive individuals False positives may occur Requires trained staff	Kabesch and Tost (2020)
Serum-specific IgE test (RAST or ImmunoCAP)	Measures the level of IgE antibodies specific to allergens in the blood.	Non-invasive Can measure multiple allergens at once Suitable for individuals with skin conditions or those unable to undergo SPT	May not distinguish between sensitization and clinical allergy Expensive	Fu et al. (2018)
Oral food challenge (OFC)	Controlled exposure to suspected food allergens under medical supervision, with observation for allergic reactions.	Gold standard for diagnosing food allergies Provides definitive diagnosis	Risk of severe allergic reaction Time-consuming Requires medical supervision	Vedanthan et al. (2021)
Patch test	Involves applying allergens on the skin with adhesive patches to detect delayed-type hypersensitivity reactions.	Useful for non-IgE mediated reactions (e.g., contact dermatitis) Can detect chronic reactions	Takes longer for results Not useful for IgE-mediated reactions	De Martinis and Sirufo (2020)
Elimination diet	Involves removing suspected foods from the diet for a period and then reintroducing them to assess for reactions.	Useful for identifying food triggers Can identify foods that do not appear on standard tests	Time-consuming Risk of inadequate nutritional intake Requires careful monitoring	Lachapelle (2009)

elimination diets, pharmacotherapy, probiotics and microbiome-based therapy, and investigation into new therapies and future research directions. This review delineates the established management approaches of food allergy-triggered skin disease and the proposed advantages of developing therapies. A graphical representation of various management strategies for food allergy-induced skin conditions is represented in Fig. 16.2.

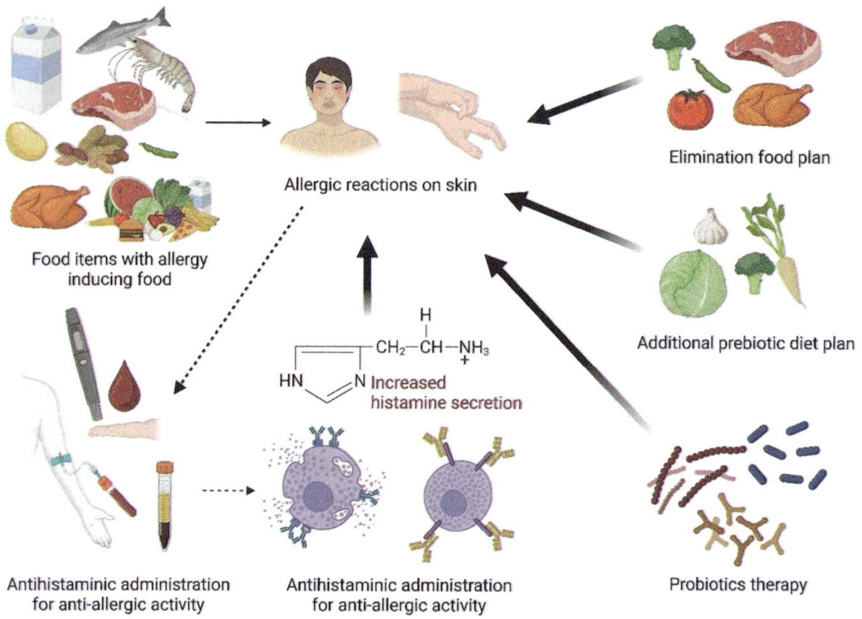

Fig. 16.2 Management strategies for food allergy-induced skin conditions

16.5.1 Role of Food Elimination Diets

Food allergy is one of the main culprits responsible for allergic skin diseases like eczema (atopic dermatitis), urticaria (hives), and angioedema (Faraz et al. 2024). The treatment of such conditions tends to be multi-dimensional, and among the most effective methods is the use of food elimination diets. Such diets aim at the detection and removal of offending foods responsible for triggering allergic responses and promoting skin inflammation (Calvani et al. 2020).

16.5.1.1 Understanding Food Allergy and Skin Conditions

Food allergies are seen when the body's immune system mistakenly recognizes a normally safe food as a threat and mounts an allergic reaction to it (Uppin et al. 2024). In food allergy-induced skin disease, this reaction is seen in the form of itching, hives, redness, and swelling (Skypala et al. 2021). Most common allergens are peanuts, tree nuts, eggs, dairy, wheat, soy, and shellfish (Muthukumar et al. 2020). Allergic reactions can either be immediate (IgE-mediated) in some or can be delayed in others (non-IgE mediated), for example, as in food protein-induced enterocolitis syndrome (Anvari et al. 2019).

In treating food allergies and skin disease, food elimination diets are a therapeutic and diagnostic intervention (Das and Panda 2021). Knowing which foods are causing allergic reactions, medical practitioners can suggest that these be excluded from the diet to avoid triggering worsening of symptoms (Gargano et al. 2021).

The procedure of starting a food elimination diet starts with identification of suspected allergens (Halken et al. 2021). One of the main strategies is to eliminate major food triggers for a duration of usually 2–6 weeks to watch for any fluctuation in the symptoms (Wechsler et al. 2022). Elimination involves avoiding the suspected foods completely from the diet so that the body may eliminate any leftover of the causative food (Vizzoto et al. 2021). Patients are, during this time, encouraged to eat a highly balanced diet in order to avoid nutritional deficiencies. Foods such as eggs, gluten, and dairy are usually eliminated but are determined by the patient's history and testing (Awuchi et al. 2020).

Once the elimination phase is finished, a reintroduction phase is undertaken, in which foods are reintroduced one at a time (Van den Houte et al. 2024). This meticulous reintroduction serves to determine which specific foods cause allergic reactions (Wang et al. 2022). Skin prick tests or blood tests can also be used to supplement this by detecting IgE responses to particular allergens (Wong and Keith 2020).

16.5.1.2 Benefits and Complication of Food Elimination Diets

The main advantage of food elimination diets is the relief of allergic skin symptoms (Skypala and McKenzie 2019). For most people, the elimination of certain allergens from the diet can result in a reduction of eczema attacks, itching, and other inflammatory skin reactions (Diotallevi et al. 2022). The diet might reinstate the balance of the immune system and dampen systemic inflammation, the common initiating cause of long-standing skin disease in food allergy patients (Kim and Kim 2019).

Additionally, food elimination diets are useful in identifying the offending foods causing allergic reactions, allowing patients to make effective decisions regarding dietary restrictions This can result in better control and long-term management of food-related skin conditions. Through identification and avoidance of specific allergens, patients can avoid unnecessary dietary restriction, eating a healthy and diverse diet (Gargano et al. 2021). While the benefits exist, food elimination diets are not problem-free. A major challenge lies in the difficulty in finding specific triggers (Gazerani 2020). Food allergies in some individuals can be multi-factorial and therefore more challenging to identify with certainty. In addition, food reintroduction needs to be carefully done under medical observation to prevent severe allergic reactions (Meyer et al. 2023).

Another issue is the risk of nutritional deficiencies (Gargano et al. 2021). Elimination of whole food groups like dairy, gluten, or nuts might leave people devoid of vital nutrients like calcium, fiber, and proteins (Davis 2019). For this reason, it is vital that people get help from a dietitian so that they eat a nutritionally balanced diet when they are undertaking the elimination period.

Finally, compliance with the diet can be challenging with the restrictive nature of the elimination process (Mirmiran et al. 2021). Social activities, eating out, and the widespread nature of hidden allergens in processed foods can be very challenging for individuals attempting to strictly follow an elimination diet (Gargano et al. 2021).

16.5.2 *Pharmacological Treatments*

Food allergy-related skin conditions like atopic dermatitis (AD), urticaria, and angioedema can affect the quality of life of patients significantly. While allergen avoidance is still the mainstay of treatment for food allergies, medication is vital for symptom control and prevention of severe reactions (Gargano et al. 2021). The following section discusses the use of antihistamines, corticosteroids, and biologics in the management of skin disorders associated with food allergies.

16.5.2.1 Antihistamines

Antihistamines are a standard part of symptomatic therapy for allergic skin reactions, which act by inhibiting histamine receptors, especially H1-receptors, to cut down pruritus, erythema, and edema of allergic reaction (Mandola et al. 2019). Histamine, when released from mast cells on exposure to an allergen, is responsible for the inflammatory cascade that results in urticaria and other allergic reactions (Stewart and Nichol 2021).

First-generation H1-antihistamines, such as diphenhydramine, chlorpheniramine, and hydroxyzine, are most frequently used because they easily pass through the blood-brain barrier, inducing sedation and cognitive impairment. Their sedative action can restrict their use during the day, particularly in cases where alertness is needed for activities like driving (Baharudin et al. 2019). Second-Generation Antihistamines Cetirizine, loratadine, and fexofenadine are drugs that have been designed to reduce sedation (Fein et al. 2019). They are more peripheral H1 receptor-selective and have less capacity to cross the blood-brain barrier. They are thus more suitable for long-term treatment of chronic urticaria and other allergic dermatoses (Ince and Ruether 2021). Second-generation antihistamines have been found to be effective in alleviating pruritus and wheal formation in chronic spontaneous urticaria (Chang et al. 2021). When doses of standard are inadequate, up-dosing (administration of greater than standard doses) of second-generation antihistamines could be considered in a supervised medical setting. Such a strategy has been found to increase efficacy with minimal augmentation of adverse effects (Podder et al. 2023).

16.5.2.2 Corticosteroids

Corticosteroids have strong anti-inflammatory and immunosuppressive activity, and are useful in the treatment of moderate to severe allergic skin reactions (Wollenberg et al. 2020). They suppress many inflammatory mechanisms, thus decreasing erythema, swelling, and pruritus (Niculet et al. 2020).

Topical Corticosteroids are the initial treatment of choice for localized allergic skin disorders. Graded according to potency (from mild to super potent), the topical corticosteroid is selected according to the severity of the condition, location of application, and patient considerations (Sreeraj et al. 2024). For example, low-potency corticosteroids such as hydrocortisone are appropriate for sensitive skin areas like the face, whereas stronger preparations such as clobetasol can be applied to thicker skin areas such as the palms and soles (Shetty and Sherje 2021). Long-term use of potent topical corticosteroids can cause skin atrophy, striae, and telangiectasia, requiring close monitoring (Mohta and Sathe 2024).

In systemic Corticosteroids, Oral corticosteroids, like prednisone, are reserved for severe or extensive allergic skin diseases that are not responsive to other therapy (Caramori et al. 2019). They are useful in the quick control of symptoms but have serious side effects, such as weight gain, osteoporosis, hyperglycemia, and adrenal suppression. Thus, their administration is usually restricted to short durations to avoid these risks (Sharma et al. 2019).

16.5.2.3 Biologic

Biologic agents have become a revolutionary method in the treatment of resistant allergic skin disorders (Ramírez-Jiménez et al. 2023). These drugs act on individual elements of the immune system responsible for allergic reactions (van de Veen and Akdis 2019).

Omalizumab is a monoclonal antibody binds to immunoglobulin E (IgE), blocking it from binding to its receptors on mast cells and basophils, thus preventing the allergic cascade (Eggel et al. 2024). Omalizumab has been approved for treating chronic spontaneous urticaria in patients who are symptomatic despite antihistamine treatment (Agache et al. 2021). Clinical trials have shown its effectiveness in treating itch severity and hives number. In February 2024, the U.S. Food and Drug Administration (FDA) approved omalizumab in the treatment of immunoglobulin E-mediated food allergy to prevent allergic reactions, including anaphylaxis, following accidental exposure (Ghouri et al. 2024).

Dupilumab to the interleukin-4 receptor alpha subunit, dupilumab suppresses signaling of IL-4 and IL-13, cytokines that are involved in the pathogenesis of atopic dermatitis (Harb and Chatila 2020). It has proven highly effective in lessening the severity of atopic dermatitis and enhancing quality of life metrics. Dupilumab recipients have reported reduction in pruritus and enhancement of skin clearance (Silverberg et al. 2021).

Other Biologics Agents including mepolizumab, reslizumab, and benralizumab that are aimed against IL-5 or its receptor have been studied mostly in eosinophilic asthma but have the potential for a role in managing food allergy-related eosinophilic dermatoses. But their use in dermatologic disorders is still being researched (Antosz et al. 2024).

16.5.3 Probiotics and Microbiome-Based Therapy

New developments in understanding the human microbiome have created new therapeutic directions in the treatment of skin disease caused by food allergy. Probiotics and microbiome therapy are new avenues being explored that involve restoring equilibrium of microbial flora in the gut and on the skin to modulate immune functions and clinical presentation (Gueniche et al. 2022). Large-scale research appearing in reputable academic journals emphasizes the complex association between microbial dysbiosis, immune dysregulation, and the pathogenesis or worsening of allergic cutaneous disorders like atopic dermatitis and urticaria.

Increasing evidence suggests that the gut–skin axis is involved in critical pathophysiology of food allergy-related skin disorders (Liu et al. 2025). Changes in gut microbiota composition and diversity have been linked to increased intestinal permeability, systemic inflammation, and abnormal immune responses (Yoo et al. 2020). These alterations can further affect skin barrier integrity and immune surveillance, leading to the development of allergic skin features. Studies have shown that atopic dermatitis patients often have a decreased presence of favorable microbes, such as different Lactobacillus and Bifidobacterium strains, as well as an overabundance of pathogenic or pro-inflammatory microorganisms. This imbalance of microbes, or dysbiosis, is thought to compromise the homeostatic processes that typically inhibit too much immune stimulation towards harmless allergens, such as food proteins (Wrześniewska et al. 2024).

Probiotics are live microorganisms that, when given in sufficient numbers, have a beneficial effect on the host. Their therapeutic potential in allergic skin disorders is diverse (Sharifi-Rad et al. 2020). One of the main ways in which probiotics act is through the modulation of the immune system. Through their interaction with intestinal epithelial cells and antigen-presenting cells, probiotics are able to induce the production of regulatory cytokines like interleukin-10 (IL-10) and transforming growth factor-beta (TGF-β) (Raheem et al. 2021). These cytokines have a central role in inducing immunological tolerance and preventing an excessive inflammatory response when challenged with food allergens (López-Enríquez et al. 2023). A number journals have pointed out that certain probiotic strains can strengthen the activity of regulatory T cells (Tregs), which play a key role in immune homeostasis and the avoidance of allergic responses.

Aside from modulating immune responses, probiotics may also play a role in enhancing the strength of the epithelial barrier. The integrity of the gut barrier is vital for avoiding the translocation of food antigens and microbial products into systemic circulation, which can initiate or exacerbate skin inflammation (Macura et al. 2024). Probiotics are found to be potent in increasing the expression of tight junction proteins and mucins, enhancing the strength of the epithelial barrier and decreasing intestinal permeability (Liu et al. 2020b). This barrier-protective action is supplemented by the synthesis of antimicrobial peptides that have the ability to suppress the growth of pathogenic bacteria (Peng and Tang 2022). When these advantages are extended to the skin, enhanced barrier function can reduce the severity of allergic dermatitis by blocking the penetration of allergens and irritants that trigger inflammatory reactions (Dębińska and Sozańska 2023).

In addition to the gut, the skin microbiome itself has been identified as a key component in the conservation of cutaneous health. A disruption in the microbe populations on the skin has the potential to predispose individuals to inflammatory dermatoses (Balato et al. 2019). Topical or systemic probiotics may serve to reconstitute normal skin microbiota, subsequently decreasing inflammation and healing (Knackstedt et al. 2020). Certain clinical studies have shown that topical probiotic preparations can mitigate the severity of atopic dermatitis lesions, decrease pruritus, and enhance the general condition of the skin. These studies indicate that both systemic and local (topical) probiotic treatment may be beneficial for modulating the immune response and the barrier function of the skin (Habeebuddin et al. 2022).

Microbiome-directed therapies extend beyond the administration of conventional probiotics. More recent developments involve the investigation of prebiotics—non-digestible food particles that selectively foster the development of healthy microorganisms—and synbiotics, which consist of probiotics and prebiotics (Chavan et al. 2023). By supplying substrates that bias toward the proliferation of health-maintaining bacteria, these therapies also augment the process of restoring a healthy microbiota (Mohr et al. 2020). Furthermore, probiotic drugs in the second-generation, namely, live biotherapeutic products developed from well-studied commensal bacteria with anti-inflammatory and immunomodulatory potential, are also under development. These have the potential to attack particular biological pathways that take part in the processes of allergic inflammation and will even provide therapeutically directed regimens tailor-made for unique patterns of host microbiota composition.

A key focus of active research is the discovery of biomarkers for the prediction of a patient's response to probiotic and microbiome-targeted therapy. Developments in genomics and metagenomics have allowed researchers to characterize the microbial communities in the gut and skin with greater precision than ever before (Mathuria et al. 2024). These investigations have discovered specific microbial signatures correlating with successful response to probiotic therapy. For instance, the presence of some Bifidobacterium strains can foresee improved clinical response in food allergy-related skin diseases in patients (Smith et al. 2021). Such biomarkers not only enable a more specific therapeutic intervention but also assist in monitoring treatment success and dosage modification.

16.5.4 Novel Therapies and Future Directions

Recent developments in the treatment of food allergy-induced skin diseases have brought new therapeutic approaches to modify disease course and enhance patient outcomes. Allergen Immunotherapy (AIT), or the stepwise introduction of allergens to desensitize the immune system, has been widely researched for food allergies, especially Oral Immunotherapy (OIT) for allergens such as peanuts (McWilliams et al. 2012). Its efficacy in treating skin manifestations is still under research (Ezhuthachan et al. 2024). Other approaches like Sublingual Immunotherapy (SLIT) and Epicutaneous Immunotherapy (EPIT) have been promising in the modulation of immune responses with a good safety profile, but more work is needed (Baker and Wang 2019). Biologic therapies have also revolutionized the management by targeting particular immune mechanisms, including monoclonal antibodies that are aimed at neutralizing free IgE antibodies and inhibiting mast cell activation, thus raising the threshold for allergic reactions and lowering skin manifestations (Ezechukwu et al. 2023). In addition, adjunct and emerging treatments, such as Chinese herbal extracts, are investigated for their prophylactic role against food-induced anaphylaxis, while cytokine-directed therapies, such as those that block IL-4 and IL-13, are examined for their capacity to regulate allergic inflammation in skin conditions caused by food allergy (Wang and Li 2012). Such new approaches have great promise but require additional clinical studies to prove their effectiveness and incorporate them into standard medical care.

The future of food allergy-induced skin disease management is in personalized medicine strategies that individualize treatments according to patient profiles, such as genetic susceptibility and unique immune reactions. Improved immunotherapy methods, like intralymphatic immunotherapy (ILIT), hold the promise of faster and longer-lasting desensitization with fewer side effects (Šošić et al. 2023). In addition, further research on the gut-skin axis and how the microbiome functions can further reveal further therapeutic targets, underlining the significance of a comprehensive understanding of immune system interactions in allergic disorders (Table 16.2).

16.6 Special Populations: Paediatrics and Adults

16.6.1 Paediatric Considerations

16.6.1.1 Early Invention and Prevention Strategies

The incidence of atopic conditions like food allergy, atopic dermatitis, and asthma has risen drastically in the past couple of decades (Šošić et al. 2023). Early childhood events, including diet, have been recognized to exert a substantial impact on the course of diseases both in children and adults (Sarkar et al. 2021). Although hereditary in nature, environmental influences such as infant nutrition in early life can also significantly influence the development of atopic disorders and, consequently, present an opportunity for prevention or deferral of onset of the disease (Haahtela et al. 2021). The diet of the pregnant woman can first have a nutritional effect on an infant's atopic condition (Greer et al. 2019).

Food allergy reactions are of concern, particularly during infancy and early childhood, and may express themselves as a wide range of clinical manifestations, including gastrointestinal, respiratory, cutaneous, and others (Solymosi et al. 2020). Allergies are of great concern from the health perspective due to the steep increase

Table 16.2 Pharmacological vs non-pharmacological treatment

Category	Treatment approach	Mechanism of action	Advantages	Limitations
Pharmacological treatments	Antihistamines (e.g., cetirizine, fexofenadine)	Block H1 receptors to reduce histamine-mediated inflammation	Rapid symptom relief, effective for urticaria	May cause drowsiness (first-generation), not curative
	Corticosteroids (topical/systemic)	Suppress inflammation and immune response	Effective for severe reactions, rapid symptom control	Long-term use can cause skin thinning, osteoporosis, adrenal suppression
	Biologic therapies (e.g., omalizumab, dupilumab)	Target immune pathways (IgE, IL-4, IL-13) to reduce allergic inflammation	Highly effective for resistant cases, long-term disease control	Expensive, requires ongoing administration, potential side effects
Non-pharmacological treatments	Food elimination diets	Identifies and removes allergenic foods to prevent immune reaction	Reduces exposure to triggers, prevents future reactions	Difficult to adhere to, risk of nutritional deficiencies
	Probiotics & Microbiome Therapy	Modulate gut and skin microbiota to improve immune tolerance	Potential long-term benefit, enhances immune regulation	Requires further research, variable effectiveness
	Skin barrier enhancement (moisturizers, emollients)	Strengthens skin integrity to reduce allergen penetration	Helps prevent eczema flares, improves hydration	Needs consistent use, does not treat underlying immune response
	Allergen immunotherapy (experimental)	Desensitization to allergens over time	Potential long-term cure, reduces severity of reactions	Still under research, risk of anaphylaxis during therapy

in the number of childhood allergies in the western world in recent decades. Reduced exposure to microbes, especially in the first years of life, has been attributed to the increase (Kallio et al. 2024). Primary prevention of allergy is defined as preventing the onset of clinical disease prior to its development (Brough et al. 2022). In a meta-analysis, sole breast-feeding for 3 months seems to decrease the prevalence of atopic dermatitis in the offspring compared with usual formula feeding (Koukou et al. 2023). With increased understanding of the immune system, it is now clear that one of the factors in the causation of allergic disease is reduced oral tolerance, a specific inhibition of the humoral and cellular immunological reaction to an antigen (Gomez-Casado et al. 2021). By minimizing exposure to environmental antigens, averting infections, stimulating the maturation of gastrointestinal mucosa and development of gut microbiota, and yielding immune-modulatory and anti-inflammatory benefits, exclusive breastfeeding can potentially prevent allergy disease (Morniroli 2024). It has been suggested that increased allergy diseases worldwide are a result of improved hygiene and reduced infant exposure of the immune system to microbial stimuli (Renz and Skevaki 2021).

Several recent meta-analyses have shown that probiotics can prevent dermatitis effectively, particularly when taken during and before pregnancy (Li et al. 2019). Several other nutrients, such as long-chain polyunsaturated fatty acids, antioxidants (e.g., vitamin C, vitamin E, beta-carotene, Zinc), folate and vitamin D, are known to have effects on immune function (Stephen et al. 2023).

16.6.2 Dermatological Food Allergy in Adults

Food allergies are common, and while they are most commonly considered in the pediatric age group, food allergies in adults can also occur, and skin findings are one of the most common presenting manifestations encountered. Food allergies in adults in which dermatoses are the primary symptom are typically immune-mediated reactions against specific food proteins. These are typically of the IgE-induced type, leading to mast cell degranulation and release of histamines that manifests in a skin-like symptomatology (Tontini and Bulfone-Paus 2021).

16.6.2.1 Clinical Manifestations

Adult dermatological food allergies can manifest with minimal to severe clinical symptoms usually restricted to skin. Urticaria (HUAS), Atopic dermatitis (Eczema), and anhiedema (swelling under the skin) these symptoms, at times, can emerge shortly after a meal (Singh et al. 2022).

16.6.2.2 Pathophysiology

Food allergic reactions arise from an immunologic reaction in which the body identifies a food protein as harmful when it is not. In an IgE-mediated food allergy, the immune system gets ready by producing IgE antibodies to the allergenic protein (Bruggink 2021). With re exposure to the offending food, the body discharges histamine and other pharmacologically active molecules from mast cells and basophils which culminates in a variety of allergic signs and symptoms ranging from skin manifestation (Gargano et al. 2021). Furthermore, there are some persons with “IgE food allergy“whose symptoms are experienced differently, and occur later, involving other immune mechanisms entirely.

16.6.2.3 Diagnosis and Management

A holistic approach to the diagnosis of dermatological food allergies is undertaken involving patient history, skin prick tests, and serum-specific IgE tests (Anvari et al. 2019). Patch tests are also helpful in the diagnosis of non-IgE mediated reactions (Cuomo et al. 2023). Besides the above investigations, elimination diets under the guidance of a physician may also be used to determine the offending allergens (Cotton et al. 2019). Once identified, the patients will have a strictly supervised elimination diet for a month followed by a progressive reintroduction of eliminated foods (Meyer et al. 2020). Management of food allergy in adults is largely a matter of avoiding the responsible allergen. For those with severe reactions, an epinephrine auto injector will be required (Cronin et al. 2024). Antihistamines and corticosteroids will suffice for prolonged symptoms when eczema or urticaria is present (Hon et al. 2019). Active research exists on some oral immunotherapy for desensitization to certain food allergens, but as yet, it is not a part of regular clinical practice (Table 16.3).

16.7 Prevention and Public Health Strategies

Allergic diseases such as atopic dermatitis (AD) and food allergies have increased dramatically in prevalence over recent decades and pose serious public health challenges. Prevention is key and should focus on interventions such as improvement of the skin barrier, early introduction of allergenic foods and promotion of public awareness and education (Sicherer and Sampson 2018).

16.7.1 Skin Barrier Enhancement Strategies

The skin functions as the primary protective barrier against various environmental allergens and pathogens. The skin barrier function which is compromised is associated with the development of atopic diseases, in particular AD and late-onset food allergies. It may be possible to avert some of these problems by improving the skin barrier from childhood. To achieve this, Emollient therapy, which consists of frequent application of moisturizers, has been attempted as a means to fortify the skin barrier. It has been shown that emollient therapy started at birth is a practical, safe, and effective means of preventing AD. There are also studies that show that early emollient therapy targeted at lowering AD development in at-risk infants is effective. Yet it remains to be seen whether Emollient therapy has any effectiveness in the prevention of food allergy. Some studies indicate that frying skin is a means to lessen allergen sensitization but there are a number of studies which indicate that there is no sufficient evidence for the use of emollients in food allergy prevention. Thus, the use of skin barrier enhancement to prevent AD is reasonably supported, but its effectiveness in the prevention of food allergies is undetermined (D'Auria et al. 2020; Kelleher et al. 2020).

Table 16.3 Summary of evidence-based allergy management approaches

Management approach	Description	Supporting evidence
Food elimination diets	Avoidance of confirmed allergenic foods to prevent reactions.	Recommended for IgE-mediated allergies; must be done under medical supervision to prevent nutritional deficiencies
Pharmacotherapy	Use of antihistamines, corticosteroids, and biologics to manage symptoms.	Antihistamines for urticaria; corticosteroids for atopic dermatitis; biologics (e.g., dupilumab, omalizumab) for severe cases
Probiotics & Microbiome-Based Therapy	Modulation of gut microbiota to enhance immune tolerance and reduce inflammation.	Studies suggest lactobacillus rhamnosus GG may reduce eczema severity in infants
Skin barrier protection	Use of emollients and skin hydration to reduce allergen penetration and inflammation.	Essential for managing atopic dermatitis and eczema, particularly in infants at high risk of allergies
Early allergen introduction	Gradual introduction of allergenic foods during infancy to promote immune tolerance.	Supported by the LEAP and EAT studies, which found early peanut and egg introduction reduces allergy risk
Allergen immunotherapy (AIT)	Gradual exposure to allergens to desensitize the immune system.	Sublingual and epicutaneous immunotherapy are emerging as effective options
Biologic therapies	Monoclonal antibodies targeting immune pathways involved in allergic reactions.	Dupilumab (IL-4/IL-13 blocker) and omalizumab (IgE blocker) are FDA-approved for severe allergic conditions
Cytokine-targeted treatments	Inhibitors targeting cytokines like IL-5, IL-33, and TSLP to reduce inflammation.	Under investigation for treating allergic dermatitis and food-induced skin reactions.
Public Health initiatives	Awareness campaigns and policies promoting early diagnosis and prevention strategies.	Educational programs in schools and healthcare settings improve allergy management

16.7.2 Early Introduction of Allergenic Foods

Standard recommendations advised against introducing allergenic foods to infants in order to prevent food allergies. In recent years, however, studies have questioned this approach, arguing that early introduction may be more advantageous. The Learning Early About Peanut Allergy (LEAP) study provided compelling evidence supporting early peanut introduction. The study found that introducing peanut-containing foods to high-risk infants between 4 to 11 months of age resulted in an 81% reduction in the development of peanut allergy by age five. This landmark finding prompted changes in dietary guidelines, now recommending early peanut introduction for infants (Allen and Koplin 2016; Sweeney et al. 2021).

16.7.3 Public Awareness and Education

Education and outreach remain critical in the implementation of the prevention measures of allergic diseases. Informing parents, guardians, and other healthcare providers on the merits of enhancing skin barrier and introducing allergens early can hasten practice uptake. Food Allergy Research and Education, commonly known as FARE, has played a major role in the initiative. These initiatives include frameworks for introducing allergens early and preventing food allergies. In the same way, the British Society for Allergy and Clinical Immunology have put out materials stressing the introduction of potentially allergenic foods in the first year of life and other children's foods in order to minimize food allergies. Misconceptions like avoidance of potentially allergic foods to prevent allergies should also be tackled in public campaigns. These outdated practices need to be countered with clear evidence proactive approach that the evidence support. Moreover, providing training to health professionals to speak in unison means that families can get clear messages that relates to the risk of the child. There are promising observations on the use of emollient therapy as a preventive measure for atopic dermatitis but their usefulness in food allergy is uncertain. At the other extreme, the introduction of certain foods known to cause allergy are established as preventive measures against food allergy (Table 16.4). Ongoing public awareness and education efforts are essential to translate these preventive strategies into widespread practice, ultimately aiming to reduce the burden of allergic diseases (Sansweet et al. 2024).

Table 16.4 Recommended early introduction of allergenic foods (D'Auria et al. 2020)

Food	Recommended age for introduction	Key consideration
Peanuts	4–6 months	Avoid whole peanuts due to choking risk. A smooth paste of peanut butter mixed with water should be introduced to infant.
Eggs	4–6 months	Well-cooked e.g. boiled, scrambled egg should be introduced to infant however raw or lightly cooked eggs should be avoided.
Milk	After 6 months	Whole cow's milk in small amounts in foods but not as a main drink.
Tree nuts	6 months	Ground nuts or nut butters mixed into foods but whole nuts should be avoided due to choking hazard.
Soy	6 months	Monitor for reactions, especially in infants with a family history of soy allergy.
Sesame	6 months	Introduce in small amounts initially.
Fish	6 months	Cooked, mashed fish (e.g., salmon, cod) can be introduced to infant however fish with high mercury levels (e.g., shark, swordfish) should be avoided.
Wheat	6 months	Soft wheat-based foods (e.g., pasta, bread) should be introduce gradually to monitor for reactions.

16.8 Conclusion

Food allergies are an emerging and substantial public health issue that has a considerable impact on individuals across the globe and plays a causative role in a variety of allergic skin disorders, such as urticaria, angioedema, atopic dermatitis, eczema, and contact dermatitis. The intricate interplay between immune processes, genetics, environmental triggers, and skin barrier dysfunction creates challenges in diagnosis and treatment. A thorough diagnostic strategy—ranging from clinical history-taking to laboratory and immunological investigations, food challenge testing, and identification of biomarkers—is necessary for proper detection and treatment of dermatological manifestations due to food allergy. Each test, however, has its limitations, and the results must be interpreted cautiously in a clinical context.

Treatment of skin diseases caused by food allergy is a multidimensional approach that unifies dietary interventions, pharmacology, and newly emerging microbiome-directed therapies. Progress in the field of biologic therapies like monoclonal antibodies (omalizumab and dupilumab) has transformed management by modulating central immune processes, and the newly emerging IL-5-directed and IL-5 receptor-targeting biologics are set to be employed in the years to come. Further, microbiome-targeted therapies and probiotics also provide new frontiers for balancing microbes and stimulating immune tolerance with the gut-skin axis as the central function in allergic cutaneous diseases. Newer therapies like allergen immunotherapy, cytokine-based therapies, and precision medicine initiatives keep emerging as more effective, personalized interventions.

In children, early nutritional therapies such as exclusive breastfeeding, probiotic supplementation, and early introduction of allergenic foods are important in preventing the development of allergic diseases. In adults, detailed diagnostic and management plans involving allergen avoidance, pharmacotherapy, and novel immunotherapies are necessary for managing food allergy-related skin diseases. Preventive strategies like skin barrier reinforcement and public education have also shown promise in reducing the increasing incidence of allergic diseases.

With research ongoing, incorporating these advances into clinical practice will be vital to enhancing patient outcomes and lessening the burden of food allergy skin diseases. As promising technology in immunotherapy and biologics molds the future of allergy care, continued research and evidence-based implementation are still important to maximize management and improve long-term well-being in all age groups.

References

Adinoff AD, Rosloniec DM, McCall LL, Nelson HS (1989) A comparison of six epicutaneous devices in the performance of immediate hypersensitivity skin testing. J Allergy Clin Immunol 84(2):168–174

Afshari M, Kolackova M, Rosecka M, Čelakovská J, Krejsek J (2024) Unraveling the skin; a comprehensive review of atopic dermatitis, current understanding, and approaches. Front Immunol 15:1361005

Agache I, Rocha C, Pereira A, Song Y, Alonso-Coello P, Solà I et al (2021) Efficacy and safety of treatment with omalizumab for chronic spontaneous urticaria: a systematic review for the EAACI biologicals guidelines. Allergy 76(1):59–70
Akdis CA (2021) Does the epithelial barrier hypothesis explain the increase in allergy, autoimmunity and other chronic conditions? Nat Rev Immunol 21(11):739–751
Allen KJ, Koplin JJ (2016) Prospects for prevention of food allergy. J Allergy Clin Immunol Pract 4(2):215–220
Antosz K, Batko J, Błażejewska M, Gawor A, Sleziak J, Gomułka K (2024) Insight into IL-5 as a potential target for the treatment of allergic diseases. Biomedicines 12(7):1531
Anvari S, Miller J, Yeh CY, Davis CM (2019) IgE-mediated food allergy. Clin Rev Allergy Immunol 57:244–260
Aquino M, Rosner G (2019) Systemic contact dermatitis. Clin Rev Allergy Immunol 56:9–18
Awuchi CG, Igwe VS, Amagwula IO (2020) Nutritional diseases and nutrient toxicities: a systematic review of the diets and nutrition for prevention and treatment. Int J Adv Acad Res 6(1):1–46
Baharudin A, Abdul Latiff AH, Woo K, Yap FBB, Tang IP, Leong KF et al (2019) Using patient profiles to guide the choice of antihistamines in the primary care setting in Malaysia: expert consensus and recommendations. Ther Clin Risk Manag 1267–75:1267–1275
Baker MG, Wang J (2019) Could this be it? Epicutaneous, sublingual, and subcutaneous immunotherapy for the treatment of food allergies. Curr Allergy Asthma Rep 19:1–12
Balato A, Cacciapuoti S, Di Caprio R, Marasca C, Masarà A, Raimondo A et al (2019) Human microbiome: composition and role in inflammatory skin diseases. Arch Immunol Ther Exp 67:1–18
Banafea GH, Bakhashab S, Alshaibi HF, Natesan Pushparaj P, Rasool M (2022) The role of human mast cells in allergy and asthma. Bioengineered 13(3):7049–7064
Barbaud A, Poreaux C, Penven E, Waton J (2015) Occupational protein contact dermatitis. Europ J Dermatol 25:527–534
Barni S, Mori F, Giovannini M, Liotti L, Mastrorilli C, Pecoraro L et al (1824) Allergic proctocolitis: literature review and proposal of a diagnostic–therapeutic algorithm. Life 2023:13
Barni S, Mori F, Pecoraro L, Saretta F, Giovannini M, Arasi S et al (2024) Food protein-induced enteropathy: a revision for the clinician. Front Pediatr 12:1417391
Bartuzi Z, Kaczmarski M, Czerwionka-Szaflarska M, Małaczyńska T, Krogulska A (2017) Position paper of food allergy section the Polish Society of Allergology on the diagnosis and management of food allergies. Alergologia Polska Polish J Allergol 4(3):109–122
Bellanti JA (2024) IgE and non-IgE food allergy: a review of immunological mechanisms. J Food Allergy 6(1):37–46
Bergmann MM, Caubet JC, Boguniewicz M, Eigenmann PA (2013) Evaluation of food allergy in patients with atopic dermatitis. J Allergy Clin Immunol Pract 1(1):22–28
Bernardino L (2021) Histamine in the crosstalk between innate immune cells and neurons: relevance for brain homeostasis and disease. In: The functional roles of histamine receptors. Springer International Publishing, pp 261–288
Bousquet J, Heinzerling L, Bachert C, Papadopoulos NG, Bousquet PJ, Burney PG et al (2012) Practical guide to skin prick tests in allergy to aeroallergens. Allergy [Internet] [cited 2025 Feb 19];67(1):18–24. Available from: https://pubmed.ncbi.nlm.nih.gov/22050279/
Brough HA, Lanser BJ, Sindher SB, Teng JMC, Leung DYM, Venter C et al (2022) Early intervention and prevention of allergic diseases. Allergy 77(2):416–441
Bruggink T (2021) Food allergy and food intolerance. In: Food safety and toxicity. CRC Press, pp 183–193
Calvani M, Anania C, Caffarelli C, Martelli A, Del Giudice MM, Cravidi C et al (2020) Food allergy: an updated review on pathogenesis, diagnosis, prevention and management. Acta Biomedica Mattioli 1885 91:1–17
Calvani M, Anania C, Bianchi A, D'auria E, Cardinale F, Votto M et al (2021) Update on Food protein-induced enterocolitis syndrome (FPIES). Acta Biomedica Mattioli 92:1885

Caramori G, Mumby S, Girbino G, Chung KF, Adcock IM. Corticosteroids. Nijkamp and Parnham's principles of Immunopharmacology 2019;661–688, Springer International Publishing
Celebi Sozener Z, Ozdel Ozturk B, Cerci P, Turk M, Gorgulu Akin B, Akdis M et al (2022) Epithelial barrier hypothesis: effect of the external exposome on the microbiome and epithelial barriers in allergic disease. Allergy 77(5):1418–1449
Çelebioğlu E, Akarsu A, Şahiner ÜM (2021) Ige-mediated food allergy throughout life. Turkish J Med Sci Turkiye Klinikleri 51:49–60
Chang J, Cattelan L, Ben-Shoshan M, Le M, Netchiporouk E (2021) Management of pediatric chronic spontaneous urticaria: a review of current evidence and guidelines. J Asthma Allergy Volume 14:187–199
Chavan AR, Singh AK, Gupta RK, Nakhate SP, Poddar BJ, Gujar VV et al (2023) Recent trends in the biotechnology of functional non-digestible oligosaccharides with prebiotic potential. Biotechnol Genet Eng Rev 39(2):465–510
Cosenza M, Sacchi S, Pozzi S (2021) Cytokine release syndrome associated with T-cell-based therapies for hematological malignancies: pathophysiology, clinical presentation, and treatment. Int J Mol Sci 22(14):7652
Cotton CC, Durban R, Dellon ES (2019) Illuminating elimination diets: controversies regarding dietary treatment of eosinophilic esophagitis. Dig Dis Sci 64:1401–1408
Cox L, Williams B, Sicherer S, Oppenheimer J, Sher L, Hamilton R et al (2008) Pearls and pitfalls of allergy diagnostic testing: report from the American College of Allergy, asthma and immunology/American Academy of allergy, asthma and immunology specific IgE test task force. Ann Allergy Asthma Immunol 101(6):580–592
Cronin C, Keohane H, O'Rourke E, O'Kelly C, Ramesh Y, Flores L et al (2024) Assessing the knowledge of anaphylaxis management and adrenaline auto-injector administration among parents of children with food allergies: a cross-sectional study. Allergie 4(3):80–93
Cuomo B, Anania C, D'Auria E, Decimo F, Indirli GC, Manca E et al (2023) The role of the atopy patch test in the diagnostic work-up of non-IgE gastrointestinal food allergy in children: a systematic review. Eur J Pediatr 182(8):3419–3431
D'Auria E, Peroni DG, Sartorio MUA, Verduci E, Zuccotti GV, Venter C (2020) The role of diet diversity and diet indices on allergy outcomes. Front Pediatr 8:545
Das A, Panda S (2021) Role of elimination diet in atopic dermatitis: current evidence and understanding. Indian J Paediatr Dermatol 22(1):21–28
Davis W (2019) Wheat belly (revised and expanded edition): lose the wheat, lose the weight, and find your path back to health. Rodale Books
De Almeida CV, Antiga E, Lulli M (2023) Oral and topical probiotics and postbiotics in skincare and dermatological therapy: a concise review. Microorganisms. 11(6):1420
de Bartolomeis A, Barone A, Begni V, Riva MA (2022) Present and future antipsychotic drugs: a systematic review of the putative mechanisms of action for efficacy and a critical appraisal under a translational perspective. Pharmacol Res 176:106078
De Martinis M, Sirufo M (2020) … MSI journal of, 2020 undefined. New perspectives in food allergy. mdpi.com M De Martinis, MM Sirufo, M Suppa, L GinaldiInternational journal of molecular sciences, mdpi.com [Internet]. [cited 2025 Feb 19]; Available from: https://www.mdpi.com/1422-0067/21/4/1474
De Martinis M, Sirufo MM, Suppa M, Ginaldi L (2020) New perspectives in food allergy. Int J Mol Sci 21(4):1474
Dębińska A, Sozańska B (2023) Epicutaneous sensitization and food allergy: preventive strategies targeting skin barrier repair—facts and challenges. Nutrients 15(5):1070
Demessant-Flavigny A, Connétable S, Kerob D, Moreau M, Aguilar L, Wollenberg A (2023) Skin microbiome dysbiosis and the role of Staphylococcus aureus in atopic dermatitis in adults and children: a narrative review. J Eur Acad Dermatol Venereol 37:3–17
Demoly P, Bousquet J, Romano A (2009) In vivo methods for the study of allergy. In: Middleton's allergy: principles and practice. Elsevier, pp 1267–1279
Dhar S, Srinivas SM (2016) Food allergy in atopic dermatitis. Indian J Dermatol 61:645–648

Diotallevi F, Campanati A, Martina E, Radi G, Paolinelli M, Marani A et al (2022) The role of nutrition in immune-mediated, inflammatory skin disease: a narrative review. Nutrients 14(3):591

Du Toit G, Tsakok T, Lack S, Lack G (2016) Prevention of food allergy. J Allergy Clin Immunol 137(4):998–1010

Eggel A, Pennington LF, Jardetzky TS (2024) Therapeutic monoclonal antibodies in allergy: targeting IgE, cytokine, and alarmin pathways. Immunol Rev 328(1):387–411

Elias PM (2022) Optimizing emollient therapy for skin barrier repair in atopic dermatitis. Ann Allergy Asthma Immunol 128(5):505–511

Ersanli C, Tzora A, Voidarou CC, Skoufos S, Zeugolis DI, Skoufos I (2023) Biodiversity of skin microbiota as an important biomarker for wound healing. Biology (Basel) 12(9):1187

Ezechukwu HC, Adegboye OA, Okunowo WO, Emeto TI (2023) Targeting IgE and Th2-cytokines in allergy: brief updates on monoclonal antibodies and antibody gene therapy. Allergie 3(2):90–104

Ezhuthachan ID, Beaudoin M, Nowak-Wegrzyn A, Vickery BP (2024) The future of food allergy management: advancements in therapies. Curr Allergy Asthma Rep 24(4):161–171

Faraz K, Seely M, Marano AL (2024) The role of the environment in allergic skin disease. Curr Allergy Asthma Rep 24(6):323–330

Fein MN, Fischer DA, O'Keefe AW, Sussman GL (2019) CSACI position statement: newer generation H 1-antihistamines are safer than first-generation H 1-antihistamines and should be the first-line antihistamines for the treatment of allergic rhinitis and urticaria. Allergy, Asthma Clin Immunol 15:1–6

Frcp P, Pepys J (1972) Laboratory methods in clinical allergy. http://dx.doi.org/101177/003591577206500320 [Internet]. 1 [cited 2025 Feb 19];65(3):271–272. Available from: https://journals.sagepub.com/doi/10.1177/003591577206500320?icid=int.sj-abstract.similar-articles.5

Friedrich M, Pohin M, Powrie F (2019) Cytokine networks in the pathophysiology of inflammatory bowel disease. Immunity 50(4):992–1006

Fu L, Wang C, Wang Y (2018) Seafood allergen-induced hypersensitivity at the microbiota-mucosal site: implications for prospective probiotic use in allergic response regulation. Crit Rev Food Sci Nutr [Internet] [cited 2025 Feb 19];58(9):1512–1525. Available from: https://www.tandfonline.com/doi/abs/10.1080/10408398.2016.1269719

Gao H, Kosins AE, Cook-Mills JM (2024) Mechanisms for initiation of food allergy by skin predisposed to atopic dermatitis. Immunol Rev 326(1):151–161

Gargano D, Appanna R, Santonicola A, De Bartolomeis F, Stellato C, Cianferoni A et al (2021) Food allergy and intolerance: a narrative review on nutritional concerns. Nutrients 13(5):1638

Gazerani P (2020) Migraine and diet. Nutrients 12(6):1658

Ghouri H, Habib A, Nazir Z, Lohana N, Akilimali A (2024) Omalizumab for the reduction of allergic reactions to foods: a narrative review. Front Allergy 5:1409342

Goleva E, Berdyshev E, Leung DYM (2019) Epithelial barrier repair and prevention of allergy. J Clin Invest 129(4):1463–1474

Gomez-Casado C, Sanchez-Solares J, Izquierdo E, Díaz-Perales A, Barber D, Escribese MM (2021) Oral mucosa as a potential site for diagnosis and treatment of allergic and autoimmune diseases. Food Secur 10(5):970

Goswami TK, Singh M, Dhawan M, Mitra S, Emran TB, Rabaan AA et al (2022) Regulatory T cells (Tregs) and their therapeutic potential against autoimmune disorders–advances and challenges. Hum Vaccin Immunother. 18(1):2035117

Greer FR, Sicherer SH, Burks A, Abrams SA, Fuchs GJ, Kim JH et al (2019) The effects of early nutritional interventions on the development of atopic disease in infants and children: the role of maternal dietary restriction, breastfeeding, hydrolyzed formulas, and timing of introduction of allergenic complementary foods. Pediatrics 143(4)

Gueniche A, Perin O, Bouslimani A, Landemaine L, Misra N, Cupferman S et al (2022) Advances in microbiome-derived solutions and methodologies are founding a new era in skin health and care. Pathogens 11(2):121

Guo C, Saltoun C (2019) Urticaria and angioedema. Allergy Asthma Proc 40(6):437–440
Gupta RS, Warren CM, Smith BM, Jiang J, Blumenstock JA, Davis MM et al (2019) Prevalence and severity of food allergies among US adults. JAMA Netw Open 2(1):e185630
Haahtela T, Alenius H, Lehtimäki J, Sinkkonen A, Fyhrquist N, Hyöty H et al (2021) Immunological resilience and biodiversity for prevention of allergic diseases and asthma. Allergy 76(12):3613–3626
Habeebuddin M, Karnati RK, Shiroorkar PN, Nagaraja S, Asdaq SMB, Khalid Anwer M et al (2022) Topical probiotics: more than a skin deep. Pharmaceutics 14(3):557
Halken S, Muraro A, de Silva D, Khaleva E, Angier E, Arasi S et al (2021) EAACI guideline: preventing the development of food allergy in infants and young children (2020 update). Pediatr Allergy Immunol 32(5):843–858
Harb H, Chatila TA (2020) Mechanisms of dupilumab. Clin Exp Allergy 50(1):5–14
Ho MHK, Wong WHS, Chang C (2014) Clinical Spectrum of food allergies: a comprehensive review. Clin Rev Allergy Immunol 46(3):225–240
Hon KL, Leung AKC, Ng WGG, Loo SK (2019) Chronic urticaria: an overview of treatment and recent patents. Recent Patents Inflamm Allergy Drug Discov 13(1):27–37
Ince M, Ruether P (2021) Histamine and antihistamines. Anaesth Intensive Care Med 22(11):749–755
Jiang S, Hinchliffe TE, Wu T (2015) Biomarkers of an autoimmune skin disease—psoriasis. Genomics Proteomics Bioinformatics [Internet] [cited 2025 Feb 19];13(4):224–233. Available from: https://doi.org/10.1016/j.gpb.2015.04.002
Johansson SGO (2004) ImmunoCAP specific IgE test: an objective tool for research and routine allergy diagnosis. Expert Rev Mol Diagn [Internet] [cited 2025 Feb 19];4(3):273–279. Available from: https://pubmed.ncbi.nlm.nih.gov/15137895/
Kabesch M, Tost J (2020) Recent findings in the genetics and epigenetics of asthma and allergy. Semin Immunopathol 42:1 [Internet]. 14 [cited 2025 Feb 19];42(1):43–60. Available from: https://link.springer.com/article/10.1007/s00281-019-00777-w, 43, 60
Kallio S, Jian C, Korpela K, Kukkonen AK, Salonen A, Savilahti E et al (2024) Early-life gut microbiota associates with allergic rhinitis during 13-year follow-up in a Finnish probiotic intervention cohort. Microbiol Spectr 12(6):e04135–e04123
Kanani A, Betschel SD, Warrington R (2018) Urticaria and angioedema. Allergy Asthma Clin Immunol 14:59
Kelleher MM, Tran L, Boyle RJ (2020) Prevention of food allergy—skin barrier interventions. Allergol Int 69:3–10
Kim JE, Kim HS (2019) Microbiome of the skin and gut in atopic dermatitis (AD): understanding the pathophysiology and finding novel management strategies. J Clin Med 8(4):444
Kim BE, Leung DYM (2018) Significance of skin barrier dysfunction in atopic dermatitis. Allergy Asthma Immunol Res 10:207–215
Kim Y, Lim KM (2021) Skin barrier dysfunction and filaggrin. Arch Pharm Res 44(1):36–48
Knackstedt R, Knackstedt T, Gatherwright J (2020) The role of topical probiotics in skin conditions: a systematic review of animal and human studies and implications for future therapies. Exp Dermatol 29(1):15–21
Konstantinou GN, Riedl MA, Valent P, Podder I, Maurer M (2023) Urticaria and angioedema: understanding complex Pathomechanisms to facilitate patient communication, disease management, and future treatment. J Allergy Clin Immunol Pract 11(1):94–106
Koukou Z, Papadopoulou E, Panteris E, Papadopoulou S, Skordou A, Karamaliki M et al (2023) The effect of breastfeeding on food allergies in newborns and infants. Children 10(6):1046
Kounis NG, Koniari I, Velissaris D, Tzanis G, Hahalis G (2019) Kounis syndrome—not a single-organ arterial disorder but a multisystem and multidisciplinary disease. Balkan Med J 36(4):212–221
Kulis MD, Smeekens JM, Immormino RM, Moran TP (2021) The airway as a route of sensitization to peanut: an update to the dual allergen exposure hypothesis. J Allergy Clin Immunol 148(3):689–693

Lachapelle JM (2009) Patch-tests dans différentes conditions climatiques. Ann Dermatol Venereol 136(8–9):621–622

Lazzarini R, Duarte I, Ferreira AL (2013) Patch tests. An Bras Dermatol [Internet]. Nov [cited 2025 Feb 19];88(6):879. Available from: https://pmc.ncbi.nlm.nih.gov/articles/PMC3900336/

Li L, Han Z, Niu X, Zhang G, Jia Y, Zhang S et al (2019) Probiotic supplementation for prevention of atopic dermatitis in infants and children: a systematic review and meta-analysis. Am J Clin Dermatol 20:367–377

Lindberg M (2011) Dermatitis MMC, 2011 undefined. Patch testing. SpringerM Lindberg, M MaturaContact dermatitis, Springer [Internet]. [cited 2025 Feb 19]; Available from: https://link.springer.com/chapter/10.1007/978-3-642-03827-3_24

Lindberg M, Matura M (2011) Patch testing. Contact dermatitis (5th ed), pp 439–64

Liu Y, Wang S, Dai W, Liang Y, Shen C, Li Y et al (2020a) Distinct skin microbiota imbalance and responses to clinical treatment in children with atopic dermatitis. Front Cell Infect Microbiol 10:336

Liu Q, Yu Z, Tian F, Zhao J, Zhang H, Zhai Q et al (2020b) Surface components and metabolites of probiotics for regulation of intestinal epithelial barrier. Microb Cell Factories 19:1–11

Liu L, Zhang Z, Xiao H, Li Z, Lin H (2025) Dietary AGEs and food allergy: insights into the mechanisms of AGEs-induced food allergy and mitigation strategies. Crit Rev Food Sci Nutr 1–18

Lopes JP, Sicherer S (2020) Food allergy: epidemiology, pathogenesis, diagnosis, prevention, and treatment. Curr Opin Immunol 66:57–64

López-Enríquez S, Múnera-Rodríguez AM, Leiva-Castro C, Sobrino F, Palomares F (2023) Modulation of the immune response to allergies using alternative functional foods. Int J Mol Sci 25(1):467

Lu HF, Zhou YC, Yang LT, Zhou Q, Wang XJ, Qiu SQ et al (2024) Involvement and repair of epithelial barrier dysfunction in allergic diseases. Front Immunol 15:1348272

Macura B, Kiecka A, Szczepanik M (2024) Intestinal permeability disturbances: causes, diseases and therapy. Clin Exp Med 24(1):232

Malling H -J, Andersen CE, Boas M -B, Holgersen F, Munch EP, Weeke B (1982) The allergy pricker. Qualitative aspects of skin prick testing using a precision needle. Allergy [Internet]. [cited 2025 Feb 19];37(8):563–567. Available from: https://pubmed.ncbi.nlm.nih.gov/7181052/

Mandola A, Nozawa A, Eiwegger T (2019) Histamine, histamine receptors, and anti-histamines in the context of allergic responses. LymphoSign J 6(2):35–51

Manuyakorn W, Tanpowpong P (2019) Cow milk protein allergy and other common food allergies and intolerances. Paediatr Int Child Health 39(1):32–40

Mathuria A, Chaudhary A, Sharma H, Mani I (2024) Multi-omics in gut microbiome. In: Multi-omics analysis of the human microbiome: from technology to clinical applications. Springer, pp 181–213

McWilliams LM, Mousallem T, Burks AW (2012) Future therapies for food allergy. Human vaccines and immunotherapeutics. Landes Biosci 8:1479–1484

Melikoglu M, Pala E, Bayraktar M (2022) Etiological causes in patients with acute urticaria and angioedema: a cross-sectional study. Allergol Immunopathol (Madr) 50(S):15–23

Messina M, Venter C (2020 Jan) Recent surveys on food allergy prevalence. Nutr Today 55(1):22–29

Meyer R, Chebar Lozinsky A, Fleischer DM, Vieira MC, Du Toit G, Vandenplas Y et al (2020) Diagnosis and management of non-IgE gastrointestinal allergies in breastfed infants—an EAACI position paper. Allergy 75(1):14–32

Meyer R, Venter C, Bognanni A, Szajewska H, Shamir R, Nowak-Wegrzyn A et al (2023) World allergy organization (WAO) diagnosis and rationale for action against cow's Milk allergy (DRACMA) guideline update–VII–Milk elimination and reintroduction in the diagnostic process of cow's milk allergy. World Allergy Organizat J 16(7):100785

Mirmiran P, Bahadoran Z, Gaeini Z (2021) Common limitations and challenges of dietary clinical trials for translation into clinical practices. Int J Endocrinol Metab 19(3):e108170

Mohr AE, Jäger R, Carpenter KC, Kerksick CM, Purpura M, Townsend JR et al (2020) The athletic gut microbiota. J Int Soc Sports Nutr 17:1–33

Mohta A, Sathe NC (2024) Topical steroid withdrawal (red skin syndrome). In: StatPearls [internet]. StatPearls Publishing

Morniroli D (2024) Neutralizing activity against SARS-COV-2 of human breast milk from COVID-19 positive and negative lactating mothers

Muthukumar J, Selvasekaran P, Lokanadham M, Chidambaram R (2020) Food and food products associated with food allergy and food intolerance–an overview. Food Res Int 138:109780

Nassau S, Fonacier L (2020) Allergic contact dermatitis. Med Clin North Am 104:61–76

Nelson HS, Rosloniec DM, McCall LI, Iklé D (1993) Comparative performance of five commercial prick skin test devices. J Allergy Clin Immunol [Internet]. [cited 2025 Feb 19];92(5):750–756. Available from: https://pubmed.ncbi.nlm.nih.gov/8227867/

Niculet E, Bobeica C, Tatu AL (2020) Glucocorticoid-induced skin atrophy: the old and the new. Clin Cosmet Investig Dermatol Volume 13:1041–1050

Pat Y, Ogulur I, Yazici D, Mitamura Y, Cevhertas L, Küçükkase OC et al (2023) Effect of altered human exposome on the skin and mucosal epithelial barrier integrity. Tissue Barriers 11(4):2133877

Patel RH, Mohiuddin SS (2020) Biochemistry, Histamine

Peng Z, Tang J (2022) Intestinal infection of Candida albicans: preventing the formation of biofilm by C. Albicans and protecting the intestinal epithelial barrier. Front Microbiol 12:783010

Piette V, Bourret E, Bousquet J, Demoly P (2002) Prick tests to aeroallergens: is it possible simply to wipe the device between tests? Allergy [Internet]. [cited 2025 Feb 19];57(10):940–942. Available from: https://pubmed.ncbi.nlm.nih.gov/12269942/

Podder I, Dhabal A, Chakraborty SS (2023) Efficacy and safety of up-dosed second-generation antihistamines in uncontrolled chronic spontaneous urticaria: a review. J Clin Aesthet Dermatol 16(3):44–50

Radonjic-Hoesli S, Hofmeier KS, Micaletto S, Schmid-Grendelmeier P, Bircher A, Simon D (2018) Urticaria and angioedema: an update on classification and pathogenesis. Clin Rev Allergy Immunol 54:88–101

Raheem A, Liang L, Zhang G, Cui S (2021) Modulatory effects of probiotics during pathogenic infections with emphasis on immune regulation. Front Immunol 12:616713

Ramírez-Jiménez F, Pavón-Romero GF, Velásquez-Rodríguez JM, López-Garza MI, Lazarini-Ruiz JF, Gutiérrez-Quiroz KV et al (2023) Biologic therapies for asthma and allergic disease: past, present, and future. Pharmaceuticals 16(2):270

Renert-Yuval Y, Thyssen JP, Bissonnette R, Bieber T, Kabashima K, Hijnen DJ et al (2021) Biomarkers in atopic dermatitis—a review on behalf of the international eczema council. J Allergy Clin Immunol 147(4):1174–1190.e1

Renz H, Skevaki C (2021) Early life microbial exposures and allergy risks: opportunities for prevention. Nat Rev Immunol 21(3):177–191

Reyes VE, Peniche AG (2019) Helicobacter pylori deregulates T and B cell signaling to trigger immune evasion. Molecular Mechanisms of Inflammation: Induction, Resolution and Escape by Helicobacter pylori. Curr Top Microbiol Immunol:229–265

Rona RJ, Keil T, Summers C, Gislason D, Zuidmeer L, Sodergren E et al (2007) The prevalence of food allergy: a meta-analysis. J Allergy Clin Immunol [Internet]. [cited 2025 Feb 19];120(3):638–646. Available from: https://pubmed.ncbi.nlm.nih.gov/17628647/

Sansweet S, Rolling C, Ebisawa M, Wang J, Gupta R, Davis CM (2024) Reaching communities through food allergy advocacy, research, and education: a comprehensive analysis. J Allergy Clin Immunol Pract 12(2):310–315

Sarasola M de la P, Taquez Delgado MA, Nicoud MB, Medina VA (2021) Histamine in cancer immunology and immunotherapy. Current status and new perspectives. Pharmacol Res Perspect 9(5):e00778

Sarkar A, Yoo JY, Valeria Ozorio Dutra S, Morgan KH, Groer M (2021) The association between early-life gut microbiota and long-term health and diseases. J Clin Med 10(3):459

Savage J, Johns CB (2015a) Food allergy. Immunol Allergy Clin N Am 35(1):45–59

Savage J, Johns CB (2015b) Food allergy: epidemiology and natural history. Immunol Allergy Clin North Am [Internet]. [cited 2025 Feb 19];35(1):45–59. Available from: https://pubmed.ncbi.nlm.nih.gov/25459576/
Severn MM, Horswill AR (2023) Staphylococcus epidermidis and its dual lifestyle in skin health and infection. Nat Rev Microbiol 21(2):97–111
Shamji MH, Valenta R, Jardetzky T, Verhasselt V, Durham SR, Würtzen PA et al (2021) The role of allergen-specific IgE, IgG and IgA in allergic disease. Allergy 76:3627–3641
Sharifi-Rad J, Rodrigues CF, Stojanović-Radić Z, Dimitrijević M, Aleksić A, Neffe-Skocińska K et al (2020) Probiotics: versatile bioactive components in promoting human health. Medicina (B Aires) 56(9):433
Sharma P, Imam S, Mehta DK (2019) Steroids in the treatment of chronic diseases. Steroids 12(7):91–97
Shetty K, Sherje AP (2021) Nano intervention in topical delivery of corticosteroid for psoriasis and atopic dermatitis—a systematic review. J Mater Sci Mater Med 32(8):88
Shulpekova YO, Nechaev VM, Popova IR, Deeva TA, Kopylov AT, Malsagova KA et al (2021) Food intolerance: the role of histamine. Nutrients 13(9):3207
Sicherer SH, Sampson HA (2014) Food allergy: epidemiology, pathogenesis, diagnosis, and treatment. J Allergy Clin Immunol [Internet] [cited 2025 Feb 19];133(2). Available from: https://pubmed.ncbi.nlm.nih.gov/24388012/
Sicherer SH, Sampson HA (2018) Food allergy: a review and update on epidemiology, pathogenesis, diagnosis, prevention, and management. J Allergy Clin Immunol 141(1):41–58
Sicherer SH, Abrams EM, Nowak-Wegrzyn A, Hourihane JO (2022) Managing food allergy when the patient is not highly allergic. J Allergy Clin Immunol Pract 10(1):46–55
Sikorska-Szaflik H, Sozańska B (2021) Primary prevention of food allergy—environmental protection beyond diet. Nutrients 13(6)
Silva EA, Bosco MRM, Mozer É (2012) Study of the frequency of allergens in cosmetics components in patients with suspected allergic contact dermatitis. An Bras Dermatol [Internet] [cited 2025 Feb 19];87(2):263–268. Available from: https://www.scielo.br/j/abd/a/CfKrKDYn95wFxb5tmBvrjzk/?lang=en
Silverberg JI, Simpson EL, Boguniewicz M, De Bruin-Weller MS, Foley P, Kataoka Y et al (2021) Dupilumab provides rapid and sustained clinically meaningful responses in adults with moderate-to-severe atopic dermatitis. Acta Derm Venereol 101(11):307
Singh AM, Anvari S, Hauk P, Lio P, Nanda A, Sidbury R et al (2022) Atopic dermatitis and food allergy: best practices and knowledge gaps—a work group report from the AAAAI allergic skin diseases committee and leadership institute project. J Allergy Clin Immunol Pract 10(3):697–706
Skypala IJ, McKenzie R (2019) Nutritional issues in food allergy. Clin Rev Allergy Immunol 57:166–178
Skypala IJ, Bartra J, Ebo DG, Antje Faber M, Fernández-Rivas M, Gomez F et al (2021) The diagnosis and management of allergic reactions in patients sensitized to non-specific lipid transfer proteins. Allergy 76(8):2433–2446
Smith AR, Knaysi G, Wilson JM, Wisniewski JA (2017) The skin as a route of allergen exposure: Part I. Immune components and mechanisms. Curr Allergy Asthma Rep 17:6
Smith NA, Germundson DL, Gao P, Hur J, Floden AM, Nagamoto-Combs K (2021) Anxiety-like behavior and intestinal microbiota changes as strain-and sex-dependent sequelae of mild food allergy in mouse models of cow's milk allergy. Brain Behav Immun 95:122–141
Solymosi D, Sárdy M, Pónyai G (2020) Interdisciplinary significance of food-related adverse reactions in adulthood. Nutrients 12(12):3725
Šošić L, Paolucci M, Flory S, Jebbawi F, Kündig TM, Johansen P (2023) Allergen immunotherapy: progress and future outlook. Expert Rev Clin Immunol 19(7):745–769
Sreeraj K, Risana VU, Mathew SR, Soman S, Charyulu RN, Chandran CS et al (2024) Optimized and safe use of topical corticosteroids in pediatric dermatology through interdisciplinary collaboration: a review. Discover Med 1(1):15

Stephen J, Manoharan D, Radhakrishnan M (2023) Immune boosting functional components of natural foods and its health benefits. Food Prod Process Nutr 5(1):61

Stewart D, Nichol A (2021) Inflammation, immunity and allergy. Anaesth Intensive Care Med 22(8):488–493

Strid J, Strobel S (2005) Skin barrier dysfunction and systemic sensitization to allergens through the skin. Curr Drug Targets Inflamm Allergy. 4:531–541

Sugita K, Akdis CA (2020) Recent developments and advances in atopic dermatitis and food allergy. Allergol Int 69:204–214

Sutton B (2024) Revealed at last: structure of the antibody-receptor complex common to all IgE-mediated allergic hypersensitivity reactions. Sci Signal 17(866):eadu0382

Sweeney A, Sampath V, Nadeau KC (2021) Early intervention of atopic dermatitis as a preventive strategy for progression of food allergy. Allergy Asthma Clin Immunol 17:30

Tan IJ, Podwojniak A, Parikh A, Cohen BA (2024) Precision dermatology: a review of molecular biomarkers and personalized therapies. Curr Issues Mol Biol 46:2975–2990 [Internet]. 2024 Mar 30 [cited 2025 Feb 19];46(4):2975–2990. Available from: https://www.mdpi.com/1467-3045/46/4/186/htm

Tedner SG, Asarnoj A, Thulin H, Westman M, Konradsen JR, Nilsson C (2022) Food allergy and hypersensitivity reactions in children and adults—a review. J Intern Med 291(3):283–302

Tennstedt D (2009) Patch tests: indications or when testing should be performed. Ann Dermatol Venereol [Internet] [cited 2025 Feb 19];136(8–9):579–583. Available from: https://europepmc.org/article/med/19686890

Tham EH, Leung DYM (2019) Mechanisms by which atopic dermatitis predisposes to food allergy and the atopic March. Allergy Asthma Immunol Res 11:4–15

Tontini C, Bulfone-Paus S (2021) Novel approaches in the inhibition of IgE-induced mast cell reactivity in food allergy. Front Immunol 12:613461

Torres T, Ferreira EO, Gonçalo M, Mendes-Bastos P, Selores M, Filipe P (2019) Update on atopic dermatitis. Acta Medica Port 32(9):606–613

Trogen B, Jacobs S, Nowak-Wegrzyn A (2022) Early introduction of allergenic foods and the prevention of food allergy. Nutrients. 14:2565

Turnbull JL, Adams HN, Gorard DA (2015a) Review article: the diagnosis and management of food allergy and food intolerances. Aliment Pharmacol Ther 41(1):3–25

Turnbull JL, Adams HN, Gorard DA (2015b) Review article: the diagnosis and management of food allergy and food intolerances. Aliment Pharmacol Ther [Internet]. [cited 2025 Feb 19];41(1):3–25. Available from: https://onlinelibrary.wiley.com/doi/full/10.1111/apt.12984

Uppin JB, Vaidya G, Mamdapur GMN (2024) Understanding food allergies and allergens: a comprehensive guide to diagnosis, treatment, and management. In: Frontiers in food biotechnology. Springer, pp 143–161

Urisu A, Ebisawa M, Ito K, Aihara Y, Ito S, Mayumi M et al (2014) Japanese guideline for food allergy 2014. Allergol Int 63(3):399–419

Valenta R, Hochwallner H, Linhart B, Pahr S (2015) Food allergies: the basics. Gastroenterology 148(6):1120–1131.e4

van de Veen W, Akdis M (2019) The use of biologics for immune modulation in allergic disease. J Clin Invest 129(4):1452–1462

Van den Houte K, Colomier E, Routhiaux K, Mariën Z, Schol J, Van den Bergh J et al (2024) Efficacy and findings of a blinded randomized reintroduction phase for the low FODMAP diet in irritable bowel syndrome. Gastroenterology 167(2):333–342

Vedanthan PK, Nelson HS, Agashe SN, Mahesh P, Katial R (eds) (2021) Textbook of allergy for the clinician. [cited 2025 Feb 19]; Available from: https://www.taylorfrancis.com/books/9781000280487, CRC Press

Vizzoto F, Testa F, Iraldo F (2021) Strategies to reduce food waste in the foodservices sector: a systematic review. Int J Hosp Manag 95:102933

Wang J, Li XM (2012) Chinese herbal therapy for the treatment of food allergy. Curr Allergy Asthma Rep 12(4):332–338

Wang LJ, Mu SC, Lin MI, Sung TC, Chiang BL, Lin CH (2022) Clinical manifestations of pediatric food allergy: a contemporary review. Clin Rev Allergy Immunol. 62:180–199
Wechsler JB, Schwartz S, Arva NC, Kim KYA, Chen L, Makhija M et al (2022) A single-food milk elimination diet is effective for treatment of eosinophilic esophagitis in children. Clin Gastroenterol Hepatol 20(8):1748–1756
Wild D (2013) The immunoassay handbook: theory and applications of ligand binding, ELISA and Related techniques. In The immunoassay handbook: theory and applications of ligand binding, ELISA and related techniques, pp 1–1013
Wollenberg A, Christen-Zäch S, Taieb A, Paul C, Thyssen JP, de Bruin-Weller M et al (2020) ETFAD/EADV eczema task force 2020 position paper on diagnosis and treatment of atopic dermatitis in adults and children. J Eur Acad Dermatol Venereol 34(12):2717–2744
Wong MM, Keith PK (2020) Presence of positive skin prick tests to inhalant allergens and markers of T2 inflammation in subjects with chronic spontaneous urticaria (CSU): a systematic literature review. Allergy, Asthma Clin Immunol 16(1):72
Woo YR, Kim HS (2024) Interaction between the microbiota and the skin barrier in aging skin: a comprehensive review. Front Physiol 15:1322205
Worth A, Sheikh A (2010) Food allergy and atopic eczema. Curr Opin Allergy Clin Immunol 10:226–230
Wrześniewska M, Wołoszczak J, Świrkosz G, Szyller H, Gomułka K (2024) The role of the microbiota in the pathogenesis and treatment of atopic dermatitis—a literature review. Int J Mol Sci 25(12):6539
Wüthrich B (2014) History of food allergy. Chem Immunol Allergy [Internet]. [cited 2025 Feb 19];100:109–19. Available from: https://pubmed.ncbi.nlm.nih.gov/24925391/
Yang G, Seok JK, Kang HC, Cho YY, Lee HS, Lee JY (2020) Skin barrier abnormalities and immune dysfunction in atopic dermatitis. Int J Mol Sci 21(8):2867
Yoo JY, Groer M, Dutra SVO, Sarkar A, McSkimming DI (2020) Gut microbiota and immune system interactions. Microorganisms 8(10):1587
Young MC (2016) Elimination diets in eczema-a cautionary tale. J Allergy Clin Immunol 4:237–238
Yu L, Li L (2022) Potential biomarkers of atopic dermatitis. Front Med (Lausanne) 9:1028694
Yu JC, Khodadadi H, Baban B (2019) Innate immunity and oral microbiome: a personalized, predictive, and preventive approach to the management of oral diseases. EPMA J 10:43–50
Zhang S, Sicherer S, Berin MC, Agyemang A (2021) Pathophysiology of non-IgE-mediated food allergy. Immunotargets Ther.:431–446
Zhou B, Li J, Liu R, Zhu L, Peng C (2022) The role of crosstalk of immune cells in pathogenesis of chronic spontaneous urticaria. Front Immunol [Internet]. [cited 2025 Feb 19];13:879754. Available from: www.frontiersin.org

Chapter 17
Integrative Physiology of Food Safety Management

Mukul Machhindra Barwant, Balwant Singh, Odangowei Inetiminebi Ogidi, Mudasir A. Dar, and Murad Muhammad

17.1 Introduction

At the dawn of the twenty-first century, the global agro-food industry stands at a critical juncture, grappling with two intertwined challenges: ensuring food security and guaranteeing food safety (Glin 2014). While food security addresses the availability and accessibility of sufficient food for all, food safety focuses on the assurance that the food consumed is free from harmful contaminants, whether microbial, chemical, or physical (Jeger et al. 2021). These challenges are not merely academic concerns but are deeply rooted in the daily lives of billions of people worldwide. The importance of food safety and quality has surged to the forefront of the agri-food sector, becoming the cornerstone of initiatives at every level—from individual farms to multinational enterprises and from regional to international governance. The stakes are high, as failures in food safety

M. M. Barwant (✉)
Department of Botany, Sanjivani Rural Education Society's, Sanjivani Arts Commerce and Science College, Kopargaon, Maharashtra, India

B. Singh
Department of Botany, K. S. Saket PG College, Ayodhya, Uttar Pradesh, India

Dr. Ram Manohar Lohia Avadh University, Ayodhya, Uttar Pradesh, India

O. I. Ogidi
Department of Biochemistry, Faculty of Basic Medical Sciences, Bayelsa Medical University, Yenagoa, Bayelsa State, Nigeria

M. A. Dar
Biofuels Institute, school of environment and safety engineering, Jiangsu University, Zhenjiang, China

M. Muhammad
University of Chinese Academy of Sciences, Beijing, People's Republic of China

T. Sarkar et al. (eds.), *Physiological Perspectives on Food Safety: Exploring the Intersection of Health and Nutrition*, https://doi.org/10.1007/978-3-031-94582-3_17

management can lead to widespread public health crises, economic losses, and an erosion of consumer trust. The Food Safety Management System (FSMS) has emerged as a pivotal framework in addressing these challenges (Ogidi et al. 2025). FSMS encompasses a comprehensive set of practices, protocols, and regulations designed to ensure the safety and quality of food throughout the entire production and supply chain (Panghal et al. 2018). Despite its critical role, the current implementation of safety and quality controls in the food chain is often deficient or poorly executed. These shortcomings result in the persistent contamination of food products by pathogens and harmful chemicals, undermining consumer confidence and posing significant risks to public health (Ogidi et al. 2023). The need for a robust and integrative approach to food safety management has never been more urgent.

Food, as a fundamental necessity of life, serves as the primary source of energy and essential nutrients for all living organisms (Ogidi and Felagha 2024). It fuels every bodily function, from basic cellular processes to complex activities such as movement, growth, and reproduction. The diversity of dietary habits across the animal kingdom ranges from omnivorous diets that include both plant and animal matter to highly specialized diets with limited variety (Muhammad et al. 2024d). Feeding behaviors vary widely, from continuous feeding and digestion to infrequent feeding episodes. This diversity underscores the complexity of the physiological mechanisms that regulate food intake, digestion, and nutrient utilization (Augustin et al. 2016).

Recent advancements in molecular biology have provided new insights into the integrative physiology of food safety and nutrition. Research has begun to unravel the complex mechanisms that underlie the body's responses to fasting, feeding, and exposure to foodborne pathogens (Leulier et al. 2017). By integrating ecological, behavioral, and physiological perspectives, researchers understand how organisms adapt to varying food availability and quality (Stipanuk and Caudill 2018). This integrative approach is essential for developing effective strategies to enhance food safety and ensure the nutritional well-being of populations worldwide. The interplay between food safety and integrative physiology is particularly evident in the context of foodborne pathogens (Muhammad et al. 2024b). These microorganisms, including bacteria, viruses, and parasites, can cause various illnesses, from mild gastrointestinal discomfort to severe, life-threatening conditions. The physiological responses to foodborne pathogens are complex and involve multiple systems, including the immune, gastrointestinal, and nervous systems (Gourama 2020). Understanding these responses is crucial for developing interventions to prevent or mitigate the impact of foodborne illnesses (Ovuru et al. 2023).

In addition to microbial contaminants, chemical hazards pose significant risks to food safety. These hazards can arise from various sources, including agricultural practices, food processing, and environmental contamination (Muhammad et al. 2024c). The physiological effects of exposure to these chemicals can be profound, leading to acute toxicity, chronic health conditions, and even cancer (Lebelo et al. 2021). Addressing these risks requires a comprehensive approach integrating food safety management with understanding the physiological mechanisms governing the body's responses to chemical exposures (Ogidi and Akpan 2023; Ogidi and Tawariowei 2023). The importance of food safety extends beyond individual health to encompass broader societal and economic impacts. Foodborne illnesses can lead to significant financial losses, including healthcare costs, lost productivity, and reduced consumer confidence in the food supply. These impacts are particularly severe in developing countries, where food safety systems are often underdeveloped or poorly enforced. Strengthening food safety management systems in these regions is essential for improving public health, reducing poverty, and promoting sustainable development (Kamboj et al. 2020).

In conclusion, the integrative physiology of food safety management represents a critical area of research and practice in the twenty-first century. By combining insights from physiology, molecular biology, and food safety management, we can develop more effective strategies to ensure the safety and quality of the global food supply. This integrative approach is essential for addressing the complex challenges of food security and food safety and safeguarding the health and well-being of populations worldwide. As we move forward, we must continue to advance our understanding of the physiological mechanisms underpinning food safety and translate this knowledge into practical solutions that can be implemented at every level of the food production and supply chain (Fig. 17.1).

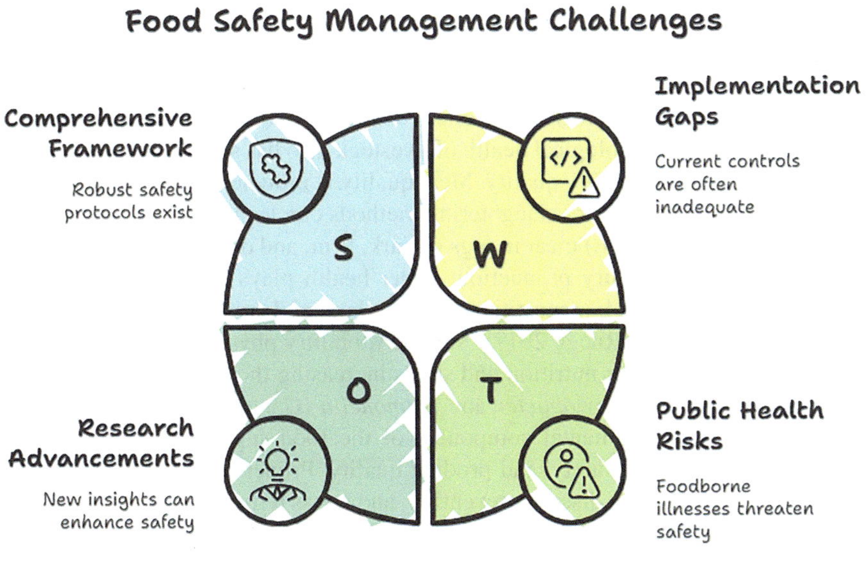

Fig. 17.1 Food safety management and challenges

17.2 Animal Physiology in the Food Supply Chain

Animal physiology plays a crucial role in the food supply chain by directly influencing livestock health, productivity, and food safety. The physiological state of food-producing animals determines their growth rates, disease resistance, and overall well-being, all of which impact the quality and safety of meat, dairy, and poultry products (Sargison 2020). Proper physiological functioning is essential for optimal metabolism, immune responses, and reproductive health, ensuring sustainable and efficient food production. Understanding the physiological mechanisms that govern animal health allows producers to implement better management strategies, reducing the risks of food contamination and enhancing food security. One of the key aspects of animal physiology in the food supply chain is its impact on livestock health and disease prevention. Healthy animals with strong immune systems are less likely to harbor and shed foodborne pathogens such as *Salmonella*, *Escherichia coli* (E. coli), and *Listeria monocytogenes* (Scallan et al. 2011). Factors such as proper nutrition, vaccination programs, and biosecurity measures help maintain physiological balance, reducing disease incidence and ensuring safer food products (FAO 2021). Conversely, poor physiological conditions, such as nutrient deficiencies and chronic illnesses, compromise immune function, increasing the risk of pathogen transmission from animals to humans through contaminated food.

Physiological stress is another critical factor affecting food safety in the livestock industry. Stressors such as transportation, overcrowding, extreme temperatures, and poor handling can trigger the hypothalamic-pituitary-adrenal (HPA) axis, leading to increased cortisol production (Carroll and Forsberg 2007). Elevated cortisol levels weaken immune function and disrupt gut microbiota, making animals more susceptible to infections and increasing pathogen shedding (Schmidt et al. 2016). Stress-induced bacterial shedding has been linked to foodborne disease outbreaks, particularly in the meat and poultry industries, where improper handling and transport practices can elevate contamination risks (Alemu et al. 2021). The physiological health of livestock also has significant implications for meat, dairy, and poultry quality. Meat quality is influenced by muscle physiology, as stress and improper slaughtering methods can lead to issues such as pale, soft, and exudative (PSE) meat in pigs or dark, firm, and dry (DFD) meat in cattle (Gregory 2010). In dairy production, udder health plays a crucial role in milk safety, with infections like mastitis leading to bacterial contamination and antibiotic residues in milk (Ruegg 2017). Similarly, poultry physiology affects egg and meat quality, with poor nutrition and stress increasing the likelihood of bacterial contamination by *Campylobacter* and *Salmonella* (Gantois et al. 2009). Animal physiology is a fundamental component of the food supply chain, influencing livestock health, food safety, and product quality. Proper physiological management through nutrition, disease prevention, and stress reduction strategies ensures optimal animal welfare and reduces contamination risks. The integration of physiological research into livestock management practices enhances food security, improves public health, and supports sustainable agricultural production. This section explores the physiological factors affecting food safety, with a focus on livestock health, the impact of physiological stress on pathogen shedding, and the implications for meat, dairy, and poultry products. Table 17.1 illustrates on the physiological factors affecting food safety.

17.2.1 Livestock Health and Food Safety

Livestock health is a critical factor in ensuring food safety, as the presence of diseases in food-producing animals can lead to the transmission of pathogens to humans. Healthy animals are less likely to harbor harmful microorganisms such as *Salmonella, Escherichia coli* (E. coli), and *Listeria monocytogenes*, which are among the leading causes of foodborne illnesses worldwide (Scallan et al. 2011). In contrast, animals with compromised immune systems due to malnutrition, stress, or infections are more susceptible to colonization by these pathogens, increasing the likelihood of contamination in meat, milk, and eggs (FAO 2021). Disease outbreaks in livestock populations not only pose a direct threat to human health but also result in economic losses for farmers and the food industry due to product recalls and trade restrictions. Preventive veterinary care plays a crucial role in maintaining livestock health and minimizing food safety risks. Vaccination programs help protect animals from infectious diseases such as foot-and-mouth disease, brucellosis, and bovine tuberculosis, which have zoonotic potential (O'Reilly and Daborn 1995). Regular veterinary check-ups and early disease detection can prevent the spread of infections within herds, reducing the presence of harmful bacteria in animal-derived food products (Grace et al. 2015). Additionally, biosecurity measures, including proper sanitation, quarantine procedures, and controlled animal movement, help limit disease transmission among livestock, ensuring that food products remain safe for consumption (FAO 2021).

Nutritional management is another key aspect of livestock health that directly impacts food safety. A well-balanced diet strengthens the immune system, reducing the risk of infections and improving overall animal welfare (Sargison 2020; Ogidi et al. 2020; Ogidi and Otenep 2020). Nutrient deficiencies, particularly in vitamins A, D, and E, as well as trace elements such as zinc and selenium, can weaken immune responses, making animals more vulnerable to bacterial and viral infections (Carroll and Forsberg 2007). Moreover, the use of contaminated feed or water sources can introduce harmful pathogens into livestock populations, increasing the likelihood of foodborne disease transmission through animal products (Oliver et al. 2011). Antibiotic use in livestock production is another factor that influences food safety. While antibiotics are essential for treating bacterial infections, their overuse can lead to antimicrobial resistance (AMR), making it harder to control infections in both animals and humans (Van Boeckel et al. 2015). Residual antibiotics in meat and dairy products can pose health risks to consumers, leading to allergic reactions and the development of resistant bacterial strains

Table 17.1 Physiological factors affecting food safety

Category	Physiological factor	Effect on food safety	Implications for meat, dairy, and poultry products
Livestock health	Immune system function	A strong immune system helps animals resist infections, reducing the presence of zoonotic and foodborne pathogens such as *salmonella*, *E. Coli*, and *listeria monocytogenes*.	Healthy animals produce safer meat, milk, and eggs with lower risks of contamination. Conversely, animals with weakened immunity shed more pathogens, increasing food safety risks.
	Nutritional status	Malnutrition weakens immune function, increases disease susceptibility, and alters gut microbiota, leading to a higher risk of pathogenic infections.	Poor nutrition results in animals more prone to foodborne infections. Deficiencies in vitamins A, D, E, and minerals like zinc and selenium impair immune responses, making animals more vulnerable to infections.
	Disease prevalence	Infectious diseases such as tuberculosis, brucellosis, and mastitis introduce bacterial contamination into animal products. Some diseases are zoonotic and can be transmitted to humans through food.	Meat from diseased animals may contain harmful bacteria. Milk from cows with mastitis has higher bacterial loads and may contain antibiotic residues. Infected poultry can carry *campylobacter* and *salmonella*, leading to foodborne illnesses in humans.
Physiological stress & pathogen shedding	Transportation stress	Stress from transport activates the hypothalamic-pituitary-adrenal (HPA) axis, leading to increased cortisol levels. Elevated stress weakens immunity and increases gut permeability, allowing pathogenic bacteria to multiply and be shed in feces.	Cattle, poultry, and swine undergoing long transport durations have been found to shed more *E. Coli* and *salmonella*. Poor transport conditions result in contamination risks during processing and slaughter.
	Overcrowding & poor housing	High stocking densities lead to increased stress, direct animal-to-animal transmission of pathogens, and reduced air quality. Stress-induced immune suppression makes animals more prone to infections.	Overcrowded poultry farms have higher incidences of *salmonella* and *campylobacter*. Inadequate ventilation in dairy farms promotes bacterial growth, increasing mastitis prevalence. Overcrowding in cattle feedlots increases the risk of *E. Coli* O157:H7 outbreaks.

(continued)

Table 17.1 (continued)

Category	Physiological factor	Effect on food safety	Implications for meat, dairy, and poultry products
	Thermal stress (heat & cold stress)	Heat stress causes dehydration, oxidative stress, and altered gut microbiota, leading to increased bacterial shedding. Cold stress suppresses immune function, making animals more susceptible to infections.	Heat-stressed dairy cows produce lower-quality milk with altered fat and protein content. Heat-stressed poultry exhibit reduced egg production and higher *salmonella* loads. Cattle exposed to extreme cold have weakened immune defenses, increasing respiratory infections and secondary bacterial complications.
Meat, dairy, & poultry safety	Meat quality & safety	Physiological stress before slaughter (e.g., transport, handling) increases metabolic disturbances, leading to poor meat quality. Stress triggers glycogen depletion, resulting in pale, soft, exudative (PSE) meat in pigs or dark, firm, dry (DFD) meat in cattle.	Poor meat quality reduces consumer acceptance and marketability. PSE meat has reduced water-holding capacity, affecting texture and flavor. DFD meat has an abnormal color, making it less desirable. Improper handling at slaughterhouses increases microbial contamination risks.
	Milk contamination & quality	Udder infections (mastitis) increase bacterial counts in milk and can introduce antibiotic residues if treated with antibiotics. Improper milking hygiene can lead to fecal contamination of raw milk.	Mastitis-causing bacteria such as *Staphylococcus aureus* and *E. Coli* can be transmitted through raw milk, posing foodborne illness risks. Antibiotic residues in milk can lead to antimicrobial resistance in consumers. Poor hygiene increases the likelihood of *listeria monocytogenes* in dairy products.
	Egg contamination & safety	Physiological stress in laying hens (due to high stocking densities, poor diet, or heat stress) weakens eggshell integrity, increasing the risk of bacterial penetration.	Eggs from stressed hens have thinner shells, making them more susceptible to *salmonella* contamination. Poor storage conditions can further promote bacterial growth, increasing the risk of foodborne illnesses in humans.
Microbial Adaptation & Food Safety Risks	Pathogen resistance to stress	Certain pathogens adapt to stressful environments (e.g., heat, acidity, cold), increasing their survival and resistance to food safety interventions.	*Listeria monocytogenes* can survive refrigeration temperatures, increasing risks in dairy and ready-to-eat meat products. *E. Coli* and *salmonella* can form biofilms on food processing surfaces, making them harder to eliminate.

(continued)

Table 17.1 (continued)

Category	Physiological factor	Effect on food safety	Implications for meat, dairy, and poultry products
	Bacterial biofilm formation	Bacteria form protective biofilms on food contact surfaces, making them resistant to cleaning and sanitization.	Biofilms contribute to cross-contamination in meat processing plants, dairy equipment, and poultry slaughterhouses, leading to persistent pathogen presence.
Food processing & storage conditions	High-pressure processing (HPP)	HPP inactivates foodborne pathogens while preserving food quality by disrupting microbial cell membranes through hydrostatic pressure.	HPP is effective in reducing *listeria* and *E. Coli* in deli meats, juices, and dairy products while maintaining freshness and nutritional value.
	Modified atmosphere packaging (MAP)	Altering oxygen levels in food packaging reduces microbial growth and extends shelf life.	MAP is widely used in fresh meat and poultry packaging to slow spoilage and maintain red meat color, reducing food waste.
	Cold chain management	Proper refrigeration and freezing conditions slow microbial growth and maintain food safety.	Breaks in the cold chain allow bacteria to multiply in perishable foods, increasing the risk of foodborne illnesses.
Chemical & residue risks	Antibiotic residues in food	Overuse of antibiotics in livestock can lead to drug residues in meat, milk, and eggs, posing health risks to consumers.	Antibiotic residues contribute to antimicrobial resistance (AMR), reducing the effectiveness of antibiotics in treating human infections. Strict withdrawal periods and residue monitoring are essential for food safety.
	Hormonal treatments	Growth hormones in livestock may influence meat quality and consumer health concerns.	Some countries regulate or ban hormone use in food animals due to potential risks, while others allow controlled use in beef production.
Environmental contaminants	Heavy metal exposure	Contaminants such as lead, cadmium, and mercury can accumulate in animal tissues through feed and water.	Heavy metals in meat, milk, and eggs pose toxicity risks to consumers, making feed quality control essential.
	Pesticide residues in animal feed	Contaminated feed can introduce pesticide residues into animal products.	Pesticide residues in meat and dairy products can pose long-term health risks, requiring strict regulatory limits.

(continued)

(Oliver et al. 2011). Regulatory agencies, such as the World Health Organization (WHO) and the U.S. Food and Drug Administration (FDA), advocate for the responsible use of antibiotics in livestock, promoting alternatives such as probiotics, prebiotics, and improved animal husbandry practices to reduce reliance on antimicrobials (WHO 2017).

17.2.2 Physiological Stress and Pathogen Shedding

Physiological stress in livestock is a significant factor influencing pathogen shedding, which directly impacts food safety. When animals experience stress due to environmental, nutritional, or handling factors, their immune system becomes compromised, increasing their susceptibility to infections and enhancing the shedding of pathogens such as *Salmonella*, *Escherichia coli* (*E. coli*), and *Campylobacter* (Schmidt et al. 2016). The activation of the hypothalamic-pituitary-adrenal (HPA) axis during stress leads to increased cortisol levels, which suppress immune function and alter gut microbiota, making animals more vulnerable to bacterial colonization (Carroll and Forsberg 2007). As a result, stressed animals not only suffer from reduced health and productivity but also become reservoirs of harmful pathogens that can contaminate meat, milk, and eggs, posing a serious risk to consumers.

Transportation stress is a major contributor to pathogen shedding in livestock. Studies have shown that animals subjected to prolonged transport exhibit elevated cortisol levels and weakened gut barriers, which facilitate the translocation of bacteria from the intestines to other parts of the body (Alemu et al. 2021). Cattle and pigs transported over long distances have been found to shed higher levels of *Salmonella* and *E. coli*, increasing the risk of contamination at slaughterhouses (Davis et al. 2008). Poultry, in particular, is highly susceptible to transport-induced stress, which has been linked to increased *Campylobacter* shedding, a major cause of foodborne illness worldwide (Skarp et al. 2016). To minimize stress-induced pathogen shedding, livestock transport practices must be optimized by reducing journey durations, ensuring proper ventilation, and providing adequate space and rest periods for animals.

Environmental stressors such as extreme temperatures, overcrowding, and poor housing conditions further exacerbate pathogen shedding in livestock. Heat stress, for instance, impairs immune function and disrupts gut integrity, leading to increased bacterial growth and shedding in dairy cattle, poultry, and swine (Collier et al. 2019). Cold stress has also been associated with increased susceptibility to infections, as animals expend more energy maintaining body temperature, leaving fewer resources for immune defense (Rostagno 2009). Overcrowded housing conditions contribute to higher stress levels, increased aggression, and elevated pathogen loads due to direct animal-to-animal transmission (Gantois et al. 2009). Implementing proper housing designs, temperature control systems, and appropriate stocking densities can help mitigate environmental stress and reduce pathogen shedding in livestock operations.

Nutritional stress is another critical factor influencing pathogen shedding in livestock. Inadequate or imbalanced diets can lead to malnutrition, weakening the immune system and making animals more prone to infections (Sargison 2020). Deficiencies in key nutrients such as vitamins A, E, and selenium have been linked to impaired immune responses and increased bacterial shedding in cattle and poultry (Carroll and Forsberg 2007). Additionally, contaminated or low-quality feed can introduce harmful bacteria into the digestive system, further increasing pathogen shedding (Oliver et al. 2011). To improve livestock health and food safety, nutritionists and farmers must ensure that animals receive balanced diets rich in essential nutrients while maintaining high-quality feed hygiene standards.

17.2.3 Implications for Meat, Dairy, and Poultry Safety

The physiological health of livestock plays a crucial role in determining the safety and quality of meat, dairy, and poultry products. The presence of pathogenic bacteria such as *Salmonella*, *Escherichia coli* (*E. coli*), and *Listeria monocytogenes* in food animals can lead to contamination of animal-derived products, posing significant risks to public health (Scallan et al. 2011). Poor animal health, stress, and inadequate hygiene practices contribute to increased pathogen shedding, making it essential to monitor and improve animal welfare to reduce foodborne disease risks (FAO 2021). Implementing proper livestock management practices, including disease prevention, stress reduction, and sanitation protocols, is essential for maintaining food safety in the meat, dairy, and poultry industries.

In the meat industry, pre-slaughter conditions significantly impact carcass contamination and meat safety. Stressful handling, poor transport conditions, and overcrowding before slaughter can lead to physiological stress, which weakens immune responses and increases bacterial shedding (Rostagno 2009). During slaughter and processing, fecal contamination from stressed animals can introduce pathogens into meat products (Koohmaraie et al. 2005). Meat safety regulations, such as the Hazard Analysis and Critical Control Points (HACCP) system, help prevent contamination by enforcing strict hygiene protocols and microbial testing (USDA 2021). Furthermore, interventions such as carcass washing, steam pasteurization, and irradiation have been developed to reduce bacterial contamination and enhance meat safety. Dairy safety is heavily influenced by udder health, as mastitis—a bacterial infection of the udder—can lead to the presence of harmful pathogens in milk (Ruegg 2017). Bacteria such as *Staphylococcus aureus*, *E. coli*, and *Streptococcus agalactiae* can enter the milk supply if infected cows are not properly monitored and treated (Oliver et al. 2011). Additionally, the use of antibiotics to treat mastitis raises concerns about antibiotic residues in milk, which can contribute to antimicrobial resistance if not properly regulated (Van Boeckel et al. 2015). Strict milk quality control measures, including somatic cell count monitoring, pasteurization, and adherence to food safety standards, are necessary to ensure that dairy products remain safe for human consumption.

In poultry production, stress-related immunosuppression increases susceptibility to bacterial infections, which can compromise food safety. Poultry products, particularly raw chicken and eggs, are commonly associated with *Salmonella* and *Campylobacter*, two of the leading causes of foodborne illness worldwide (Skarp et al. 2016). Stressors such as high stocking densities, heat stress, and poor nutrition contribute to increased pathogen shedding, leading to higher contamination risks during processing (Gantois et al. 2009). Implementing biosecurity measures, improving housing conditions, and using probiotics in feed are effective strategies for reducing bacterial loads in poultry and minimizing contamination risks (Mead 2000).

17.3 Innovative Interventions Based on Physiology

Innovative interventions based on animal physiology are transforming the food production industry by enhancing disease management, food safety, and overall product quality. Understanding the physiological mechanisms that regulate immune function, stress responses, and metabolic pathways allows researchers to develop targeted solutions that improve animal health and minimize contamination risks (Van Boeckel et al. 2015). These interventions include tailored antimicrobial treatments, optimized food processing techniques, and physiological insights for risk reduction. By integrating physiology-based strategies, food producers can enhance sustainability, efficiency, and public health protection.

One of the most significant advancements in livestock management is the development of tailored antimicrobial treatments that align with individual animal physiology. The widespread use of antibiotics in food animals has contributed to antimicrobial resistance (AMR), which threatens both animal and human health (Oliver et al. 2011). Targeted treatments, based on physiological markers and microbiome analysis, enable veterinarians to administer precise antimicrobial therapies that combat infections while preserving beneficial gut bacteria (Mehdi et al. 2018). Alternative approaches, such as probiotics, bacteriophage therapy, and immunomodulators, have emerged as promising solutions to reduce antibiotic dependence and maintain livestock health (Clavijo and Flórez 2018).

Innovations in food processing also leverage physiological insights to improve product safety and quality. High-pressure processing (HPP) and biopreservation techniques utilize physiological principles to inactivate pathogens without compromising the nutritional integrity of food (Zhang et al. 2019). HPP disrupts microbial cell membranes through pressure-induced physiological stress, eliminating bacteria while preserving the texture and flavor of meat and dairy products (Tiwari et al. 2009). Similarly, biopreservation techniques use naturally occurring antimicrobial compounds, such as bacteriocins and organic acids, to extend shelf life and prevent spoilage (Gálvez et al. 2007). These advancements align with consumer preferences for minimally processed foods while enhancing safety.

Physiological insights are also driving improvements in risk reduction strategies within livestock production. Stress-induced pathogen shedding is a major concern in food safety, as stress hormones like cortisol weaken immune defenses and increase bacterial loads in the gut (Rostagno 2009). By implementing stress-reducing interventions—such as improved housing conditions, dietary modifications, and precision feeding—producers can mitigate these risks and reduce contamination in meat, dairy, and poultry products (Schmidt et al. 2016). Additionally, biosensors based on physiological markers have been developed to detect early signs of contamination in processing facilities, allowing for real-time intervention (Velusamy et al. 2010).

Physiology-based innovations are revolutionizing food production by improving animal health, enhancing processing techniques, and reducing contamination risks. Tailored antimicrobial treatments help combat disease while minimizing antibiotic resistance, advanced food processing methods ensure safer products, and physiological insights guide effective risk management strategies. As research continues to deepen our understanding of animal physiology, these interventions will play an increasingly vital role in ensuring sustainable, safe, and high-quality food production. By integrating these scientific advancements, the food industry can meet global demands for safer and more sustainable agricultural practices (Ogidi and Izah 2024). These interventions include tailored antimicrobial treatments, optimized food processing techniques, and physiological insights for risk reduction. By integrating physiological knowledge into livestock management and food production practices, these innovations contribute to safer and more sustainable food systems. Innovative interventions based on physiology for food safety and management is shown in Table 17.2.

17.3.1 Tailored Antimicrobial Treatments

The increasing prevalence of antimicrobial resistance (AMR) in food-producing animals has necessitated the development of tailored antimicrobial treatments that align with individual animal physiology. Traditional blanket antibiotic treatments have contributed to the spread of resistant bacterial strains, posing a major threat to both animal and human health (Van Boeckel et al. 2015). Tailored antimicrobial treatments focus on precision medicine approaches, utilizing diagnostic tools such as microbial culture testing, genetic sequencing, and biomarker analysis to determine the most effective and targeted treatment for specific infections (Laxminarayan et al. 2016). By customizing antimicrobial therapy, this strategy minimizes unnecessary antibiotic use while improving treatment efficacy and food safety.

One approach in tailored antimicrobial treatments is the use of microbiome analysis to assess an animal's gut health before prescribing antibiotics. The gut microbiota plays a crucial role in immune function and digestion, and disrupting its balance through indiscriminate antibiotic use can lead to dysbiosis and

Table 17.2 Innovative interventions based on physiology for food safety and management

Category	Innovative intervention	Physiological basis	Impact on food safety & management
Tailored antimicrobial treatments	Microbiome-based therapy	Uses gut microbiota analysis to determine antimicrobial treatments, avoiding disruption of beneficial microbes.	Reduces unnecessary antibiotic use, lowers antimicrobial resistance (AMR), and improves livestock gut health, decreasing pathogen shedding. Enhances digestion and feed efficiency.
	Phage therapy	Utilizes bacteriophages (viruses that target specific bacterial pathogens) to treat infections in livestock.	Selectively kills harmful bacteria like *salmonella*, *E. Coli*, and *campylobacter* without harming beneficial gut flora. Prevents bacterial resistance and offers an eco-friendly alternative to antibiotics.
	Probiotics & prebiotics	Enhances gut microbiota by promoting beneficial bacteria that outcompete pathogens and improve digestion.	Reduces pathogen colonization in the digestive tract, minimizes bacterial contamination in meat, dairy, and poultry, and supports animal growth and health.
	Immunomodulation strategies	Uses natural immune stimulants like beta-glucans, essential oils, and plant extracts to enhance immune function.	Strengthens natural disease resistance in livestock, reducing infection rates and the need for antibiotics, thereby minimizing antibiotic residues in food.
	Targeted antibiotic treatments	Utilizes molecular diagnostics to identify bacterial infections, allowing for precision antibiotic use.	Ensures effective treatment while reducing overuse of broad-spectrum antibiotics, decreasing antibiotic resistance in foodborne pathogens.
Optimized food processing techniques	High-pressure processing (HPP)	Uses hydrostatic pressure to disrupt microbial cell membranes while preserving food nutrients and structure.	Eliminates pathogens like *listeria monocytogenes* and *E. Coli* in ready-to-eat meats, dairy, and juices, while maintaining texture, flavor, and nutritional value.

(continued)

Table 17.2 (continued)

Category	Innovative intervention	Physiological basis	Impact on food safety & management
	Biopreservation (bacteriocins & organic acids)	Employs antimicrobial compounds naturally produced by bacteria to inhibit pathogen growth and spoilage organisms.	Extends shelf life of dairy, meat, and fermented foods. Reduces reliance on synthetic preservatives, aligning with consumer demand for natural products.
	Active & intelligent packaging	Uses antimicrobial coatings, oxygen scavengers, and smart sensors that detect spoilage and microbial contamination.	Extends shelf life, reduces bacterial contamination, improves food quality monitoring, and provides early warnings of spoilage or safety issues.
	Cold plasma technology	Uses ionized gas (plasma) to sterilize food surfaces without the need for heat or chemicals.	Effectively kills bacteria, viruses, and fungi on meat, poultry, seafood, and fresh produce while preserving food texture and taste.
	Electrolyzed water treatment	Uses electrically charged water to create antimicrobial properties that kill pathogens.	Provides a chemical-free, residue-free method for sanitizing meat, dairy, and food processing equipment.
Physiological insights for risk reduction	Stress management interventions	Reduces livestock stress through improved transport, housing, and handling practices to lower cortisol levels.	Minimizes stress-induced pathogen shedding, reducing contamination risks during slaughter and processing. Improves animal welfare and meat quality.
	Precision nutrition programs	Provides targeted diets rich in essential vitamins, minerals, and amino acids to enhance immune function and digestive health.	Strengthens livestock disease resistance, reduces the likelihood of infections and pathogen transmission, and improves overall feed conversion efficiency.
	Biosensors for pathogen detection	Uses rapid biosensors to detect physiological markers of microbial contamination in real time at various food production stages.	Allows early detection of foodborne pathogens like *salmonella* and *E. Coli*, improving food safety monitoring, quality control, and contamination prevention.

(continued)

Table 17.2 (continued)

Category	Innovative intervention	Physiological basis	Impact on food safety & management
	Genetic selection for disease resistance	Uses genetic screening and breeding programs to enhance disease resistance traits in livestock.	Reduces the need for antibiotics, decreases the prevalence of foodborne diseases in animal production, and improves the sustainability of livestock farming.
	Modified atmosphere packaging (MAP)	Alters oxygen and carbon dioxide levels inside food packaging to slow microbial growth and oxidation.	Prolongs freshness and shelf life of meat, poultry, and seafood while maintaining safety and quality.
	Thermal stress reduction strategies	Uses climate-controlled housing, misting systems, and dietary adjustments to reduce heat stress in livestock.	Prevents stress-related immunosuppression, reduces pathogen shedding, and improves meat and milk production efficiency.
Environmental & chemical risk management	Antibiotic residue monitoring programs	Implements strict regulations and surveillance to detect antibiotic residues in food products.	Prevents antibiotic contamination in meat, milk, and eggs, reducing AMR risks and ensuring compliance with food safety regulations.
	Heavy metal & pesticide monitoring	Regularly tests feed, water, and soil for contaminants such as lead, cadmium, and pesticides.	Prevents accumulation of toxic substances in animal tissues, reducing risks of long-term health effects for consumers.
	Water quality & feed safety programs	Ensures proper water sanitation and high-quality feed sources free from fungal toxins and contaminants.	Reduces exposure to mycotoxins and pathogens that could affect animal health and lead to foodborne illnesses.

increased pathogen susceptibility (Mehdi et al. 2018). Advanced sequencing technologies allow veterinarians to analyze an animal's microbiome and determine whether an antibiotic is necessary, or if alternative therapies such as probiotics, prebiotics, or competitive exclusion strategies would be more beneficial (Clavijo and Flórez 2018). This shift towards microbiome-based interventions reduces AMR risks while supporting overall livestock health and productivity.

Phage therapy has emerged as another promising alternative to conventional antibiotics in tailored antimicrobial treatments. Bacteriophages are viruses that specifically target and destroy bacterial pathogens without affecting beneficial microbes or contributing to resistance (Doss et al. 2017). Studies have shown that phage therapy can effectively reduce *Salmonella*, *Escherichia coli*, and *Campylobacter* infections in poultry and livestock, providing a targeted and sustainable solution to bacterial control (Clavijo and Flórez 2018). Additionally, phage therapy can be used in combination with antibiotics to enhance treatment efficacy while minimizing antibiotic resistance development.

Immunomodulation is another tailored approach that enhances an animal's natural defenses against infections. Immunostimulants such as beta-glucans, essential oils, and plant-derived compounds have been shown to boost immune function, allowing animals to resist infections without the need for excessive antibiotic use (Kogut et al. 2018). Vaccination programs tailored to specific herd or flock needs also reduce the reliance on antibiotics by preventing infections before they occur (Mehdi et al. 2018). These tailored interventions help maintain livestock health while ensuring that food products remain free of excessive antibiotic residues and resistant bacteria.

17.3.2 Optimized Food Processing Techniques

Optimized food processing techniques leverage physiological insights to enhance food safety, extend shelf life, and preserve the nutritional integrity of animal-derived products. Traditional food processing methods, such as thermal pasteurization and chemical preservatives, can sometimes degrade the sensory and nutritional qualities of meat, dairy, and poultry products (Zhang et al. 2019). Advances in food science have introduced innovative processing methods that maintain microbial safety while preserving the physiological properties of food. These techniques include high-pressure processing (HPP), biopreservation, and active packaging, which are increasingly being adopted by the food industry to meet consumer demands for minimally processed, high-quality foods. High-pressure processing (HPP) is an advanced, non-thermal technique that effectively inactivates foodborne pathogens while preserving the texture, flavor, and nutritional content of food (Tiwari et al. 2009). Unlike heat-based pasteurization, which can denature proteins and alter sensory attributes, HPP applies hydrostatic pressure to food products to disrupt microbial cell membranes without compromising product quality (Zhang et al. 2019). This technique has been widely used in meat and dairy industries to eliminate pathogens such as *Listeria monocytogenes* and *Escherichia coli*, reducing the risk of foodborne illnesses while maintaining product freshness (Patterson et al. 2011). HPP offers a sustainable solution to food safety by minimizing the need for chemical preservatives and thermal treatments.

Biopreservation techniques utilize naturally occurring antimicrobial agents, such as bacteriocins, organic acids, and probiotics, to inhibit the growth of spoilage microorganisms in food (Gálvez et al. 2007). Bacteriocins, such as nisin and pediocin, are peptides produced by lactic acid bacteria that selectively target and kill foodborne pathogens without affecting beneficial microbes (Silva et al. 2018). Additionally, probiotic strains, such as *Lactobacillus* and *Bifidobacterium*, can be incorporated into dairy products to enhance gut health while preventing the proliferation of harmful bacteria (Rul 2017). The use of biopreservation aligns with consumer preferences for clean-label products, reducing reliance on synthetic additives while improving food safety.

Active packaging systems have emerged as another innovative food processing technique designed to enhance food safety and shelf life. Unlike traditional packaging, which merely protects food from external contamination, active packaging incorporates antimicrobial agents, oxygen scavengers, and moisture regulators to actively inhibit microbial growth (Han 2014). For example, packaging infused with silver nanoparticles or essential oils has been shown to reduce bacterial contamination in meat and poultry products (Mahmoud et al. 2019). By leveraging physiological principles of microbial inhibition, active packaging helps maintain food freshness, reduce waste, and extend the distribution range of perishable products.

17.3.3 Physiological Insights for Risk Reduction

Understanding physiological responses in livestock and microorganisms has led to significant advancements in risk reduction strategies in food production. Physiological factors such as stress, immune function, and microbial adaptation play critical roles in determining food safety risks, particularly concerning pathogen shedding and contamination (Rostagno 2009). By studying these physiological mechanisms, researchers and food producers can develop targeted interventions to mitigate risks associated with foodborne illnesses. Insights into stress-induced microbial shedding, immune modulation, and biosensor-based detection systems have enhanced efforts to reduce contamination and improve food safety across the meat, dairy, and poultry industries. One major application of physiological insights is in reducing stress-induced pathogen shedding in livestock. Stressors such as transportation, overcrowding, and extreme temperatures trigger the hypothalamic-pituitary-adrenal (HPA) axis, leading to elevated cortisol levels that weaken immune defenses and increase bacterial proliferation in the gut (Carroll and Forsberg 2007). Studies have shown that cattle and poultry experiencing high stress levels shed greater amounts of *Salmonella* and *Escherichia coli*, increasing the risk of contamination in processing facilities (Schmidt et al. 2016). By implementing strategies such as low-stress handling, environmental enrichment, and optimized nutrition, food producers can minimize stress responses in animals, thereby reducing pathogen shedding and improving food safety.

Advancements in immunomodulation have also contributed to risk reduction in food production. Natural immune stimulants such as probiotics, prebiotics, and essential oils have been shown to enhance livestock immune responses, reducing their susceptibility to infections (Mehdi et al. 2018). For example, supplementing poultry diets with *Lactobacillus* probiotics has been found to lower *Campylobacter* colonization in the gut, reducing contamination risks in poultry processing plants (Clavijo and Flórez 2018). Similarly, dietary modifications in dairy cattle, including the use of selenium and vitamin E supplements, have been shown to improve immune function and lower the incidence of mastitis, decreasing bacterial contamination in milk (Sargison 2020). These physiological-based interventions offer sustainable alternatives to antibiotic use while enhancing overall food safety.

Biosensor technology, which utilizes physiological markers for real-time pathogen detection, has emerged as an essential tool for risk reduction in food processing. Biosensors detect microbial contaminants by analyzing physiological indicators such as ATP bioluminescence, enzyme activity, and metabolic byproducts (Velusamy et al. 2010). These rapid detection systems enable food producers to identify contamination risks at critical control points, allowing for immediate corrective actions before products reach consumers. Advances in nanotechnology have further improved biosensor sensitivity, enabling the detection of foodborne pathogens at extremely low concentrations (Velusamy et al. 2010). By integrating biosensor technology into food safety monitoring programs, producers can enhance risk management and reduce foodborne disease outbreaks.

17.4 Physiological Basis of Foodborne Pathogens

Foodborne pathogens present a challenge to global food safety since they can survive, grow, and cause disease in various food environments. Knowledge of the physiological mechanisms involved in their behavior is critical to the effective management of food safety risks. These include their stress responses, biofilm formation, virulence factors, antimicrobial resistance, and cold tolerance, all of which contribute to their persistence in the food supply chain and their ability to infect human hosts.

One of the important physiological features of foodborne pathogens is their ability to adapt to environmental stresses encountered during food processing and storage. For example, most pathogens, like Escherichia coli and Listeria monocytogenes, react to high heat with the synthesis of heat shock proteins (HSPs) that minimize thermal injury to cellular components, like proteins and membranes. They serve as molecular chaperons that ensure proper folding and stability in the cellular structure at high temperatures (Jay et al. 2021). Acid stress, found frequently in fermented foods or in an acidic environment, is regulated through the expression of acid tolerance responses (ATRs). Both Salmonella enterica and E. coli employ various enzymatic mechanisms including glutamate decarboxylase as well as the arginine deiminase pathway for maintaining internal pH, ensuring improved survival at low pH values (Bhunia 2018). Similarly, osmotic stress, caused by salting or drying, is balanced by the accumulation of osmoprotectants like betaine and proline, which stabilize cellular structures and prevent dehydration (Huang et al. 2022).

Another important physiological characteristic of foodborne pathogens is their ability to form biofilms. Biofilms are a complex community of microorganisms encapsulated in an extracellular polymeric substance (EPS) matrix, allowing pathogens to adhere to surfaces and resist environmental stresses, including sanitation efforts. Biofilm formation is especially worrisome in food processing facilities, where pathogens such as Listeria monocytogenes, Pseudomonas aeruginosa, and Salmonella spp. survive even with stringent cleaning protocols. Regulation of biofilm development is carried through cell-to-cell communication mechanism, also known as quorum sensing; it involves a signaling molecule which uses an acyl homoserine lactone (AHL) and coordinates microbial functions, such as EPS production, and virulence (Vu et al. 2020).

Virulence factors are key mediators of pathogenicity by foodborne pathogens. Toxins are proteins secreted by several pathogens, such as Clostridium botulinum, which secretes neurotoxins, and Bacillus cereus, which is known for its emetic and diarrheal toxins (Cahill et al. 2022). On the other hand, adhesion to host cells represents one of the important virulence mechanisms, where in, for instance, E. coli expresses fimbriae and adhesins to attach themselves to epithelial surfaces (Crofts et al. 2018). In addition, the invasive pathogens such as Salmonella and Listeria monocytogenes use the secretion systems to invade host cells and evade the immune defenses. These include the type III and type IV secretion systems that deliver effector proteins directly into host cells, thus enabling intracellular survival and immune evasion (Radoshevich and Cossart 2018).

The antimicrobial resistance among foodborne pathogens is an increasing threat to public health. AMR mechanisms, including efflux pumps, enzymatic degradation of antibiotics, and changes in drug targets, make it difficult for pathogens to be killed by antimicrobial treatments. Horizontal gene transfer through plasmids, transposons, and integrons is a main pathway for spreading resistance genes through microbial communities (Hu et al. 2021). Antimicrobial-induced mutation accelerates resistance evolution, especially at sub-lethal antibiotic concentrations, thereby complicating the control of foodborne diseases (Blázquez et al. 2021).

Psychrotrophic pathogens, such as Listeria monocytogenes and Yersinia enterocolitica, have specific physiological adaptations that permit psychrotrophy, the ability to grow at refrigeration temperatures. Such adaptations include the production of cryoprotectants, such as trehalose, and modifications in the composition of membrane lipids that allows them to maintain fluidity at low temperatures (Tirumalai et al. 2020). This ability of survival and growth in cold conditions emphasizes the need for a strict temperature control process through the cold chain to evade outbreaks linked to ready-to-eat foods.

The physiological basis of foodborne pathogens includes adaptive mechanisms that enable these pathogens to persist in food systems and cause disease. The physiological traits involved in these mechanisms are essential for understanding the development of effective food safety management strategies that will mitigate risks associated with these pathogens (Table 17.3).

17.4.1 Microbial Behavior and Adaptations in Foodborne Pathogens

The survival and proliferation of foodborne pathogens in different environmental conditions are significant factors that affect food safety. Foodborne microorganisms consist of bacteria, viruses, and fungi, which show an excellent ability to adapt with fluctuations in the environment to provide sustainability for these

Table 17.3 Physiological basis of foodborne pathogens

Physiological term	Cause mechanism	References
Microbial behavior and adaptations	Pathogens exhibit complex behavior and adaptations to survive in food environments, including temperature, pH, and nutrient availability. Adaptations such as heat shock proteins aid survival.	Jay et al. (2021)
Environmental stressors	Temperature, pH, and nutrient availability affect microbial survival. Pathogens, including *Salmonella* and *E. coli*, exhibit stress responses to adapt to cold and acidic environments.	Silva et al. (2021)
Acid tolerance mechanisms	*Salmonella* and *E. coli* utilize acid tolerance mechanisms to survive in low pH conditions, often via glutamate decarboxylase and arginine deiminase.	Bhunia (2018)
Heat shock proteins (HSPs)	Heat shock proteins help pathogens like *Listeria monocytogenes* survive high temperatures by repairing damaged proteins.	Tirumalai et al. (2020)
Toxin production	Toxins like botulinum toxin, enterotoxins, and TSST-1 are key virulence factors in pathogens like *Clostridium botulinum* and *Staphylococcus aureus*. These toxins are regulated by environmental factors.	Zong et al. (2021)
Immune evasion mechanisms	Pathogens such as *Salmonella* and *Listeria* evade host immune responses by manipulating immune signaling pathways, contributing to their virulence.	Ibarra and Steele-Mortimer (2021)
Pathogen survival in the gastrointestinal tract	Pathogens survive in the gastrointestinal tract by using mechanisms such as acid tolerance and proteases to resist stomach acid and bile.	Allen et al. (2021)
Biofilm formation	Biofilm formation by pathogens like *Salmonella* enhances resistance to environmental stresses and contributes to persistent infections in food environments.	Liu et al. (2021)
Immune system interactions	The immune system's response to foodborne pathogens involves both innate and adaptive immune mechanisms, including inflammation, phagocytosis, and antibody production.	Miyamoto et al. (2020)
Vulnerable populations	Immunocompromised, elderly, pregnant, and chronically ill populations are at higher risk for severe foodborne illnesses due to weakened or altered immune systems.	Riley et al. (2021)

microorganisms in food systems and enhance their pathogenic potential. Such behaviors are mediated by physiological mechanisms, where microorganisms are able to perceive and respond to different types of stressors. These stressors include stress that causes adverse effects on their cells, resulting in damage, hindered growth, and decreased ability to reproduce. Therefore, the study of microbial behavior and responses to these stressors is fundamental in controlling foodborne diseases and the associated risk of public health hazard.

Microorganisms in food environments are constantly exposed to environmental stressors, including temperature fluctuations, pH changes, and nutrient availability. Their adaptation to these challenges is mediated by complex biochemical and molecular pathways. Pathogens such as Salmonella enterica, Escherichia coli, and Listeria monocytogenes have evolved specific stress responses that enable them to survive in a wide range of food matrices, from refrigerated to ambient temperatures. These adaptive changes serve the basic function of facilitating the survival of these pathogens; additionally, it serves in increasing virulence, such that their infection occurs readily in predisposed populations.

17.4.2 Environmental Stressors: Temperature, pH, and Nutrient Availability

Temperature, pH, and nutrient availability are major environmental stressors that have significant effects on microbial survival and growth. Both high and low temperature fluctuations can cause severe physiological disruption in foodborne pathogens. The survival of pathogens at high temperatures is associated with heat shock responses, including the production of heat shock proteins (HSPs). These proteins help in the refolding of denatured proteins and stabilization of cellular structures (Jay et al. 2021). Conversely, cold stress, experienced during refrigeration and freezing, activates pathways that preserve membrane integrity and protein function at low temperatures. For instance, Listeria monocytogenes and Yersinia enterocolitica can grow at refrigeration temperatures due to the synthesis of cryoprotectants like trehalose, which protects against ice crystal formation in cells (Tirumalai et al. 2020).

The pH of food products is also an important stressor for pathogens, especially in fermented foods or acidified environments. Pathogens like Salmonella enterica and E. coli have acid tolerance mechanisms that enable them to survive in low-pH environments. These adaptations are controlled by acid tolerance response (ATR) systems that include the induction of specific genes encoding glutamate decarboxylase and arginine deiminase proteins, which aid in the buffering of intracellular pH (Bhunia 2018). In the same way, the survival of pathogens in high-pH environments, such as in alkaline foods, is enabled by mechanisms that control ion homeostasis and improve protein stability.

The availability of nutrients is a significant factor that influences microbial growth. The existence or absence of key nutrients such as nitrogen, carbon, and sulfur will decide whether a pathogen grows or rests. In this case, for example, if the levels of available nutrients are low, E. coli and Salmonella enter a stationary phase where their metabolic activities decrease, and they increase their resistance to environmental stresses (Silva et al. 2021). This metabolic shift serves as a survival strategy that permits these pathogens to survive under such adverse conditions as nutrient-limited environments encountered during food storage.

17.4.3 Mechanisms of Toxin Production

Toxin production is an important mechanism of virulence in several foodborne pathogens and may contribute directly to the development of disease in humans. *Clostridium botulinum, Bacillus cereus,* and *Staphylococcus aureus*, for example, produces a range of powerful toxins that cause illness from nausea and vomiting to potentially fatal neurological damage. Toxin production is tightly regulated by a variety of genetic and environmental factors that include temperature, pH, and nutrient availability. The molecular mechanisms controlling toxin synthesis follow primarily microbial growth phase patterns. Toxins are generally synthesized during the stationary phase, when pathogens abandon a more resource-intensive metabolic state.

For instance, Clostridium botulinum is known to produce botulinum toxin when under anaerobic conditions. This toxin production has been tightly regulated by the response of the bacterium to the environmental cues like pH and temperature (Cahill et al. 2022). Likewise, Bacillus cereus produces two primary forms of enterotoxins that are associated with diarrhea and emetic symptoms. These toxins are regulated by genetic systems sensitive to nutrient availability, temperature, and oxygen levels (Jensen et al. 2021). Staphylococcus aureus is another notorious pathogen that produces a variety of enterotoxins and exotoxins, such as the super antigenic toxic shock syndrome toxin (TSST-1), under conditions of nutrient abundance, high temperature, or low pH (Zong et al. 2021).

Regulation of toxin production is typically governed by quorum sensing, a mechanism where microorganisms interact with signaling molecules to coordinate group behaviors, including toxin synthesis. For example, Vibrio cholerae, the causative agent of cholera, synthesizes cholera toxin as a response to quorum sensing mechanisms activated by autoinducer molecules' accumulation (Chen et al. 2021). Quorum sensing also plays a significant role in the regulation of biofilm formation and other virulence factors, which altogether enhance the ability of the pathogen to cause disease.

The behavior and survival adaptations of the foodborne pathogen are ultimately determined by their interaction with specific environmental factors or intrinsic microbial abilities. Temperature and pH significantly regulate the physiological level of responses enabling pathogens to be viable and multiply well in food matrices. Understanding mechanism stress responses of pathogens, related to toxin formation, is fundamentally important in appraising the hazard of foodborne diseases. Further study into these physiological processes will require establishing improved food safety management practices, the design of novel antimicrobial treatments, and prevention of foodborne illness outbreaks.

17.5 Human Physiology and Food Safety Risks

Susceptibility to foodborne pathogens as well as response to them significantly depends on human physiology. Key factors include the digestive system, immune responses, and host factors that may be vulnerability due to an immunocompromised state or chronic conditions in determining how well the body prevents, manages, and recovers from foodborne infections. These physiological mechanisms explain why the outcomes of foodborne illness are sometimes more severe in some individuals and populations than in others. It is crucial to understand the nexus between human physiology and food safety risks in developing effective food safety management strategies.

17.5.1 Digestive System and Foodborne Pathogen Interactions

The human digestive system is the prime site where a foodborne pathogen first has contact with the body. That system includes both the mouth and esophagus, stomach and small intestine and large intestine in which foodstuffs and nutrition pass but may also serve to act as front-line defense against any pathogens. The stomach creates a very acidic environment (pH 1.5–3.5) that prevents the proliferation or survival of many of the pathogens entering it. Certain pathogens, such as Helicobacter pylori, Salmonella enterica, and Escherichia coli O157:H7, can adapt to survive in this highly acidic environment of the stomach (Jay et al. 2021). Pathogens, such as Salmonella, have acid tolerance mechanisms that allow them to survive the acidity of the stomach and infect the intestines where they can adapt and multiply further.

Once pathogens enter the small intestine, digestive enzymes, bile salts, and a complex microbiome are available to either inhibit or promote the colonization of these pathogens. Defense against harmful microorganisms is also balanced with nutrient absorption and food digestion in the small intestine. Some pathogens, including Campylobacter jejuni and Vibrio cholerae, have specialized mechanisms to interact with the surface of the mucosa in the intestinal region, making them adhere and enter host cells (Allen et al. 2021). For example, Vibrio cholerae induces the sequestering of fluids into the intestines, causing diarrhea, a primary symptom of infection. Similarly, Clostridium perfringens and Bacillus cereus produce enterotoxins that cause an intestinal permeability disorder, thus causing gastrointestinal distress (Cahill et al. 2022).

The large intestine also has a very significant role in pathogen survival limitation, which is supported by a resident microbiota that is acting like a barrier to colonization by pathogenic microorganisms. In cases where this balance of such microbiota is disturbed, like after antibiotic treatment, pathogenic bacteria cause overgrowth and conditions for diseases like Clostridium difficile infection (C. difficile) (Penders et al. 2021). The inflammatory responses from the pathogen interaction with the intestinal lining also cause further complications, such as systemic infections when pathogens translocate into the bloodstream.

17.5.2 Immune Responses to Foodborne Hazards

The human immune system acts as the first line of defense against foodborne pathogens, involving both innate and adaptive immune mechanisms to detect, neutralize, and eliminate harmful microorganisms. It provides the first line of defense through physical barriers such as epithelial cells lining the gut and the mucus layer that traps pathogens. The gut epithelial cells are secreted defensins and cathelicidins that act directly on pathogens to kill them or inhibit their multiplication (Miyamoto et al. 2020). Another central function would be phagocytic cells, which includes neutrophils and macrophages that could recognize and internalize pathogens with an initiation of inflammation in attempts to contain the infection and to eliminate it.

In addition to these physical and chemical barriers, the innate immune system is also mediated through pattern recognition receptors (PRRs), including Toll-like receptors (TLRs) that recognize conserved molecular patterns derived from pathogens termed pathogen-associated molecular patterns (PAMPs). The stimulation of these PRRs on immune cells triggers the production of cytokines that are signaling molecules to modulate the immune response and facilitate clearance of pathogens (Liu et al. 2021). For instance, in Salmonella enterica, the activation of PRR leads to the production of pro-inflammatory cytokines, which facilitate coordinated immune response and enhance neutrophil cytotoxic activity against the pathogen (Ibarra and Steele-Mortimer 2021).

The adaptive immunity is more specific against foodborne pathogens through the generation of antigen-specific antibodies and memory cells that provide long-lasting immunity against future infections. B cells produce antibodies, such as IgA, that are secreted into the gastrointestinal tract to neutralize pathogens before they can adhere to epithelial cells (Barrett et al. 2021). Eliminating infected cells and orchestrating the immune response itself also involves T cells. For example, CTLs can detect and lysates infected cells, preventing the spread of pathogens such as Listeria monocytogenes, which can invade host cells and replicate intracellularly (Radoshevich and Cossart 2018). However, some foodborne pathogens can evade the immune system's detection or suppress its responses, as in the case of Salmonella, which manipulates host immune signaling; this complicates the body's ability to mount an effective defense (Rao et al. 2020).

17.5.3 Vulnerable Populations: Immunocompromised and Chronic Conditions

Certain populations are at a greater risk of developing severe outcomes of foodborne illness, including immunocompromised patients, elderly people, pregnant women, and patients with chronic conditions such as diabetes. These populations are at an increased risk due to their compromised immune defenses or the physiological changes associated with their conditions. For instance, people with impaired immune systems caused by HIV/AIDS, cancer chemotherapy, or organ transplants cannot respond to pathogens as effectively. In such groups, even less virulent pathogens can cause life-threatening infections (Riley et al. 2021).

Pregnant women face special food safety risks, primarily because physiological alterations in pregnancy-including changes to the immune function and hormonal levels-sometimes increase vulnerability to specific types of foodborne infections. Commonly implicated in foodborne infections in pregnant women are Listeria monocytogenes, Toxoplasma gondii, and Campylobacter jejuni, which cause infections that sometimes lead to miscarriage, premature birth, or neonatal infection (Gerba et al. 2020). Additionally, chronic conditions such as diabetes can impair immune responses and alter the gastrointestinal environment, making it easier for pathogens like Salmonella and Clostridium difficile to cause infection (Zong et al. 2021).

The elderly are also at a higher risk, due to the effects of aging on the immune system, including reduced T-cell activity and slower production of antibodies, leading to decreased immune responses to foodborne pathogens. Consequently, infections caused by foodborne pathogens are often more severe in older adults and may lead to prolonged hospitalization or death (Tremblay et al. 2022). This will include chronic conditions, which are also common in the elderly, such as cardiovascular diseases and obesity, that further compromise the immune system and contribute to worse outcomes after foodborne illnesses.

Food safety risks are intricately linked to human physiology, primarily through the roles of the digestive system and immune responses in the pathogenesis of foodborne infections. The interactions of foodborne pathogens with the digestive system, the immune defenses mounted in response, and the susceptibility of vulnerable populations all contribute to the complexity of foodborne illnesses. Understanding these physiological mechanisms is crucial for developing effective public health strategies to protect high-risk groups and reduce the incidence of foodborne diseases.

17.6 Case Studies

17.6.1 Application of Physiological Principles in Food Safety

Integrating physiological principles into food safety management has significantly improved the control of foodborne pathogens. One prominent example is using probiotics to combat Salmonella contamination in poultry (Muhammad et al. 2024a). Researchers studied the physiological environment of the avian gut, focusing on factors such as pH, microbial competition, and nutrient availability, which influence Salmonella survival (Dunkley et al. 2009). Introducing probiotic feed additives enhanced the gut health of chickens, creating an unfavorable environment for Salmonella colonization. This approach reduced the prevalence of the pathogen in poultry and minimized the risk of transmission to humans through contaminated meat (Khan et al. 2020; Mkangara 2023). Another application involves simulating the human digestive system to study pathogen behavior. Laboratory models of the gastrointestinal tract, which replicate conditions like stomach acidity and enzyme activity, have been used to assess the survival of pathogens such as *Listeria* and *E. coli* in processed foods (Ayivi et al. 2020). These studies have informed better food processing techniques, such as heat treatment and fermentation, to ensure pathogen inactivation (Unruh 2018). By leveraging physiological insights, food safety experts can develop targeted interventions that address the root causes of contamination, ultimately improving public health outcomes.

17.6.2 Success Stories in Contamination Control

Implementing Hazard Analysis and Critical Control Points (HACCP) in the seafood industry is a landmark success in contamination control. In the 1990s, the U.S. seafood industry faced a significant challenge with Vibrio contamination in raw oysters, which caused severe gastrointestinal illnesses (Ahmad et al. 2023). Producers identified critical control points, such as water temperature during harvest and storage conditions, by adopting HACCP principles to minimize bacterial growth. This systematic approach dramatically reduced Vibrio-related illnesses and set a benchmark for other food industries. Similarly, the dairy industry has successfully controlled Cronobacter contamination in infant formula through advanced filtration and pasteurization techniques. Cronobacter, a pathogen that can cause life-threatening infections in infants, was a significant concern for formula manufacturers (Kamboj et al. 2020). By implementing rigorous testing protocols and improving processing technologies, the industry has ensured the safety of its products for vulnerable populations. These success stories demonstrate the effectiveness of proactive contamination control measures and highlight the importance of continuous innovation in food safety management.

17.6.3 Lessons Learned from Outbreaks

The 2011 Listeria outbreak in cantaloupes in the U.S. is a stark reminder of the consequences of lapses in food safety protocols. The outbreak, which resulted in 33 deaths, was traced back to unsanitary conditions at a single processing facility (Clapp 2016). Investigators found that the facility's equipment was difficult to clean, allowing Listeria to thrive and contaminate the fruit (Avoids and Malbec 2018). This tragedy underscored the critical importance of sanitation in food processing and led to stricter regulations and improved industry practices. Another key lesson comes from the 2008 melamine contamination in infant formula in China, where melamine, a chemical used in plastics, was added to milk to inflate protein content artificially (Rajpoot et al. 2021). This led to kidney damage in thousands of infants and exposed weaknesses in the oversight of food safety systems. Governments and industries worldwide strengthened their monitoring and enforcement mechanisms to prevent such incidents (Mahardika et al. 2017). These outbreaks highlight the need for robust food safety management systems, continuous vigilance, and a commitment to learning from past mistakes to protect public health.

17.7 Future Directions in Integrative Physiology and Food Safety

Nutritional science has played a pivotal role in advancing human health by elucidating essential nutrient requirements and mitigating deficiencies (Davy and Davy 2019). Emerging protein sources, such as cultivated meat and precision-fermented dairy, will serve as complementary additions to traditional diets, offering a pathway to reduce environmental impact while maintaining nutritional diversity. Simultaneously, advancements in personalized nutrition and diagnostic technologies will enable individuals to make informed dietary choices, while innovations in smart farming and gene editing will enhance resource efficiency (FAO and WHO 2020). Although conventional agriculture will remain predominant, biotechnological innovations will play an increasingly significant role in shaping the future of food production. Recent advancements in systems-level metabolic analysis have illuminated the complex communication networks between tissues, particularly in the context of exercise-induced adaptations and maternal-fetal metabolic interactions. Meanwhile, precision medicine approaches are being integrated with deep phenotyping to tailor nutritional strategies based on individual microbiome profiles, genetic predispositions, and metabolic responsiveness (Wu et al. 2020). The evolving nexus of integrative physiology and food safety is poised to address pressing global challenges emerging zoonoses, chemical contaminants, and climate-driven nutritional insecurity through interdisciplinary innovation (Wang et al. 2024). By employing multi-omics platforms, AI-driven predictive modeling, and bioengineered *in vitro* systems, researchers can decode complex interactions between foodborne hazards (e.g., pathogens, toxins) and host physiology, enabling personalized risk assessments and interventions tailored to genetic, metabolic, and microbiome variability (Leulier et al. 2018). Concurrently, advances in biomarker discovery and nutraceutical design will empower precision strategies to mitigate chronic exposure effects, such as endocrine disruption or immunotoxicity (FAO and WHO 2020).

Climate resilience demands physiological insights into stress-adapted crops and livestock, optimizing nutrient retention and thermoregulation under environmental extremes, while one health frameworks integrate human, animal, and ecosystem data to preempt zoonotic spillover. The integration of cellular agriculture with regenerative farming practices presents a balanced approach that harmonizes tradition with innovation (Jiang et al. 2021). By embracing these scientific advancements, we can develop a resilient and sustainable food system capable of meeting the demands of a growing global population while safeguarding environmental health. The coming decade represents a pivotal opportunity to reimagine the food supply chain, fostering a more humane, sustainable, and adaptable model. Translational success is mainly based on harmonizing global policies with evidence-based contaminant thresholds and ethical governance of emerging technologies, ensuring equitable access to innovations like CRISPR-engineered crops or AI-enabled supply-chain monitoring (Liu et al. 2023). Gene editing, particularly through technologies such as CRISPR, will be central to these advancements. By engineering crops with enhanced resistance to pests, diseases, and climate variability, gene editing has the potential to increase yields while promoting sustainable agricultural practices. Furthermore, in the realm of food biotechnology, CRISPR facilitates precise genetic modifications in microorganisms and cellular systems, optimizing their ability to produce high-value proteins and bioactive compounds (Shen et al. 2024). This targeted genetic refinement enhances the efficiency, scalability, and sustainability of food production, paving the way for a more secure and environmentally responsible global food system. Ultimately, the field must bridge molecular mechanisms to systemic outcomes, fostering sustainable food systems through collaborative, physiology-informed strategies that balance safety, nutrition, and planetary health an imperative requiring sustained investment in cross-sector research, education, and policy co-creation (Nordhagen et al. 2022). Lastly, the future of integrative physiology and food safety will depend on our ability to harmonize cutting-edge technologies with sustainability imperatives, such as climate-resilient food systems and personalized nutrition safety protocols, to create a healthier and safer food environment for all.

17.8 Conclusion

The integrative approach to food safety management, rooted in physiological principles, offers a comprehensive framework for understanding and mitigating risks associated with foodborne pathogens. By examining the physiological adaptations of microbes, human and animal physiological responses, and the technological advancements in food safety monitoring, this chapter has highlighted the importance of a multidisciplinary perspective in ensuring food safety across the supply chain. A key takeaway from this discussion is the interplay between environmental stressors and microbial adaptations, which directly impact pathogen survival and toxin production. Understanding these mechanisms allows for the development of targeted interventions, including optimized antimicrobial treatments and improved food processing techniques. Furthermore, human and animal physiology play critical roles in food safety risk assessments, emphasizing the need for tailored strategies to protect vulnerable populations and ensure livestock health. The integration of real-time physiological monitoring tools and predictive microbiology has revolutionized food safety management, enabling proactive hazard identification and control. Additionally, case studies have demonstrated the successful application of physiological insights in contamination control, reinforcing the effectiveness of science-driven approaches in minimizing foodborne risks. As food safety challenges continue to evolve, future research should focus on advancing physiological-based interventions and refining data-driven risk assessment models. Innovations in biotechnology, microbiome analysis, and personalized food safety strategies hold promise for further enhancing food safety standards. By leveraging integrative physiology, stakeholders across the food industry can work collaboratively to safeguard public health and ensure the sustainability of food production systems. In conclusion, the application of physiological principles in food safety management is an indispensable component of modern food protection strategies. A holistic understanding of microbial behavior, host interactions, and technological advancements will continue to shape the future of food safety, driving improvements in contamination control and risk mitigation.

References

Ahmad RS, Munawar H, Saima H, Siddique F (2023) Introductory chapter: food safety. In: Food safety-new insights. IntechOpen

Alemu RG, Getabalew M, Akeberegn D (2021) Effects of transportation stress on welfare and meat quality of food animals: a review. Vet Med Int 2021:5561189. https://doi.org/10.1155/2021/5561189

Allen D, Jones M, Williams P (2021) *Vibrio cholerae* interactions with the human gastrointestinal tract: mechanisms of colonization and virulence. Microorganisms 9(6):1251. https://doi.org/10.3390/microorganisms9061251

Augustin MA, Riley M, Stockmann R, Bennett L, Kahl A, Lockett T, Osmond M, Sanguansri P, Stonehouse W, Zajac I (2016) Role of food processing in food and nutrition security. Trends Food Sci Technol 56:115–125

Avoids PT, Malbec SU (2018) FOO LAW

Ayivi RD, Gyawali R, Krastanov A, Aljaloud SO, Worku M, Tahergorabi R, Silva RC d, Ibrahim SA (2020) Lactic acid bacteria: food safety and human health applications. Dairy 1(3):202–232

Barrett JM, Johnson RE, Howard LM (2021) IgA-mediated immunity and its role in gut protection against foodborne pathogens. Front Immunol 12:732903. https://doi.org/10.3389/fimmu.2021.732903

Bhunia AK (2018) Foodborne microbial pathogens: mechanisms and pathogenesis, 2nd edn. Springer. https://doi.org/10.1007/978-1-4939-8848-0

Blázquez J, Couce A, Rodríguez-Beltrán J (2021) Antimicrobial resistance mutations: mechanisms and selection. Ann Rev Microbiol 75:287–314. https://doi.org/10.1146/annurev-micro-090817-062455

Cahill SM, Li J, McClelland M (2022) Mechanisms of toxin production in foodborne pathogens. Toxins 14(6):423. https://doi.org/10.3390/toxins14060423

Carroll JA, Forsberg NE (2007) Influence of stress and nutrition on cattle immunity. Vet Clin N Am Food Anim Pract 23(1):105–149. https://doi.org/10.1016/j.cvfa.2007.01.003

Chen X, Duan W, Zhang Y (2021) Quorum sensing regulation in *Vibrio cholerae* and its effect on virulence factors. Microorganisms 9(3):659. https://doi.org/10.3390/microorganisms9030659

Clapp S (2016) Fixing the food system. Bloomsbury Publishing

Clavijo V, Flórez MJ (2018) The gastrointestinal microbiome and its association with the control of pathogens in broiler chicken production: a review. Poult Sci 97(3):1006–1021. https://doi.org/10.3382/ps/pex308

Collier RJ, Baumgard LH, Zimbelman RB, Xiao Y (2019) Heat stress: physiological effects and consequences for dairy cattle. J Dairy Sci 102(4):3781–3796. https://doi.org/10.3168/jds.2018-15799

Crofts AA, Giovanetti SM, Rubin EJ, Poly FM, Gutiérrez RL, Faherty CS (2018) Enterotoxigenic *Escherichia coli* pathogenesis. Nat Rev Microbiol 16(8):485–496. https://doi.org/10.1038/s41579-018-0022-7

Davis MS, Mader TL, Holt SM, Parkhurst AM (2008) Stress responses of cattle during road transport. J Anim Sci 86(6):1650–1657. https://doi.org/10.2527/jas.2007-0744

Davy KP, Davy BM (2019) Advances in nutrition science and integrative physiology: insights from controlled feeding studies. Front Physiol 10:486989. https://doi.org/10.3389/fphys.2019.01341

Doss J, Culbertson K, Hahn D, Camacho J, Barekzi N (2017) Phage therapy: an alternative to antibiotics in the age of multi-drug resistance. World J Gastrointest Pharmacol Ther 8(3):162–173. https://doi.org/10.4292/wjgpt.v8.i3.162

Dunkley K, Callaway T, Chalova V, McReynolds J, Hume M, Dunkley C, Kubena L, Nisbet D, Ricke S (2009) Foodborne salmonella ecology in the avian gastrointestinal tract. Anaerobe 15(1–2):26–35

FAO (2021) The state of food and agriculture 2021: making agrifood systems more resilient to shocks and stresses. Food and Agriculture Organization of the United Nations

FAO and WHO (2020) The future of food safety: transforming knowledge into action for people, economies and the environment—transforming knowledge into action for people, economies and the environment. Technical summary by FAO and WHO, Rome. https://doi.org/10.4060/ca8386en

Gálvez A, Abriouel H, Benomar N, Lucas R (2007) Bacteriocin-based strategies for food biopreservation. Int J Food Microbiol 120(1–2):51–70. https://doi.org/10.1016/j.ijfoodmicro.2007.06.001

Gantois I, Ducatelle R, Pasmans F, Haesebrouck F, Van Immerseel F, P. (2009) Mechanisms of egg contamination by *Salmonella Enteritidis*. FEMS Microbiol Rev 33(4):718–738. https://doi.org/10.1111/j.1574-6976.2008.00161.x

Gregory TR (2010) Understanding natural selection: Essential concepts and common misconceptions.Evolution: Education and Outreach 3(2):145–162. https://doi.org/10.1007/s12052-010-0224-0

Gerba CP, Bradeen JM, McKendry MP (2020) Foodborne pathogens and risks during pregnancy. J Food Prot 83(5):904–915. https://doi.org/10.4315/JFP-19-446

Glin LC (2014) Governance of global organic agro-food networks from Africa. Wageningen University and Research

Gourama H (2020) Foodborne pathogens. In: Food safety engineering. Springer, pp 25–49

Grace D, Mutua F, Ochungo P, Kruska R, Jones K, Brierley L, Dominguez-Salas P (2015) Mapping of poverty and likely zoonoses hotspots. International Livestock Research Institute
Han JH (2014) Innovations in food packaging. Academic
Hu Y, Zhang T, Gao G (2021) Antimicrobial resistance genes in foodborne pathogens. Annu Rev Food Sci Technol 12(1):161–182. https://doi.org/10.1146/annurev-food-062320-101942
Huang L, Zhang Q, Xu W, Sun W (2022) Osmotic stress adaptation in foodborne pathogens. Front Microbiol 13:854872. https://doi.org/10.3389/fmicb.2022.854872
Ibarra JA, Steele-Mortimer O (2021) *Salmonella enterica* interactions with the intestinal immune system. Front Immunol 12:691084. https://doi.org/10.3389/fimmu.2021.691084
Jay JM, Loessner MJ, Golden DA (2021) Modern food microbiology, 8th edn. Springer. https://doi.org/10.1007/978-1-4939-3047-2
Jeger M, Beresford R, Bock C, Brown N, Fox A, Newton A, Vicent A, Xu X, Yuen J (2021) Global challenges facing plant pathology: multidisciplinary approaches to meet the food security and environmental challenges in the mid-twenty-first century. CABI Agric Biosci 2(1):1–18
Jensen BB, Rødbotten M, Hansen LH (2021) Toxin production in *Bacillus cereus*: a review of environmental regulation and metabolic pathways. J Appl Microbiol 131(5):1420–1431. https://doi.org/10.1111/jam.15143
Jiang S, Wang F, Li Q, Sun H et al (2021) Environment and food safety: a novel integrative review. Environ Sci Pollut Res 28:54511–54530. https://doi.org/10.1007/s11356-021-16069-6
Kamboj S, Gupta N, Bandral JD, Gandotra G, Anjum N (2020) Food safety and hygiene: a review. Int J Chem Stud 8(2):358–368
Khan S, Moore RJ, Stanley D, Chousalkar KK (2020) The gut microbiota of laying hens and its manipulation with prebiotics and probiotics to enhance gut health and food safety. Appl Environ Microbiol 86(13):e00600–e00620
Kogut MH, Arsenault RJ, Johansen LM (2018) Innate immune response to *Salmonella enterica* in poultry: immune function and organ microbiomes. Poult Sci 97(2):479–485. https://doi.org/10.3382/ps/pex407
Koohmaraie M, Scanga JA, De La Zerda MJ, Koohmaraie M, Elder RO (2005) Pre-harvest strategies to reduce foodborne pathogens in food animals. Meat Sci 71(1):79–91. https://doi.org/10.1016/j.meatsci.2005.04.002
Laxminarayan R, Van Boeckel T, Teillant A (2016) The economics of antibiotic resistance: a theory of optimal use. J Environ Econ Manag 78:157–166. https://doi.org/10.1016/j.jeem.2016.03.009
Lebelo K, Malebo N, Mochane MJ, Masinde M (2021) Chemical contamination pathways and the food safety implications along the various stages of food production: a review. Int J Environ Res Public Health 18(11):5795
Leulier F, MacNeil LT, Lee W-J, Rawls JF, Cani PD, Schwarzer M, Zhao L, Simpson SJ (2017) Integrative physiology: at the crossroads of nutrition, microbiota, animal physiology, and human health. Cell Metab 25(3):522–534
Leulier F, MacNeil LT, Rawls JF, Cani PD, Schwarzer M, Zhao L, Simpson SJ (2018) Integrative physiology: at the crossroads of nutrition, microbiota, animal physiology and human health. Cell Metab 25(3):522. https://doi.org/10.1016/j.cmet.2017.02.001
Liu Q, Zhang C, Chen Y (2021) The role of pattern recognition receptors in detecting foodborne pathogens. Microorganisms 9(10):2154. https://doi.org/10.3390/microorganisms9102154
Liu Z, Wang S, Zhang Y, Feng Y, Liu J, Zhu H (2023) Artificial intelligence in food safety: a decade review and bibliometric analysis. Food Secur 12(6):1242. https://doi.org/10.3390/foods12061242
Mahardika K, Mastuti I, Zafran Z (2017) Histopathological study on nephropathy caused by oral administration with melamine and cyanuric acid in humpback grouper (Cromileptes altivelis). Aquac Aquar Conserv Legis 10(2):328–334
Mahmoud BSM, Lin HT, Ismail AA (2019) Application of natural antimicrobial agents in food packaging. Food Sci Technol Int 25(3):213–227. https://doi.org/10.1177/1082013218823838

Mead GC (2000) Prospects for "competitive exclusion" treatment to control *Salmonella* and other foodborne pathogens in poultry. Vet J 159(2):111–123. https://doi.org/10.1053/tvjl.1999.0420

Mehdi Y, Létourneau-Montminy MP, Gaucher ML, Chorfi Y, Suresh G, Rouissi T, Godbout S (2018) Use of antibiotics in broiler production: global impacts and alternatives. Anim Nutr 4(2):170–178. https://doi.org/10.1016/j.aninu.2018.03.002

Miyamoto S, Murakami S, Inoue M (2020) Antimicrobial peptides and their role in pathogen control. J Food Saf 40(1):e12914. https://doi.org/10.1111/jfs.12914

Mkangara M (2023) Prevention and control of human salmonella enterica infections: an implication in food safety. Int J Food Sci 2023(1):8899596

Muhammad M, Ahmad J, Basit A, Mohamed HI, Khan A, Kamel EA (2024a) Antimicrobial activity of Penicillium species metabolites. In: Fungal secondary metabolites. Elsevier, pp 369–383

Muhammad M, Li W-J, Li L, Liu Y-H, Ali K, Ahmed I (2024b) Introduction of microbiomes, viromes and biofilms. In: Microbial ecology. CRC Press, pp 1–30

Muhammad M, Wahab A, Waheed A, Mohamed HI, Hakeem KR, Li L, Li W-J (2024c) Harnessing bacterial endophytes for environmental resilience and agricultural sustainability. J Environ Manag 368:122201

Muhammad M, Waheed A, Wahab A, Majeed M, Nazim M, Liu Y-H, Li L, Li W-J (2024d) Soil salinity and drought tolerance: an evaluation of plant growth, productivity, microbial diversity, and amelioration strategies. Plant Stress 11:100319

Nordhagen S, Lambertini E, DeWaal CS, McClafferty B, Neufeld LM (2022) Integrating nutrition and food safety in food systems policy and programming. Glob Food Sec 32:100593. https://doi.org/10.1016/j.gfs.2021.100593

O'Reilly LM, Daborn CJ (1995) The epidemiology of *Mycobacterium bovis* infections in animals and man: a review. Tuber Lung Dis 76(1):1–46. https://doi.org/10.1016/0962-8479(95)90591-X

Ogidi OI, Akpan UM (2023) Impacts of chemical use in agricultural practices: perspectives of soil microorganism and vegetation. In: One health implications of agrochemicals and their sustainable alternative. Springer Nature Singapore Private Ltd.

Ogidi OI, Felagha I (2024) Gut microbiome and diet interaction. In: Nutrition controversies and advances in autoimmune disease. IGI Global

Ogidi IO, Izah SC (2024) Emerging technologies in the global south food industry: prospects and challenges. In: Food safety and quality in global south. Springer Nature Singapore Private Ltd. https://doi.org/10.1007/978-981-97-2428-4_7

Ogidi OI, Otenep EJ (2020) Determination of nutritional properties of honey from *Apis mellifera*. Int J Food Sci Nutr 5(1):23–26

Ogidi OI, Tawariowei AM (2023) Multi-elemental analysis and potential human health risk assessment of selected canned foods sold in Nigeria. Food Env Saf J 22(4):268–276. https://doi.org/10.4316/fens.2023.025

Ogidi OI, Charles EE, Onimisi AM, Amugeh R (2020) Assessment of nutritional properties and heavy metal composition of African giant land snails (*Archachatinamarginata*) and Clams (*Mercenariamercenaria*) from Ekowe Community. European J Nutr Food Saf 12(6):99–108. https://doi.org/10.9734/EJNFS/2020/v12i630242

Ogidi OI, Izah SC, Akpan UM (2023) Food quality and agrochemical use: integrated monitoring, assessment, and management policies. In: One health implications of agrochemicals and their sustainable alternative. Springer Nature Singapore Private Ltd.

Ogidi OI, Wenapere CM, Chukwunonso OA (2025) Enhancing food safety and quality control with computer vision systems. In: Bora D, Bania R (eds) Computer vision techniques for agricultural advancements. IGI Global Scientific Publishing, pp 51–88. https://doi.org/10.4018/979-8-3693-8019-2.ch003

Oliver SP, Murinda SE, Jayarao BM (2011) Impact of antibiotic use in dairy cattle on antimicrobial resistance of infectious bacteria. Curr Opin Microbiol 14(5):573–579. https://doi.org/10.1016/j.mib.2011.07.005

Ovuru KF, Izah SC, Ogidi OI, Imarhiagbe O, Ogwu MC (2023) Slaughterhouse facilities in developing nations: sanitation and hygiene practices, microbial contaminants and sustainable management system. Food Sci Biotechnol 33(3):519–537. https://doi.org/10.1007/s10068-023-01406-x

Panghal A, Chhikara N, Sindhu N, Jaglan S (2018) Role of food safety management systems in safe food production: a review. J Food Saf 38(4):e12464

Patterson MF, McKay AM, Connolly M, Linton M (2011) The effect of high pressure on the survival of *Escherichia coli* and *Listeria monocytogenes* in milk and fruit juices. J Food Prot 74(9):1523–1530. https://doi.org/10.4315/0362-028X.JFP-11-133

Penders J, Vink C, Dijkshoorn L (2021) Dysbiosis and its role in gastrointestinal infections. Front Microbiol 12:726513. https://doi.org/10.3389/fmicb.2021.726513

Radoshevich L, Cossart P (2018) *Listeria monocytogenes*: towards a complete picture of its physiology and pathogenesis. Nat Rev Microbiol 16(1):32–46. https://doi.org/10.1038/nrmicro.2017.126

Rajpoot M, Bhattacharya R, Sharma S, Gupta S, Sharma V, Sharma AK (2021) Melamine contamination and associated health risks: gut microbiota does make a difference. Biotechnol Appl Biochem 68(6):1271–1280

Riley LW, Ko G, Goldberg J (2021) Immunocompromised populations and foodborne diseases. Front Immunol 12:715655. https://doi.org/10.3389/fimmu.2021.715655

Rostagno MH (2009) Can stress in farm animals increase food safety risk? Foodborne Pathog Dis 6(7):767–776. https://doi.org/10.1089/fpd.2009.0315

Ruegg PL (2017) A 100-year review: mastitis detection, management, and prevention. J Dairy Sci 100(12):10381–10397. https://doi.org/10.3168/jds.2017-13023

Rul F (2017) Probiotics and dairy products. Dairy Sci Technol 97(1):67–79. https://doi.org/10.1007/s13594-016-0301-5

Sargison N (2020) Sheep flock health: a planned approach. Wiley

Scallan E, Hoekstra RM, Angulo FJ, Tauxe RV, Widdowson MA, Roy SL, Jones JL, Griffin PM (2011) Foodborne illness acquired in the United States—major pathogens. Emerg Infect Dis 17(1):7–15. https://doi.org/10.3201/eid1701.P11101

Schmidt JW, Brichta-Harhay DM, Bosilevac JM, Kalchayanand N, Shackelford SD, Wheeler TL (2016) Prevalence, enumeration, serotypes, and antimicrobial resistance phenotypes of *salmonella* enterica isolates from carcasses at two large United States commercial beef processing plants. Appl Environ Microbiol 82(15):4470–4479. https://doi.org/10.1128/AEM.00890-16

Shen J, Zhang D, He Y, Shen Y, Duan M, Zhao Y, Liu Z, Jia F (2024) Evolving CRISPR/Cas system for food safety monitoring across the food supply chain. TrAC Trends Anal Chem 181:118050. https://doi.org/10.1016/j.trac.2024.118050

Silva CCG, Silva SP, Ribeiro SC (2018) Application of bacteriocins and protective cultures in dairy food preservation. Front Microbiol 9:594. https://doi.org/10.3389/fmicb.2018.00594

Silva C, Oliveira C, Pereira S (2021) Stress adaptation in *Escherichia coli* and *Salmonella* and its implications for food safety. Int J Food Microbiol 349:1–11. https://doi.org/10.1016/j.ijfoodmicro.2021.109174

Skarp CPA, Hänninen ML, Rautelin HIK (2016) *Campylobacteriosis*: the role of poultry meat. Clin Microbiol Infect 22(2):103–109. https://doi.org/10.1016/j.cmi.2015.11.019

Stipanuk MH, Caudill MA (2018) Biochemical, physiological, and molecular aspects of human nutrition-e-book: biochemical, physiological, and molecular aspects of human nutrition-e-book. Elsevier Health Sciences

Tirumalai R, Salim AS, Weng Y (2020) Cold tolerance mechanisms in foodborne pathogens. Food Res Int 136:109444. https://doi.org/10.1016/j.foodres.2020.109444

Tiwari BK, O'Donnell CP, Patras A, Cullen PJ (2009) Anthocyanin and ascorbic acid degradation in sonicated strawberry juice. J Agric Food Chem 57(20):10047–10052. https://doi.org/10.1021/jf902606j

Tremblay A, Martin G, Vallée T (2022) Age-related factors influencing foodborne illness severity. J Am Geriatr Soc 70(1):116–125. https://doi.org/10.1111/jgs.17135

Unruh DA (2018) Addressing and anticipating food safety challenges: microbiology and policy frameworks for Shiga toxin-producing Escherichia coli (STEC) and salmonella. Kansas State University

USDA (2021) Food safety and inspection service regulations. United States Department of Agriculture

Van Boeckel TP, Brower C, Gilbert M, Grenfell BT, Levin SA, Robinson TP, Teillant A, Laxminarayan R (2015) Global trends in antimicrobial use in food animals. Proc Natl Acad Sci 112(18):5649–5654. https://doi.org/10.1073/pnas.1503141112

Velusamy V, Arshak K, Korostynska O, Oliwa K, Adley C (2010) An overview of foodborne pathogen detection: in the perspective of biosensors. Biotechnol Adv 28(2):232–254. https://doi.org/10.1016/j.biotechadv.2009.12.004

Vu B, Chen M, Crawford RJ, Ivanova EP (2020) Bacterial biofilms: regulation, formation, and strategies for eradication. Crit Rev Microbiol 46(3):277–299. https://doi.org/10.1080/1040841X.2020.1729693

Wang Y, Gu H, Yin X, Geng T, Long W, Fu H, She Y (2024) Deep leaning in food safety and authenticity detection: an integrative review and future prospects. Trends Food Sci Technol 146:104396. https://doi.org/10.1016/j.tifs.2024.104396

WHO (2017) Guidelines on use of medically important antimicrobials in food-producing animals. World Health Organization

Wu Q, Chen T, El-Nezami H, Savidge TC (2020) Food ingredients in human health: ecological and metabolic perspectives implicating gut microbiota function. Trends Food Sci Technol 100:103–117. https://doi.org/10.1016/j.tifs.2020.04.007

Zhang H, Kong B, Xiong YL, Sun X (2019) Antioxidant activity of peptides from meat protein hydrolysates as measured by electron spin resonance spectroscopy. Food Chem 301:125289. https://doi.org/10.1016/j.foodchem.2019.125289

Zong Z, Wang Y, Lu J (2021) Staphylococcus aureus: toxin production and its control by environmental factors. J Appl Microbiol 131(6):1827–1836. https://doi.org/10.1111/jam.14864

Chapter 18
Immunological Responses to Foodborne Threats

Anindya Sundar Ray, Tapas Pal, Subhendu Dhibar, Swtarka Das, Sreelakshmi R.S., and Animesh Sen

18.1 Introduction

Foodborne infections have become increasingly prevalent over time and now constitute a significant threat to public health globally. Foodborne pathogens are responsible for thousands of infections and pose a risk to both human health and the economy due to their adverse health effects. Each year, unsafe food is responsible for approximately 600 million cases of illness and 420,000 deaths worldwide. This public health burden leads to the loss of an estimated 33 million Disability-Adjusted Life Years (DALYs) annually (World Health Organization 2025). Children are disproportionately affected; every year, approximately 1.25 million children under the age of

A. S. Ray
Ethnopharmacology Laboratory, Department of Biological Sciences and Microbiology, Biswa Bangla Biswabidyalay (University), Bolpur, Shibpur, Birbhum, West Bengal, India
e-mail: anindyasundar.ray@bbb.ac.in

Ethnopharmacology Laboratory, Department of Botany, Visva-Bharati University, Santiniketan, Birbhum, West Bengal, India

T. Pal
Raiganj University, Raiganj, West Bengal, India

S. Dhibar
Colloid Chemistry Laboratory, Department of Chemistry, The University of Burdwan, Golapbag, Burdwan, West Bengal, India
e-mail: sdhibar@scholar.buruniv.ac.in

S. Das
Division of Microbiology and Immunology, College of Veterinary Medicine, Cornell University, Ithaca, NY, USA

Sreelakshmi R.S. · A. Sen
Regional Ayurveda Research Institute, Gangtok, India

T. Sarkar et al. (eds.), *Physiological Perspectives on Food Safety: Exploring the Intersection of Health and Nutrition*, https://doi.org/10.1007/978-3-031-94582-3_18

five die. A diverse range of pathogenic microorganisms can contaminate food products throughout the manufacturing and processing stages, as well as during storage and transportation prior to consumption. Widespread gastrointestinal disorders triggered by foodborne human pathogens are estimated to cause substantial financial and health burdens. According to the World Health Organization (WHO), nearly 30% of foodborne mortality occurs in children under five years of age (World Health Organization 2025). Foodborne microorganisms may cause mild to severe symptoms, such as diarrhea, or debilitating diseases, such as meningitis. This is attributed to the underdeveloped immune systems of young children, rendering them more susceptible to foodborne illnesses (Gourama 2020). They are also more likely to engage in hand-to-mouth behaviors or consume food contaminated with bacteria, viruses, or parasites. Several studies have found that the incidence and significance of various foodborne disorders are influenced by the relationships between pathogenic organisms, humans, foodstuffs, and the environment (Gourama 2020). As a consequence of foodborne illnesses, pathogenic microorganisms, including bacteria, viruses, fungi, and parasites, are implicated in the development of these diseases, although bacteria appear to be the most common etiological agents and are capable of exhibiting diverse characteristics and functions. Certain bacteria, such as *Clostridium botulinum*, *Clostridium perfringens*, *Bacillus subtilis*, and *Bacillus cereus*, can produce spores and are highly heat-resistant. For instance, *Staphylococcus aureus* and *Clostridium botulinum* are heat-tolerant bacteria capable of producing toxins. In most cases, they are mesophilic, with optimal growth temperatures ranging between 20 and 45 °C. Furthermore, certain foodborne pathogens, such as *Listeria monocytogenes* and *Yersinia enterocolitica*, can survive in refrigerated conditions or at temperatures below 10 °C and can be transmitted to humans (Bintsis 2017).

In this chapter, we provide a comprehensive overview of foodborne pathogens, including bacteria, viruses, parasites, and their associated toxins that contribute to foodborne illnesses. We delve into the mechanisms through which these pathogens invade the host, produce toxins, and disrupt normal physiological functions. Furthermore, the chapter explores the host's defense systems, highlighting both innate and adaptive immune responses that are activated to combat these infections. Understanding these interactions is crucial for developing effective prevention and control strategies to reduce the global burden of foodborne diseases.

18.2 Foodborne Illness Outbreaks: A Public Health Perspective

18.2.1 Bacteria

Foodborne infections caused by bacterial pathogens represent a significant global health concern. In the United States alone, it is expected that foodborne diseases result in between 128,000 and 325,000 hospitalizations and 3000 to 5000 deaths annually, alongside approximately 76 million cases of illness. The economic burden associated with these illnesses is substantial, reaching up to $83 billion per year (Aladhadh 2023). There are many different varieties, forms, and characteristics of bacteria, which are the most frequent cause of illnesses related to food. Certain harmful bacteria, such as *Clostridium botulinum*, *Clostridium perfringens*, *Bacillus subtilus*, and *Bacillus cereus*, are extremely heat-resistant due to their capacity to produce spores. Some, like *Clostridium botulinum* and *Staphylococcus aureus*, can produce poisons that are resistant to heat. A majority of infections grow best at temperatures between 20 and 45 degrees Celsius because they are mesophilic. Nonetheless, some foodborne pathogens (psychrotrophs), such as Yersinia enterocolitica and Listeria monocytogenes, can grow in cold environments or at temperatures lower than 10 °C (Bintsis 2017; Aladhadh 2023).

Salmonella, particularly in eggs and egg products, emerged as the most frequently identified bacterial agent responsible for these outbreaks. Other notable pathogens included *Campylobacter*, *Yersinia*, *Shiga* toxin-producing *Escherichia coli* (STEC), and *Listeria monocytogenes*, which collectively contributed to a significant proportion of foodborne illnesses in the EU during the same year. Notably, illnesses associated with *L. monocytogenes* were among those leading to the highest fatality rates (Aladhadh 2023).

A common spore-forming, Gram-positive bacterium in the environment is *Bacillus cereus*. Because its spores are so sticky and difficult to remove, it can survive in settings where food is produced. *B. cereus* can survive in various kinds of diets because it can withstand a broad range of temperatures (4–55 °C), pH levels (4.9–9.3), and water activity levels. It produces two types of toxins that cause foodborne illness: diarrheal toxins that cause diarrhea and abdominal discomfort and an emetic toxin that causes nausea and vomiting. Diarrheal syndrome is associated with meat, soups, and dairy products, whereas emesis usually happens after eating starchy meals like rice. *B. cereus* is still a serious foodborne pathogen even though symptoms are usually mild and temporary. *Bacillus cereus* and *Bacillus anthracis* are remarkably similar, according to genetic and genomic investigations, and some strains of *B. cereus* carry plasmids that resemble *Bacillus anthracis'* toxin plasmids. There is little data on incidence because food poisoning caused by *B. cereus* is not a disease that must be reported in the majority of nations. Although fatal cases have been reported, it is acknowledged that there may be a considerable underreporting of *B. cereus* disease because the symptoms are often moderate, transient, and self-limiting. Out of the 31 main pathogens in the US, *B. cereus* is thought to be responsible for 0.7% of foodborne illnesses (Bintsis 2017; Aladhadh 2023).

Gram-negative, spiral-shaped *Campylobacter* species, especially *Campylobacter jejuni,* are a major cause of diarrheal disease globally. Contaminated food, especially undercooked poultry, unpasteurized milk, and contaminated water, are the primary methods that *C. jejuni* infects humans, leading to gastrointestinal and often systemic disorders. These bacteria thrive in a limited range of oxygen and temperature because they are thermophilic, microaerophilic, and genetically unstable. *Campylobacters* can go into a viable but non-culturable condition under stress, which allows for ongoing transmission (Bintsis 2017).

Clostridium spp. are spore-forming, Gram-positive (in early growth stages), anaerobic or aerotolerant bacteria belonging to the family Bacillaceae. Vegetative cells vary in shape, from short coccoid rods to long filamentous forms, occurring singly, in pairs, or in chains. Endospores, which form under unfavorable conditions, give *Clostridium* spp. their characteristic shape and enable them to survive extreme conditions, such as boiling and exposure to air. *Clostridium* spp. are widespread in soil, animal intestines, and the environment. The most powerful known toxins, known as botulinum neurotoxins, are produced by the motile *Clostridium botulinum*. These toxins are classified into seven types (A–G), with types A, B, E, and F causing botulism in humans. Foodborne botulism is frequently linked to improperly processed or stored foods, such as canned goods, and necessitates strict thermal processing to inactivate spores. The genome and toxin production of the bacterium have been extensively investigated using modern sequencing techniques, and recent outbreaks have brought to light the dangers of improper thermal processing and temperature abuse in food storage (Bintsis 2017).

Escherichia coli (*E. coli*) is a diverse group of Gram-negative, facultative anaerobic bacteria, typically harmless but with pathogenic strains that pose public health risks. A varied genus of facultative anaerobic, Gram-negative bacteria, *Escherichia coli* (*E. coli*) is generally benign but can have pathogenic strains that are dangerous to the public's health. Enterohemorrhagic *E. coli* (EHEC), which generates Shiga toxin (STEC), is one of six families of pathogenic *E. coli* that are divided according to disease processes. Hemolytic uremic syndrome (HUS), a potentially fatal illness that results in kidney failure, and severe diarrhea are two symptoms of STEC infections, especially those caused by strain O157:H7. Cattle are the main reservoir for *E. coli* O157:H7, a disease that was discovered in 1982 and associated with outbreaks from meat that is not properly cooked. Contaminated fruit, water, or food can spread the disease. The 2011 *E. coli* O104:H4 outbreak in Europe, which resulted in more than 4000 illnesses and 50 fatalities, is one example of an outbreak that has raised concerns around the world. Interventions have been implemented to reduce contamination in food production, yet non-O157 serogroups and antibiotic resistance remain emerging challenges (Bintsis 2017; Aladhadh 2023).

Staphylococcus aureus are nonmotile, gram-positive cocci that grow in diverse arrangements and are facultative anaerobes, thriving better under aerobic conditions. They can endure for a longer period of time in harsh environments and are found across nature, invading the skin, mucosal surfaces, and internal tissues of birds and mammals. These bacteria are extremely salt-tolerant, can thrive in a broad range of temperatures (7–47.8 °C) and pH (4.5–9.3), and under some circumstances, can produce enterotoxins. Skin lesions, bacteremia, pneumonia, and toxic shock syndrome are among the infections that *S. aureus* is mostly responsible for. Staphylococcal food poisoning, which is typified by gastrointestinal symptoms, can be caused by foodborne contamination by *S. aureus*, which is frequently brought on by incorrect handling of food, refrigeration, or hygiene. Foodborne outbreaks are associated with heat-resistant superantigens called *Staphylococcal* enterotoxins, especially types A and D (Bintsis 2017).

18.2.2 Fungi

Fungi represent a unique category of foodborne pathogens, distinct from bacteria and viruses, with outbreaks of related illnesses being comparatively rare. Among the approximately 1.5 million known fungal species, only about 300 are recognized as pathogenic to humans (Aladhadh 2023). Despite their relative rarity, notable outbreaks have occurred, primarily attributed to the consumption of fungal secondary metabolites, including various toxins. These incidents often affect vulnerable populations, such as individuals with compromised immune systems due to conditions like organ transplants or immunosuppressive therapies. However, there are instances where immunocompetent individuals also experience foodborne fungal illnesses (Aladhadh 2023).

One significant outbreak occurred in 2013 in the United States, where over 200 individuals developed gastroenteritis characterized by vomiting, nausea, and diarrhea after consuming yogurts contaminated with *Mucor circinelloides*. Another incident involved *Rhizopus microsporus*, which was linked to food poisoning in seven hospital patients in Hong Kong due to contaminated pre-packaged meals or cornstarch used in the production of allopurinol tablets. A literature review has identified key filamentous fungi such as *Aspergillus*, *Fusarium*, and *Mucor* as potential agents of localized infections when ingested or inhaled, even in immunocompetent hosts. While invasive pulmonary diseases caused by these pathogens predominantly affect immune-compromised patients—often through inhalation—the gastrointestinal route is also significant. This underscores the necessity for such patients to avoid foods that may harbor fungal contaminants (Aladhadh 2023). The potential for invasive fungal infections (IFIs) arising from contaminated food or nutritional supplements is an area requiring further exploration, particularly regarding the effectiveness of dietary restrictions for high-risk individuals. Although the fungal foodborne illnesses are less common than those caused by bacterial agents, they present a critical area for public health surveillance and intervention.

18.2.3 Virus

Viruses such as Hepatitis A and Norovirus are major causes of foodborne illnesses, often transmitted through the consumption of contaminated shellfish, including oysters, clams, mussels, and cockles. These viruses are resilient and can survive under extreme environmental conditions like freezing, heat, and chemical exposure. Hepatitis A, in particular, is known for its stability in the environment, and although it cannot replicate outside of a host, it remains infectious for long periods in water and food. Shellfish, which filter feed on water, can accumulate viruses in their digestive tracts. Unlike bacteria, these viruses are difficult to remove through cleaning processes such as depuration, and because shellfish are often consumed raw or lightly cooked, they can harbor the virus, making them a significant vector for transmission. The consumption of contaminated shellfish has led to several outbreaks of Hepatitis A, such as the large-scale outbreak linked to clams in Shanghai, which affected hundreds of thousands of people (Bintsis 2017).

Norovirus, another significant foodborne virus, causes the majority of acute viral gastroenteritis cases globally. This highly contagious virus is spread primarily through fecal-oral transmission and can persist on food and surfaces even after cleaning. Norovirus outbreaks are common in environments like nursing homes, hospitals, and cruise ships, where the close quarters and shared food sources facilitate rapid transmission. Contaminated food, including raw or undercooked oysters and even frozen berries, has been linked to numerous outbreaks. For instance, a series of outbreaks involving frozen raspberries highlighted the challenge of identifying the source, with multiple separate outbreaks traced back to a single contaminated batch from Serbia. Additionally, asymptomatic food handlers have been identified as a key factor in spreading Norovirus, contributing to up to one-quarter of outbreaks. Effective prevention strategies rely on maintaining good personal hygiene, particularly handwashing, which remains the most effective method for preventing the spread of Norovirus. Enhanced hygiene measures, such as using hand sanitizers with urea and citric acid, may also help in combating these resilient viruses (Bintsis 2017).

18.2.4 Parasites

Foodborne parasites are microorganisms that require a living host for nourishment and reproduction. Unlike bacteria, they do not multiply in food but instead in the tissues of infected hosts, typically animals or humans. The infectious stage of these parasites is often a cyst, which can be transmitted through contaminated food or water. Some common foodborne parasites include *Cyclospora cayetanensis*, *Toxoplasma gondii*, and *Trichinella spiralis* (Bintsis 2017).

Cyclospora cayetanensis is a protozoan parasite that primarily affects the human small intestine. It is often linked to contaminated produce, such as raspberries, lettuce, and basil, and can cause prolonged illness with symptoms like non-bloody diarrhea, bloating, nausea, fatigue, and abdominal cramping. Outbreaks have been reported in the US and Canada, and it is estimated to cause around 15,000 foodborne illnesses annually in the US (Bintsis 2017).

Toxoplasma gondii is another protozoan parasite, which uses cats as its primary host but can infect almost all warm-blooded animals, including humans. It is typically acquired through the consumption of undercooked meat, especially pork and lamb, contaminated water, or contact with infected cat feces. In humans, it may cause toxoplasmosis, which can lead to symptoms like fever, muscle aches, and swollen lymph nodes. The parasite is of particular concern to pregnant women, as it can be transmitted to the fetus, potentially causing severe birth defects or even death (Bintsis 2017).

Table 18.1 Common foodborne pathogens and their associated health effects

No.	Name of the pathogens	Quick facts on pathogens	Hallmark symptoms
Bacteria			
1.	*Staphylococcus aureus*	Gram-positive, cocci, facultative anaerobe. **Details**: Produces heat-stable enterotoxins that cause food poisoning. Symptoms include vomiting, abdominal cramps, and diarrhea, typically appearing within hours of ingestion. Commonly found in improperly stored foods.	Nausea, Vomiting, Diarrhea, Abdominal discomfort syndromes.
2.	*Bacillus cereus*	**Nature**: Gram-positive rod, spore-forming, facultative anaerobe. **Details**: Causes two types of foodborne illnesses: The emetic type (vomiting, caused by a heat-stable toxin) and the diarrheal type (caused by a heat-labile toxin). Found in rice, pasta, and starchy foods.	Diarrhea Vomiting Abdominal pain Nausea
3.	*Clostridium perfringens*	**Nature**: Gram-positive rod, spore-forming, anaerobic. **Details**: Produces enterotoxins that cause abdominal cramps and diarrhea. Symptoms occur 8–12 hours after ingestion. Associated with improperly cooked or reheated meat and poultry.	Abdominal cramps Diarrhea Sometimes nausea and vomiting
4.	*Salmonella* spp.	**Nature**: Gram-negative rod, facultative anaerobe. **Details**: Causes salmonellosis, characterized by diarrhea, fever, and abdominal cramps. Transmitted through contaminated food, water, or contact with infected animals.	Fever Abdominal cramps Diarrhea Vomiting Headache
5.	*Shigella* spp.	**Nature**: Gram-negative rod, facultative anaerobe. **Details**: Causes shigellosis (bacillary dysentery), with symptoms including severe diarrhea, often with blood, fever, and abdominal cramps. Spread through contaminated food, water, or person-to-person contact.	Fever Abdominal cramps Diarrhea Vomiting Headache
6.	*Escherichia coli*	**Nature**: Gram-negative rod, facultative anaerobe. **Details**: Pathogenic strains include: Enterotoxigenic *E. coli* (**ETEC**): Causes traveler's diarrhea. Enterohemorrhagic *E. coli* (**EHEC**): Produces Shiga toxins, leading to bloody diarrhea and possible hemolytic uremic syndrome (HUS). Found in contaminated food, water, and undercooked meat.	Fever Abdominal cramps Diarrhea Vomiting Headache

(continued)

Table 18.1 (continued)

No.	Name of the pathogens	Quick facts on pathogens	Hallmark symptoms
7.	*Campylobacter* spp.	**Nature**: Gram-negative curved rod, microaerophilic. **Details**: Causes campylobacteriosis, with symptoms including diarrhea (sometimes bloody), fever, and abdominal pain. Commonly associated with raw or undercooked poultry and unpasteurized milk.	Diarrhea (often bloody) Abdominal pain Nausea Vomiting Malaise Fever
8.	*Clostridium botulinum*	**Nature**: Gram-positive rod, spore-forming, obligate anaerobe. **Details**: Produces botulinum toxin, causing botulism, a rare but serious illness characterized by muscle paralysis, respiratory failure, and possible death. Associated with improperly canned or preserved foods.	Vertigo Double or blurred vision Loss or light reflex Difficulty in swallowing Dry mouth Weakness Respiratory paralysis
9.	*Salmonella tympi*	**Nature**: Gram-negative rod, facultative anaerobe. **Details**: Causes typhoid fever, a systemic infection with symptoms such as high fever, abdominal pain, weakness, and rash. Spread through contaminated food or water.	Headache Fever (high fever, often with a stepwise increase) Rose spot (small, pink spots on the abdomen or chest) Nausea Vomiting Abdominal pain
10.	*Listeria monocytogenes*	**Nature**: Gram-positive rod, facultative anaerobe. **Details**: Causes listeriosis, which may lead to fever, muscle aches, diarrhea, and severe complications such as meningitis or miscarriage. Found in contaminated dairy products, ready-to-eat foods, and raw produce.	Fever Chills Headache Arthralgia (joint pain) Prostration (extreme fatigue or weakness) Malaise (general feeling of illness) Swollen lymph nodes
11.	*Campylobacter jejuni*	**Nature**: Gram-negative, spiral-shaped bacterium. **Details**: A leading cause of bacterial gastroenteritis, often associated with consuming raw or undercooked poultry. Causes diarrhea, abdominal cramps, fever, and nausea.	Fever Chills Headache Arthralgia (joint pain) Prostration (extreme weakness or fatigue) Malaise (general feeling of discomfort or illness) Swollen lymph nodes Muscle aches

(continued)

Table 18.1 (continued)

No.	Name of the pathogens	Quick facts on pathogens	Hallmark symptoms
12.	*Yersinia enterocolitica*	**Nature**: Gram-negative, rod-shaped bacterium. **Details**: Causes yersiniosis, typically through contaminated food (e.g., pork). Symptoms include diarrhea, abdominal pain (often mimicking appendicitis), fever, and vomiting.	Fever Vomiting Fatigue
13.	*Vibrio cholearae*	**Nature**: Gram-negative, comma-shaped bacterium. **Details**: Causes cholera, characterized by severe watery diarrhea (rice-water stools) and dehydration. Transmitted via contaminated water or food.	Abdominal cramps Diarrhea Vomiting Fever Nausea Headache Dehydration
14.	*Vibrio parahaemolyticus*	**Nature**: Gram-negative, rod-shaped bacterium. **Details**: Associated with consuming raw or undercooked seafood. Symptoms include watery diarrhea, abdominal cramps, nausea, and vomiting.	Abdominal cramps Diarrhea Vomiting Fever Nausea Headache Dehydration
15.	*Aeromonas hydrophila*	**Nature**: Gram-negative, rod-shaped, facultative anaerobe. **Details**: Commonly found in freshwater and is associated with aquatic environments. It can cause foodborne illness through the consumption of contaminated food, especially undercooked or contaminated seafood, or through direct contact with infected water. It is also a known pathogen for various gastrointestinal and wound infections.	Diarrhea Abdominal cramps Nausea & vomiting Fever (in some cases) In severe cases, it can lead to septicemia or wound infections, particularly in immunocompromised individuals.
16.	*Pseudomonas aeruginosa*	**Nature**: Gram-negative, rod-shaped, opportunistic pathogen. **Details**: Commonly found in soil, water, and on human skin; can cause infections in immunocompromised individuals. While it is not a frequent cause of foodborne illness, it can contaminate food and cause gastroenteritis, particularly in environments where hygiene is poor or food is stored improperly. It is resistance to many antibiotics and disinfectants.	Diarrhea Abdominal cramps Nausea Vomiting Fever (in some cases) Severe cases may lead to pneumonia or bloodstream infections

(continued)

Table 18.1 (continued)

No.	Name of the pathogens	Quick facts on pathogens	Hallmark symptoms
17.	*Bacillus anthracis*	**Nature**: Bacterial pathogen (Gram-positive, spore-forming, rod-shaped). **Details**: It is the causative agent of anthrax, a zoonotic disease that can affect both humans and animals. Although it is not commonly a cause of foodborne illness, it can be contracted through consumption of contaminated meat, especially from infected animals. The bacteria produce spores that are resistant to harsh environmental conditions and can persist in soil for long periods. Foodborne anthrax is rare, but can be highly fatal if not treated promptly.	**Ingestion**: Nausea Vomiting Diarrhea (may be bloody) Severe systemic illness, including sepsis and shock **Inhalation (inhalation anthrax)**: Fever Difficulty breathing Shock and respiratory failure (often fatal if untreated) **Cutaneous (cutaneous anthrax)**: Painless ulcer with a black center Swelling around the site of infection
18.	*Neisseria meningitidis*	**Nature:** Gram-negative, encapsulated diplococcus bacterium. **Details**: It has several pathogenicity factors, including a polysaccharide capsule that protects against host immune responses, lipooligosaccharide, an endotoxin that triggers strong inflammation, pili and adhesins that facilitate mucosal attachment, and IgA protease, which helps evade mucosal immunity. The bacterium is transmitted person-to-person via respiratory secretions. It primarily affects children, adolescents, and immune-compromised individuals, with outbreaks occurring worldwide.	Meningitis Meningococcemia
Viruses			
19.	Norovirus	**Nature**: Non-enveloped, single-stranded RNA virus. **Details**: A leading cause of acute gastroenteritis, characterized by vomiting, diarrhea, abdominal pain, and fever. Highly contagious and transmitted through contaminated food, water, or surfaces.	Nausea Vomiting Watery non-bloody diarrhea Dehydration
20.	Rotavirus	**Nature**: Double-stranded RNA virus. **Details**: A major cause of severe diarrhea in infants and young children. Transmitted via the fecal-oral route. Vaccines are available.	Fever Vomiting Watery non-inflammatory diarrhea

(continued)

Table 18.1 (continued)

No.	Name of the pathogens	Quick facts on pathogens	Hallmark symptoms
21.	Astrovirus	**Nature**: Non-enveloped, single-stranded RNA virus. **Details**: Causes mild gastroenteritis, especially in children and immunocompromised individuals. Symptoms include diarrhea, nausea, and low-grade fever.	Fever Vomiting Watery non-inflammatory diarrhea
22.	Hepatitis A	**Nature**: RNA virus (Picornaviridae family). **Details**: Transmitted primarily through the fecal-oral route via contaminated food, water, or close contact with an infected person. Causes hepatitis A, an acute liver infection. The virus is highly contagious but does not lead to chronic infection. A vaccine is available for prevention.	Fatigue Nausea Vomiting Abdominal pain Jaundice (yellowing of the skin and eyes) Dark urine Pale stool Fever Loss of appetite
23.	Enteric adenovirus	**Nature**: Non-enveloped, double-stranded DNA virus. **Details:** Causes gastroenteritis, particularly in young children. Symptoms include prolonged diarrhea, vomiting, and fever.	Fever Vomiting Watery non-inflammatory diarrhea
Parasites			
24.	*Taenia saginata*	**Nature**: Parasitic flatworm (cestode). **Details**: Acquired by consuming undercooked beef containing larvae. Causes mild gastrointestinal symptoms or asymptomatic infections.	Nervousness Insomnia Hunger pains Anorexia Weight loss Abdominal pain
25.	*Taenia solium*	**Nature**: Parasitic flatworm (cestode). **Details**: Transmitted through undercooked pork. Can cause cysticercosis if larvae invade tissues, including the brain (neurocysticercosis).	Nervousness Insomnia Hunger pains Anorexia Weight loss Abdominal pain
26.	*Entamoeda histolytica*	**Nature**: Amoebic protozoan parasite. **Details**: Causes amoebiasis, characterized by diarrhea, abdominal cramps, and potentially severe complications like liver abscesses. Transmitted through contaminated food or water.	Abdominal pain Diarrhea Constipation Headache Drowsiness Ulcers
27.	*Toxoplasma gondii*	**Nature**: Intracellular protozoan parasite. **Details**: Causes toxoplasmosis, which can be asymptomatic in healthy individuals but dangerous for immunocompromised individuals and pregnant women. Transmitted through undercooked meat or contaminated cat feces.	Fever Headache Myalgia Rash

(continued)

Table 18.1 (continued)

No.	Name of the pathogens	Quick facts on pathogens	Hallmark symptoms
28.	*Cyclospora cayetanensis*	**Nature**: Parasitic protozoan (coccidian). **Details**: Transmitted through ingestion of food or water contaminated with sporulated oocysts. Causes cyclospora infection (cyclosporiasis), leading to watery diarrhea, abdominal cramps, bloating, and fatigue. Sporulation occurs in the environment, and oocysts are resistant to standard disinfectants. Often associated with fresh produce like berries, herbs, and lettuce.	Watery diarrhea Abdominal cramps Bloating Nausea Fatigue Loss of appetite Weight loss
29.	*Toxoplasma gondii*	**Nature**: Parasitic protozoan (apicomplexan). **Details**: Transmitted through ingestion of undercooked meat containing tissue cysts, contaminated water or food with oocysts, or via congenital transmission. Causes toxoplasmosis, which is often asymptomatic but can lead to severe complications in immunocompromised individuals and during pregnancy. Cats are the definitive hosts, shedding oocysts in their feces.	Flu-like symptoms (fever, muscle aches, fatigue) Swollen lymph nodes, Neurological symptoms (encephalitis, seizures) or damage to the eyes (chorioretinitis). Congenital infection can result in miscarriage, stillbirth, or developmental abnormalities.
30.	*Trichinella spiralis*	**Nature**: Parasitic nematode (roundworm). **Details**: Causes trichinellosis, with symptoms including diarrhea, muscle pain, fever, and swelling of the face. Transmitted through consumption of raw or undercooked meat (especially pork) containing larvae.	Gastroenteritis Fever Oedema around eyes Perspiration Muscular pain, Prostration Laboured breathing
31.	*Trichinella spiralis*	**Nature**: Parasitic roundworm (nematode). **Details**: Transmitted through consumption of undercooked or raw meat (commonly pork or wild game) containing encysted larvae. Causes trichinellosis (or trichinosis), where larvae migrate to and encyst in muscle tissue after being released in the intestines.	**Initial stage** (intestinal phase): Diarrhea Abdominal pain Nausea **Muscle invasion phase**: Muscle pain Fever Swelling of the face (especially around the eyes) Weakness Difficulty breathing (in severe cases) **Complications** (in severe cases): Myocarditis (inflammation of the heart muscle) Neurological symptoms (headaches, confusion) Pneumonia

Trichinella spiralis is a parasitic roundworm that primarily infects carnivorous mammals, including humans. It is commonly contracted by consuming undercooked pork or wild game. The adult worms reside in the small intestine, where they produce larvae that travel through the bloodstream and encyst in muscles. Symptoms of trichinosis include nausea, vomiting, fever, muscle pain, and swelling. While cases of trichinosis have significantly decreased in the US, it still causes some foodborne illnesses and is preventable through proper cooking temperatures. These parasites present significant health risks, especially in areas with poor sanitation or inadequate food safety measures (Bintsis 2017). Table 18.1 represents common foodborne pathogens and their associated health effects.

18.3 Unveiling the Body's Arsenal: Immune Mechanisms Against Foodborne pathogen Invasion

The human body is in a perpetual state of vigilance, defending against the relentless threat posed by bacterial invaders that seek to colonize tissues and disrupt homeostasis. To combat these microbial attacks, the immune system employs a complex array of mechanisms that intricately evolve to detect, neutralize, and eliminate pathogens. At the forefront of this defense lies the innate immune system, which acts as a rapid-response unit, recognizing conserved bacterial signatures, such as lipopolysaccharides (LPS), peptidoglycans, or flagellin, through pattern recognition receptors (PRRs) such as toll-like receptors (TLRs). These molecular sentinels trigger cascading signals that ignite inflammatory responses, recruiting neutrophils, macrophages, and other immune cells to the site of infection. Complement proteins amplify the attack, opsonize bacteria for phagocytosis, or directly lyse the microbial membranes. Simultaneously, the adaptive immune system refines this defense, generating pathogen-specific antibodies and memory T cells to ensure long-term protection. However, the battle is not without consequence; excessive inflammation can lead to tissue damage, underscoring the delicate balance between eradication and regulation. This article explores the multifaceted immune strategies employed against microbial invasion, from initial detection and inflammatory signaling to the coordinated cellular and humoral responses that safeguard human health. By unraveling these mechanisms, we gain critical insights into host-pathogen interactions, paving the way for novel therapeutic approaches to combat antibiotic resistance and enhance immune resilience. Figures 18.1 and 18.2 represent the general immune response against different foodborne pathogens.

18.3.1 Innate Immunity

This system can always be present but is normally dormant and non-specific to any particular antigen (Tian et al. 2024). Foodborne intoxication is often said to be caused by foodborne pathogenic microorganisms or their toxins, and the symptoms,

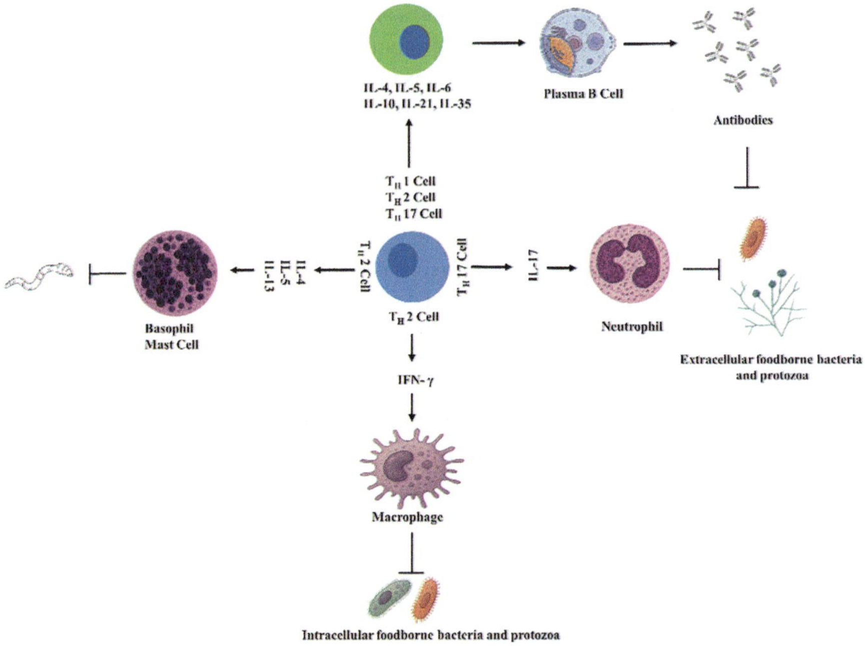

Fig. 18.1 A schematic illustration of the differentiation of naïve CD4⁺ T-helper (T_H) cells into distinct subsets (T_H1, T_H2, and T_H17) and their respective roles in coordinating immune responses against foodborne pathogens. Upon activation by specific cytokines, naïve T_H cells differentiate into T_H1 cells, which secrete IFN-γ to activate macrophages for eliminating intracellular pathogens such as Listeria monocytogenes and Toxoplasma gondii. TH2 cells produce IL-4, IL-5, and IL-13 to activate basophils and mast cells. T_H17 cells secrete IL-17, enhancing neutrophil recruitment and defense against extracellular bacteria and protozoa. T_H1, T_H2, and T_H17 cells assist B cell maturation into plasma cells through various interleukins (IL-4, IL-5, IL-6, IL-10, IL-11, IL-21, IL-35) that produce antibodies targeting various foodborne pathogens

such as nausea and vomiting, usually appear within 30 minutes to 1 hour of ingestion. The natural enteric defense system is vital in the struggle against intoxication or infection. The complex system entails many factors that work in harmony to safeguard the intestinal tract from pathogens. In the same way, there is the epithelial barrier of the intestines, the mucus, the complement proteins, the antimicrobial peptides, the resident microbiota, the pattern recognition receptors, among others (Li et al. 2023). The latter are designed to detect the pathogens or the so-called pathogen-associated molecular patterns, which include but are not limited to the toll-like receptors and the protein known as nucleotide-binding oligomerization domain, or NOD. Many molecules, including these PAMPs (pathogen-associated molecular patterns) as well as TLRs and the NOD proteins, can start signaling events that result in more phagocytosis, getting rid of the pathogen in the body (Albiger et al. 2007; Kubelkova and Macela 2019; Tian et al. 2024). Both TLRs and NLRs (NOD-like receptors, NLRs) have been demonstrated to be essential for both maintaining the homeostasis of the colonizing microflora and protecting the host against microbial

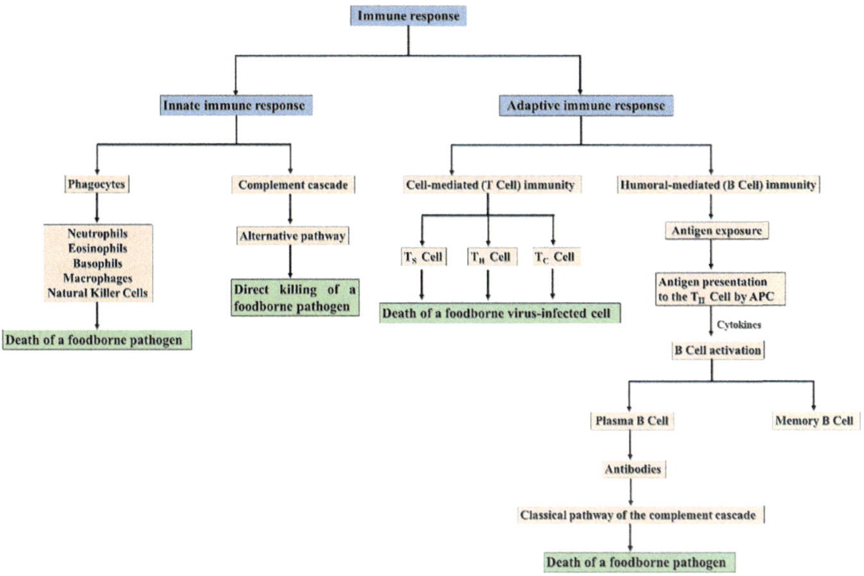

Fig. 18.2 Flow chart illustrating the immune response, divided into innate and adaptive responses. The innate response includes phagocytes like neutrophils and macrophages, leading to the death of a foodborne pathogen. The adaptive response is split into cell-mediated (T Cell) immunity, involving T_S, T_H, and T_C cells, resulting in the death of a virus-infected cell, and humoral-mediated (B Cell) immunity, involving antigen exposure, B cell activation, and antibody production, leading to pathogen death. Key terms: immune response, phagocytes, T cells, B cells, antibodies

infections (Kubelkova and Macela 2019). They can respond quickly to bacterial infections and start the long-lasting adaptive immunity process by recognizing particular microbial ligands and using unique adaptor molecules to activate various signaling pathways that in turn cause subsequent inflammatory and immune responses.

18.3.2 Through Complement-Mediated Destruction of Food-Borne Pathogens

When pathogens like *Neisseria meningitidis* enter the body, they face an assault from immune system components known as complement proteins (Schneider et al. 2007). These proteins facilitate bacterial elimination through three distinct mechanisms: the classical complement pathway, the alternative complement pathway, or the lectin pathway. The initial stages of the classical complement pathway involve antibodies that attach to the surface of invading bacteria. These attached antibodies then become targets for a specific complement protein complex called C1, which binds to the antibody's Fc region (tail). Once attached, C1 triggers a series of cleavage and reformation events among complement complexes. This process culminates in either the formation of a membrane attack complex (MAC) on the bacterial surface or the generation of opsonins that mark the bacterium for destruction. While MAC can penetrate the cell membrane of Gram-negative bacteria, it cannot do so for Gram-positive bacteria. In Gram-negative bacteria, MAC produces pores that allow the entry of membrane-damaging substances like lysozyme, rendering the bacterium vulnerable to osmotic lysis (Schneider et al. 2007).

The adaptive immune response usually takes 4–7 days to emerge. Only a small group of pathogens that require an extended incubation time and induce systemic infection may be significantly impacted by the adaptive immune response during foodborne pathogen infection. Hepatitis A virus, *Shigella* species, *Salmonella enterica*, *Yersinia enterocolitica*, *Listeria monocytogenes*, *Toxoplasma gondii*, *Trichinella* species, and a few more viral and parasitic diseases are among them. Reactive arthritis and Guillain-Barré syndrome are two adverse outcomes that might result from an adaptive immune response to some pathogenic bacterial infections (Schneider et al. 2007).

18.3.3 Killing of Pathogens by Phagocytosis

Phagocytes are also capable of destroying bacterial populations. The process of **opsonization** involves immune proteins, such as acute-phase proteins, including complement and antibodies, binding to bacteria. The coating of opsonized bacteria is identified by phagocytic cells, which subsequently surround and destroy them by phagocytosis. This method of elimination by immune cells is termed **opsonophagocytic** killing. Opsonins, such as antibodies, attach to antigens on the surface of pathogens, increasing their susceptibility to phagocytes. Complement C3b is crucial in this opsonization strategy to limit bacterial infections, especially for Gram-positive ones like *Staphylococcus* spp., which are not easily killed by MAC (membrane attack complex) (Dunkelberger and Song 2010). Once bacteria are taken in and surrounded by the phagosome, a series of intracellular processes takes place, and enzymes degrade the bacteria into smaller parts. These pieces are then presented on the surface of the phagocyte with molecules of class II major histocompatibility (MHC class II) (Dunkelberger and Song 2010).

18.3.4 Helper T Cells and Cell-Mediated Destruction of Pathogens

Helper T cells and Cell-mediated destruction of pathogens present in blood circulation capture these parts of bacterial origin and, in response, activate a process that leads to the secretion of proteins known as **cytokines**. The principal types of helper T cells are grouped into two, TH1 and TH2. The two types of cells differ in the type of cytokines they produce. Th1 cells mainly secrete interferon-g (IFN-g), which augments cell-mediated immune mechanisms. Th2 cells mostly secrete interleukin-4 (IL-4) to enhance humoral immunity through B cell activation. Extracellularly associated B cells produce antibodies that inhibit bacterial growth and survival (Zheng et al. 2020). Some of the bacteria that are taken in via phagocytosis utilize phagocyte's surviving killing mechanisms (Kaufmann 1999). Macrophages are some of the most common intracellular bacteria targets (for example, *Salmonella* spp.) that reside within the compartments of cells. They cannot be reached by complement or antibody but rather are removed by cell selection. Infected macrophages display the bacterial peptide on their cell surface with the help of MHC class II molecules. It is referred to as antigen presentation (Kaufmann 1999; Masud et al. 2019).

A helper T cell scans the surface of MHC class II molecules using the T cell receptor (TCR) for the peptides they carry. If a peptide of bacterial origin is displayed, the Th1 cell secretes IFN-g. This particular cytokine induces effector (killing) mechanisms (for example, lysozyme production) within the infected macrophage to digest and destroy the invading bacterium. IFN-g also increases antigen presentation by cells, making the bacterium more noticeable to the immune system and more prone to attack (Masud et al. 2019).

18.3.5 Humoral-Mediated Destruction of Pathogens

The humoral immune response is a vital component of the adaptive immune system in combating foodborne pathogens, particularly those that remain extracellular. It primarily involves B lymphocytes, which, upon activation, differentiate into plasma cells that produce specific antibodies. Secretory IgA (sIgA), present in the mucosal linings of the gastrointestinal tract, plays a key role by preventing the adherence of pathogens to the intestinal epithelium and neutralizing their toxins. Systemic antibodies like IgM and IgG also contribute significantly—IgM acts as the first responder during infection, while IgG provides long-lasting immunity. These antibodies help neutralize pathogens and their toxins, promote opsonization, and activate the complement system. The complement system, activated during the humoral immune response against foodborne pathogens, primarily involves the classical pathway. This pathway is initiated when IgG or IgM antibodies bind to antigens present on the surface of the pathogen. Once antibodies bind to the pathogen, the C1 complex (comprising C1q, C1r, and C1s) is activated, leading to a cascade of proteolytic activations involving C4, C2, and ultimately the formation of the C3 convertase. This enzyme cleaves C3 into C3a and C3b—C3b acts as an opsonin, enhancing phagocytosis, while C3a acts as an anaphylatoxin, promoting inflammation. Further activation leads to the formation of the membrane attack complex (MAC), which creates pores in the pathogen's membrane, causing cell lysis (Tan and Coussens 2007; Elgueta et al. 2010).

The host immune response against *Bacillus cereus*, is primarily mediated through innate immune mechanisms, particularly the activation of the NLRP3 inflammasome. Upon infection, the pore-forming toxins hemolysin BL (HBL) and non-hemolytic enterotoxin (NHE) induce potassium efflux from host cells, a danger signal that is sensed by the cytosolic innate immune receptor NLRP3. This leads to the assembly of the inflammasome complex, which includes the adaptor protein ASC and caspase-1. Activated caspase-1 then processes pro-inflammatory cytokines IL-1β and IL-18 into their active forms and cleaves gasdermin D, resulting in pyroptosis—an inflammatory form of cell death. Additionally, Toll-like receptors such as TLR2 and potentially TLR4 contribute to the immune response by recognizing bacterial components and triggering inflammatory cytokine production, including TNF, IL-6, and IFN-γ, particularly in the context of ocular infections like endophthalmitis. This multi-pronged innate immune response aims to contain the infection but can also contribute to tissue damage and systemic inflammation if dysregulated (Enosi Tuipulotu et al. 2021).

The immune response to *Vibrio cholerae* begins with the activation of innate immunity, where intestinal epithelial cells detect bacterial components like LPS, flagellin, cholera toxin (CTX), and OmpU through pattern recognition receptors (PRRs), leading to the secretion of proinflammatory cytokines such as IL-1β, IL-6, IL-8, and TNF-α. This recruits innate immune cells including neutrophils, macrophages, dendritic cells (DCs), and mucosal-associated invariant T (MAIT) cells to the site of infection. The adaptive immune response follows, with activated DCs promoting Th1, Th17, and Tfh cell differentiation, which in turn stimulates B cell activation and antibody production. Antibodies, especially secretory IgA (sIgA), IgG, and IgM, target cholera toxin and bacterial surface antigens to prevent adhesion and facilitate bacterial clearance. Vibriocidal antibody titers, particularly against O-specific polysaccharide (OSP), correlate with long-term immunity, although full protection involves both cellular and humoral mechanisms (Montero et al. 2023).

The host immune response during *Taenia solium* neurocysticercosis (NCC) is highly dynamic and depends on the cyst stage and location in the brain. Initially, viable cysts evade immune detection by releasing immunomodulatory molecules that promote anti-inflammatory cytokines like IL-10 and TGF-β, inducing regulatory T cells and alternatively activated macrophages. As cysts degenerate—naturally or through treatment—they lose this ability, triggering a strong inflammatory Th1 response characterized by cytokines such as TNF-α, IFN-γ, and IL-1β. This leads to disruption of the blood-brain barrier, immune cell infiltration, granuloma formation, and eventually calcification. The immune profile transitions to a mixed Th1/Th2 response with continued fibrosis and tissue remodeling. Extraparenchymal infections often elicit stronger antibody and cytokine responses and are associated with severe symptoms like hydrocephalus and intracranial hypertension (Prodjinotho et al. 2020).

The immune response against hepatitis viruses (HAV, HBV, and HCV) involves both innate and adaptive mechanisms, which vary by virus. Innate immunity initiates with the recognition of viral components via pattern recognition receptors like RIG-I, MDA5, and TLR3, leading to the production of type I and III interferons and induction of interferon-stimulated genes (ISGs). HCV elicits strong IFN responses, while HAV and HBV trigger minimal ISG induction. Adaptive immunity plays a critical role, especially virus-specific CD8+ and CD4+ T cell responses, which are essential for viral clearance in acute infections. In HAV, CD4+ T cells and neutralizing antibodies dominate the response. In HBV, broad T cell responses and antibodies to HBsAg ensure resolution in most adults. HCV, however, often progresses to chronic infection due to delayed T cell responses, viral escape mutations, and T cell exhaustion, despite initial ISG upregulation and interferon activity (Stapleton 1995; Park and Rehermann 2014).

18.4 Future Prospects

Understanding the immune response against foodborne pathogens remains a critical area of research with far-reaching implications for public health, food safety, and disease prevention. In the future, advancements in molecular immunology, genomics, and bioinformatics are expected to revolutionize our ability to detect, monitor, and mitigate foodborne illnesses. The development of next-generation vaccines, Immunotherapeutics, and rapid diagnostic tools tailored to specific pathogens holds promise for significantly reducing the global burden of foodborne diseases. Moreover, the integration of microbiome research can enhance our knowledge of how commensal gut microbes influence immune responses to ingested pathogens. Personalized nutrition and precision medicine approaches may offer new avenues to modulate host immunity and enhance resistance to infection through diet and functional foods. Nanotechnology and biosensor development may also improve early detection and real-time monitoring of pathogens in food systems, further preventing outbreaks. Collaboration between immunologists, microbiologists, food technologists, and public health agencies will be essential to translating laboratory findings into effective strategies for food safety and disease control.

18.5 Conclusion

The immune system plays a vital role in protecting the host from foodborne pathogens, employing both innate and adaptive mechanisms to detect and eliminate harmful microorganisms. Despite the body's natural defense strategies, various pathogens have evolved mechanisms to evade immune detection and establish infection. A comprehensive understanding of these host-pathogen interactions is crucial for developing effective preventive and therapeutic strategies. Ongoing research has significantly improved our grasp of immune responses at the cellular and molecular levels. However, the dynamic nature of emerging and re-emerging foodborne pathogens necessitates continuous scientific innovation. Strengthening immune-based interventions, enhancing food hygiene practices, and promoting public awareness can collectively contribute to reducing the incidence of foodborne illnesses and improving global health outcomes.

References

Aladhadh M (2023) A review of modern methods for the detection of foodborne pathogens. Microorganisms 11:1111. https://doi.org/10.3390/microorganisms11051111

Albiger B, Dahlberg S, Henriques-Normark B, Normark S (2007) Role of the innate immune system in host defence against bacterial infections: focus on the Toll-like receptors. J Intern Med 261:511–528. https://doi.org/10.1111/j.1365-2796.2007.01821.x

Bintsis T (2017) Foodborne pathogens. AIMS Microbiol 3:529–563. https://doi.org/10.3934/microbiol.2017.3.529

Dunkelberger JR, Song W-C (2010) Complement and its role in innate and adaptive immune responses. Cell Res 20:34–50. https://doi.org/10.1038/cr.2009.139

Elgueta R, De Vries VC, Noelle RJ (2010) The immortality of humoral immunity. Immunol Rev 236:139–150. https://doi.org/10.1111/j.1600-065X.2010.00924.x

Enosi Tuipulotu D, Mathur A, Ngo C, Man SM (2021) Bacillus cereus: epidemiology, virulence factors, and host–pathogen interactions. Trends Microbiol 29:458–471. https://doi.org/10.1016/j.tim.2020.09.003

Gourama H (2020) Foodborne pathogens. In: Food Engineering Series, pp. 25–49

Kaufmann SHE (1999) Cell-mediated immunity: dealing a direct blow to pathogens. Curr Biol 9:R97–R99. https://doi.org/10.1016/S0960-9822(99)80059-1

Kubelkova K, Macela A (2019) Innate immune recognition: an issue more complex than expected. Front Cell Infect Microbiol 9. https://doi.org/10.3389/fcimb.2019.00241

Li L, Dickinson MS, Coers J, Miao EA (2023) Pyroptosis in defense against intracellular bacteria. Semin Immunol 69:101805. https://doi.org/10.1016/j.smim.2023.101805

Masud S, Prajsnar TK, Torraca V et al (2019) Macrophages target *Salmonella* by Lc3-associated phagocytosis in a systemic infection model. Autophagy 15:796–812. https://doi.org/10.1080/15548627.2019.1569297

Montero DA, Vidal RM, Velasco J et al (2023) Vibrio cholerae, classification, pathogenesis, immune response, and trends in vaccine development. Front Med (Lausanne) 10. https://doi.org/10.3389/fmed.2023.1155751

Park S-H, Rehermann B (2014) Immune responses to HCV and other hepatitis viruses. Immunity 40:13–24. https://doi.org/10.1016/j.immuni.2013.12.010

Prodjinotho UF, Lema J, Lacorcia M et al (2020) Host immune responses during Taenia solium Neurocysticercosis infection and treatment. PLoS Negl Trop Dis 14:e0008005. https://doi.org/10.1371/journal.pntd.0008005

Schneider MC, Exley RM, Ram S et al (2007) Interactions between Neisseria meningitidis and the complement system. Trends Microbiol 15:233–240. https://doi.org/10.1016/j.tim.2007.03.005

Stapleton JT (1995) Host immune response to hepatitis a virus. J Infect Dis 171:S9–S14. https://doi.org/10.1093/infdis/171.Supplement_1.S9

Tan T-T, Coussens LM (2007) Humoral immunity, inflammation and cancer. Curr Opin Immunol 19:209–216. https://doi.org/10.1016/j.coi.2007.01.001

Tian J, Liu H, Che J, Song L (2024) Editorial: innate immunity against intracellular bacteria: mechanisms and strategies. Front Immunol 15. https://doi.org/10.3389/fimmu.2024.1396114

World Health Organization (2025) Estimating the burden of foodborne diseases

Zheng D, Liwinski T, Elinav E (2020) Interaction between microbiota and immunity in health and disease. Cell research 30:492–506. https://doi.org/10.1038/s41422-020-0332-7

Index

T. Sarkar et al. (eds.), *Physiological Perspectives on Food Safety: Exploring the Intersection of Health and Nutrition*, https://doi.org/10.1007/978-3-031-94582-3

9783031945816